Biofilm Applications to Revolutionize Food Technology

Tanmay Sarkar • Dibyajit Lahiri
Moupriya Nag • Debasmita Bhattacharya
Editors

Biofilm Applications to Revolutionize Food Technology

Springer

Editors
Tanmay Sarkar
Department of Food Processing Technology
Malda Polytechnic, West Bengal State
Council of Technical Education,
Government of West Bengal
Malda, West Bengal, India

Moupriya Nag
Department of Biotechnology
University of Engineering and Management
Kolkata, West Bengal, India

Dibyajit Lahiri
Department of Biotechnology
University of Engineering and Management
Kolkata, West Bengal, India

Debasmita Bhattacharya
Department of Basic Science and
Humanities
Institute of Engineering and Management
Kolkata, West Bengal, India

ISBN 978-3-031-85204-6 ISBN 978-3-031-85205-3 (eBook)
https://doi.org/10.1007/978-3-031-85205-3

This Springer imprint is published by the registered company Springer Nature Switzerland AG
The registered company address is: Gewerbestrasse 11, 6330 Cham, Switzerland

Contents

The Enigmatic World of Biofilm

1

Mohan Das, Devalina Khamaru, Sayantan Santra, Rudrak Gupta, Sarobi Das, and Rintu Banerjee

Abstract

Biofilms are complex microbial structures that are encapsulated by self-produced matrices and are considered as one of the most fascinating phenomena in the world of microbiology. The formation of biofilm primarily relies on the ability of adhesion of microorganisms on a substrate to form clusters and finally to develop an intricate structure. This structure plays a key role by providing media for communication, by absorbing nutrition and by protecting the microorganisms that reside inside. Thorough research has been dedicated to the elucidation of biochemical pathways that are involved in biofilm formation. These extensive investigations also established the understanding of attachment proteins, quorum sensing (QS) and extracellular polymeric substances (EPS). Research has also shed light on understanding the structural and functional aspects of microorganisms within biofilms which are versatile and dynamic. Microorganisms residing inside this complex architecture possess the ability for reciprocal coordination as well as competition for resources and niches which improve their organizational capacity to withstand adverse environmental conditions and antimicrobial agents. Besides, biofilm promotes microbial diversity and evolution by enhancing the flow of genes among microbes. Biofilm is an enigmatic architecture that provides a distinctive interface among biotechnology, microbial ecology and environment. Extensive research on their development, function, evolution and

M. Das (✉) · D. Khamaru · S. Santra · R. Banerjee
Agricultural and Food Engineering Department, Indian Institute of Technology Kharagpur, Kharagpur, India
e-mail: mohan@agfe.iitkgp.ac.in

R. Gupta
Virus Research and Diagnostic Laboratory, ICMR-National Institute of Cholera and Enteric Diseases, Kolkata, India

S. Das
Department of Biology, University of Texas at Arlington, Arlington, TX, USA

T. Sarkar et al. (eds.), *Biofilm Applications to Revolutionize Food Technology*,
https://doi.org/10.1007/978-3-031-85205-3_1

technological applications could clarify both the advantages and disadvantages of these microorganisms on human civilization and nature. This chapter highlights biofilm formation, its functional aspects, the relationship with the environment, its application in biotechnology, technical advancements in research and its negative influences.

Keywords

Biofilm · Quorum sensing · Microorganisms · Genesis · Adhesion

1.1 Introduction

Bacterial biofilms that cling to surfaces, crafting intricate formations, are fascinating happenings obscured in mystery. Whether clinical, industrial or natural, this ubiquitous phenomenon wields extensive influence. Its arising involves multifaceted phases. Initially, cells affix to substrates, micro-clusters start to form and extracellular polymeric substances (EPS) exude. This polymeric matrix supports structural cohesion, sheltering embedded microbes. Gram-positive and negative varieties utilize distinct adhesion mechanisms during development. Gram-positives leverage proteins and teichoic acids, while negatives depend on lipopolysaccharides and flagella (Bucher et al. 2015). As a result of microbial propagation, biomass proliferates and biofilms grow more complex. Metabolic and physiological profiles of biofilm residents diverge from freely floating counterparts. Mature configurations exhibit three-dimensional designs permitting nutritional influx and waste removal upon attaining climax growth. Microbes in biofilms carry out many responsibilities, such as degrading natural substances, recycling nutrients and undergoing biochemical changes. Quorum sensing permits microbial cells to collaborate and correspond during the final stage. It is an important regulatory mechanism in biofilm progression which enables bacterial cells to create and react to signalling molecules. This system controls the manifestation of genes and numerous biofilm formation factors. Cyclic-di-GMP (c-di-GMP) is a signalling molecule that plays a part in both the maturation and dispersion of biofilms. It directs the production of EPS and other factors that comprise the biofilm architecture, impacting the arrangement and durability of the biofilm (Nair et al. 2017). The fabrication and structure of biofilms are influenced by hydrodynamic conditions, temperature and pH. Flow rhythms impact the process of bacterial adhesion and detachment, whereas environmental parameters regulate the activity and proliferation of microorganisms. While biofilms are environmentally significant, they can also have negative effects such as biofouling, corrosion and the persistent existence of disease-causing microorganisms in clinical and industrial environments. It provides heightened resistance to antimicrobial agents, rendering them difficult to eliminate. The biofilm-associated cells' altered physiology and protective matrix are responsible for their antibiotic resistance. The development of biofilms on both host tissues and medical devices increases the risk of long-term infections and treatment failures. Chronic wounds, infections linked to prosthetic implants and catheters are among examples. Surfaces and equipment in the food industry can be contaminated with biofilm microorganisms, which in turn causes food to decompose and have a reduced shelf life. Food safety and hygiene can be compromised by those microorganisms that attach spontaneously to biofilms. The versatile and enigmatic world of biofilms has significant implications for several fields of study. To effectively regulate and utilize the potential of biofilms in various habitats, it is important to understand how biofilms originate, the responsibilities they hold in the environment and what can be the utilization of them.

1.2 Genesis of Biofilm

Studies have suggested that the initial step of biofilm formation is triggered by environmental factors that encourage the shift to a life-forming state on a surface. Distinct organisms exhibit variability in their response to environmental inputs. In some cases, certain strains of *Escherichia coli* and *Vibrio cholerae* demonstrate an incapacity to develop biofilms in a minimal medium until additional amino acids are supplied. Conversely, *E. coli* O517:H7 is documented to form biofilm, especially in conditions with reduced nutritional levels. While environmental conditions such as temperature, osmolarity, pH, iron and oxygen levels impact biofilm progression, the composition of media initiates their assembly and influences the whole development process (Dewanti and Wong 1995; O'Toole et al. 2000). Bacteria adhere to various surfaces, forming microcolonies that mature into biofilms through distinct developmental stages and ultimately biofilm maturation and dispersion take place. Research investigating the genetic mechanisms underlying biofilm production in *Escherichia coli*, *Pseudomonas aeruginosa*, *Pseudomonas putida*, *Clostridium difficile* and *Vibrio cholerae* suggests that an elevation in the intracellular second messenger c-di-GMP triggers biofilm formation. The observed phenomenon involves the enhancement of biofilm matrix synthesis and the inhibition of swimming motility. Therefore, the bacterium's ability to detect and react to surfaces following its first attachment represents a crucial stage for the development of biofilms (Hinsa et al. 2003; Römling et al. 2013). The binding process requires the microbe to conquer the repellent pulls generated by the surface and the negative-charged bacterial membrane. Plastic, being hydrophobic, exhibits a diminished repellent effect in comparison to hydrophilic substrates such as glass and metal. The decrease in repellent pull is directly proportional to the enhancement of adhesion strength. The outer surface of the bacterial cell is adorned with flagella, fimbriae and pili, which allow for attachment. Bacteria with functioning flagella can circulate and withstand electrostatic forces which are unfavourable when they come into contact with certain surfaces, such as the conditioning film of the host or other materials (Jamal et al. 2018; O'Toole and Kolter 1998; Palmer et al. 2007). Recently, attached cells are loosely connected to a surface, easily capable of detaching. The stage in question is referred to as the reversible attachment stage, which is frequently distinguished by cells that are polarly linked (Caiazza and O'Toole 2004; Hinsa et al. 2003). Over a period of time, certain cells will subsequently transition into the irreversible adhesion phase, during which they establish a flat position on the surface and exhibit resistance against any physical efforts to displace them. After undergoing irreversible attachment, bacteria undergo proliferation and initiate the production of biofilm matrix constituents, resulting in the formation of microcolonies, which are tiny aggregations of bacteria. Ultimately, they transform into enormous cellular clusters enveloped by a matrix.

Contrasting Adhesion Strategies in Bacteria

The outermost covering of gram-negative bacteria is the lipopolysaccharide (LPS) layer which constitutes three components, O- antigen, core and lipid A, whereas in gram-positive bacteria there is no LPS layer, but teichoic acid is an essential component of their cell wall. One similar characteristic of LPS and teichoic acid is that they impart charges to the exterior surface of bacteria, which might affect their adhesion to surfaces and thus lead to the development of biofilms. The variations in different regions of LPS of gram-negative bacteria could have a significant impact on the beginning of adhesion and subsequent production of biofilms (Lee et al. 2016). Core lipopolysaccharide structures remain consistent across diverse genera, implying opportunities for cross-species therapies targeting this conserved region (Bennett-Guerrero et al. 2000). However, lipopolysaccharide modifications by mutation vary in their effects on biofilm development, with alterations to core regions and O-antigens sometimes enhancing and other times inhibiting biofilm formation depending on the bacteria (Chalabaev et al. 2014). The testing revealed that mutants with shorter lipopolysaccharides (LPS) possessed an increased potential to develop biofilms (Heacock-Kang et al. 2018). The pathogenic microbe *Pseudomonas aeruginosa* incorporates A and B bands in its O-antigen region and alterations in the B band area enhance biofilm formation (Makin and Beveridge 1996). The formation of biofilms among such variants is equally impacted by transient and experimental elements. Hence, alterations in the sticky characteristics observed in device-mediated infections might potentially be attributed to fluctuations in the surface features of bacteria, including hydrophobicity and cell surface charges, which could be influenced by lipopolysaccharide. Unlike gram-negative bacteria, gram-positive bacteria do not possess a lipopolysaccharide layer on their external surface. However, these microorganisms possess substantial peptidoglycan and external glycol polymers that exhibit variability, as exemplified by the presence of teichoic acid in the cell wall glycol polymers of Pneumococci. Just as LPS influences gram-negative bacteria, teichoic acid plays a role in gram-positive bacterial colonization (Bucher et al. 2015).

Unveiling the Phenomenon of Population Growth

When the microbial cells undergo division, the resulting new cells proliferate in the exterior direction from the place of the attachment site, forming clusters

(Hall-Stoodley and Stoodley 2002). Usually, these relationships and development occur inside the growing biofilm and give rise to an architecture resembling a mushroom. It is hypothesized that the mushroom shape facilitates the transportation of essential nutrients to microorganisms located deep inside a biofilm. Subsequent to an earlier period of delay, a notable surge in population is noted, commonly referred to as the exponential growth phase. The outcome is contingent upon the characteristics of the surroundings, encompassing both its physical and chemical attributes. The phenomenon of rapid development is accompanied by a depletion of the ambient nutrients derived from the surroundings. At this point, factors that initially contributed to the attachment start to decrease while biological mechanisms take over. The elimination of polysaccharide intercellular adhesion (PIA) polymers and the existence of divalent cations combine to establish more robust intercellular interaction (Dunne Jr 2002). In non-motile species, the formation of surface extensions is hindered due to limited mobility and the lack of necessity for them. At the same time, several genes that code for proteins on cell surfaces and excretory products are turned on. Opr C and Opr E are two examples of surface proteins (porins) that make it easier for external substances to enter into cells and outflow of excretory products such as polysaccharides from the cell (Hancock et al. 1990). The structural composition of many polysaccharides found in gram-negative bacteria is characterized by a degree of simplicity, encompassing either homopolysaccharides or heteropolysaccharides (Sutherland 2001). These substances have a crucial role in providing mechanical integrity and are essential for the adherence and cohesiveness of biofilms, as well as their capacity to evade severe fluctuating surroundings.

Final Phase of Development

The stationary developmental phase refers to a stage in which the pace of cellular replication is equivalent to the pace of cellular apoptosis. Quorum sensing refers to the utilization of many cell signalling pathways by the biofilm when there is an abundance of cells. Autoinducers are signalling compounds produced and released by microbes that sense cell density. These compounds stimulate the genetic expression of alginates processing enzymes and mechanical components that are essential components of the extracellular matrix. The biofilm degrades during the dying period. In order to decompose the polysaccharides that keep the biofilm integrated, the microorganisms produce their own unique enzymes. This proactively releases exposed bacteria so that they can colonize new substrates (Sutherland 2001). In order for bacteria to be able to move about, a genetic cycle is completed by up-regulating operons that code for flagellar proteins and down-regulating genes that code for many porins. This allows the biofilm to adhere and stay together (Garrett et al. 2008).

1.3 Functional Aspects of Microbes

Chemical Applications

Elevated phenol levels pose a significant threat to human health and the environment, potentially leading to severe ecological and public health consequences. As a result of industrial activities, there has been an urgency to minimize the amount of phenol in the atmosphere, which presents a challenge that calls for the development of suitable solutions. Multiple investigations have demonstrated the function that microbes play in breaking down phenols, therefore using them as a countermeasure against the phenol issue is one approach. In reality, since microbes break down phenols into innocuous chemicals and subsequent mineral contaminants, biological water treatment has gained popularity (Bai et al. 2007). Many research efforts have previously endeavoured to verify the significance of biofilms of microbes in mitigating the deleterious effects of phenols. This process includes the aerobic decomposition of phenols by microbes such as *Acetobacter* sp. and *Candida* sp. The experimental research demonstrating the time-dependent and time-dependent effectiveness of several pathogenic strains in the aerobic breakdown of phenols by biofilms of microbes was conducted (Abd-El-Haleem et al. 2003; Ojumu et al. 2005; Unell et al. 2008). In an attempt to comprehend the rheological activities of fluids, attempts were undertaken to inspect the mechanical attributes of microbial biofilms. Examining and grasping these characteristics offer valuable data on the biofilms of microorganisms that are significant for both commercial and medicinal applications. Investigations have demonstrated that *Pseudomonas aeruginosa* biofilms possess viscoelastic properties, and they aim to emphasize the significance of the surface tension of rough surface areas. Current accessible studies and scientific information indicate the existence of inherently occurring polymer traits in *Streptococcus* mutans biofilms on dental plaque (Wloka et al. 2004).

Quorum Sensing

The method of quorum sensing synchronizes the biofilm production pathway. It has been demonstrated that quorum sensing, which is characterized as cell-to-cell signalling including cell-to-cell interaction, is important for the development of biofilms in harmful bacteria. It is an essential procedure that keeps track of the number of bacteria cells. If the necessary quantity of cells is reached, it triggers messenger molecules called autoinducers to emerge, which supports the quorum sensing process (Federle and Bassler 2003). Greater microbial cell numbers allow the microbe to release autoinducers into the surroundings, thereby turning off the quorum sensing process. Enzymatic breakdown alters the quorum sensing system, either by preventing the production of new biofilms or by weakening those that have already formed. Along with the cell population-based signalling system, quorum sensing is additionally important for the control of gene expression. Whenever the necessary quantity of cells is reached, the autoinducers are released, which attach to transcriptional factors and trigger either the induction or inhibition of some genes which might prove advantageous to the pathogen. There are notable variations between gram-positive and gram-negative microbes regarding the autoinducers that initiate the quorum sensing procedure. N-homoserine lactone is an autoinducer utilized by gram-negative bacteria. Its length varies with the cell density of gram-negative bacteria. On the other hand, the quorum sensing mechanism is initiated by the gram-positive bacteria using peptides as signals (Vasudevan 2014). Studies have confirmed that through quorum sensing, gene regulation occurs to expand the range of nutrients available and increase competency on competitor bacteria and environment (Thoendel et al. 2011). Contemporary research endeavour and scientific inquiries support the role of quorum sensing in biofilm synthesis among bacteria. For this purpose, mutant isolates of bacteria, which do not contain the gene required for the synthesis of the cell signalling molecules involved in the quorum sensing process, have been examined for biofilm formation and the results obtained are monitored (Schaber et al. 2004). Though recent research has proved that quorum sensing plays a critical role in the establishment of bacterial biofilms, some studies failed to support the notion that quorum sensing is essential for the formation of biofilms. However, appreciation of signalling chemicals enables the recognition of a compound that alters the mechanisms associated with quorum sensing in order to influence biofilm formation more (Annous et al. 2009). It is here pertinent to appreciate that comprehension of the role of cell communication and how quorum sensing occurs in the formation of bacterial biofilms rely fundamentally on the chemical identity of the signalling molecules. Identifying appropriate objectives is critical because QS and biofilm formation are two interrelated concepts that are present in today's literature and are the focus of several studies and tests that can be employed to mitigate the adverse effects of biofilms (Cvitkovitch et al. 2003). However, further research is required to understand the part played by quorum sensing in the formation of bacterial biofilms, and the importance of cell signalling in the virulence and antibiotic resistance of the pathogens (Fig. 1.1).

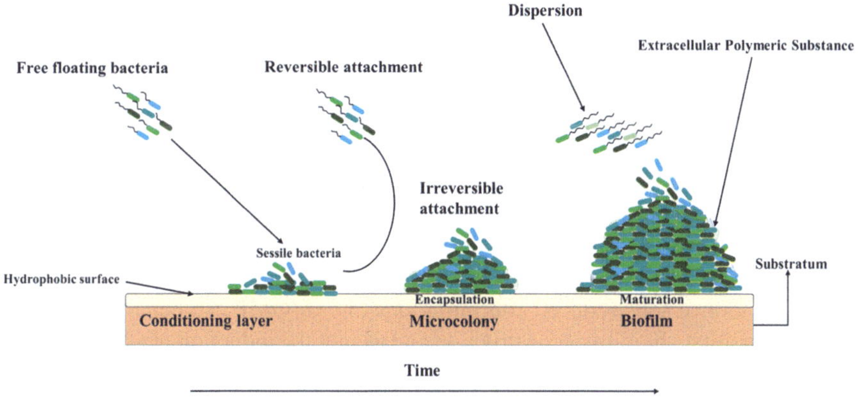

Fig. 1.1 Stages of biofilm development

1.4 Relevance to the Environment

Some factors that may interfere with the attachment of bacteria include temperature, pH, food availability and hydrokinetic conditions are some of the environment aspects that are believed to affect the attachment, there are also some aspects observed in the bacterial cell surface, which include flagellation/mobility, hydrophobicity and surface roughness. One of the reasons that may play a role in becoming privileged inside a heterogeneous population of microorganisms is the properties of cells and their appendages, such as fimbriae and flagella, as well as cell junctions that would allow them to interact with one other and produce extracellular polysaccharides and proteins (Donlan 2002; Oder et al. 2017). Bacterial species exhibiting hydrophobic characteristics are more inclined to adhere to surfaces compared to those with hydrophilic properties. Nevertheless, the formation of biofilms is prone to transpire on surfaces that include rough textures, hydrophobic features and are covered with surface conditioning layers. The pace of bacterial adhesion in aqueous environments could be enhanced by boosting the speed of the flow, water temperature or level of nutrients, as long as these elements do not surpass critical thresholds (Donlan 2002; Pagán and García-Gonzalo 2015; Prakash et al. 2003).

Cyclic-di-GMP

C-di-GMP in bacteria is of significant importance in the process of the development of biofilm. The compound was initially recognized as an allosteric stimulator of cellulose production in *Gluconacetobacter xylinus*. Since then, this molecule has been identified as a crucial compound that controls the shift from the motile, planktonic form to a sessile, biofilm-associated form (Cotter and Stibitz 2007; Nair et al. 2017). Multiple species of bacteria are capable of producing c-di-GMP, a molecule that has been shown to affect the modulation of multiple physiological processes, including cell surface attachment and biofilm formation, EPS production, bacterial motility and expression of virulence factors. Cyclic-di-GMP has been reported to have effects on several basic bacterial processes, such as cell cycle increase, development, fimbriae synthesis, type III secretion system regulation, RNA synthesis/turnover and stress response (Sisti et al. 2013; Srivastava and Waters 2012; Toyofuku et al. 2016). Diguanylate cyclases (DGCs) and phosphodiesterase (PDEs)are enzymes that synthesize, degrade as well as regulate the concentration of cyclic-di-GMP within cells (Cruz et al. 2012). They are synthesized by converting two GTP molecules by DGCs and degraded by breaking into smaller molecules either pGpG or two GMP molecules by PDEs. Hyp bromide concentrations are often associated with surface attachment, EPS synthesis, bacterial biofilm formation and a state of dormancy. Conversely, reduced concentrations of c-di-GMP can enhance bacterial mobility, facilitate biofilm deconstruction and trigger the induction of virulence cascades (Cruz et al. 2012). The presence of GGDEF and HY domains can be observed either independently or in combination as hybrid proteins that encompass both domains. Yet, hybrid proteins often exhibit either PDE or DGC action exclusively, even so, there are instances when both activities may coexist (Cruz et al. 2012). The

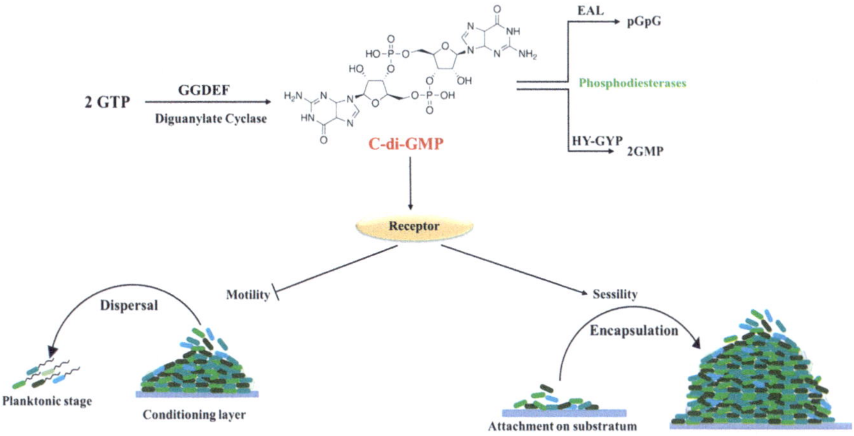

Fig. 1.2 Effect of C-di-GMP level in biofilm formation

control of many phases in the biofilm-building pathway is influenced by the concentrations of c-di-GMP within the cell. When it comes to *Pseudomonas aeruginosa*, the initial stage of developing a biofilm, known as reversible attachment, is controlled by flagellar motion. This motion is influenced by FleQ, a transcriptional regulator that governs the expression of flagellar genes. Yet when c-di-GMP binds to FleQ, it triggers a conformational alteration that diminishes the aquatic mobility of the bacteria (Nair et al. 2017). C-di-GMP was found to modulate the production of Pel and Psl as well as other biofilm matrix biopolymers in *P. aeruginosa* and other bacteria during the various stages of biofilm development. Cyclic-di-GMP might also be involved in controlling biofilm dispersion. For example, exposure to agents that produce nitric oxide (NO) is known to result in the dispersion of *P. aeruginosa* biofilms. It is for this reason that contact with the substance increases the activity of peroxidase (PDE) leading to reduced levels of intracellular c-di-GMP in the affected biofilms (Nair et al. 2017). The cyclic di-GMP signalling cycle depends on the activators that are crucial in synthesizing and degrading this particular molecule. Further, we can also identify effector proteins that get involved in c-di-GMP coupling and target components that represent the final end product of the regulation module. Because it increases peroxidase (PDE) activity, which causes c-di-GMP within the targeted biofilms to be reduced. Consequently, DGCs and PDEs sense exogenous and endogenous cues and tune the cells in c-di-GMP levels. These, in turn, regulate the activity of compounds that interact with c-di-GMP leading to physiological and behavioural changes within the cell (Mills et al. 2011) (Fig. 1.2).

Hydrodynamic Conditions

This matrix is intrusive to a broad range of hydrodynamic conditions and seems to differ across habitats (Lembre et al. 2012). They affect biofilm formation because these factors change nutrient and air availability and cause shear stresses that affect the adhesion of cells on surfaces. It also demonstrated that fluid hydrodynamics controls the rate at which microbial cells, nutrients and air are transported from the fluid to the biofilm (Gomes et al. 2018). Furthermore, all stated variables can influence the properties of biofilm, such as its density and hardness. As a result, these aspects can subsequently affect the transport of nutrients and signals within the biofilm (Purevdorj et al. 2002). The phenomenon of dissociation appears to be high when exterior shear stresses acting on a biofilm outstrip the interior strain of the typically rigid matrix responsible for embracing the biofilm collectively (Fu et al. 2017). Cell detachment from biofilm can either be enhanced by an increase in exterior shear stress or a decrease in interior strength such as the breakdown of the polymeric biofilm matrix. There are two contrasting events associated with an elevated flow rate. Firstly, it facilitates the transportation of nutrients toward the surface, thereby promoting the development of cells in the bacterial layer and the generation of exopolysaccharides. Secondly, as the speed of flow increases, shear rates also increase, leading to further deterioration and disengagement of biofilm components. Consequently, the volume of biomass bound to the solid substrate decreases. Nevertheless, the separation of the biofilm due to hydrodynamic pressures results in a drop in live biomass, thereby leading to a drop in EPS release, thereby elevated shear stresses may result in the formation of denser and thinner biofilms (Liu and Tay 2001). Cells growing in environments with high shear stress exhibit improved attachment and reduced dissociation frequencies.

Other Factors

External variables have the ability to impact both the characteristics of bacteria (via alterations in gene regulation and/or the physical and chemical properties of the cell surface) and the properties of the surface (mostly via changes in physicochemical qualities) (Pagán and García-Gonzalo 2015). In addition, they have the ability to modulate the levels of the second messenger c-di-GMP, which contributes to controlling several aspects linked to biofilm formation, including cell extensions, surface proteins, EPS and cell mobility (Toyofuku et al. 2016; Pompilio et al. 2008). Biofilm production may be greatly impacted by the pH of the surroundings. This has already been shown that this element influences the initial stage of biofilm development and bacterial adherence to substrates. pH is regarded to be a crucial factor in the initial attachment of microorganisms, including *Staphylococcus epidermidis* to substrates (Nostro et al. 2012). Furthermore, it was also demonstrated that the pH of the environment influences the formation of microbial biofilm slime, which in turn influences the efficiency of bio-catalysts because every enzyme requires a specific pH (Chaieb et al. 2007). The ideal pH for exopolysaccharide production is understood to vary depending on the species in question, although, for the majority of bacteria, it is about 7 (Tilahun et al. 2016). The synthesis of exopolysaccharides contributes to biofilm defence toward stressful conditions such as pH. Because of this, bound microbial cells in biofilms are more resistant to pH fluctuations than free cells. For instance, in lower pH environments, the slimy matrix of microbial biofilm may aid in slowing down the quick passage of ions and permit the creation of a pH gradient inside the extracellular matrix. But in circumstances as in higher pH, aberrantly delicate and unorganized biofilms due to impaired biofilm development were identified as well, coupled with attachment restriction for some bacteria including *S. aureus* and *S. epidermidis* (Nostro et al. 2012). Temperature may additionally have an impact on the development of bacterial biofilm. Because the breakdown of foods is believed to be reliant upon the availability and turnover of enzymes that control the establishment of different physiological and biochemical systems in bacteria, the ideal temperature is believed to encourage bacterial growth and hasten the generation of biofilm (Tilahun et al. 2016). On the other hand, microbial development may be inhibited when the temperature is raised over the optimal range because of a drop in response rates. This might have an impact on the formation of biofilms (Garrett et al. 2008). Temperature may impact the physical characteristics of molecules inside and outside cells, as well as enzymes. It was demonstrated that the existence of microbial cell exterior extensions, such as flagella, pili and fimbriae, which aid in substrate adhesion, is temperature-dependent (Garrett et al. 2008). It was discovered that when EPS temperature rises, a gel-like material is produced. This substance progressively gets stronger up to a threshold degree, at the point when the gel turns into a liquid. Therefore, reduced temperatures can result in greater homogeneous exopolysaccharide characteristics, which frequently enhance the likelihood of microbial biofilm attachment (Garrett et al. 2008). However, it was shown that elevated temperatures in certain bacteria increased the biofilm's adherence to the substrate. Perhaps oxygen availability is necessary for the initiation of biofilm formation and microbial adhesion to underwater structures. To be more concrete, oxygen availability may not only influence biofilm formation but also the yield of microorganisms' energy. Specifically, low oxygen may decrease microbial metabolic activities of microbes attached to biofilms. Consequently, oxygen is assumed to be one of the external factors that might influence the biochemical nature and development of biofilms at large (Keleştemur et al. 2018). Dynamic dispersion, a critical step in biofilm formation, is frequently initiated in response to a reduction in oxygen availability (Toyofuku et al. 2016). For example, microbial cells are located at the bottom of a biofilm and receive less oxygen than cells above them, which enhances their efficiency in siphoning off the subcutaneous layers of a biofilm. Oxygen is required during the production of biofilm for a number of bacteria including *E. coli* provided there is no separation signal. Nutrients may affect the movement of free-floating microorganisms to non-motile microorganisms since the nutritional status decides whether the bacteria are going to form a biofilm or remain in the solution (Tilahun et al. 2016). In addition, as described above, the concentration of nutrients can affect biofilm formation and displacement of cells within them. Besides, proving that fluctuations within the availability of a crucial nutrient influence the biological processes concerning microbes in biofilms (Bühler et al. 1998). Numerous studies have attempted to understand the role of nutritional status in the formation of microbial biofilm. For example, high nutrient concentrations in potable water supply systems contributed to an increase of cells in biofilms. The influence of glucose concentrations on the production of biofilms is somewhat understood, but the effects of varying concentrations of nitrogen throughout the same operation are not as well understood (Rochex and Lebeault 2007). However, variations in the concentrations of yeast extract and peptone, two excellent suppliers of nitrogen, showed no discernible effect on the development of *E. coli* biofilms.

1.5 Biotechnological Utilization

Unique Features of Biofilm Microorganisms Used in Biotechnological Applications

Microorganisms that form biofilms maintain structural integrity, cell-to-cell communication among themselves, surface attachment, nutrient absorption, and resistance to environmental stresses and antibiotics. All these are maintained by the extracellular matrix (ECM) secreted by the biofilm microorganisms. ECM is a complex mixture of proteins, polysaccharides and nucleic acids. Some components of the ECM are useful biotechnologically in fields such as cosmetics, food and pharmaceutical industries. Some examples of such components are Rhamnolipids which is a biosurfactant and BslA. Ramnolipids and BslA (Biosurfactant Lipopeptide A) are both commercially useful due to their unique properties and applications: Rhamnolipids produced by *P. aeruginosa* are best studied as an extracellular secondary metabolite. They consist of a hydrophilic (water-attracting) sugar head group (rhamnose) and a hydrophobic (water-repellent) fatty acid tail. Rhamnolipids exhibit excellent surface activity, emulsification and foaming properties. Rhamnolipids are good for bioremediation (Dusane et al. 2010). They enhance the solubility and bioavailability of hydrophobic pollutants, aiding in the cleanup of contaminated soil and water. It helps in oil sludge recovery and enhance oil recovery by reducing the interfacial and surface tension between water and oil easing oil flow (Mishra et al. 2022; Alahmad 2015). It is used in personal care products such as shampoo, cosmetics and soaps for its mildness and foaming ability. They can be used as natural pesticides and soil conditioners. As detergents and cleaning agents, rhamnolipids are eco-friendly alternatives to synthetic surfactants. In the food industry they can stabilize emulsions and improve food texture. Rhamnose is a precursor of many high-quality flavour components, rhamnolipids produced by biofilm-forming bacteria *P. aeruginosa* are a good source of this rhamnose. In the pharmaceutical industry, rhamnolipids have potential antimicrobial, antifungal and antiviral properties. It increases the membrane permeability by forming aggregates on the membrane surface forming transmembrane pores leading to cell death (Dusane et al. 2010). Thus, regarding biotechnological utilization, it has a role in wound healing treatment and organ transplants. As a whole its low-level toxicity and biodegradability, milder production conditions as compared to other synthetic surfactants make it gain extra attention.

Quorum Sensing Mechanisms Controlling Rhamnolipids Production and Its Commercial Utilization

Quorum sensing is a cell-to-cell communication process used by bacteria to coordinate gene expression in response to population density. Rhamnolipids production can be influenced by quorum sensing mechanisms, specifically the las and rhl, which regulate the expression of genes involved in rhamnolipid biosynthesis in *Pseudomonas aeruginosa* (Kumar and Das 2018). The las and rhl quorum sensing systems in *Pseudomonas aeruginosa* are interlinked and collaborate to control the synthesis of rhamnolipids by producing specific autoinducers and activating essential transcriptional activator proteins. Autoinducers, such as N-acyl homoserine lactones, play a significant role in regulating the production of rhamnolipids. The las and rhl quorum sensing systems in *Pseudomonas aeruginosa* are interlinked and collaborate to control the synthesis of rhamnolipids by producing specific autoinducers and activating essential transcriptional activator proteins (Berlanga and Guerrero 2016). In *Pseudomonas aeruginosa*, specific autoinducers such as N-butyryl homoserine lactone (C4- HSL) (PAI -2) and N-3-oxododecanoyl homoserine lactone (3-oxo-C12-HSL) (PAI-1) are involved in quorum sensing, which controls the expression of genes related to rhamnolipid biosynthesis. The interaction of the above-mentioned autoinducers with transcriptional activator proteins such as RhlR in the rhl QS system induces transcription of rhlAB operon genes and rhlC genes encoding enzymes such as ramnosyltranferase1 and ramnosyltranferase2 essential for rhamnolipid production. Bacterial cell density increases, induces the las system and increases the concentration of PAI-1. PAI-1 binds to the transcriptional activator protein LasR forming the LasR-PAI-1 complex inducing las QS system-controlled genes. The las system controls the expression of RhlR, a key regulator in the rhl system, indicating a hierarchical relationship between the two QS systems. LasR and LasI in the las system influence the expression of genes in the rhl system, ultimately impacting rhamnolipid biosynthesis (Dusane et al. 2010). The coordinated action of both QS systems ensures the proper regulation of rhamnolipid production in response to changes in bacterial population density and environmental cues. Commercially, understanding and manipulating quorum sensing pathways can be leveraged to enhance rhamnolipid production for various applications. By modulating the quorum sensing systems, researchers and biotechnologists can potentially increase rhamnolipid yields in microbial cultures. This can be achieved through strategies such as genetic engineering, modifying the genetic pathways involved in quorum sensing to optimize rhamnolipid production in bacterial strains used for industrial purposes, also by bioreactor designing, engineering bioreactors that provide optimal conditions for quorum sensing-mediated rhamnolipid production, such as nutrient availability, oxygen levels and population density control. By harnessing the regulatory role of quorum sensing in rhamnolipid biosynthesis, researchers can potentially upscale the production of these valuable biosurfactants for various industrial applications, including bioremediation, pharmaceuticals and cosmetics. BslA, the surface layer protein of *Bacillus subtilis* biofilms, another example of biosurfactant, holds significant commercial value owing to its distinctive characteristics and wide-ranging potential applications across diverse industries. BslA is another protein of the ECM of *B. subtilis*. It can self-polymerize forming structural polymers. This protein is amphiphilic in nature and forms a hydrophobic coat on the surface of the biofilm. Commercial aspects of BslA which falls under the hydrophobin group of proteins include its role as a good stabilizer in the food industry as in ice cream manufacturing. BslA aids in the proper mixing of air, milk proteins, sugar, fat and water enhancing its texture, and giving a stable mixture. It helps ice cream to be in frozen condition for a longer period of time even in hot weather. It also slows down ice crystal formation thus giving a smooth texture. Biofilms represent promising "cell factories" for producing a variety of chemicals via fermentation processes, wastewater treatment and bioremediation. These biofilm-centred cell factories provide numerous benefits for industrial use (Berlanga and Guerrero 2016).

1.6 Technological Advances in Biofilm Research

In view of the upgradation in the biotechnological field, there are various advancements in biofilm research. Different models and imaging techniques are there to delve into the biological aspects of biofilm so that new strategies can be put forward to control pathogens and treat various infectious diseases (Alhede et al. 2012).

Biofilm Growth Models

There are various growth models that can be used to know various biological facts about biofilms. The models can be actually engineered according to the system that is being investigated.

Static Growth Model

This is a simple biofilm experimental model where the biofilms are allowed to form at the bottom of a microtitre plate. One can also add substrates such as glass beads and hydroxyapatite discs to the bottom for attachment of the biofilms . These substrates can later be taken out for direct imaging (Cleaver and Garnett 2023). Biofilms can also be visualized directly under a microscope when grown within an 8-well chamber slide (Jurcisek et al. 2011). Early biofilm formation can be detected in the static growth model examination. It is cheaper and simpler to execute. One disadvantage is that the cultures are not continuously supplied with nutrients thus mature biofilms cannot be obtained easily (Merritt et al. 2011). Another static biofilm growth model is colony biofilm. Bacterial colonies are grown over filters and placed on agar plates. Transferring the filters at regular intervals to a fresh medium allows a semi-continuous flow of nutrients (Wentland et al. 1996). This model of growth is economical, large biomass can be grown in a short amount of time. This model helps us to study biofilms as they are thick and are used for a variety of experiments.

It allows cryo thin-sectioning of biofilms to obtain vertical cross-sections. Differential staining methods are carried out to study gene expression heterogeneity within biofilms (Franklin et al. 2015).

Dynamic Growth Models

Flow Cell System

A flow cell is a device that permits the microscopic observation of the initiation and growth of a biofilm in real-time. It is constructed simply from a piece of square capillary tubing mounted on a base. The flow cell is placed on the stage of a microscope, and a source of microorganisms and nutrients is permitted to pass through the tube. It is a more advanced technique; it overcomes the limitation of the static growth model. In this flow cell system, there is a continuous flow of nutrients and air allowing biofilms to reach maturation.

Biofilm Reactor System

Biofilm bioreactors fall under the dynamic growth model which is used to cultivate and analyse biofilms under controlled conditions. These are specialized tanks or vessels that are meant to create an environment for whole cells or enzymes to convert biochemical into products. Biofilm reactors are tailored to deliver highly reproducible biofilm growth conditions, ensuring increase in reaction rate for effective and rapid conversion of biochemical into products. These have ports for nutrient entry and gaseous exchange. There are additional ports for inhibitors such as antimicrobials and the substrate upon which biofilms grow can be removed for analysis. This type of bioreactor model is best suited for oral biofilm modelling. Here the continuous flow of nutrients replicates the saliva covering the tooth surface (Cleaver and Garnett 2023).

Microfluidic System

Microfluidic platforms play a crucial role in studying bacterial adhesion and biofilm formation. They offer dynamic real-time analysis and precise control over relevant parameters compared to traditional static and flow chamber assays. These are miniaturized systems ideal for high-throughput studies and visualizing biofilm behaviour at the microscale. This system provides controlled flow conditions, enabling real-time analysis. It includes devices such as Bio flux (Fluxion Biosciences) and Biofilm Chip, this system uses small channels of micron diameter to maintain constant and low volume flow of fluid facilitating small-scale experiments. Key advantages include flow control enabling the precise manipulation of fluid flow mimicking physiological conditions. It enables label-free observation of bacterial proliferation at single-cell resolution and also provides a closer representation of biological conditions compared to other techniques. Another advantage of microfluidic devices is the ability to image the biofilm in situ in real-time (Kim et al. 2012). A recent study has been done using *E. coli* as a model organism to investigate biofilm adhesion and biofilm formation (Straub et al. 2020).

Microcosm, Single and Mixed Specie Modelling

Traditionally, biofilm investigations focused on studying single species of bacteria. In these models, researchers examine the ability of a single bacterial species to form biofilms. The biofilm consists of cells from a single microbial species adhering to a surface. These models help us understand the fundamental processes of biofilm formation, adhesion and growth. They are useful for studying the impact of specific genes, proteins, antimicrobials or environmental factors on biofilm development as there is less variability in studying single species (Cleaver and Garnett 2023). We all know that biofilms do not grow as a single species. They thrive in a symbiotic state in natural environments. In this more complex scenario, researchers study mixed-species biofilms. These biofilms involve multiple microbial species interacting within the same community. Different species may cooperate, compete or exhibit synergistic effects, sharing nutrients and antimicrobial resistance, making them more resilient. Mixed-species biofilms are relevant in natural environments, clinical settings and industrial processes. Understanding interactions between species helps us address biofilm-related challenges (Cleaver and Garnett 2023). Microcosm biofilms represent complex ecosystems within biofilms. These models aim to mimic the intricate microbial communities found in natural environments. They include various microbial species, such as bacteria, fungi, protozoa and viruses. Different species interact within the biofilm, forming intricate networks. Microcosms allow researchers to study nutrient cycling and material exchange. These models capture the resilience and adaptability of biofilm communities. For example, in vitro models such as *S. aureus* and *Candida aureus* dual species, as well as *E. coli* and *Enterococcus faecalis* gut biofilm model, are used to investigate biofilm related to different infections at different body locations (Lee et al. 2014).

Advanced Imaging Techniques Used to Study Biofilms

Biofilms are fascinating structures formed by bacterial and fungal cells that adhere to surfaces and proliferate within a protective extracellular matrix. Understanding biofilm formation and development is crucial for medical research, environmental control and disease treatment. Some of the techniques are discussed below.

Light Microscopy

Light microscopy is a versatile tool for observing biofilms. It is the most basic and cheap technique for visualizing and confirming the biofilm material on the substrate being analysed. Though cheap and easy to perform, this technique provides far lower resolution (Rodrigues et al. 2022).

Fluorescence Microscopy

This technique is a step forward in technological aspects from light microscopy. There are many fluorophores used in this technique, and probes with DNA tags are used for species-specific staining. Fluorescence in situ hybridization (FISH) has been widely used for in situ analysis of the spatial distribution of biofilms. It combines fluorescently labelled nucleic acid probes with confocal laser scanning microscopy (CLSM) to visualize biofilm architecture in three dimensions. By targeting specific microbial species or functional genes, FISH provides insights into the social interactions and spatial arrangement of different species within biofilms. FISH combined with CLSM allows researchers to visualize biofilms in their native, hydrated state. This 3D imaging capability provides a deeper understanding of biofilm structure and organization. Unlike traditional sectioning methods, CLSM captures images through the entire thickness of the biofilm without disrupting its integrity, i.e., it is non-invasive (Barbosa et al. 2023).

Electron Microscopic Techniques

Electron microscopy techniques for biofilm imaging are methods that provide high-resolution insights into biofilm structure, ultra-structure and 3D architecture.

Scanning Electron Microscopy

Scanning electron microscopy (SEM) uses a focused beam of electrons to scan the surface of a sample. It provides detailed images of the sample's topography. It requires minimum sample preparation ensuring preservation of the actual sample structure. SEM offers excellent resolution, allowing visualization of biofilm surface features. SEM is a well-established technique to assess the interaction between bacteria within the biofilm and with the substrate surface. Disadvantages include, SEM only captures surface information; deeper layers remain inaccessible. Non-conductive samples may require additional coating with conductive materials. Variable pressure SEM (VP-SEM) operates at lower vacuum levels, allowing imaging of non-conductive samples without coating. VP-SEM eliminates the need for conductive coating. It provides better depth penetration than conventional SEM but it has a slightly lower resolution compared to high vacuum SEM (Relucenti et al. 2021).

Transmission Electron Microscopy

Transmission electron microscopy (TEM) is a powerful imaging technique that provides detailed insights into the ultrastructure of biofilms. TEM uses a beam of electrons to illuminate a thin section of the biofilm sample. Electrons passing through the sample create an image based on their interactions with the sample's components. The resulting image reveals fine details at the nanometer scale, including individual cells, extracellular matrix and other structural elements. This technique is much less frequently used than SEM because a very thin section of samples using an ultramicrotome is required. TEM reveals the arrangement of bacterial cells, EPS and other components in biofilms. TEM captures interactions between neighbouring cells, such as cell adhesion and communication. Researchers can identify and characterize the extracellular polymeric substances (EPS) that form the biofilm matrix. TEM assesses the impact of antimicrobial agents on biofilm structure and integrity. There are some limitations to using TEM. Sample preparation can introduce artifacts, affecting the native biofilm structure. TEM provides 2D images of thin sections, limiting the understanding of 3D biofilm architecture. TEM cannot visualize deeper layers within thick biofilms. Sample preparation and imaging require expertise and time (Relucenti et al. 2021).

1.7 Negative Aspect of Biofilms

Role of Antibiotic Resistance

One of the most significant global health issues is the establishment and propagation of antibiotic resistance in microorganisms (Cepas et al. 2019). The microbial biofilm colonies that cause persistent infections can lead to a number of effects, including antibiotic resistance. *Klebsiella pneumoniae* is one of the opportunistic pathogens and resistant to multiple antibiotics. It infects people and also a significant source of hospital-acquired infections and has a higher risk of morbidity and mortality because treatment options are limited (Navon-Venezia et al. 2017). As summarized from various articles, a bacterium is capable of resisting adverse effects such as antibiotics and antimicrobial chemicals by forming a biofilm (Høiby et al. 2010). According to a study, it is firmly stated that biofilm is not vulnerable to antibiotic intervention and it carries notable responsibility for recurrent and chronic infections caused by pathogenic bacteria from across the world (Verderosa et al. 2019). It is so since cells of bacteria may become more difficult to kill by antibiotics and the sterilizing chemicals due to biofilm formation and getting enclosed in a complex structure hard to control microbes (Khatoon et al. 2018; Lajhar et al. 2018). Much more about antibiotic resistance is experienced by the microbes living in a biofilm than the free-floating bacteria (Høiby et al. 2010). It is important to note that bacteria exposed to a lifestyle in biofilms use a specific type of antibiotic resistance. Other research has established that the microbes within biofilms produce several molecular defence systems that protect cells in hostile environments. These mechanisms include reactions of the biofilm matrix to antibiotics such as the ability or the extent to which the biofilm slows down bacterial activity or metabolic rates to render antibiotics ineffective, genetic resistance and production of long-lived cells that are highly resistant to various antibiotics (Pinto et al. 2020). High variation rates in the specific genera of microbes that form biofilms are able to devise strategies to develop resistance to antibiotics, and in the course of doing so, provide an opportunity for the genetic material to create proteins that counteract or destroy antibiotics using efflux pumps effectively (Ciofu and Tolker-Nielsen 2019). Among the defences against antibiotics, microbes within biofilms form metabolically dormant persisters that can survive at high antibiotic concentrations. Since there are too many bacteria in the matrix, they come into close proximity with one another and may share resistant genes, which could eventually lead to the acquisition of that resistance gene by the whole population (Balcázar et al. 2015). Thus, the highest genetic variability of bacteria within a biofilm is the main determinant of antibiotic resistance. Due to this, the nature of the interaction between cells within this multimodal framework has shown that biofilm is essential in the transfer of conjugative plasmids as pointed out by Lécuyer et al. (2018). Therefore, one of the major causes of the development of antibiotic resistance is the biofilm formed by the bacteria (Pinto et al. 2020).

Clinical Consequences of Biofilm Formation

In clinical practices, civilization has faced challenges due to the proliferation of several illnesses caused by bacterial biofilms. Research conducted at the National Institute of Sciences (NIH) has identified biofilms as the primary outcome of infections, accounting for around 60% of cases (Lewis 2001). Research and experiments have additionally proven that biofilms play a significant role in grownup gum infections, accounting for 40–50% of cases. Research has also demonstrated that 15–20% of newborns with cerebrospinal (CSF) fluid shunts have diseases linked to biofilms. A large percentage of people who are sent to hospitals for treatment of infection of the urinary tract are susceptible to biofilm infections, which can have detrimental effects. This risk was seen to be greater in patients who are receiving urinary catheters. Research has demonstrated that catheter-associated biofilms had a high level of antibiotic resistance and proved challenging to cure Since the bacteria that are contributing to the infection have been found to be capable of developing biofilms, biofilm infestation contributed to 95% of the patients with urinary tract infections. Tests for antibiotic susceptibility that are often used might not be effective in treating illnesses linked to biofilms. Because these biofilm diseases are so persistent, an enormous sum of resources is actually spent on treating them. The length of therapy is an additional consideration since patients who are hospitalized for an extended period of time are more susceptible to infection and because prolonged hospital stays raise treatment costs. Thus, effective methods are required to counteract the negative effects of biofilms (Danhorn and Fuqua 2007).

Biofilm Formation in Food Industries

Infections caused by food can result from infections resulting from microbial biofilm on food or in machines used in industries. Bacteria associated with biofilm may discharge toxins when it is present in factories that process food. Those bacteria may infect a food matrix, leading to disease outbreaks. Regardless of the cause, biofilms in food factories are dangerous for people's well-being. The types of bacteria that make up this three-dimensional biological architecture determine how hazardous it is.

Bacillus cereus

Preventing the persistence of *Bacillus cereus* in food processing facilities is essential for safeguarding human health. Additionally, because these bacteria produce endospores, they may withstand industrialized pasteurization operations. These bacteria can create two different types of toxins: one called heat-resistant cereulide, which triggers nausea, and another called diarrheal enterotoxins, which induce diarrhoea and abdominal discomfort. It is recognized that they can all result in food poisoning. It is especially significant as this frequent food-associated infection produces haemolysis that can cause severe dehydration and potentially death (López et al. 2015; Tschiedel et al. 2015). The preconditioning process begins with the initial adhesion of *B. cereus* to food surfaces, resulting in making it easier for further attachment of different types of secondary bacteria to quickly attach themselves. This kind of bacterium can often be detected in food and drink companies as well as dairy industries (Ehling-Schulz et al. 2015). The majority of biofilms, developed by *B. cereus,* are found in the juncture of gas and liquid and a characteristic ring that is linked to the deposited wall in dairy industries. On the contrary, around submerged surfaces such as pipelines, a few strains of it can also form a biofilm (Hayrapetyan et al. 2015; Wijman et al. 2007). Bacteria found in such biofilms may readily travel great distances through the supply lines of food factories, creating major hazards to health if they make it to the food samples that are given to customers (Hayrapetyan et al. 2015).

Escherichia coli

Within the human gut microbes, the majority of *E. coli* strains do not pose a health risk. Some strains, in contrast, are harmful pathogens transmitted through food that may infect humans and are spread by raw milk, fresh meat, fruits and vegetables, as well as by drinking water. These goods can become tainted during the food production process or at the point of origin. When growing vegetables, the use of tainted water supplies can lead to infection in the food sector prior to the harvesting stage. In addition to occurring in later settings following the cleaning and handling of the unprocessed product (vegetables, dead animals, etc.), this contamination can also occur as a result of conditions of storage that promote the rapid development of the current bacterial contaminants (Carter et al. 2019). *E. coli* strains have been shown in several investigations to be able to adhere to a wide range of substrates, such as Teflon coating glass, stainless steel, polystyrene, polypropylene, PVC and biological surfaces. This species forms biofilms in large part because of the outermost material's hydrophobic properties. For instance, the *E. coli* O157:H7 strain formed thick biofilms on stainless steel and borosilicate glass, while barely any biofilms were seen on polypropylene, most likely because of the material's hydrophobic properties. Due to its ability to withstand excessive acidity, *Escherichia coli* is likewise recognized for its role in the emergence of pathogenic strains such as O26. These strains are typically produced by procedures used in food processing that involve acetic acid, citric acid, propionic acid and lactic acid (Lajhar et al. 2017). In terms of clinical problems, enterohaemorrhagic gastroenteritis (which manifests as watery stool with bloodshed in the stools) is caused by Shiga toxin (Stx)-producing *E. coli* (STEC) secreting Stx1 and Stx2 toxins. This type of bacteria can cause infection as well as death when these microbes get contaminated with soups, sauce, cooked poultry, cooked beef, salads and other fresh products. In some cases, this condition can lead to haemolytic uremic syndrome, which is dangerous to the kidneys and causes thrombocytopenia (Bowen and Coward 2018).

1.8 Conclusion

Therefore, the intricate environment of biofilm has a complex population of microbial organisms with applications in different fields. In this chapter, we have discussed various attributes to coincide with the formation, role, importance, application and drawbacks of biofilms. The formation of biofilm begins by portraying the difference in adhesion mechanisms between gram-positive and gram-negative bacteria and also the growth of the population and the final growth stage of it. Understanding these steps yields valuable information about the formation and regulation of biofilms. The features that point out the role of biofilms in the context of ecology include cyclic-di-GMP signalling, hydrodynamic conditions and other features such as temperature and pH since they determine the structure and function of biofilms in various environments. Among the functional activities, chemical application and communication mechanism called quorum sensing are some of the most intriguing functions of microorganisms in biofilms. These processes influence the optimality of the biofilm and its capacity to evolve in its environment. In this chapter, advances in technology are described as having expanded biochemical knowledge of biofilms and are said to ease the generation of novel strategies for biofilm control and management. However, it is important not to lose sight of the fact that there are definite disadvantages with biofilms, which are noted for life threats in clinical settings and for antibiotic resistance. Moreover, in the food industry, biofilm formation by *E. coli* and *Bacillus cereus* is a key factor in public health and food hygiene issues. Biofilm is filled with challenge and mystery and still, there is so much that remains a secret. Understanding more about its origin, functions, importance and impacts enables us to learn more about the microbial community and introduce new approaches to fight biofilms in different environments.

References

Abd-El-Haleem D, Beshay U, Abdelhamid AO, Moawad H, Zaki S (2003) Effects of mixed nitrogen sources on biodegradation of phenol by immobilized Acinetobacter sp. strain W-17. Afr J Biotechnol 2(1):8–12. https://doi.org/10.5897/AJB2003.000-1002

Alhede M, Qvortrup K, Liebrechts R, Høiby N, Givskov M, Bjarnsholt T (2012) Combination of microscopic techniques reveals a comprehensive visual impression of biofilm structure and composition. FEMS Immunol Med Microbiol 65(2):335–342. https://doi.org/10.1111/j.1574-695X.2012.00956.x

Annous BA, Fratamico PM, Smith JL (2009) Scientific status summary: quorum sensing in biofilms: why bacteria behave the way they do. J Food Sci 74(1):R24–R37. https://doi.org/10.1111/j.1750-3841.2008.01022.x

Bai J, Wen JP, Li HM, Jiang Y (2007) Kinetic modeling of growth and biodegradation of phenol and m-cresol using Alcaligenes faecalis. Process Biochem 42(4):510–517. https://doi.org/10.1016/j.procbio.2006.10.004

Balcázar JL, Subirats J, Borrego CM (2015) The role of biofilms as environmental reservoirs of antibiotic resistance. Front Microbiol 6:1216. https://doi.org/10.3389/fmicb.2015.01216

Barbosa A, Miranda S, Azevedo NF, Cerqueira L, Azevedo AS (2023) Imaging biofilms using fluorescence in situ hybridization: seeing is believing. Front Cell Infect Microbiol 13:1195803. https://doi.org/10.3389/fcimb.2023.1195803

Bennett-Guerrero E, McIntosh TJ, Barclay GR, Snyder DS, Gibbs RJ, Mythen MG, Poxton IR (2000) Preparation and preclinical evaluation of a novel liposomal complete-core lipopolysaccharide vaccine. Infect Immun 68(11):6202–6208. https://doi.org/10.1128/iai.68.11.6202-6208.2000

Berlanga M, Guerrero R (2016) Living together in biofilms: the microbial cell factory and its biotechnological implications. Microb Cell Factories 15:1–11. https://doi.org/10.1186/s12934-016-0569-5

Bowen EE, Coward RJ (2018) Advances in our understanding of the pathogenesis of hemolytic uremic syndromes. Am J Physiol Renal Physiol 314:454–461. https://doi.org/10.1152/ajprenal.00376.2017.-Hemo

Bucher T, Oppenheimer-Shaanan Y, Savidor A, Bloom-Ackermann Z, Kolodkin-Gal I (2015) Disturbance of the bacterial cell wall specifically interferes with biofilm formation. Environ Microbiol Rep 7(6):990–1004. https://doi.org/10.1111/1758-2229.12346

Bühler T, Ballestero S, Desai M, Brown MRW (1998) Generation of a reproducible nutrient-depleted biofilm of Escherichia coli and Burkholderia cepacia. J Appl Microbiol 85(3):457–462. https://doi.org/10.1046/j.1365-2672.1998.853501.x

Caiazza NC, O'Toole GA (2004) SadB is required for the transition from reversible to irreversible attachment during biofilm formation by Pseudomonas aeruginosa PA14. J Bacteriol 186(14):4476–4485. https://doi.org/10.1128/JB.186.14.4476-4485.2004

Carter MQ, Feng D, Li HH (2019) Curli fimbriae confer Shiga toxin-producing Escherichia coli a competitive trait in mixed biofilms. Food Microbiol 82:482–488. https://doi.org/10.1016/j.fm.2019.03.024

Cepas V, López Y, Muñoz E, Rolo D, Ardanuy C, Martí S, Xercavins M, Horcajada JP, Bosch J, Soto SM (2019) Relationship between biofilm formation and antimicrobial resistance in gram-negative bacteria. Microb Drug Resist 25(1):72–79. https://doi.org/10.1089/mdr.2018.0027

Chaieb K, Chehab O, Zmantar T, Rouabhia M, Mahdouani K, Bakhrouf A (2007) In vitro effect of pH and ethanol on biofilm formation by clinical ica-positive Staphylococcus epidermidis strains. Ann Microbiol 57:431–437. https://doi.org/10.1007/BF03175085

Chalabaev S, Chauhan A, Novikov A, Iyer P, Szczesny M, Beloin C, Caroff M, Ghigo JM (2014) Biofilms formed by gram-negative bacteria undergo increased lipid a palmitoylation, enhancing in vivo survival. MBio 5(4). https://doi.org/10.1128/mBio.01116-14

Ciofu O, Tolker-Nielsen T (2019) Tolerance and resistance of Pseudomonas aeruginosa biofilms to antimicrobial agents—how P. aeruginosa can escape antibiotics. Front Microbiol 10:913. https://doi.org/10.3389/fmicb.2019.00913

Cleaver L, Garnett JA (2023) How to study biofilms: technological advancements in clinical biofilm research. Front Cell Infect Microbiol 13:1335389. https://doi.org/10.3389/fcimb.2023.1335389

Cotter PA, Stibitz S (2007) c-di-GMP-mediated regulation of virulence and biofilm formation. Curr Opin Microbiol 10(1):17–23. https://doi.org/10.1016/j.mib.2006.12.006

Cruz DP, Huertas MG, Lozano M, Zárate L, Zambrano MM (2012) Comparative analysis of diguanylate cyclase and phosphodiesterase genes in Klebsiella pneumoniae. BMC Microbiol 12. https://doi.org/10.1186/1471-2180-12-139

Cvitkovitch DG, Li YH, Ellen RP (2003) Quorum sensing and biofilm formation in Streptococcal infections. J Clin Invest 112(11):1626–1632. https://doi.org/10.1172/JCI200320430

Danhorn T, Fuqua C (2007) Biofilm formation by plant-associated bacteria. Ann Rev Microbiol 61(1):401–422. https://doi.org/10.1146/annurev.micro.61.080706.093316

Dewanti R, Wong AC (1995) Influence of culture conditions on biofilm formation by Escherichia coli O157: H7. Int J Food Microbiol 26(2):147–164. https://doi.org/10.1016/0168-1605(94)00103-D

Donlan RM (2002) Biofilms: microbial life on surfaces. Emerg Infect Dis 8(9):881. https://doi.org/10.3201/eid0809.020063

Dunne WM Jr (2002) Bacterial adhesion: seen any good biofilms lately? Clin Microbiol Rev 15(2):155–166. https://doi.org/10.1128/CMR.15.2.155-166.2002

Dusane DH, Zinjarde SS, Venugopalan VP, Mclean RJ, Weber MM, Rahman PKSM (2010) Quorum sensing: implications on Rhamnolipid biosurfactant production. Biotechnol Genet Eng Rev 27(1):159–184. https://doi.org/10.1080/02648725.2010.10648149

Ehling-Schulz M, Frenzel E, Gohar M (2015) Food–bacteria interplay: pathometabolism of emetic Bacillus cereus. Front Microbiol 6:704. https://doi.org/10.3389/fmicb.2015.00704

Federle MJ, Bassler BL (2003) Interspecies communication in bacteria. J Clin Invest 112(9):1291–1299. https://doi.org/10.1172/jci200320195

Franklin MJ, Chang C, Akiyama T, Bothner B (2015) New technologies for studying biofilms. In: Microbial biofilms, pp 1–32. https://doi.org/10.1128/9781555817466.ch1

Fu L, Wu C, Zhou Y, Zuo J, Ding Y (2017) Investigation on evaluation criteria of backwashing effects for a pilot-scale BAF treating petrochemical wastewater. Environ Technol (United Kingdom) 38(20):2523–2533. https://doi.org/10.1080/09593330.2016.1269838

Garrett TR, Bhakoo M, Zhang Z (2008) Bacterial adhesion and biofilms on surfaces. Prog Nat Sci 18(9):1049–1056. https://doi.org/10.1016/j.pnsc.2008.04.001

Gomes IB, Meireles A, Gonçalves AL, Goeres DM, Sjollema J, Simões LC, Simões M (2018) Standardized reactors for the study of medical biofilms: a review of the principles and latest modifications. Crit Rev Biotechnol 38(5):657–670. https://doi.org/10.1080/07388551.2017.1380601

Hall-Stoodley L, Stoodley P (2002) Developmental regulation of microbial biofilms. Curr Opin Biotechnol 13(3):228–233. https://doi.org/10.1016/S0958-1669(02)00318-X

Hancock REW, Siehnel R, Martin N (1990) Outer membrane proteins of pseudomonas. Mol Microbiol 4(7):1069–1075. https://doi.org/10.1111/j.1365-2958.1990.tb00680.x

Hayrapetyan H, Muller L, Tempelaars M, Abee T, Nierop Groot M (2015) Comparative analysis of biofilm formation by Bacillus cereus reference strains and undomesticated food isolates and the effect of free iron. Int J Food Microbiol 200:72–79. https://doi.org/10.1016/j.ijfoodmicro.2015.02.005

Heacock-Kang Y, Zarzycki-Siek J, Sun Z, Poonsuk K, Bluhm AP, Cabanas D, Fogen D, McMillan IA, Chuanchuen R, Hoang TT (2018) Novel dual regulators of Pseudomonas aeruginosa essential for productive biofilms and virulence. Mol Microbiol 109(3):401–414. https://doi.org/10.1111/mmi.14063

Hinsa SM, Espinosa-Urgel M, Ramos JL, O'Toole GA (2003) Transition from reversible to irreversible attachment during biofilm formation by Pseudomonas fluorescens WCS365 requires an ABC transporter and a large secreted protein. Mol Microbiol 49(4):905–918. https://doi.org/10.1046/j.1365-2958.2003.03615.x

Høiby N, Bjarnsholt T, Givskov M, Molin S, Ciofu O (2010) Antibiotic resistance of bacterial biofilms. Int J Antimicrob Agents 35(4):322–332. https://doi.org/10.1016/j.ijantimicag.2009.12.011

Jamal M, Ahmad W, Andleeb S, Jalil F, Imran M, Nawaz MA et al (2018) Bacterial biofilm and associated infections. J Chin Med Assoc 81(1):7–11. https://doi.org/10.1016/j.jcma.2017.07.012

Jurcisek JA, Dickson AC, Bruggeman ME, Bakaletz LO (2011) In vitro biofilm formation in an 8- well chamber slide. J Vis Exp 47. https://doi.org/10.3791/2481

Kamal-Alahmad (2015) The definition, preparation and application of Rhamnolipids as biosurfactants. Int J Nutr Food Sci 4(6):613–623. https://doi.org/10.11648/j.ijnfs.20150406.13

Keleştemur S, Avci E, Çulha M (2018) Raman and surface-enhanced Raman scattering for biofilm characterization. Chemosensors 6(1):5. https://doi.org/10.3390/chemosensors6010005

Khatoon Z, McTiernan CD, Suuronen EJ, Mah T-F, Alarcon EI, Alarcon Bacterial EI (2018) Bacterial biofilm formation on implantable devices and approaches to its treatment and prevention. Heliyon 4:e01067. https://doi.org/10.1016/j.heliyon.2018

Kim J, Park H-D, Chung S (2012) Microfluidic approaches to Bacterial biofilm formation. Molecules 17:9818–9834. https://doi.org/10.3390/molecules17089818

Kumar R, Das AJ (2018) Quorum sensing: its role in Rhamnolipid production. Rhamnolipid Biosurfactant 125–135. https://doi.org/10.1007/978-981-13-1289-2_11

Lajhar SA, Brownlie J, Barlow R (2017) Survival capabilities of Escherichia coli O26 isolated from cattle and clinical sources in Australia to disinfectants, acids and antimicrobials. BMC Microbiol 17(1):47. https://doi.org/10.1186/s12866-017-0963-0

Lajhar SA, Brownlie J, Barlow R (2018) Characterization of biofilm-forming capacity and resistance to sanitizers of a range of E. coli O26 pathotypes from clinical cases and cattle in Australia. BMC Microbiol 18(1). https://doi.org/10.1186/s12866-018-1182-z

Lécuyer F, Bourassa J-S, Gélinas M, Charron-Lamoureux V, Burrus V, Beauregard PB (2018) Biofilm formation drives transfer of the conjugative element ICE Bs1 in Bacillus subtilis. MSphere 3(5). https://doi.org/10.1128/msphere.00473-18

Lee KWK, Periasamy S, Mukherjee M, Xie C, Kjelleberg S, Rice SA (2014) Biofilm development and enhanced stress resistance of a model, mixed-species community biofilm. ISME J 8(4):894–907. https://doi.org/10.1038/ismej.2013.194

Lee KJ, Lee MA, Hwang W, Park H, Lee KH (2016) Deacylated lipopolysaccharides inhibit biofilm formation by Gram-negative bacteria. Biofouling 32(7):711–723. https://doi.org/10.1080/08927014.2016.1193595

Lembre P, Lorentz C, Di Martino P (2012) Exopolysaccharides of the biofilm matrix: a complex biophysical world. In: The complex world of polysaccharides, pp 371–392. https://doi.org/10.5772/51213

Lewis KIM (2001) Riddle of biofilm resistance. Antimicrob Agents Chemother 45(4):999–1007. https://doi.org/10.1128/AAC.45.4.999-1007.2001

Liu Y, Tay JH (2001) Metabolic response of biofilm to shear stress in fixed-film culture. J Appl Microbiol 90(3):337–342. https://doi.org/10.1046/j.1365-2672.2001.01244.x

López AC, Minnaard J, Pérez PF, Alippi AM (2015) A case of intoxication due to a highly cytotoxic Bacillus cereus strain isolated from cooked chicken. Food Microbiol 46:195–199. https://doi.org/10.1016/j.fm.2014.08.005

Makin SA, Beveridge TJ (1996) The influence of A-band and B-band lipopolysaccharide on the surface characteristics and adhesion of Pseudomonas aeruginosa to surfaces. Microbiology 142(2):299–307. https://doi.org/10.1099/13500872-142-2-299

Merritt JH, Kadouri DE, O'Toole GA (2011) Growing and analyzing static biofilms. Curr Protoc Microbiol 22(1):1B. https://doi.org/10.1002/9780471729259.mc01b01s22

Mills E, Pultz IS, Kulasekara HD, Miller SI (2011) The bacterial second messenger c-di-GMP: mechanisms of signalling. Cell Microbiol 13(8):1122–1129. https://doi.org/10.1111/j.1462-5822.2011.01619.x

Mishra S, Huang Y, Li J, Wu X, Zhou Z, Lei Q et al (2022) Biofilm-mediated bioremediation is a powerful tool for the removal of environmental pollutants. Chemosphere 294:133609. https://doi.org/10.1016/j.chemosphere.2022.133609

Nair HAS, Periasamy S, Yang L, Kjelleberg S, Rice SA (2017) Real time, spatial, and temporal mapping of the distribution of c-di-GMP during biofilm development. J Biol Chem 292(2):477–487. https://doi.org/10.1074/jbc.M116.746743

Navon-Venezia S, Kondratyeva K, Carattoli A (2017) Klebsiella pneumoniae: a major worldwide source and shuttle for antibiotic resistance. FEMS Microbiol Rev 41(3):252–275. https://doi.org/10.1093/femsre/fux013

Nostro A, Cellini L, Di Giulio M, D'Arrigo M, Marino A, Blanco AR, Favaloro A, Cutroneo G, Bisignano G (2012) Effect of alkaline pH on staphylococcal biofilm formation. APMIS 120(9):733–742. https://doi.org/10.1111/j.1600-0463.2012.02900.x

O'Toole GA, Kolter R (1998) Flagellar and twitching motility are necessary for Pseudomonas aeruginosa biofilm development. Mol Microbiol 30(2):295–304. https://doi.org/10.1046/j.1365-2958.1998.01062.x

Oder M, Fink R, Bohinc K, Torkar KG (2017) The influence of shear stress on the adhesion capacity of legionella pneumophila. Arh Hig Rada Toksikol 68(2):109–115. https://doi.org/10.1515/aiht-2017-68-2904

Ojumu TV, Bello OO, Solomon. (2005) Evaluation of microbial systems for bioremediation of petroleum refinery effluents in Nigeria. Afr J Biotechnol 4(1):31–35

O'Toole GA, Gibbs KA, Hager PW, Phibbs PV Jr, Kolter R (2000) The global carbon metabolism regulator Crc is a component of a signal transduction pathway required for biofilm development by Pseudomonas aeruginosa. J Bacteriol 182(2):425–431. https://doi.org/10.1128/jb.182.2.425-431.2000

Pagán R, García-Gonzalo D (2015) Influence of environmental factors on bacterial biofilm formation in the food industry: a review. Postdoc j., (ART-2015-95845). https://doi.org/10.14304/surya.jpr.v3n6.2

Palmer J, Flint S, Brooks J (2007) Bacterial cell attachment, the beginning of a biofilm. J Ind Microbiol Biotechnol 34(9):577–588. https://doi.org/10.1007/s10295-007-0234-4

Pinto RM, Soares FA, Reis S, Nunes C, Van Dijck P (2020) Innovative strategies toward the disassembly of the EPS matrix in bacterial biofilms. Front Microbiol 11:952. https://doi.org/10.3389/fmicb.2020.00952

Pompilio A, Piccolomini R, Picciani C, D'Antonio D, Savini V, Di Bonaventura G (2008) Factors associated with adherence to and biofilm formation on polystyrene by Stenotrophomonas maltophilia: the role of cell surface hydrophobicity and motility. FEMS Microbiol Lett 287(1):41–47. https://doi.org/10.1111/j.1574-6968.2008.01292.x

Prakash B, Veeregowda BM, Krishnappa G (2003) Biofilms: a survival strategy of bacteria. Curr Sci:1299–1307. https://www.jstor.org/stable/24108133

Purevdorj B, Costerton JW, Stoodley P (2002) Influence of hydrodynamics and cell signaling on the structure and behavior of Pseudomonas aeruginosa biofilms. Appl Environ Microbiol 68(9):4457–4464. https://doi.org/10.1128/AEM.68.9.4457-4464.2002

Relucenti M, Familiari G, Donfrancesco O, Taurino M, Li X, Chen R, Artini M, Papa R, Selan L (2021) Microscopy methods for biofilm imaging: focus on SEM and VP-SEM pros and cons. Biology 10:51. https://doi.org/10.3390/biology10010051

Rochex A, Lebeault JM (2007) Effects of nutrients on biofilm formation and detachment of a Pseudomonas putida strain isolated from a paper machine. Water Res 41(13):2885–2892. https://doi.org/10.1016/j.watres.2007.03.041

Rodrigues CF, Allkja J, Mendes L, Azevedo AS (2022) Methods for the visualization of multispecies biofilms. In: Multispecies biofilms: technologically advanced methods to study microbial communities. Springer International Publishing, Cham, pp 35–78. https://doi.org/10.1007/978-3-031-15349-5_2

Römling U, Galperin MY, Gomelsky M (2013) Cyclic di-GMP: the first 25 years of a universal Bacterial second messenger. Microbiol Mol Biol Rev 77(1):1–52. https://doi.org/10.1128/mmbr.00043-12

Schaber JA, Carty NL, McDonald NA, Graham ED, Cheluvappa R, Griswold JA, Hamood AN (2004) Analysis of quorum sensing-deficient clinical isolates of Pseudomonas aeruginosa. J Med Microbiol 53(9):841–853. https://doi.org/10.1099/jmm.0.45617-0

Sisti F, Ha DG, O'Toole GA, Hozbor D, Fernández J (2013) Cyclic-di-GMP signalling regulates motility and biofilm formation in Bordetella bronchiseptica. Microbiology 159(Pt_5):869–879. https://doi.org/10.1099/mic.0.064345-0

Srivastava D, Waters CM (2012) A tangled web: regulatory connections between quorum sensing and cyclic Di-GMP. J Bacteriol 194(17):4485–4493. https://doi.org/10.1128/JB.00379-12

Straub H, Eberl L, Zinn M, Rossi RM, Maniura-Weber K, Ren Q (2020) A microfluidic platform for in situ investigation of biofilm formation and its treatment under controlled conditions. J Nanobiotechnol 18:1–12. https://doi.org/10.1186/s12951-020-00724-0

Sutherland IW (2001) Microbial polysaccharides from Gram-negative bacteria. Int Dairy J 11(9):663–674. https://doi.org/10.1016/S0958-6946(01)00112-1

Thoendel M, Kavanaugh JS, Flack CE, Horswill AR (2011) Peptide signaling in the Staphylococci. Chem Rev 111(1):117–151. https://doi.org/10.1021/cr100370n

Tilahun A, Haddis S, Teshale A, Hadush T (2016) Review on biofilm and microbial adhesion. Int J Microbiol Res 7(3):63–73. https://doi.org/10.5829/idosi.ijmr.2016.63.73

Toyofuku M, Inaba T, Kiyokawa T, Obana N, Yawata Y, Nomura N (2016) Environmental factors that shape biofilm formation. Biosci Biotechnol Biochem 80(1):7–12. https://doi.org/10.1080/09168451.2015.1058701

Tschiedel E, Rath PM, Steinmann J, Becker H, Dietrich R, Paul A et al (2015) Lifesaving liver transplantation for multi-organ failure caused by Bacillus cereus food poisoning. Pediatr Transplant 19(1):E11–E14. https://doi.org/10.1111/petr.12378

Unell M, Nordin K, Jernberg C, Stenström J, Jansson JK (2008) Degradation of mixtures of phenolic compounds by Arthrobacter chlorophenolicus A6. Biodegradation 19(4):495–505. https://doi.org/10.1007/s10532-007-9154-2

Vasudevan R (2014) Biofilms: microbial cities of scientific significance. J Microbiol Experiment 1(3). https://doi.org/10.15406/jmen.2014.01.00014

Verderosa AD, Totsika M, Fairfull-Smith KE (2019) Bacterial biofilm eradication agents: a current review. Front Chem 7:495483. https://doi.org/10.3389/fchem.2019.00824

Wentland EJ, Stewart PS, Huang CT, McFeters GA (1996) Spatial variations in growth rate within Klebsiella pneumoniae colonies and biofilm. Biotechnol Prog 12(3):316–321. https://doi.org/10.1021/bp9600243

Wijman JGE, De Leeuw PPLA, Moezelaar R, Zwietering MH, Abee T (2007) Air-liquid interface biofilms of Bacillus cereus: formation, sporulation, and dispersion. Appl Environ Microbiol 73(5):1481–1488. https://doi.org/10.1128/AEM.01781-06

Wloka M, Rehage H, Flemming HC, Wingender J (2004) Rheological properties of viscoelastic biofilm extracellular polymeric substances and comparison to the behavior of calcium alginate gels. Colloid Polym Sci 282(10):1067–1076. https://doi.org/10.1007/s00396-003-1033-8

Architectural Marvel: Biofilm Formation

2

Mohan Das, Devalina Khamaru, Sayantan Santra, Siddu Lalsangi, Rudrak Gupta, Sarobi Das, and Rintu Banerjee

Abstract

The first stage in microcolony formation and biofilm development is the attachment of bacteria to the surfaces, involving the aggregation of microbes beneath a self-assembled structure of extracellular polymeric substances (EPS). Exopolysaccharides, different kinds of structural proteins, lipids, enzymes and extracellular DNA make up the bulk of this organized structure. It imparts rigidity and also serves as an external organ for digestion. A crucial regulatory compound, cyclic-di-GMP modulates its activities and promotes the shift from the free planktonic stage to biofilm-forming sessile stage. The EPS of Enterobacteriaceae primarily consists of curli amyloid fibres. Disruption of EPS structural components with enzymes such as glycoside hydrolase and DNase is possible for bacterial biofilm. Microorganisms living inside biofilms are significantly more resistant to antibiotics when compared to motile free-living microbes and this resistance has become a threat to human specially in the clinical environments as well as food industries, regardless of whether there are strict protocols to mitigate their development and propagation. For better observation and understanding of the intricate architecture of biofilm, modern methods of imaging SEM, VPSEM and FIBSEM have been proven very beneficial. Advanced research on the architectural framework of biofilm and advanced techniques for its destruction remains an important concern for human welfare.

M. Das (✉) · D. Khamaru · S. Santra · S. Lalsangi · R. Banerjee
Agricultural and Food Engineering Department, Indian Institute of Technology Kharagpur, Kharagpur, India
e-mail: mohan@agfe.iitkgp.ac.in

R. Gupta
Virus Research and Diagnostic Laboratory, ICMR-National Institute of Cholera and Enteric Diseases, Kolkata, India

S. Das
Department of Biology, University of Texas at Arlington, Arlington, TX, USA

T. Sarkar et al. (eds.), *Biofilm Applications to Revolutionize Food Technology*,
https://doi.org/10.1007/978-3-031-85205-3_2

Keywords

Biofilm development · Bacteria attachment · Extracellular polymeric substances · Cyclic-di-GMP · Biofilm resistance and architectural framework

2.1 Introduction

In the vast domain of natural marvels, biofilms are captivating and complex wonders of architecture. These organized communities of microorganisms adhere to the substratum and form intricate frameworks that present a challenge to our understanding of microbial life. Research into the initial attachment of microbes to the surfaces, microcolony formation, compounds that make up extracellular polymeric substances, the 3D structure of biofilm framework and advanced imagining techniques is crucial for a better understanding of intricate biofilm development. Derjaguin–Landau–Verwey–Overbeek (DLVO) as well as thermodynamic theory are able to explain the adhesion mechanism of bacteria. DLVO theory focuses on the equilibrium between attractive van der Waals forces and repulsive electrostatic interactions. According to this, the attractive force required for adhesion is administrated by electrostatic interaction and Van der Waal forces. Conversely, thermodynamic theory contemplates free energy shifts during adhesion. Factors such as surface charge, hydrophobicity and roughness determine the free energy change during adhesion according to this theory. Microbial interaction with substratum can be better understood with an understanding of these theories. Followed by adhesion, spatial structure is established, resulting in the development of microcolonies with these microorganisms. Based on the arrangement of bacterial cells, biofilm's function and behaviour are influenced.

Extracellular polymeric substances (EPS), an important compound of extracellular matrix function as glue to bind the biofilm together. EPS includes various elements such as polysaccharides, proteins, extracellular DNA and enzymes. It also confers protective measures, structural integrity and nutrient accessibility to biofilm. For maintaining the function and development of biofilm, EPS architecture is crucial. Cells within the matrix are enveloped by EPS, which provides a microenvironment to them. A special type of EPS that creates capsules shows more significant affinity towards cell exteriors when compared to other types (Sutherland 2001). To understand the 3D architecture of biofilm, advanced imaging techniques are required. Light microscopy, confocal laser scanning microscopy, scanning electron microscopy and variable pressure scanning electron microscopy are some of the examples that offer insights into biofilm organization and structure. These advanced techniques also facilitate the study of biofilms at different magnifications, from micro to nano levels. They allow us to investigate from surface interactions to biofilm development and mature structure. Biofilms constitute complex microbial communities that possess a sophisticated organization formed by attachment to microbial cells to substratum, synthesis of EPS and spatial organization. A significant understanding of these processes can mitigate biofilm-associated problems in different sectors such as the food industry and clinical environment.

2.2 Microbial Adhesion and Surface Interactions

A process model comprising five phases has been proposed to describe the development of biofilms. The methodology comprised four distinct phases: (i) the initial cellular adhesion to the substratum; (ii) the cells' subsequent permanent adherence; (iii) the formation of biofilm microcolonies; (iv) the biofilm's growth and maturation and (v) dispersal of bacterial cells. It is hypothesized that physiological changes occur in the cells and EPS is produced concurrently with the initial stages of cell attachment to the surface. EPS functions as a cohesive agent, guaranteeing the cells' connection to the exterior. Extracellular polymeric substance (EPS) is a hydrogel primarily esteemed for its protein, polysaccharide and nucleic acid components. It supplies the biofilm with its fundamental mechanical integrity. Following the initial division that occurs during attachment, the cells proceed to develop into microcolonies. As the cells mature, they endure additional division and generate a higher amount of EPS. This phase of biofilm development possesses the capability to proliferate beyond the epidermis and develop into a three-dimensional architecture. Once cells have reached maturation, they are capable of escaping the biofilm and establishing colonies in various locations (Asally et al. 2012; Hall-Stoodley and Stoodley 2002; Stoodley et al. 2002).

The process by which microbes accumulate on an acquiring surface can be separated into three distinct phases: (1) adsorption, which refers to the deposition of a cell onto the adsorbent from a substratum; (2) attachment, in which cells' connection with the collector is fortified, usually using the formation of polymer linkages and (3) colonization, which denotes the expansion and proliferation of cells on the collector's surface (Fletcher 1996). A surface-forming organic or inorganic stratum known as the conditioning film serves as a foundation site for the initial reversibly associated cells; however, the cohesiveness of the conditioning layer determines the matrix's robustness. Aqueous media, such as water, can cause this layer to develop in minutes and then continue to expand for hours (Donlan 2002; Garrett et al. 2008; Perni et al. 2014).

The presence of any substance within the bulk fluid can result in its settling onto a substrate and subsequent incorporation into a conditioning layer due to gravitational force or flow movement. This layer alters the underlying materials to enhance the ability of germs to reach them. This layer alters the underlying materials to enhance the ability of microbes to reach them. Bacterial communities can flourish on substrates because they offer a place to anchor and nutrients that promote their development (Garrett et al. 2008). The duration required for the development of conditioning film might range from an instant to several minutes (Garrido et al. 2014). After the process of surface conditioning, microorganisms undergo attachment, adhesion and subsequent retention on a given substrate (Whitehead and Verran 2015). The potential impact of the preliminary chemical composition of the surface on microbial adhesion and retention suggests that the presence of a conditioning layer may influence the development of biofilms (Moreira et al. 2017; Whitehead and Verran 2015).

Bacterial Adhesion Theory

Biofilm mostly consists of bacteria with sizes ranging from 0.5 to 2μm, which lies in the colloidal particle range. Thus, the DLVO theory details bacterial attachment (Marshall et al. 1971). The DLVO theory elucidates the process by which bacteria adhere to substrates, attributing it to Lifshitz–Van der Waals forces, Coulomb forces within this bilayer, and, in its expanded formulation, acid–base binding. Because of the prevalence of Van der Waals attraction in surface regions, particles cannot disengage off that region by Brownian motion, resulting in irreversible adhesion. Van der Waals interaction drops dramatically with greater separation, making Coulomb interaction increasingly significant. The overall connections (ΔGadh) of the bacterium with its substrate interface are dictated by the equilibrium of attracting van der Waals interactions (ΔGvdW) and repellent electrostatic interactions (ΔGdl) caused by the proximity of two electrical layers of the organism and surface of the material. Derjaguin–Landau–Verwey–Overbeek (DLVO) theory is generally mathematically formulated by

$$\Delta G_{\text{adh}}\left(d\right) = \Delta G_{\text{vdW}}\left(d\right) + \Delta G_{\text{dl}}\left(d\right) \tag{2.1}$$

The above-mentioned equation demonstrates how the gap between the cellular structure and the substratum affects both forces. If the bacteria are near the exterior, the van der Waals interactions get stronger and cells remain stuck to the surface permanently. Conversely, when the space separating the microbial cell from the matrix increases, the van der Waals interactions significantly diminish and Coulomb interactions take a greater role, leading to the cell's repulsive effect on the substratum (Hori and Matsumoto 2010). Furthermore, a critical component of bacterial attachment is the cell's ionicity. Positively and negatively charged particles will naturally gravitate towards one another if the ionic strength is strong. If the ionic strength is modest, the particles will naturally repel one another because of their similar charges. Nonetheless, reversible adhesion might take place when a microbial cell comes close to the interface as a result of its mobility or Brownian motion. Here, the bacterial cell must surpass a specific energy threshold. Upon approaching the surface, a microbial cell can employ its extensions, such as pili and flagella, to adhere to the substrate. These extensions can penetrate the energy threshold and promote bacterial adherence (Hori and Matsumoto 2010). The DLVO theory elucidates the likelihood of a living entity surmounting an electrostatic obstacle for clarity. A recent study has demonstrated a correlation between ionic strength and attachment of bacteria, aligning with the DLVO theory.

Extended DLVO Theory

One way to look at hydrogen bonding is as an example of the Lewis acid–base interaction (Van Oss 1991). Consequently, the expanded DLVO theory was developed, which introduced both hydrophobic/hydrophilic forces and osmotic interactions into the model. However, the impact of osmotic interactions on bacterial attachment can be disregarded since they have little effect. Consequently, the result may be expressed by the subsequent equation.

$$\Delta G_{\mathrm{adh}} = \Delta G_{\mathrm{vdW}} + \Delta G_{\mathrm{dl}} + \Delta G_{\mathrm{ab}} \tag{2.2}$$

The symbol ΔG_{vdW} represents the Lifshitz–van der Waals interaction, ΔG_{dl} represents the two-layer electric interaction and ΔG_{ab} represents acid–base interactions. The extended DLVO theory, which considers hydrophobic and repulsive hydration impact, provides more accurate predictions of adherence compared to the ordinary DLVO theory. The interplay between acids and bases also has a significant impact on reducing the energy barrier. Nevertheless, the acid–base interaction only becomes active when the interval from the substrate is modest, specifically 5 nm (Bos et al. 1999).

Thermodynamic Theory

A physicochemical method was employed to define microbial adhesion with substrates and to quantify the van der Waals, electrostatic and dipole forces in regard to free energy. This is formulated as

$$\Delta G_{\mathrm{adh}} = \gamma_{\mathrm{sm}} - \gamma_{\mathrm{sl}} - \gamma_{\mathrm{ml}} \tag{2.3}$$

γ_{sm}, γ_{sl} and γ_{ml} represent the free energies of the solid–microorganism, solid–liquid and microorganism–liquid interfaces, respectively. If the value of free energy is negative, it indicates that attachment is preferred (Silva-Dias et al. 2015).

2.3 Microcolony Formation and Spatial Organization

For bacteria to adhere successfully, it is necessary to be near interfaces. This might be accomplished by either active or passive motions in the aqueous phase. To aid cell adhesion, flagellated Gram-type bacteria often use movement and secrete polysaccharides to nourish the substrate. As an illustration, Gram-negative bacteria like *P. aeruginosa* (PA) move to the exterior of the surface, using the Type IV-based secretion system twitching movement, which releases Psl polysaccharide and leaves routes to encourage adhesion between cells. After using flagellar movement (swimming or swarming) to get onto surfaces, PA down-regulates its motility (Zhao et al. 2013). The MSHA pill in *V. cholera* has a role in substrate screening and in the formation of numerous smaller colonies (Teschler et al. 2015). Mechano-sensing with modulation using a two-component structure, mediated by c-di-GMP, contributes to the majority of Gram-negative bacteria. External polysaccharide production facilitates the attachment of Gram-positive flagellated bacterium *B. subtilis* with substrates which reduces the development of flagella (Vlamakis et al. 2013).

S. aureus bacterium, which lacks flagella, utilizes Brownian motion to move towards substrates. Polysaccharide intercellular adhesins (PIA) and extracellular DNA (eDNA) assist during first attachment, whereas cell wall anchored protein (CWAP) produced in the exterior aids in adhering to substrates in macro colonies by a process that does not rely on PIA (Speziale et al. 2014). Regarding endocarditis infection, Gram-positive non-flagellated *Enterococcus* employs multiple surface adhesins, such as biofilm-associated pilus (Ebp), aggregation substances (Agg) and enterococcal surface protein (Esp), to facilitate early attachment (Barbu et al. 2014). A typical method for initiating irreversible attachment involves the release of polysaccharides to produce flagellar attachment. In non-flagellated microbes, surface adhesins may aid in this process. However, it is undeniable that extreme cases or alternative patterns of microcolony development can be found. Flagella can participate actively in the production of biofilms in Gram-positive as well as Gram-negative bacteria. Hence, the most effective approach to inhibit the development of biofilms is to specifically target flagella. Increased quantities of c-di-GMP serve as an intriguing regulator that inhibits movement and function during biofilm development (Valentini and Filloux 2016). Decreased c-di-GMP levels can result in the restoration of movement and in the dispersion of organisms. The involvement of c-di-GMP in the movement was extensively seen among Gram-negative bacteria, although there are indications of its indirect influence among Gram-positive bacteria (Valentini and Filloux 2016). Nevertheless, there are documented cases of Gram-positive bacteria without locomotory appendages, like *S. aureus*, where surface sensing triggers the emergence of adhesion molecules and leads to permanent adherence (Harapanahalli et al. 2015).

Extracellular Matrix

The bacteria that make up the majority of biofilms typically make up a maximum of 10 percent of the dry mass, while the matrix might constitute more than 90%. The matrix refers to the extracellular matter that is primarily synthesized by the microbes alone. This substance serves as a medium within which the biofilm bacteria are enclosed. The biofilm is composed of many biological polymers, collectively known as extracellular polymeric substances (EPS). The aforementioned compounds serve as the framework for the 3D framework of the biofilm and are accountable for both attachments to substrates and cohesion within the biofilm. The development of a biofilm enables an identity that is completely separate from the free-floating condition. Extracellular polymeric substances (EPS) attach biofilm microorganisms and maintain their intimate contact, facilitating intensive connections such as communication between cells and the development of symbiotic micro consortia. Due to the presence of enzymes outside cells, a highly adaptable extrinsic gastric apparatus is formed. This structure effectively captures soluble and colloidal nutrients from the water, enabling their use as resources of both nutrition and energy. The structure of the matrix serves as a recycling site by preserving all the constituents of ruptured cells. This encompasses DNA, which potentially serves as a storage of genes for the process of horizontal gene transmission. EPS can also function as a nutrition supply; however, certain elements of EPS are recyclable at low rates. Due to the intricate nature of EPS, the full breakdown of all its components necessitates a diverse array of enzymes. Dehydration, oxidative or charged biocides, UV rays, host immunological defence, protozoan grazers, and certain antibiotics and metal ions are all shielded from bacteria by the matrix. Within the ecologically constrained EPS matrix, population viability is continuously adjusted through a combination of rivalry and collaborative efforts. Not all biofilm bacteria, especially those farthest from the outside world, may benefit ecologically from the matrix. Biofilm competitive models showed that polymer manufacturers evolved to have an evolutionary advantage over non-producers. This might be because polymers bring the progeny of polymer manufacturers nearer to abundant oxygen habitats (Xavier and Foster 2007). Due to the wide variety of substrate biopolymers and the difficulties in analysing them, EPS is often termed 'the dark matter of biofilms'(Flemming et al. 2007). Factors such as temperature, shear stresses, nutrition accessibility, microbes and biofilm type can significantly affect EPS. In the beginning, EPS was referred to as 'extracellular polysaccharides'. However, they were later renamed due to the realization that the matrix also includes lipids, nucleic acids, proteins and other biological polymers including humic compounds. EPS is a significant portion of the reduced-carbon pool found in soils, sediments and dispersed particulates in both marine and freshwater ecosystems worldwide. In the ecology of bacteria, nutrients are considered essential elements and hence play a crucial role (Decho et al. 2005; Ortega-Morales et al. 2007).

Architecture of EPS

The microbial cells inside a biofilm are enveloped by EPS, which provide the immediate surroundings for those cells. Certain EPS, namely those that create capsules, have a stronger association with cell exteriors compared to other types. The establishment and sustenance of organized multicellular populations of different microbes critically depend upon the synthesis and abundance of EPS (Sutherland 2001). The biofilm can exhibit many morphologies including smooth as well as rough, fluffy, filamentous or fat. Additionally, the biofilm may differ in porousness, with certain macrocolonies resembling mushrooms and including water-filled cavities. Every one of these architectures has a common outcome: to temporarily suspend biofilm populations and enable the formation of persistent mixed-species microconsortia, along with their relationships and variations. This creates highly varied environments on a microscopic scale, promoting biodiversity. The structure of biofilms is shaped by various aspects, such as hydrodynamic conditions, nutrient availability, bacterial movement, interaction between cells, and the presence of extracellular polysaccharides and proteins. This is evident from the changes in the appearance of biofilms created by mutants without certain elements of EPS. The scaffolding of biofilm by *Bacillus subtilis* is composed of an exopolysaccharide and the protein TasA. Each of these components is necessary for maintaining the structural stability of the framework and for the formation of biofilm building construction, which resembles fruiting body-like patterns (Branda et al. 2006). The biofilms of mucoid isolates of *Pseudomonas aeruginosa* include alginate, one of the most extensively researched exopolysaccharides that facilitate the formation of biofilm (Franklin and Ohman 1993; Tielen et al. 2005).

When faced with circumstances that cause alginate growers to create biofilms with varying structural components, non-mucoid strains tend to generate biofilms that are flatter and more uniform. The bacterial clustering into microbial colonies and the variation in the structure of developed biofilms are both impacted by the addition of acetyl groups with alginate (Franklin and Ohman 1993; Tielen et al. 2005). Negatively charged EPS, which includes carboxylic groups, may react with multivalent cations to significantly impact biofilm construction. When calcium ions link polyanionic alginate particles, it promotes the growth of dense biofilms that are more resistant to mechanical stress (Körstgens et al. 2001).

Amyloid-Type Proteins

Specific proteins are only present throughout the production of biofilms. Additionally, several proteins situated on the exterior in different bacteria show similarity to biofilm-associated proteins (Bap) discovered in *S. aureus*. The core area of Bap is intriguing due to its predominant composition of protein with several similar repetitions that might include amyloid-like peptide sequences. Various species exhibit various structural and variation patterns due to the presence of multiple repetitions. These patterns play a crucial role in facilitating attachment and evading the immune system. An instance of this is the Fap (functional amyloids in *Pseudomonas*) protein in *P. aeruginosa*, which exhibits amyloid-like properties. Evolutionary investigations suggest the presence of similar proteins in other bacterial species (Ruhal and Kataria 2021). Subsequent research is anticipated to uncover more homologs in several microorganisms. These proteins have functions at several stages of biofilm development, encompassing the initial phase, development and disintegration (Jiang et al. 2018; Sun et al. 2018). Additionally, they function as routes inside biofilms, facilitating nutrient availability, while also possessing surfactant characteristics that aid in dispersion. Comparably, the *E. coli* biofilm matrix contains the amyloid fibre protein curli, which is a structurally important amyloid protein that is built on the surface of the bacterium, particularly in areas where nutrients are very less (Smith et al. 2017). Due to their prevalence in both types of bacterial species, amyloid-like proteins may provide the most fruitful avenues for further investigation into matrix proteins. Biofilm development cannot take place without amyloid proteins, which might potentially be targeted for medicinal use. Proteins such as *S. aureus* amyloids, which are sensitive to pH and Ca^{2+}, are examples of surface-secreted, unstructured proteins that may need treatment and polymerization to achieve their full functional potential. Biofilm can be inhibited by a variety of compounds that attack amyloid proteins. For instance, the compound FNO7, which is a tiny molecule consisting of conjugated 2-pyridones, inhibited the polymerization of primary curli component

CsgA in *E. coli* both in laboratory conditions as well as inside living organisms (Lim et al. 2012). The compounds have been selected from a large chemical library using a screening process. To successfully focus on amyloid proteins for antibiofilm objectives, it is essential to possess a comprehensive understanding of the structural biology of such proteins. An instance of this is the antibiotic rifapentine, which hindered the production of curli-dependent biofilms on various surfaces without affecting the proliferation of bacteria (Maher 2016). The antibiotic has a benefit due to its multifaceted strategy, as it also reduces genes associated with cellulose. Certain proteins that possess amyloid domains tend to aggregate. These proteins can infiltrate bacteria and induce the clustering of cytosolic proteins, leading to the destruction of the bacterial cell without impacting the host cells (Bednarska et al. 2016). To understand the process of biofilm development, it is necessary to thoroughly examine functionally relevant homologous proteins in bacteria that are not closely linked in terms of evolutionary history.

Polysaccharides

In the field of biofilm research, the release of polysaccharides appears to be a necessary condition in several types of bacteria, particularly for the early attachment process. Nevertheless, these polysaccharides also serve a crucial function in the preservation of fully developed biofilms. Biofilm development in *E. coli* relies on cellulose; however, exopolysaccharides may additionally control biofilms and prevent them through feedback mechanisms since an excess of them can reduce biofilm size (Gualdi et al. 2008). A surface polysaccharide that is released by *S. aureus* and involved in immunological resistance and biofilm development improves the bacteria's ability to adhere to one another intercellularly (Cramton et al. 1999; Mack et al. 1996). Various functions are displayed by exopolysaccharides released by *P. aeruginosa*, such as alginate, Psl and Pel (Franklin et al. 2011). *P. aeruginosa* undergoes morphological transformation to a mucoid form after a prolonged invasion. It is possible that the primary exopolysaccharide causing this mucoid structure is alginate, whereas Psl and Pel are likely liable for biofilm adherence and preservation (Ruhal and Kataria 2021). Some of these results may lead to the development of a technique for inhibiting biofilms by concentrating on polysaccharides. Thus, more research is needed to determine if the excessive production of polysaccharides by a certain strain of bacteria suppresses the biofilm formation of a different Gram-type bacterium. In situations when there are two biofilms, the polysaccharide synthesis locus (Psl) facilitates communication between different bacteria. Consequently, it is crucial to investigate the molecular specifics of anti-biofilm features that modify the abiotic or bacterial surfaces' physical attributes or function as signalling molecules. Research showed how Psl of *P. aeruginosa* was suppressed by protein-A released by *S. aureus*, and as a result, biofilm formation (Armbruster et al. 2016). The dispersal of developed biofilms with glycosidase hydrolase, which targets every variety of biofilm polysaccharides, is an additional alternative strategy. This method was successful in degrading both Gram kinds of biofilms (Flemming and Wingender

2010). Testing is necessary to determine whether the immune system reacts to these polysaccharides and whether antibodies can bind the epitopes of biofilm-specific polysaccharides. Antibodies produced by humans that target the polysaccharide Psl, which is found in *P. aeruginosa*, can boost the immune system's reaction to biofilms and ultimately destroy them (Ray et al. 2017). A vaccine against *Staphylococci* was also developed using poly-N-acetyl beta-glucosamine (Soliman et al. 2018). A vaccination composed of a typical mixture of polysaccharides could potentially be effective, but further research is required to produce vaccines founded on matrix polysaccharides.

Extracellular DNA

The significance of DNA found outside of cells (eDNA) in bacteria of both Gram types is significant (Whitchurch et al. 2002). The results unequivocally showed DNase therapy significantly decreased biofilms. Additionally, electrostatic properties of eDNA may contribute to its ability to withstand antibiotics (Wilton et al. 2016). Potential modes of extracellular DNA (eDNA) release involve autolysis, death caused by prophages and active discharge. Thus, eDNA is a prevalent component that is present in both Gram types of bacteria.

Outer Membrane Vesicle

Outer membrane vesicles (OMVs) present in Gram-negative bacteria conduct an essential function in the formation of biofilms. This is because such vesicles are integral components of the structure of a biofilm, and OMVs are necessary to transport matrix proteins inside biofilms (Wang et al. 2015). Bacterial outer membrane vesicles (OMVs) play a crucial role in promoting bacterial survival, particularly in challenging circumstances, by facilitating the release of OMVs. For instance, proteins with folding errors can be eliminated by OMVs, and the administration of antibiotics causes DNA damage, resulting in alterations in lipopolysaccharide (LPS) production and subsequent modifications in OMVs. The involvement of OMVs is crucial in biofilm development. This observation was made by comparing the proteomics of OMVs with matrix proteins (Couto et al. 2015). By comparing their protein profiles with those in the biofilm matrices, researchers have shown that OMVs make about 20% of the matrix's proteins. Proteomic examinations have also verified the presence of prevalent proteins inside membrane vesicles obtained from Gram-positive *S. aureus*. Furthermore, the internalization of these membrane vesicles into the cytosol is essential for the cytotoxic effects of *S. aureus* (Gurung et al. 2011; Lee et al. 2009). Significantly, membrane vesicles (MVs) of *S. aureus* have the ability to transport active β-lactamase, which grants tolerance to certain antibiotics (J. Lee et al. 2013). Thus, OMVs can serve to be promising vehicles to deliver biofilm-restricting elements, such as natural chemicals. Recent research indicates that OMVs have promising potential as vaccine adjuvants because they can stimulate both innate and adaptive immune responses. Additionally, OMVs can be produced in large quantities and can be bioengineered (Bitto and Kaparakis-Liaskos 2017).

Enzymes

A wide range of extracellular enzymes were identified in biofilms, with a significant number of them playing a role in breaking down biopolymers. The scaffolding acts as an extrinsic digestive system by including enzymes responsible for breaking down EPS molecules. This process converts biopolymers into smaller molecules, which may be absorbed and used as resources for carbon and energy. Furthermore, certain enzymes may engage in the decomposition of EPS involved in biofilm structure formation, facilitating the separation of microbes from biofilms. Many extracellular enzymes derived from fungi and bacteria are manufactured commercially on a huge scale for industrial use. Furthermore, extracellular enzymes provide their own purification functions in sediments, water and soil. These functions were adapted for the biological treatment of wastewater and drinking water by utilizing flocs and biofilms to break down organic materials (Flemming and Wingender 2010). Additionally, extracellular enzymes are employed in the breakdown of synthetic compounds by targeting the polymer skeleton or breaking down chemicals such as antioxidants or plasticizers (like terephthalates). Additionally, exogenous redox enzymes contribute to corrosion that is affected by microbes (Busalmen et al. 2002) (Fig. 2.1).

2.4 Three-Dimensional Architecture and Imaging Techniques

Biofilm enables bacteria to thrive in many challenging conditions with restricted resources, while their free-floating counterparts cannot live. Nutrient-depleted hostile situations are plentiful in the natural world. Thus, biofilms are the predominant mode of bacterial growth in natural environments (Costerton et al. 1987). The structural makeup of a biofilm is directly related to the movement of nutrients, the migration of mass, the rate of growth and the administration of different species inside the biofilm (De Beer and Stoodley 1995; Yang et al. 2000). Biofilms exhibit self-organization in a certain arrangement that increases the speed at which they consume nutrients or energy (Beyenal and Tanyolaç 1998). For instance, biofilms proliferate in a filamentous configuration across water bodies. The filamentous form of the biofilms serves two purposes: it increases the surface area available for absorbing substrates and reduces the drag caused by the flow of the waterway or river, thus facilitating the detachment of the biofilms (Hall-Stoodley et al. 2004).

The biofilm architecture refers to the complete 3D surface that is filled with biomass in a biofilm. Advancements in imaging techniques, such as confocal scanning laser microscopy (CSLM), have made it possible to collect photographs at various

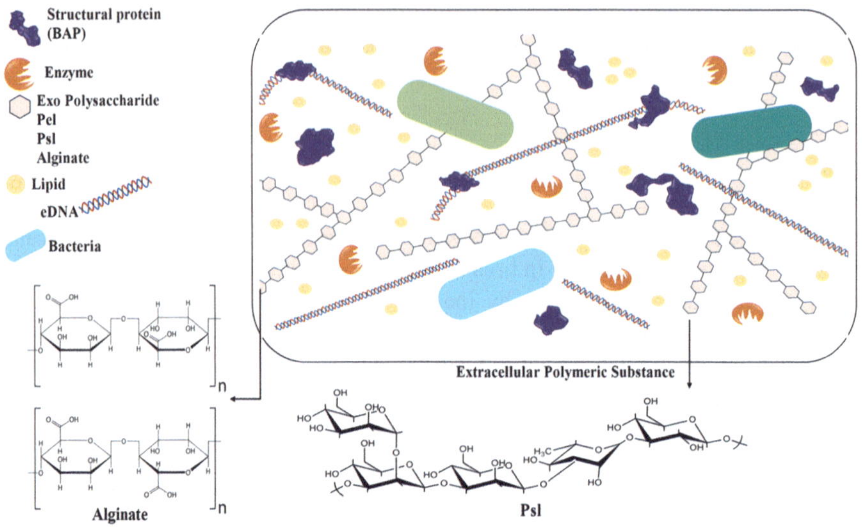

Fig. 2.1 Composition of biofilm matrix inside organized biofilm community

depths. This enables the generation of pictures that may be used to determine the 3D structure. The assessment of biofilm architecture is achieved by the utilization of numerous 2D as well as 3D biofilm structural factors, which are derived from biofilm photographs. Several software tools dedicated to measuring the architecture of biofilms are already published (Neu and Lawrence 2014).

Utilization of Microscopic Methods for Biofilm Visualization

Light Microscopy

Light microscopy (LM) is a fundamental visualization method that is beneficial in visually identifying the presence of biofilms and has significant predictive importance (Hong et al. 2014). The method may be utilized to quantitatively evaluate the amount of biofilm biomass. It is a simple and cost-effective procedure (De Carvalho and Da Fonseca 2007). Light microscopy, although useful, has inherent limitations in terms of magnification and resolution, making it unsuitable for accurately describing the complex characteristics of biofilm cell morphology or EPS construction. During research on teeth microbial communities, the use of light microscopy to investigate semi-thin slices of demineralized teeth proved to be the most effective method for obtaining a comprehensive view of the root canal. This technique allowed for the examination of greater regions at a low level of magnification. While the specimens were inspected employing a scanning electron microscope (SEM), no microbes were located inside teeth tubules. But when those exact specimens were demineralized and embedded in resin, microorganisms were identified in their semi-thin sections. Light microscopy (LM) photos detected the existence of microbes, which were then verified by transmission electron microscopy (TEM) images (Richardson et al. 2009).

Confocal Laser Scanning Microscopy (CLSM)

Quantitative assessment of architectural metrics such as biovolume (cells total volume in the observing region), dimension and abrasion is made possible using confocal laser scanning microscopy (CLSM). A real-time 4-D depiction of a specimen's structure that changes over time is also possible (Klausen et al. 2003). The technique of CLSM was employed in conjunction with a fluorescent dye to effectively analyse various kinds of biofilms (Bridier et al. 2010; Guilbaud et al. 2015; Sun et al. 2015; Villacorte et al. 2015).

CLSM involves the use of pathogen-specific probes that are tagged with distinct fluorescent dyes. This technique, known as FISH followed by CLSM, allows for the detection of only one species in specimens containing many species (Thornton et al. 2011). Using the same method, we examined disturbance across species and evaluated competition among species (Bridier et al. 2014). Experiments evaluating the antimicrobial properties of drugs utilized CLSM in conjunction with certain fluorophores to identify viable and non-viable microbial cells, as well as how they were distributed (Bridier et al. 2012; Doroshenko et al. 2014; Hope et al. 2002; Verma et al. 2010).

For the purpose of observing and quantifying biofilms, CLSM is the preferred approach. Using fluorophores, there is a finite amount of reporter compounds, and the signal of concern could be obscured by the interaction between the probe and intrinsic biofilm fluorescence; these are some of the constraints of CLSM biofilm research.

Scanning Electron Microscopy

Electron microscopy is exceptional in its ability to visualize microbial biofilms in all their granularity, which is essential for investigations that link variations in biofilm architecture with functioning. Nevertheless, to prevent the development of artefacts, it is essential to adjust and modify the sample-specific methods for sample preparation, equipment selection and acquisition settings. If high-resolution pictures and high magnification are required to precisely portray biofilm structure, traditional SEM and field emission scanning electron microscopy (FESEM) are the ideal approaches for imaging (Bossù et al. 2020b; Hung et al. 2013; Rodrigues et al. 2013). When doing comparative studies, such as evaluating the effectiveness of a medicine or therapy in preventing biofilm formation, SEM is a valuable tool. This is because the data obtained from SEM imaging show a strong correlation with the results obtained from other analytical methods (Di Bonaventura et al. 2006, 2008; Hasan et al. 2015; Li et al. 2015, 2020; Orsinger-Jacobsen et al. 2013; Relucenti et al. 2021; Van Laar et al. 2015). The undeniable benefits of SEM and FESEM include the capability to capture images at various magnifications (ranging from 20 to 30,000x) with exceptional resolution (between 50 to 100 nm) and a significant depth of field. In addition, 3D imaging software enables the collection and measurement of intricate morphological discoveries. Nevertheless, several drawbacks may arise in the investigation of biofilms. Dehydration and application of a conductive substance can lead to the breakdown of EPS and a reduction in the size of the biofilm (Alhede et al. 2012). The constraints of traditional SEM and FESEM might be solved by employing specialized methods for studying biofilms or by utilizing innovative SEM techniques such as variable pressure SEM (VP-SEM), cryo-SEM and environmental SEM (ESEM).

Variable Pressure Scanning Electron Microscopy

VP-SEM enables superior resolution imaging in difficult experimental environments, as it enables the observation of completely moistened and out-gassing materials using minimal or no sample preparation processes (Bossù et al. 2019; Bossù et al. 2020a; Redler et al. 2019). When studying biofilm characteristics, it is crucial to characterize the hydrated biofilm. However, employing traditional high-vacuum SEM, which requires dehydration and drying, might lead to a breakdown of the matrix EPS and result in a distorted form and altered structure of the biofilm. VP-SEM, on the other hand, does not need the removal of water or drying. When combined with suitable fixation and subsequent heavy metal staining to improve contrast and resolution, it may be employed for the characterization of bacterial (Ishii et al. 2004; Priester et al. 2007; Weber et al. 2014; Weimer et al. 2006) and fungal biofilms (Joubert et al. 2017). The main fixatives commonly utilized are aldehydes that create cross-links with proteins, and osmium tetroxide (OsO_4) which prefers to attach to lipids, resulting in the preservation of intricate details. It is utilized as both a fixative and a differentiating reagent in research on moist fungal biofilms (Joubert et al. 2017), specifically to emphasize the presence of lipid droplet inclusion. Heavy metal staining is a well-established technique that was employed in electron microscopy since the beginning.

2.5 Engineering Approaches for Modulating Biofilm Architecture

The early 90s saw the introduction of the term Biofilm Engineering by the Center for Biofilm Engineering at Montana State University. Recently in 2019 also the National Biofilm Innovation Center (NBIC) organized a workshop on biofilm engineering where the researchers and industrial people discussed some strategies to prevent, detect, manage, and engineer detrimental biofilms and utilize the beneficial ones (Mukherjee and Cao 2021). Organizations such as the National Biofilms Innovation Centre (NBIC) are dedicated to advancing biofilm research and delivering impactful innovations for society. By understanding biofilms better, we can harness their potential and address critical challenges. Biofilms play a central role in global challenges, including infections, antimicrobial resistance, food safety, water security, marine biofouling and corrosion. Their economic, social and environmental impacts are estimated at a staggering $5 trillion per annum. Research on biofilms can lead to practical applications, from fighting antibiotic resistance to engineering anti-clogging pipes and creating long-lasting boat paints.

Different stages of the biofilm life cycle are controlled by regulatory networks. As we all know about the term Quorum Sensing (QS). Quorum sensing is intercellular communications. It regulates the formation of biofilm through signalling molecules. Thus, QS serves as a target for bioengineering biofilms. Biofilm engineering can be utilized as a technology for sustainable use of water, food and energy. During wastewater treatment, membrane systems are being used increasingly, even during the recycling of water from wastewater, energy production and food processing membrane systems are in use extensively. Multiplication of biofilms forming microorganisms in the membrane system results in the formation of thick extra polymeric substances clogging the whole system. This is known as biofouling. Biofouling poses a threat to the sustainable use of these resources. Biofouling reduces the efficiency of the membrane system. We can solve this problem considerably by engineering quorum-sensing circuits. Engineered beneficial biofilms can be produced and applied over a membrane system. This protective biofilm layer can prevent other harmful bacteria and also can regulate its growth. One example of this type of bioengineered strain is *E. coli* which after devising a genetic circuit can sense its presence and limit the formation of biofilms (Mukherjee and Cao 2021). The LasI/LasR QS system of *Pseudomonas aeruginosa* is being used. In this system, LasR is the transcriptional activator protein. LasI along with LasR constitutes the Las system. LasI/LasR module directs the production of an autoinducer PAI-1 [N-(3-oxododecanoyl)-L-homoserine Lactone] (Pesci et al. 1997). This autoinducer is expressed at basal levels at lower cell densities. The concentration of this diffusible molecule increases with an increase in cell density until the cell density reaches saturation. At this point of concentration, the autoinducer binds to a specific target protein known as LasR (activator protein). This autoinducer-activator complex activates many genes that it controls. Autoinducer–protein complex helps in activating LasI promoter leading to increased production of BdcAE50Q (dispersal protein) as cell density starts increasing and reaches a threshold value.

There are many strategies employed to control biofouling of membrane bioreactor for example application of ultrasonic vibration, alteration of the membrane surface properties and chemical cleaning of the membranes with acid or alkali. Cleaning by chemicals is very effective in killing microbes and removing biofouling but overuse of chemicals alters the membrane properties. One of the strategies of nitric oxide production by bioengineering can be promising in preventing biofouling. Nitric oxide is an intracellular signalling molecule that plays an important role in various biological processes. It has been shown to be involved in biofilm dispersal by inducing phosphodiesterse activity resulting in the degradation of cyclic di-guanylate monophosphate(c-diGMP) triggering a change in gene expression promoting the transition from the biofilm mode of growth to the free-swimming planktonic state (Luo et al. 2015). Exogenous addition of NO, using NO donor compounds, has been shown to disperse biofilms. For example, sodium nitroprusside: at low concentrations (25–500 nM) induces *Pseudomonas aeruginosa* biofilm dispersal. By promoting biofilm dispersal, NO can reduce fouling on surfaces. NO is a quorum quencher, meaning it disrupts biofilm signals through metabolic rewiring. Interestingly, NO is also secreted by probiotics (beneficial bacteria). If probiotic biofilms that secrete NO are used, they could mitigate the growth of other microbes on the filtration membrane. Essentially, these NO-secreting probiotics act as multifunctional bioweapons for membrane remediation (Wood et al. 2016). This is a promising approach because of safety as unlike harsh chemicals they will not harm the environment or human health if released in the atmosphere during filtration. The use of probiotics brings sustainability in membrane cleaning and also provides a green alternative to other traditional membrane cleaning methods (Fig. 2.2).

Cyclic di-GMP (c-di-GMP) and its role in controlling biofilm formation: cyclic di-GMP is a second messenger molecule present in bacteria playing a crucial role in regulating many bacterial behaviours including the transition between motile (free-living) cells and biofilm formers. It regulates biofilm development in a wide variety of bacteria; thus, bioengineering of this target is compatible. c-di-GMP is a signalling molecule synthesized by diguanylate cyclases (DGCs) and degraded by phosphodiesterases (PDEs). An indispensable part of the biofilm life cycle is biofilm dispersion. Conversion of biofilms mode back to the planktonic mode of life is known as dispersion. Low levels of c-di-GMP are associated with dispersion which is in contrast to the formation of biofilms. Environmental factors such as the presence of glutamate and nitric oxide trigger dispersion which requires the presence of c-di-GMP degrading enzymes. C-di-GMP and its role in controlling biofilm formation are studied in model organism *P. aeruginosa* (Valentini and Filloux 2016). We will discuss the mechanism in this review further. In *P. aeruginosa*, c-di-GMP degrading enzymes are DipA, RbdA, NbdA and MucR. NbdA is a phosphodiesterase enzyme that triggers dispersal in response to environmental factors such as nitric oxide but the other phosphodiesterases present such as DipA and RbdA act regardless of the dispersion factors (Park and Sauer 2022). c-di-GMP is not the sole enzyme that plays a key role in dispersion, but two guanylate cyclases, namely,

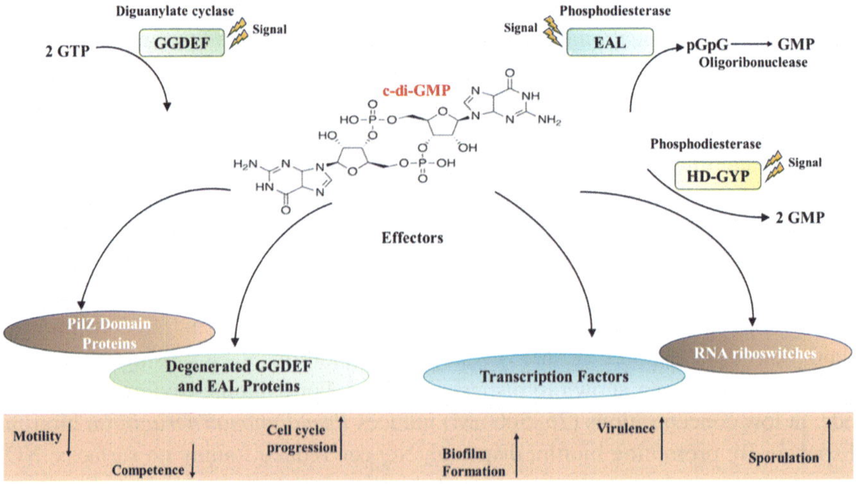

Fig. 2.2 Role of c-di-GMP in controlling biofilm formation

GcbA and NicD and another chemotaxis-like MCP homolog BdlA. Under favourable environmental conditions for dispersal to occur, BdlA gets activated after phosphorylation in a c-di-GMP-dependent manner. At first, there is a sudden burst in the c-di-GMP level, this rise is due to the activity of two diguanylate cyclase enzymes GcbA (PA 4843) and NicD (PA 4929). A rise in c-di-GMP level is followed by non-processive proteolysis of BdlA. After being phosphorylated and activated, BdlA triggers phosphodiesterase enzymes such as RbdA and DipA resulting in a decrease of c-di-GMP levels leading to dispersion (Park and Sauer 2022) (Fig. 2.3).

The role of c-di-GMP in controlling biofilm formation and dispersal has made researchers focus on discovering strategies capable of reducing the intracellular level of c-di-GMP enabling suppression of biofilm formation in membrane systems resulting in sustainability in utilizing resources, disabling many biofilms related phenotypes such as antibiotic resistance and membrane clogging. Diguanylate cyclases (DGC) are the key factor for influencing c-di-GMP for the production of biofilm. Many researchers have searched for inhibitors of the WspR protein a DGC in *Pseudomonas aeruginosa*. Their findings revealed an immunosuppressive drug azathioprine to block WspR-dependent c-di-GMP formation in bacterial cells but in an indirect way. Azathioprine has been shown to inhibit 5-aminoimidazole-4-carboxamide ribotide (AICAR) transformylase involved in purine biosynthesis. This disturbance of the intracellular nucleotide pool inhibits c-di-GMP biosynthesis (Antoniani et al. 2013).

Quorum quenching (QQ) is a fascinating strategy that disrupts bacterial communication within biofilms. QQ disrupts quorum sensing by interfering with the autoinducers (AIs) produced or by interfering with their downstream effects. QQ enzymes cleave AIs, preventing their accumulation. For example, lactonases are being applied to break down the lactone rings of N-acyl homoserine lactones along with acylase cleaving acyl groups (Fang et al. 2020). This breakdown of signalling

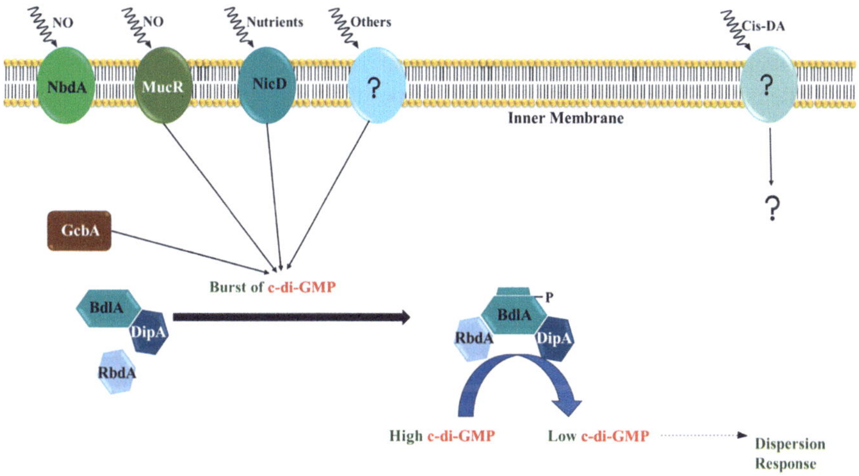

Fig. 2.3 Mechanism for the control of biofilm formation and dispersal

molecules disrupts the communication network within the microbial community, hindering biofilm development. QQ enzymes have been applied in antifouling strategies, particularly in membrane bioreactors for wastewater treatment. By reducing AHL-producing bacteria on membrane surfaces, QQ enzymes help prevent biofilm formation and microbial attachment, thereby improving the efficiency of wastewater treatment processes. An example of an antifouling strategy is bioengineered *E. coli* producing lactonases AiiA and AiiO from *Bacillus thuringiensis* and *Agrobacterium tumifaciens* reduces AHL-producing bacteria on the membrane surface (Fang et al. 2020).

Enzymatic QQ treatments have been shown to reduce the production of extracellular polymeric substances (EPS) and the expression of genes related to microbial attachment and agglomeration. This reduction in microbial attachment helps prevent biofilm formation and promotes cleaner surfaces. High intracellular levels of c-di-GMP promote biofilm formation by the production of adhesins and extrapolymeric substances while low levels trigger biofilm dispersion and the development of structures which help in motility returning to planktonic growth. One of the beneficial aspects of bioengineering is the enhanced degradation of organic pollutants by the bioengineered biofilm of *Comamonas testoteroni* can be achieved by constitutively expressing a c-di-GMP synthase from *E. coli* called YedQ. Thus, this also provides a suitable example of engineering approaches for modulating biofilm architecture (Cao et al. 2023).

Lack of knowledge about biofilm development and some drawbacks of biological toolkits have hindered biofilm eradication and utilization. Growing interest in bioengineering approaches for modulating biofilm architecture will shed light on biofilm-controlling strategies and their application in medical, foodprocessing, agricultural, industrial and environmental fields.

2.6 Conclusions

In the complex process of biofilm development, adherence of microorganisms to the substrate is the foundational step, which shapes the complex structure and arrangements of microbial communities. This adhesion can be better understood with the assistance of theories such as thermodynamic, DLVO and extended DLVO theories, which shed light on the complex interactions between the substrate and microbial cell. As a result of adherence, microcolonies originate as attached microbes organize themselves spatially within the biofilm matrix. A broad range of elements including extracellular DNA, exopolysaccharides, enzymes and different types of proteins are crucial to organizing the biofilm structure. EPS is one of the most abundant components of this matrix and plays a key role in maintaining integrity by enclosing microbial cells. Depending upon the composition and structure, biofilm's lifespan and adaptability also vary.

3D architectural intricacies of biofilm can be revealed using advanced imaging techniques. Complex structures and spatial arrangements can be revealed by traditional light microscopy or by advanced CLSM to SEM. Their three-dimensional architecture is now more beautifully visualized as variable pressure scanning electron microscopy offers the visualization of specimens in their own hydrated conditions.

References

Alhede M, Qvortrup K, Liebrechts R, Høiby N, Givskov M, Bjarnsholt T (2012) Combination of microscopic techniques reveals a comprehensive visual impression of biofilm structure and composition. FEMS Immunol Med Microbiol 65(2):335–342. https://doi.org/10.1111/j.1574-695X.2012.00956.x

Antoniani D, Rossi E, Rinaldo S, Bocci P, Lolicato M, Paiardini A et al (2013) The immunosuppressive drug azathioprine inhibits biosynthesis of the bacterial signal molecule cyclic-di-GMP by interfering with intracellular nucleotide pool availability. Appl Microbiol Biotechnol 97:7325–7336. https://doi.org/10.1007/s00253-013-4875-0

Armbruster CR, Wolter DJ, Mishra M, Hayden HS, Radey MC, Merrihew G, MacCoss MJ, Burns J, Wozniak DJ, Parsek MR, Hoffman LR (2016) *Staphylococcus aureus* protein a mediates interspecies interactions at the cell surface of *Pseudomonas aeruginosa*. MBio 7(3). https://doi.org/10.1128/mBio.00538-16

Asally M, Kittisopikul M, Rué P, Du Y, Hu Z, Çağatay T, Robinson AB, Lu H, Garcia-Ojalvo J, Süel GM (2012) Localized cell death focuses mechanical forces during 3D patterning in a biofilm. Proc Natl Acad Sci USA 109(46):18891–18896. https://doi.org/10.1073/pnas.1212429109

Barbu EM, Mackenzie C, Foster TJ, Höök M (2014) SdrC induces staphylococcal biofilm formation through a homophilic interaction. Mol Microbiol 94(1):172–185. https://doi.org/10.1111/mmi.12750

Bednarska NG, van Eldere J, Gallardo R, Ganesan A, Ramakers M, Vogel I, Baatsen P, Staes A, Goethals M, Hammarström P, Nilsson KPR, Gevaert K, Schymkowitz J, Rousseau F (2016) Protein aggregation as an antibiotic design strategy. Mol Microbiol 99(5):849–865. https://doi.org/10.1111/mmi.13269

Beyenal H, Tanyolaç A (1998) The effects of biofilm characteristics on the external mass transfer coefficient in a differential fluidized bed biofilm reactor. Biochem Eng J 1(1):53–61. https://doi.org/10.1016/S1369-703X(97)00010-7

Bitto NJ, Kaparakis-Liaskos M (2017) The therapeutic benefit of bacterial membrane vesicles. Int J Mol Sci 18(6):1287. https://doi.org/10.3390/ijms18061287

Bos R, van der Mei HC, Busscher HJ (1999) Physico-chemistry of initial microbial adhesive interactions – its mechanisms and methods for study. FEMS Microbiol Rev 23(2):179–230. https://doi.org/10.1111/j.1574-6976.1999.tb00396.x

Bossù M, Saccucci M, Salucci A, Di Giorgio G, Bruni E, Uccelletti D, Sarto MS, Familiari G, Relucenti M, Polimeni A (2019) Enamel remineralization and repair results of biomimetic hydroxyapatite toothpaste on deciduous teeth: an effective option to fluoride toothpaste. J Nanobiotechnol 17(1). https://doi.org/10.1186/s12951-019-0454-6

Bossù M, Matassa R, Relucenti M, Iaculli F, Salucci A, Di Giorgio G, Familiari G, Polimeni A, Di Carlo S (2020a) Morpho-chemical observations of human deciduous teeth enamel in response to biomimetic toothpastes treatment. Materials 13(8). https://doi.org/10.3390/MA13081803

Bossù M, Selan L, Artini M, Relucenti M, Familiari G, Papa R, Vrenna G, Spigaglia P, Barbanti F, Salucci A, Di Giorgio G, Rau JV, Polimeni A (2020b) Characterization of scardovia wiggsiae biofilm by original scanning electron microscopy protocol. Microorganisms 8(6). https://doi.org/10.3390/microorganisms8060807

Branda SS, Chu F, Kearns DB, Losick R, Kolter R (2006) A major protein component of the Bacillus subtilis biofilm matrix. Mol Microbiol 59(4):1229–1238. https://doi.org/10.1111/j.1365-2958.2005.05020.x

Bridier A, Dubois-Brissonnet F, Boubetra A, Thomas V, Briandet R (2010) The biofilm architecture of sixty opportunistic pathogens deciphered using a high throughput CLSM method. J Microbiol Methods 82(1):64–70. https://doi.org/10.1016/j.mimet.2010.04.006

Bridier A, de Sanchez-Vizuete MP, Le Coq D, Aymerich S, Meylheuc T, Maillard JY, Thomas V, Dubois-Brissonnet F, Briandet R (2012) Biofilms of a Bacillus subtilis hospital isolate protect Staphylococcus aureus from biocide action. PLoS One 7(9):e44506. https://doi.org/10.1371/journal.pone.0044506

Bridier A, Briandet R, Bouchez T, Jabot F (2014) A model-based approach to detect interspecific interactions during biofilm development. Biofouling 30(7):761–771. https://doi.org/10.1080/08927014.2014.923409

Busalmen JP, Vazquez M, De Sánchez SR (2002) New evidences on the catalase mechanism of microbial corrosion. Electrochim Acta 47(12):1857–1865. https://doi.org/10.1016/S0013-4686(01)00899-4

Cao Z, Xia W, Wu S et al (2023) Bioengineering *Comamonas testosteroni* CNB-1: a robust whole-cell biocatalyst for efficient PET microplastic degradation. Bioresour Bioprocess 10:94. https://doi.org/10.1186/s40643-023-00715-7

Costerton W, Cheng KJ, Geesey GG, Ladd T, Nickel C, Dasgupta M, Marrie TJ (1987) Bacterial biofilms in nature and disease. Ann Rev Microbiol 41:435–464. https://doi.org/10.1146/annurev.mi.41.100187.002251

Couto N, Schooling SR, Dutcher JR, Barber J (2015) Proteome profiles of outer membrane vesicles and extracellular matrix of *Pseudomonas aeruginosa* biofilms. J Proteome Res 14(10):4207–4222. https://doi.org/10.1021/acs.jproteome.5b00312

Cramton SE, Gerke C, Schnell NF, Nichols WW, Götz F (1999) The intercellular adhesion (ica) locus is present in *Staphylococcus aureus* and is required for biofilm formation. Infect Immun 67(10):5427–5433. https://doi.org/10.1128/iai.67.10.5427-5433.1999

De Beer D, Stoodley P (1995) Relation between the structure of an aerobic biofilm and transport phenomena. Water Sci Technol 32(8):11–18. https://doi.org/10.1016/0273-1223(96)00002-9

De Carvalho CCCR, Da Fonseca MMR (2007) Assessment of three-dimensional biofilm structure using an optical microscope. BioTechniques 42(5):616–620. https://doi.org/10.2144/000112403

Decho AW, Visscher PT, Reid RP (2005) Production and cycling of natural microbial exopolymers (EPS) within a marine stromatolite. Palaeogeogr Palaeoclimatol Palaeoecol 219(1–2):71–86. https://doi.org/10.1016/j.palaeo.2004.10.015

Di Bonaventura G, Pompilio A, Picciani C, Iezzi M, D'Antonio D, Piccolomini R (2006) Biofilm formation by the emerging fungal pathogen *Trichosporon asahii*: development, architecture, and antifungal resistance. Antimicrob Agents Chemother 50(10):3269–3276. https://doi.org/10.1128/AAC.00556-06

Di Bonaventura G, Piccolomini R, Paludi D, D'Orio V, Vergara A, Conter M, Ianieri A (2008) Influence of temperature on biofilm formation by Listeria monocytogenes on various food-contact surfaces: relationship with motility and cell surface hydrophobicity. J Appl Microbiol 104(6):1552–1561. https://doi.org/10.1111/j.1365-2672.2007.03688.x

Donlan RM (2002) Biofilms: microbial life on surfaces. Emerg Infect Dis 8(9):881. https://doi.org/10.3201/eid0809.020063

Doroshenko N, Tseng BS, Howlin RP, Deacon J, Wharton JA, Thurner PJ, Gilmore BF, Parsek MR, Stoodley P (2014) Extracellular DNA impedes the transport of vancomycin in *Staphylococcus epidermidis* biofilms preexposed to subinhibitory concentrations of vancomycin. Antimicrob Agents Chemother 58(12):7273–7282. https://doi.org/10.1128/AAC.03132-14

Fang K, Park OJ, Hong SH (2020) Controlling biofilms using synthetic biology approaches. Biotechnol Adv 40:107518. https://doi.org/10.1016/j.biotechadv.2020.107518

Flemming HC, Wingender J (2010) The biofilm matrix. Nat Rev Microbiol 8(9):623–633. https://doi.org/10.1038/nrmicro2415

Flemming HC, Neu TR, Wozniak DJ (2007) The EPS matrix: the "house of biofilm cells". J Bacteriol 189(22):7945–7947. https://doi.org/10.1128/JB.00858-07

Fletcher M (1996) Bacterial attachment in aquatic environments: a diversity of surfaces and adhesion strategies. Bacterial adhesion: molecular and ecological diversity 19(1):1–24

Franklin MJ, Ohman DE (1993) Identification of algF in the alginate biosynthetic gene cluster of Pseudomonas aeruginosa which is required for alginate acetylation. J Bacteriol 175(16):5057–5065. https://doi.org/10.1128/jb.175.16.5057-5065.1993

Franklin MJ, Nivens DE, Weadge JT, Howell PL (2011) Biosynthesis of the *Pseudomonas aeruginosa* extracellular polysaccharides, alginate, Pel, and Psl. Front Microbiol 2:167. https://doi.org/10.3389/fmicb.2011.00167

Garrett TR, Bhakoo M, Zhang Z (2008) Bacterial adhesion and biofilms on surfaces. Prog Nat Sci 18(9):1049–1056. https://doi.org/10.1016/j.pnsc.2008.04.001

Garrido KD, Palacios RJS, Lee C, Kang S (2014) Impact of conditioning film on the initial adhesion of *E. coli* on polysulfone ultrafiltration membrane. J Ind Eng Chem 20(4):1438–1443. https://doi.org/10.1016/j.jiec.2013.07.029

Gualdi L, Tagliabue L, Bertagnoli S, Ieranò T, De Castro C, Landini P (2008) Cellulose modulates biofilm formation by counteracting curli-mediated colonization of solid surfaces in *Escherichia coli*. Microbiology 154(7):2017–2024. https://doi.org/10.1099/mic.0.2008/018093-0

Guilbaud M, Piveteau P, Desvaux M, Brisse S, Briandet R (2015) Exploring the diversity of Listeria monocytogenes biofilm architecture by high-throughput confocal laser scanning microscopy and the predominance of the honeycomb-like morphotype. Appl Environ Microbiol 81(5):1813–1819. https://doi.org/10.1128/AEM.03173-14

Gurung M, Moon DC, Choi CW, Lee JH, Bae YC, Kim J, Lee YC, Seol SY, Cho DT, Kim SI, Lee JC (2011) *Staphylococcus aureus* produces membrane-derived vesicles that induce host cell death. PLoS One 6(11):e27958. https://doi.org/10.1371/journal.pone.0027958

Hall-Stoodley L, Stoodley P (2002) Developmental regulation of microbial biofilms. Curr Opin Biotechnol 13(3):228–233. https://doi.org/10.1016/S0958-1669(02)00318-X

Hall-Stoodley L, Costerton JW, Stoodley P (2004) Bacterial biofilms: from the natural environment to infectious diseases. Nat Rev Microbiol 2(2):95–108. https://doi.org/10.1038/nrmicro821

Harapanahalli AK, Chen Y, Li J, Busscher HJ, Van Der Mei HC (2015) Influence of adhesion force on icaA and cidA gene expression and production of matrix components in *Staphylococcus aureus* biofilms. Appl Environ Microbiol 81(10):3369–3378. https://doi.org/10.1128/AEM.04178-14

Hasan S, Danishuddin M, Khan AU (2015) Inhibitory effect of zingiber officinale towards *Streptococcus mutans* virulence and caries development: in vitro and in vivo studies. BMC Microbiol 15(1):1. https://doi.org/10.1186/s12866-014-0320-5

Hong SD, Dhong HJ, Chung SK, Kim HY, Park JO, Ha SY (2014) Hematoxylin and eosin staining for detecting biofilms: practical and cost-effective methods for predicting worse outcomes after endoscopic sinus surgery. Clin Exp Otorhinolaryngol 7(3):193–197. https://doi.org/10.3342/ceo.2014.7.3.193

Hope CK, Clements D, Wilson M (2002) Determining the spatial distribution of viable and nonviable bacteria in hydrated microcosm dental plaques by viability profiling. J Appl Microbiol 93(3):448–455. https://doi.org/10.1046/j.1365-2672.2002.01703.x

Hori K, Matsumoto S (2010) Bacterial adhesion: from mechanism to control. Biochem Eng J 48(3):424–434. https://doi.org/10.1016/j.bej.2009.11.014

Hung C, Zhou Y, Pinkner JS, Dodson KW, Crowley JR, Heuser J, Chapman MR, Hadjifrangiskou M, Henderson JP, Hultgren SJ (2013) *Escherichia coli* biofilms have an organized and complex extracellular matrix structure. MBio 4(5):e00645. https://doi.org/10.1128/mBio.00645-13

Ishii S, Koki J, Unno H, Hori K (2004) Two morphological types of cell appendages on a strongly adhesive bacterium, *Acinetobacter sp.* strain Tol 5. Appl Environ Microbiol 70(8):5026–5029. https://doi.org/10.1128/AEM.70.8.5026-5029.2004

Jiang W, Wang Y, Luo J, Li X, Zhou X, Li W, Zhang L (2018) Effects of antimicrobial peptide GH12 on the cariogenic properties and composition of a cariogenic multispecies biofilm. Appl Environ Microbiol 84(24). https://doi.org/10.1128/AEM.01423-18

Joubert LM, Ferreira JA, Stevens DA, Nazik H, Cegelski L (2017) Visualization of *aspergillus fumigatus* biofilms with scanning electron microscopy and variable pressure-scanning electron microscopy: a comparison of processing techniques. J Microbiol Methods 132:46–55. https://doi.org/10.1016/j.mimet.2016.11.002

Klausen M, Heydorn A, Ragas P, Lambertsen L, Aaes-Jørgensen A, Molin S, Tolker-Nielsen T (2003) Biofilm formation by *Pseudomonas aeruginosa* wild type, flagella and type IV pili mutants. Mol Microbiol 48(6):1511–1524. https://doi.org/10.1046/j.1365-2958.2003.03525.x

Körstgens V, Flemming HC, Wingender J, Borchard W (2001) Influence of calcium ions on the mechanical properties of a model biofilm of mucoid *Pseudomonas aeruginosa*. Water Sci Technol 43(6):49–57. https://doi.org/10.2166/wst.2001.0338

Lee EY, Choi DY, Kim DK, Kim JW, Park JO, Kim S, Kim SH, Desiderio DM, Kim YK, Kim KP, Gho YS (2009) Gram-positive bacteria produce membrane vesicles: proteomics-based characterization of Staphylococcus aureus-derived membrane vesicles. Proteomics 9(24):5425–5436. https://doi.org/10.1002/pmic.200900338

Lee J, Lee EY, Kim SH, Kim DK, Park KS, Kim KP, Kim YK, Roh TY, Gho YS (2013) *Staphylococcus aureus* extracellular vesicles carry biologically active β-lactamase. Antimicrob Agents Chemother 57(6):2589–2595. https://doi.org/10.1128/AAC.00522-12

Li YF, Sun HW, Gao R, Liu KY, Zhang HQ, Fu QH, Qing SL, Guo G, Zou QM (2015) Inhibited biofilm formation and improved antibacterial activity of a novel nanoemulsion against cariogenic *Streptococcus mutans* in vitro and in vivo. Int J Nanomedicine 10:447–462. https://doi.org/10.2147/IJN.S72920

Li Y, Xiao P, Wang Y, Hao Y (2020) Mechanisms and control measures of mature biofilm resistance to antimicrobial agents in the clinical context. ACS omega 5(36):22684–22690. https://doi.org/10.1021/acsomega.0c02294

Lim JY, May JM, Cegelski L (2012) Dimethyl sulfoxide and ethanol elicit increased amyloid biogenesis and amyloid-integrated biofilm formation in Escherichia coli. Appl Environ Microbiol 78(9):3369–3378. https://doi.org/10.1128/AEM.07743-11

Luo J, Zhang J, Barnes RJ, Tan X, McDougald D, Fane AG, Zhuang G, Kjelleberg S, Cohen Y, Rice SA (2015) Nitric oxide-based control of membrane biofouling. Microb Biotechnol 8:549–560. https://doi.org/10.1111/1751-7915.12261

Mack D, Fischer W, Krokotsch A, Leopold K, Hartmann R, Egge H, Laufs R (1996) The intercellular adhesin involved in biofilm accumulation of *Staphylococcus epidermidis* is a linear beta-1, 6-linked glucosaminoglycan: purification and structural analysis. J Bacteriol 178(1):175–183. https://doi.org/10.1128/jb.178.1.175-183.1996

Maher MC (2016) Identification and characterization of inhibitors of amyloids and amyloid-based biofilm in E. Coli through cell-based high-throughput screening (Doctoral dissertation, Stanford University)

Marshall KC, Stout R, Mitchell R (1971) Mechanism of the initial events in the sorption of marine bacteria to surfaces. Microbiology 68(3):337–348. https://doi.org/10.1099/00221287-68-3-337

Moreira JMR, Gomes LC, Whitehead KA, Lynch S, Tetlow LA, Mergulhão FJ (2017) Effect of surface conditioning with cellular extracts on *Escherichia coli* adhesion and initial biofilm formation. Food Bioprod Process 104:1–12. https://doi.org/10.1016/j.fbp.2017.03.008

Mukherjee M, Cao B (2021) Engineering controllable biofilms for biotechnological applications. Microb Biotechnol 14(1):74. https://doi.org/10.1111/1751-7915.13715

Neu TR, Lawrence JR (2014) Investigation of microbial biofilm structure by laser scanning microscopy. Productive Biofilms 1–51. https://doi.org/10.1007/10_2014_272

Orsinger-Jacobsen SJ, Patel SS, Vellozzi EM, Gialanella P, Nimrichter L, Miranda K, Martinez LR (2013) Use of a stainless steel washer platform to study *Acinetobacter baumannii* adhesion and biofilm formation on abiotic surfaces. Microbiology (United Kingdom) 159(PART 12):2594–2604. https://doi.org/10.1099/mic.0.068825-0

Ortega-Morales BO, Santiago-García JL, Chan-Bacab MJ, Moppert X, Miranda-Tello E, Fardeau ML, Carrero JC, Bartolo-Pérez P, Valadéz-González A, Guezennec J (2007) Characterization of extracellular polymers synthesized by tropical intertidal biofilm bacteria. J Appl Microbiol 102(1):254–264. https://doi.org/10.1111/j.1365-2672.2006.03085.x

Park S, Sauer K (2022) Controlling biofilm development through cyclic di-GMP signaling. In: Pseudomonas aeruginosa: biology, pathogenesis and control strategies, pp 69–94. https://doi.org/10.1007/978-3-031-08491-1_3

Perni S, Preedy EC, Prokopovich P (2014) Success and failure of colloidal approaches in adhesion of microorganisms to surfaces. Adv Colloid Interf Sci 206:265–274. https://doi.org/10.1016/j.cis.2013.11.008

Pesci EC, Pearson JP, Seed PC, Iglewski BH (1997) Regulation of las and rhl quorum sensing in *Pseudomonas aeruginosa*. J Bacteriol 179(10):3127–3132. https://doi.org/10.1128/jb.179.10.3127-3132.1997

Priester JH, Horst AM, Van De Werfhorst LC, Saleta JL, Mertes LAK, Holden PA (2007) Enhanced visualization of microbial biofilms by staining and environmental scanning electron microscopy. J Microbiol Methods 68(3):577–587. https://doi.org/10.1016/j.mimet.2006.10.018

Ray VA, Hill PJ, Stover KC, Roy S, Sen CK, Yu L, Wozniak DJ, Digiandomenico A (2017) Anti-Psl targeting of *Pseudomonas aeruginosa* biofilms for neutrophil-mediated disruption. Sci Rep 7(1):16065. https://doi.org/10.1038/s41598-017-16215-6

Redler A, Miglietta S, Monaco E, Matassa R, Relucenti M, Daggett M, Ferretti A, Familiari G (2019) Ultrastructural assessment of the anterolateral ligament. *Orthopaedic*. Journal of Sports Medicine 7(12). https://doi.org/10.1177/2325967119887920

Relucenti M, Familiari G, Donfrancesco O, Taurino M, Li X, Chen R et al (2021) Microscopy methods for biofilm imaging: focus on SEM and VP-SEM pros and cons. Biology 10(1):51. https://doi.org/10.3390/biology10010051

Richardson N, Mordan NJ, Figueiredo JAP, Ng YL, Gulabivala K (2009) Microflora in teeth associated with apical periodontitis: a methodological observational study comparing two protocols and three microscopy techniques. Int Endod J 42(10):908–921. https://doi.org/10.1111/j.1365-2591.2009.01594.x

Rodrigues D, Bañobre-López M, Espiña B, Rivas J, Azeredo J (2013) Effect of magnetic hyperthermia on the structure of biofilm and cellular viability of a food spoilage bacterium. Biofouling 29(10):1225–1232. https://doi.org/10.1080/08927014.2013.834893

Ruhal R, Kataria R (2021) Biofilm patterns in gram-positive and gram-negative bacteria. Microbiol Res 251:126829. https://doi.org/10.1016/j.micres.2021.126829

Silva-Dias A, Miranda IM, Branco J, Monteiro-Soares M, Pina-Vaz C, Rodrigues AG (2015) Adhesion, biofilm formation, cell surface hydrophobicity, and antifungal planktonic susceptibility: relationship among *Candida spp*. Front Microbiol 6:205. https://doi.org/10.3389/fmicb.2015.00205

Smith DR, Price JE, Burby PE, Blanco LP, Chamberlain J, Chapman MR (2017) The production of curli amyloid fibers is deeply integrated into the biology of *Escherichia coli*. Biomol Ther 7(4). https://doi.org/10.3390/biom7040075

Soliman C, Walduck AK, Yuriev E, Richards JS, Cywes-Bentley C, Pier GB, Ramsland PA (2018) Structural basis for antibody targeting of the broadly expressed microbial polysaccharide poly-N-acetylglucosamine. J Biol Chem 293(14):5079–5089. https://doi.org/10.1074/jbc.RA117.001170

Speziale P, Pietrocola G, Foster TJ, Geoghegan JA (2014) Protein-based biofilm matrices in Staphylococci. Front Cell Infect Microbiol 4:171. https://doi.org/10.3389/fcimb.2014.00171

Stoodley P, Sauer K, Davies DG, Costerton JW (2002) Biofilms as complex differentiated communities. Ann Rev Microbiol 56(1):187–209. https://doi.org/10.1146/annurev.micro.56.012302.160705

Sun L, Liao K, Wang D (2015) Effects of magnolol and honokiol on adhesion, yeast-hyphal transition, and formation of biofilm by *Candida albicans*. PLoS One 10(2). https://doi.org/10.1371/journal.pone.0117695

Sun X, Lin Z-W, Hu X-X, Yao W-M, Bai B, Wang H-Y, Li D-Y, Chen Z, Cheng H, Pan W-G, Deng M-G, Xu G-J, Tu H-P, Chen J-W, Deng Q-W, Yu Z-J, Zheng J-X (2018) Biofilm formation in erythromycin-resistant *Staphylococcus aureus* and the relationship with antimicrobial susceptibility and molecular characteristics. Microb Pathog 124:47–53. https://doi.org/10.1016/j.micpath.2018.08.021

Sutherland IW (2001) The biofilm matrix–an immobilized but dynamic microbial environment. Trends Microbiol 9(5):222–227. https://doi.org/10.1016/S0966-842X(01)02012-1

Teschler JK, Zamorano-Sánchez D, Utada AS, Warner CJ, Wong GC, Linington RG, Yildiz FH (2015) Living in the matrix: assembly and control of *Vibrio cholerae* biofilms. Nat Rev Microbiol 13(5):255–268. https://doi.org/10.1038/nrmicro3433

Thornton RB, Rigby PJ, Wiertsema SP, Filion P, Langlands J, Coates HL, Vijayasekaran S, Keil AD, Richmond PC (2011) Multi-species bacterial biofilm and intracellular infection in otitis media. BMC Pediatr 11. https://doi.org/10.1186/1471-2431-11-94

Tielen P, Strathmann M, Jaeger KE, Flemming HC, Wingender J (2005) Alginate acetylation influences initial surface colonization by mucoid *Pseudomonas aeruginosa*. Microbiol Res 160(2):165–176. https://doi.org/10.1016/j.micres.2004.11.003

Valentini M, Filloux A (2016) Biofilms and cyclic di-GMP (c-di-GMP) signalling: lessons from *Pseudomonas aeruginosa* and other bacteria. J Biol Chem 291(24):12547–12555. https://doi.org/10.1074/jbc.R115.711507

Van Laar TA, Chen T, You T, Leung KP (2015) Sublethal concentrations of carbapenems alter cell morphology and genomic expression of *Klebsiella pneumoniae* biofilms. Antimicrob Agents Chemother 59(3):1707–1717. https://doi.org/10.1128/AAC.04581-14

Van Oss CJ (1991) The forces involved in bioadhesion to flat surfaces and particles – their determination and relative roles. Biofouling 4(1–3):25–35. https://doi.org/10.1080/08927019109378192

Verma V, Harjai K, Chhibber S (2010) Structural changes induced by a lytic bacteriophage make ciprofloxacin effective against older biofilm of *Klebsiella pneumoniae*. Biofouling 26(6):729–737. https://doi.org/10.1080/08927014.2010.511196

Villacorte LO, Ekowati Y, Neu TR, Kleijn JM, Winters H, Amy G, Schippers JC, Kennedy MD (2015) Characterisation of algal organic matter produced by bloom-forming marine and freshwater algae. Water Res 73:216–230. https://doi.org/10.1016/j.watres.2015.01.028

Vlamakis H, Chai Y, Beauregard P, Losick R, Kolter R (2013) Sticking together: building a biofilm the *Bacillus subtilis* way. Nat Rev Microbiol 11(3):157–168. https://doi.org/10.1038/nrmicro2960

Wang W, Chanda W, Zhong M (2015) The relationship between biofilm and outer membrane vesicles: a novel therapy overview. FEMS Microbiol Lett 362(15). https://doi.org/10.1093/femsle/fnv117

Weber K, Delben J, Bromage TG, Duarte S (2014) Comparison of SEM and VPSEM imaging techniques with respect to *Streptococcus mutans* biofilm topography. FEMS Microbiol Lett 350(2):175–179. https://doi.org/10.1111/1574-6968.12334

Weimer PJ, Price NPJ, Kroukamp O, Joubert LM, Wolfaardt GM, Van Zyl WH (2006) Studies of the extracellular glycocalyx of the anaerobic cellulolytic bacterium *Ruminococcus albus* 7. Appl Environ Microbiol 72(12):7559–7566. https://doi.org/10.1128/AEM.01632-06

Whitchurch CB, Tolker-Nielsen T, Ragas PC, Mattick JS (2002) Extracellular DNA required for bacterial biofilm formation. Science 295(5559):1487–1487. https://doi.org/10.1126/science.295.5559.1487

Whitehead KA, Verran J (2015) Formation, architecture and functionality of microbial biofilms in the food industry. Curr Opin Food Sci 2:84–91. https://doi.org/10.1016/j.cofs.2015.02.003

Wilton M, Charron-Mazenod L, Moore R, Lewenza S (2016) Extracellular DNA acidifies biofilms and induces aminoglycoside resistance in *Pseudomonas aeruginosa*. Antimicrob Agents Chemother 60(1):544–553. https://doi.org/10.1128/AAC.01650-15

Wood TL, Guha R, Tang L, Geitner M, Kumar M, Wood TK (2016) Living biofouling-resistant membranes as a model for the beneficial use of engineered biofilms. Proc Natl Acad Sci 113(20):E2802–E2811. https://doi.org/10.1073/pnas.1521731113

Xavier JB, Foster KR (2007) Cooperation and conflict in microbial biofilms. Proc Natl Acad Sci 104(3):876–881. https://doi.org/10.1073/pnas.0607651104

Yang X, Beyenal H, Harkin G, Lewandowski Z (2000) Quantifying biofilm structure using image analysis. J Microbiol Methods 39(2):109–119. https://doi.org/10.1016/S0167-7012(99)00097-4

Zhao K, Tseng BS, Beckerman B, Jin F, Gibiansky ML, Harrison JJ, Luijten E, Parsek MR, Wong GCL (2013) Psl trails guide exploration and microcolony formation in *Pseudomonas aeruginosa* biofilms. Nature 497(7449):388–391. https://doi.org/10.1038/nature12155

Nutritional Fortification: Biofilm Contribution

3

Monisha Nath, Ankita Gon, Arnab Chakraborty, and Soma Das

Abstract

Biofilms are defined as communities of microorganisms that attach to surfaces and are a prevalent mode of growth for microorganisms in nature. Biofilms are currently identified as an assemblage of surface-associated microbial cells that are enclosed in hydrated extracellular polymeric substances (EPS). Food fortification is defined as the practice of adding vitamins and minerals to commonly consumed foods during processing to increase their nutritional value. During production, surfaces and equipment used in food processing might get contaminated with bacteria. These microorganisms have the potential to cause foodborne disease or loss of food quality if they form biofilms and act as reservoirs for pathogens or spoilage organisms. Biofilms may unintentionally encapsulate extra nutrients or fortifiers in certain situations. Because of their encapsulation, these substances may be less bioavailable and less effective in strengthening the finished product when released into the food matrix. Various types of probiotic bacteria help produce various nutraceuticals and micronutrients that help in the in situ mechanism of fortification of meat and meat-associated products. This results in enhancing the nutritional quality of the product. Enzymes or metabolites that can break down or change the chemical structure of fortificants found in the food matrix can be produced by biofilms. The bioavailability or effectiveness of the fortified nutrients may decrease as a result. Foods that have been fortified may degrade more quickly due to biofilms created by spoiling microbes, resulting in a lower shelf life. Over time, the existence of biofilms may alter the

M. Nath · A. Gon · A. Chakraborty · S. Das (✉)
Department of Allied Health Sciences: School of Health and Medical Sciences, Adamas University, Barasat, India

Department of Pharmaceutical Technology: School of Health and Medical Sciences, Adamas University, Barasat, India

T. Sarkar et al. (eds.), *Biofilm Applications to Revolutionize Food Technology*,
https://doi.org/10.1007/978-3-031-85205-3_3

taste, texture, and nutritional value of food by fostering microbial growth and enzymatic activity.

Keywords
Food fortification · Biofilm · Bacteria · Fortificants · Contamination · Encapsulation of nutrients

3.1 Overview of Biofilm

Biofilms are described as microbial communities that adhere to surfaces and are a common way for microorganisms to proliferate in the natural world. Currently, biofilms are defined as a collection of microbial cells that adhere to surfaces and are encased in hydrated extracellular polymeric substance (EPS). Retrospectively, it has been shown that the first biofilm observation was performed by the Dutch scientist Antonie van Leeuwenhoek in the seventeenth century. He observed microbial communities in dental plaque and other habitats using a basic handmade microscope. Identifying these particles as "animalcules," he pointed out that they were made up of a variety of microorganisms. In the early 1970s, microbiologist J. William Costerton and associates coined the term "biofilm" in their investigation of the bacterial communities that grow on surfaces in aquatic environments. They discovered that these communities were highly ordered structures with specialized adaptive activities, such as food intake, waste elimination, and defense against environmental stresses, rather than just spontaneous assemblages of individual cells. It has varied morphologies depending on constituent bacteria and the condition under which it was formed. It adheres to surfaces and forms a slimy, protective layer. They are present in a variety of settings, including soil, water, and the human body. Biofilm is used in bioremediation, biofuel generation, and wastewater treatment that causes infections on medical implants. Scientists are looking at cutting-edge tactics including developing antimicrobial substances, surface coatings, and enzymes that specifically target biofilms to prevent the formation of biofilms.

3.2 Classification of Biofilm

1. *Bacteria*
   - Cell surface hydrophobicity.
   - Adhesins.
   - Quorum sensing.
   - Extracellular polysaccharide.
2. *Host*
   - Cell receptor.
   - PH.
   - Proteins.
   - Platelets.

3. *Implant*
   - Hydrophobicity.
   - Surface chemistry.
   - Roughness.
   - Porosity.

3.3 Cell Surface Hydrophobicity

The hydrophobicity of the bacterial cell surface is critical to many physiological processes and interactions with their environment (Donlan 2002). The following are some essential facts about the hydrophobicity of bacterial cell surfaces:

- Hydrophobic interactions have the potential to facilitate the adhesion of bacterial cells to surfaces. Hydrophobic bacteria are more likely to attach themselves to biotic surfaces, such as those found in hosts, as well as abiotic surfaces, such as those found in ambient surfaces and medical equipment. This adhesion is often a crucial initial step for activities such as biofilm formation and infection.
- *Quorum Sensing*: Quorum sensing is a sophisticated mechanism used by bacterial cells to synchronize their responses to changes in population density (Sakuragi and Kolter 2007). Typically, it functions as described below.

Bacteria produce and release small compounds into their environment called autoinducers, which function as signaling molecules. The distribution of autoinducers in the environment is directly proportional to the number of bacteria present. It is possible for bacteria to identify the presence and amount of these signaling molecules, thanks to a class of receptors known as autoinducer receptors or LuxR-type receptors. Regulation of gene expression: Receptor activation occurs when autoinducer concentrations cross a particular threshold. Additionally, inside the bacterial population, this activation sets off alterations in gene expression.

- *Extracellular polysaccharide*: EPS is an extracellular polymeric substance that encases structured colonies of bacteria, which is the process of biofilm formation (Li et al. 2022).

Bacterial cell adhesion to surfaces and subsequent production of biofilms are influenced by the hydrophobicity of the cell surface. Surface-adhering bacteria may be more robust when their surfaces are hydrophobic.

- *Cell receptor*: Proteins on a cell's surface or interior that receive signals are called cellular receptors. This is a chemical signal in normal physiology that is produced when a protein ligand interacts with a protein receptor. A chemical message called a ligand is released by one cell to communicate with another cell or with itself (Flemming et al. 2016a).

- *PH*: The pH of the various host environments ranges from extremely acidic ($pH < 2$) in the stomach to mildly acidic ($6.5 < pH < 4.5$) on the skin and plant surfaces, neutral (pH 7.4) in the blood, and basic ($pH < 8.5$) in some portions of the intestine.
- *Proteins*: Proteins play crucial roles in the host's body.

Proteins called enzymes serve as biological catalysts, quickening the pace at which certain chemical reactions take place inside cells. DNA replication, metabolism, digestion, and cellular signaling are among the processes that enzymes are involved in. DNA polymerases, metabolic enzymes such as ATP synthase and hexokinase, and digestive enzymes such as amylase, lipase, and protease are a few examples.

Structural proteins provide stability and support to cells, tissues, and organs. Structural proteins such as collagen, elastin, keratin, and actin form the structure of connective tissues, skin, hair, nails, muscles, and bones. These proteins function to strengthen, stretch, and retain these tissues.

Chemicals such as ions, nutrients, gases, and hormones are transported throughout the body by proteins known as transport proteins, which are able to pass through cell membranes. Examples of carrier proteins include hemoglobin and albumin, as well as membrane transporters like glucose transporters.

Hormones and proteins that operate as signaling molecules: Proteins regulate a variety of physiological processes and cellular responses.

Specific receptors on target cells attach to hormones including insulin, glucagon, adrenaline, and growth factors, which then initiate signaling pathways that regulate growth, development, metabolism, and homeostasis (Vert et al. 2012).

Antibodies and Immunoglobulins: Proteins are essential for the immune system's protection against foreign invaders and infections. Specialized proteins produced by B cells in response to antigens are called antibodies or immunoglobulins. They identify antigens and attach them to either neutralize their deleterious effects or mark them for elimination by immune cells (López et al. 2010).

- *Platelets*: Although platelets actively bind, confine, and kill germs, they can also harm the host due to their proinflammatory effector functions and interactions with the coagulation system (e.g., acute lung injury and sepsis).

3.4 Example of Biofilm

Escherichia coli biofilm consists of a bacterial colony embedded in a matrix of extracellular polymeric substances (EPS).

- *How to produce biofilm*: Producing biofilms involves a series of steps where microorganisms adhere to a surface, form a matrix of extracellular polymeric substances (EPS), and develop into structured communities. Here is a general overview of the process.

Attachment

First, microorganisms adhere to the surface. This surface may be biotic (such as tissues or cells) or abiotic (such as ambient surfaces or medical implants). The facilitation of attachment is influenced by adhesion molecules, surface features, and environmental factors (Hall-Stoodley et al. 2004).

Colonization

Attached microorganisms begin to proliferate and disperse throughout the surface, forming microscopic colonies. During this phase, the microbial population grows and diversifies, and numerous species begin to contribute to the biofilm community.

Microbes in the biofilm generate and release extracellular polymeric substances (EPS), which surround the microbial cells in a matrix. EPS is composed of proteins, lipids, nucleic acids, and polysaccharides and provides structural support and defense for the biofilm population (Aslam et al. 2012).

Biofilm Architecture

To promote the circulation of nutrients and waste materials throughout the community, biofilms have an intricate architecture that includes characteristics such as towers, canals, and mushroom-like structures. The robustness and flexibility of the biofilm are facilitated by the spatial organization and variety of the microbial cells inside it (Lear and Lewis 2012).

Quorum Sensing

Through the creation and detection of signaling molecules, quorum sensing allows microorganisms within the biofilm to coordinate and communicate with one another.

Biofilm Dispersal

It is the process by which bacteria migrate to occupy uncharted regions or surfaces after separating from the biofilm. Dispersal processes include the enzymatic breakdown of the EPS matrix, active cell separation, and environmental cues that induce cell movement (Shrout et al. 2006) (Fig. 3.1).

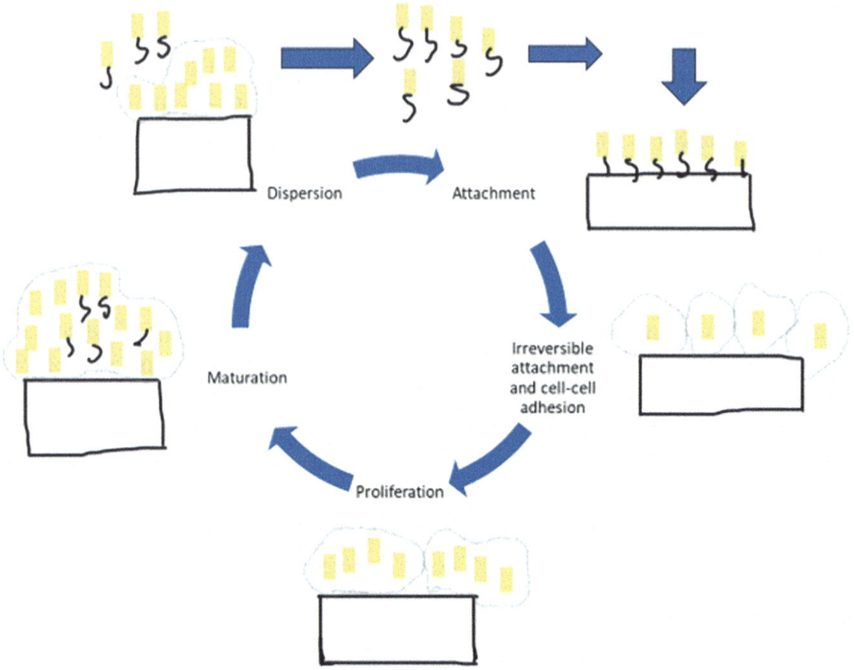

Fig. 3.1 The stages of biofilm formation

3.5 Application of Biofilm in Food Industry

One or more species can form complex microbial communities termed biofilms by immersing themselves in an extracellular matrix with different compositions, depending on the kind of food-producing environment and the colonizing species. These biofilms can be composed of a variety of microorganisms, such as fungi and bacteria. When many bacterial species coexist in a biofilm, it may cling to a surface more easily, which has significant ecological benefits (Meyer 2015). Polysaccharides such as cellulose, proteins, or foreign DNA make up the majority of the extracellular matrix. This matrix can be attached to either biological structures (vegetables, meat, bones, fruits, etc.) or hard surfaces (food industry equipment, transport, dispensing and storage surfaces, soil, etc.) (Flemming et al. 2016b). These biofilms' high durability in the food industry is due to a structural function played by the extracellular matrix. It includes extracellular enzymes for nutritional reasons, creates intricate gradients for the diffusion of nutrients and oxygen, permits the transmission of cell communication molecules, and shields the embedded cells from harmful substances. To sum up, the formation of biofilms offers several benefits to the microbial cells in the food industry, including chemical protection from chemicals, antimicrobials, and disinfectants, as well as physical resistance against desiccation and mechanical resistance against liquid streams in pipelines (Ferreira et al. 2014).

3.6 Role of Biofilm in Food Fortification

The biofilm lifestyle has improved food quality dramatically in recent years, and its applications in the pharmaceutical and nutraceutical industries have grown tremendously. The substances that are present in the EPS that have the ability to lower cholesterol, function as prebiotic agents, and have antibacterial and antioxidant properties are known as bioactive compounds (Prete et al. 2021).

Therefore, the food sector has been interested in the exopolysaccharide generation by the LAB group. The application of EPS as a thickening agent and causing rheological alterations in food has elevated it to a significant area of study (Jurášková et al. 2022). One of the many functional qualities that the extracellular polymer (EPS) generated by sessile colonies contributes to is preventing different types of pathogenic pathogens from adhering to animal cell lines (González-Cancho et al. 1980). Because of their diverse physicochemical characteristics, the EPS produced by the LAB serves as substitute food additives. Because the EPS alters the rheological properties of food, it is a valuable area for research and useful to the food sectors (Korcz and Varga 2021). It is more advantageous to use biofilm-bound bacterial cells in the wine-making industries because the biofilm-embedded *Lactobacillus plantarum* bacterial cells demonstrated the ability of malolactic fermentation and that their efficacy for L-malic acid conversion was higher in sessile cells than in planktonic form (Pannella et al. 2020). Likewise, in unfavorable circumstances, such as a low pH environment during malolactic fermentation, the biofilm-bound cells of other LABs, such as *Lactobacillus casei* ATCC 334, showed enhanced survival capacity (Wang et al. 2018). The amount of vital nutrients available in food and its digestibility are the two main factors that determine its nutritional quality. Fermentation may improve nutritional qualities and also make food more digestible. Microbial enzymes that are fermenting initially begin to break down the macronutrients throughout the fermentation process (Xiang et al. 2019). Increasing the amount and bioavailability of nutrients results in an increase in nutritional density, which is one of the many ways that fermentation may improve the nutritional quality of food. By stimulating the production of promoters for the absorption of antinutritional component degradation, nutritional density may be raised. This can impact the mucosa's ability to absorb nutrients and the predigestive phases of dietary elements (Nkhata et al. 2018).

Biofilms are communities of microorganisms that adhere to surfaces and produce a protective matrix, which can play a significant role in various aspects of food science and technology, including food fortification. In various forms, the biofilms are considered in fermented food formation, e.g., cheese, vinegar, kombucha, kefir, wine, lambic beer, miso, and kimchi (Wolfe and Dutton 2015).

On the other hand, fermentation leads to the disintegration of organic macromolecules through microbial activity into simple molecules. For example, enzymes from yeast convert starch and sugar to alcohol, whereas the proteins get converted to amino acids/peptides. The microbial enzymatic activities on food components lead to food fermentation producing desirable biochemical alterations, which lead to the noticeable modification of the food. Fermentation is a natural process of improving essential amino acids, vitamins, proteins, antinutrients, flavors, food appearance, and enriched aroma. Thus, fermentation is a process to reduce the energy required to cook and make the food safer. Hence, microbial actions play an important role in food fermentation by enhancing both the physical as well as chemical properties of food. Fermented foods have longer shelf life, improved organoleptic properties, eliminated unwanted/ harmful components from food, and improved antioxidant properties. Examples of fermented foods include cheese, garri, bread, ogi, milk, and yogurt.

Substrate composition and fermented microbes are the two main factors influencing fermented food quality. The treatment of food and the period of fermentation during food processing also influence fermented food quality. For most fermented beverages and foods, lactic acid bacteria (LAB) have been found to be the most dominant group of microbes and they impart beneficial effects to fermented foods (Mokoena 2017). The group of LAB mainly consists of *Streptococcus, Enterococcus, Lactobacillus, Leuconostoc*, and *Pediococcus* and also yeasts and molds such as *Kluyveromyces, Debaryomyces, Geotrichum, Saccharomyces, Penicillium, Mucor*, and *Rhizopus* species (Anukam and Reid 2009).

Here is how biofilms are relevant in this context:

- *Probiotic Delivery*: Biofilms can be used to deliver probiotics more effectively in fortified foods. Probiotics are beneficial bacteria that can support gut health, and embedding these bacteria in biofilms can enhance their survival through the digestive tract, ensuring that a higher number of viable probiotics reach the gut.
- *Nutrient Encapsulation*: Biofilms can encapsulate and protect sensitive nutrients, vitamins, or minerals during food processing and storage. This can ensure that the food that has been fortified retains its nutritious worth throughout time by maintaining the potency and efficacy of these nutrients.
- *Enhanced Nutrient Utilization*: Nutrient absorption and digestion can be facilitated by certain microbes that produce biofilms. It could be able to improve the bioavailability of vital elements in fortified meals by adding these microbes to food items.
- *Food Safety*: Biofilms have the potential to improve food safety by preventing the development of harmful microbes. Benefiting bacteria can reduce the risk of contamination and deterioration in fortified food items by outcompeting harmful pathogens when they form biofilms.
- *Functional Ingredients*: Food items can employ biofilms as transporters for beneficial substances. The effectiveness of food fortification is increased overall by the specific health advantages that these biofilm-encapsulated substances can offer, such as better digestion or heightened immune function.
- *Innovation in Food Processing*: A novel strategy for raising the nutritional value of food is the fortification of food with biofilms. The study of biofilm technology

may contribute to the creation of fresh approaches to adding and maintaining nutrients in food items.

In general, the thoughtful application of biofilms in food fortification can boost nutrient delivery and effectiveness, raise food safety standards, and result in healthier food items.

3.7 Challenges with Biofilms in the Fortification of Food

Although biofilms have a number of potential benefits for food fortification, there are drawbacks and difficulties with their use as well.

Complexity of Control

Controlling the production of biofilms can be challenging. Nutrient or probiotic distribution and efficacy in fortified meals might vary depending on irregular or uncontrolled biofilm growth.

Risk of Contamination

If biofilms are not correctly controlled, they may harbor harmful microbes. This may increase the chance of contamination in food items, particularly if cross-contamination occurs during processing or the biofilm matrix is not well maintained.

Processing Challenges

The processing and formulation of food items may become more difficult if biofilms are included. Biofilms can alter the final product's texture, flavor, or appearance, thus rigorous testing and optimization are necessary.

Regulatory and Safety Concerns

Food items containing biofilms may give rise to safety and regulatory issues. It is important to make sure that the biofilm-forming bacteria used in food fortification are safe and compliant with regulations; this may include a lot of testing and paperwork.

Potential for Unintended Interactions

Complex interactions between biofilms and other substances may have unexpected consequences for the food's safety or nutritional value. To make sure that the biofilm components do not have a detrimental effect on the product, thorough formulation and stability testing are needed.

Economic Costs

Biofilm-based technology development and deployment can be expensive. This covers costs for quality control procedures, production procedures, and research and development. These expenses could be passed along to customers or have an effect on how reasonably priced fortified goods are.

Consumer Acceptance

Consumers may not fully comprehend the idea of biofilms, and there may be opposition or doubt regarding food items derived from biofilms. Addressing any issues and fostering acceptance requires effective communication as well as education.

Limited Research

It is still early in the field of study to apply biofilms to food fortification. In order to completely comprehend the advantages and long-term impacts of biofilm inclusion in food items, additional research is necessary.

3.8 Microbe-Assisted Nutritional Quality Improvement in Fermented Foods

Because fermenting bacteria are present, foods that have undergone fermentation have enhanced nutritional benefits. Anabolic or catabolic fermentation techniques can be used to synthesize vitamins and growth factors or to break down complicated molecules. Nutrients from plant cells and structures—mostly in grains and seeds—are released by undigested materials. Plant materials have more nutritional value when they are broken down by enzymes, which makes it possible to produce cellulose-rich substrates. Fermentation by yeast and lactic acid bacteria (LAB) can improve food's digestibility and nutritional content. For dairy products, alcoholic drinks, and olives to ferment, yeast biofilms are essential. They facilitate the release of inorganic phosphorus and the breakdown of phytic substances. The co-agglomeration of yeast and *Candida boidinii* results in polymicrobial biofilms that improve olive oil fermentation. Biofilms of yeast enrich the taste of wine (Hasan et al. 2014; Sarkar et al. 2021; Zara et al. 2020; Arroyo-López et al. 2012; Santona et al. 2018; Fancello et al. 2020; Perpetuini et al. 2018; Legras et al. 2016; Ángeles Pozo-Bayón and Victoria 2011).

Fermentation even improves microbial enzymatic activity by creating an acidic environment at a temperature of around 22–25 °C (Mokoena et al. 2005). The reduction of antinutrients phytic acid (phytase-mediated degradation) and tannins, which increases the bioavailability of polysaccharides (amylases), or simple sugars, free fatty acids (lipase), proteins (protease), and iron, is one of the main purposes of enzyme-mediated hydrolysis in fermented food products.

3.9 Role of Biofilm in Media Lactic Acid Fermentation Upon Nutritional Qualities of Food

The ability of *Lactobacillus plantarum* biofilm-embedded bacterial cells to perform malolactic fermentation was demonstrated (Pannella et al. 2020). Sessile cells were more effective in converting L-malic acid than planktonic cells in this regard. This suggests that biofilm-bound cells could be used more often in the wine-making industry. Similar to this, biofilm-bound cells of other LABs, such as *Lactobacillus casei* ATCC 334, demonstrated enhanced resilience to unfavorable circumstances, such as a low pH environment during malolactic fermentation (Wang et al. 2018).

The amount of vital elements in food and how easily it may be digested affect its nutritional quality. By increasing the amount and bioavailability of nutrients, fermentation can improve both aspects and increase nutritional density. In order to influence nutrient uptake by mucosa and predigestive stages of dietary elements, nutritional density can be increased by promoting synthesis for antinutritional component breakdown absorption. Protein solubility and the bioavailability of minerals and amino acids can both be enhanced by lactic acid fermentation. By reducing phytates, 50% of tannins, and 90% of oligosaccharides, this mechanism affects disorders associated with nutrition deficiencies either directly or indirectly. Food fermentation has a direct therapeutic effect on consumers' health by making vitamins such as thiamine, niacin, riboflavin, and folic acid more readily available (Xiang et al. 2019; Nkhata et al. 2018; Rollán et al. 2019; Samtiya et al. 2020; Melini et al. 2019).

By reducing indigestible plant material such as cellulose, hemicelluloses, and polygalacturonic and glucuronic acids, the fermentation in food enhances the bioavailability of minerals and trace elements (Gupta et al. 2015). By inhibiting the liver's production of cholesterol and the intestine's absorption of both endogenous and dietary cholesterol, it even lowers blood cholesterol (Jesch and Carr 2017). Because of its stability, durability, and safety for the food product, it helps to ward off illnesses such as diarrhea and salmonellosis (Minh 2014).

Alterations and Improvement of Biological Constituents of Fermented Foods.

Bio Enrichment with Vitamins

Marine Roseobacter Clade biofilm aids in the production of essential vitamins and auxins, as well as the synthesis of vitamins such as folic acid, biotin, and vitamin K. Lactobacillus biofilm also contributes to the synthesis of vitamin K and water-soluble vitamins such as pyridoxine, riboflavin, thiamine, cobalamin, folates, and thiamine. Indonesian dish "tempe" is prepared by soaking, dehulling, and partially cooking soya beans in the presence of molds or *Rhizopus oligosporus*. This fermentation process increases riboflavin, fatty acids, niacin, and vitamin B12, which are not found in vegetarian foods. This process reduces cooking time and enhances the digestibility of legume or cereal mixtures. Mexican pulque, the oldest alcoholic beverage in America, is made from Agave juices, a traditional beverage made from unmalted and malted kafficorn. Kaffir beer, a traditional beverage in South Africa, has a gruel and thin consistency with a sour but pleasant taste. Kaffir beer can be substituted with maize or millet, providing a higher content of riboflavin and nicotinic acid or niacin while maintaining a constant thiamine level during fermentation (Dang and Lovell 2015; Braga et al. 2016; Patel et al. 2013; Nout and Kiers 2005; Solange et al. 2014).

Enrichment with Antioxidants

The antioxidant activity of fermented foods containing Asian enzyme soybean products is analyzed using methods such as reducing electricity analysis, general evaluation of phenol content, and the possibility of DPH radical absorption. Some Asian soybean products such as Bikan, Pace, Kinema, Chongkoku can, Chan, NATO, Tunao, and Souls have antioxidant activity. Kimchi and yogurt also indicate antioxidant characteristics. Dairy pools selected from bacteria were used to ferment cereal flour starter culture for the synthesis of antioxidant peptides. DPPH analysis showed that lactic acid bacteria (LAB) during skim milk fermentation had free radical scavenging capacity ranging from 14.7% to 50.8% (v/v) after 24 to 72 h. This study showed that different LAB types and fermentation times had a significant effect on the antioxidant properties. *Lacticaseibacillus paracasei* and *Lactobacillus curvatus* increased the antioxidant activity of fermented sausage, improving its safety and taste. *L. curvatus* had better DPPH reducing and scavenging activity, whereas *L. paracasei* had better antilipid properties and OH scavenging ability (Karamoko et al. 2016; Coda et al. 2012; Abubakr et al. 2012; Zhang et al. 2017).

Enrichment with Peptides

Bioactive peptides (BAP) are produced from proteolytic microbes, mainly Bacillus, when food is fermented. BAPs have antihypertensive properties and are immunomodulatory and antithrombic agents. However, due to a lack of large-scale facilities and high enzyme costs, BAP production is not yet reaching the benchmark.

Fermentation with microorganisms is a cost-effective method for BAP production. Four classes of cell envelope proteinases (CEP) have been found, with *L. helveticus* being the species with the highest proteolytic capacity of the genus Lactobacillus. BAPs can inhibit angiotensin-converting enzyme (ATE) in fermented milk products such as kefir, koumiss, cheese, yogurt, fermented fish products, and fermented camel milk. BAPs are also generated by soybean product fermentation, which helps in treating and preventing metabolic diseases (Fong and Yildiz 2015; Miyamoto et al. 2015; Moslehishad et al. 2013).

3.10 Microbial Enzyme Production

Fermentation and Enzymes in Asian Foods (Tamang et al. 2007; Mine et al. 2005; Zeng et al. 2013; Montriwong et al. 2015).

Fermented microbes, mainly Bacillus, produce enzymes such as proteinase, amylase, catalase, mannase, and cellulase
Carbohydrate-producing enzymes include α-amylase, amyloglucosidase, pectinase, maltase, cellulase, invertase, lipase, alkaline proteases, and β-galactosidase
Enzymes from mycelia fungi such as *Actinomucor*, *Amylomyces*, *Mucor*, *Aspergillus*, *Rhizopus*, *Monascus*, and *Neurospora* are used in fermented beverages and foods
Aspergillus oryzae in koji has industrial applications
Starter cultures in Himalayas use dry, stable, and cake-like amylolytic starter culture for alcohol production
Bacillus subtilis' nattokinase enzyme has fibrinolytic capacity
Other fermented food bacteria, including *Vagococcus carniphilus*, *Bacillus amyloliquefaciens*, *Pediococcus acidilactici*, *Vagococcus lutrae*, *Enterococcus faecium*, *Enterococcus faecalis*, and *Enetrococcus gallinarum*, also possess fibrinolytic activities

3.11 Increased Production of Poly-Glutamate, Isoflavone, and Saponin Values

The four chemical components of soybeans yield isoflavones, including daidzein, genistein, and glycitein. The fermentation process of fermented foods such as tempe, chungkokjang, miso, natto, sufu, douche, and thuanao converts glucosidic isoflavones to aglycones. Following tempe fermentation, the isoflavone level rises, and the isoflavones in doenjang activate the LDL-C receptor, which may help avoid vascular illnesses such as hyperlipoproteinemia. *Bacillus subtilis* plays a major role for saponin fortification thus regulating constipation (Stanley and Lazazzera 2005).

Degradation of Antinutrients

Antinutrients in dietary items are harmful and reduce the bioavailability of important nutrients. Microbes in fermented foods break down these antinutrients, making them edible. Dewatering, grating, peeling, washing, roasting, and fermentation all minimize the amount of cyanide in cassava products such as fufu and gari. Streptococcus, Lactobacillus, and Leuconostoc may detoxify the cyanogenic glycosides linamarinis and lotaustralin found in cassava tubers, resulting in hydrocyanic acid with a low boiling point. *R. oligosporus* in tempe transforms nondigestible oligosaccharides into disaccharides and monosaccharides, which are easily absorbed. *B. subtilis* eliminates antinutritive compounds from kinema during rabadi and idli fermentation (Bamidele et al. 2015).

Biochemical Alterations in Fermentation of Cereals

Cereals are an important source of dietary proteins, carbs, fiber, minerals, and vitamins globally. However, they encounter hurdles in consumer appeal due to their low protein content, lack of key amino acids, limited starch bioavailability, the presence of antinutritive chemicals, and coarse texture. Several ways have been explored to improve cereal nutritional qualities, including genetic modification, amino acid addition, heating, sprouting, milling, and fermentation.

Natural cereal fermentation reduces carbohydrate content, indigestible polysaccharides, and oligosaccharides while increasing vitamin B group bioavailability and amino acid production. Fermentation also aids in the enzymatic biodegradation of phytates, creating ideal circumstances for the bioavailability of soluble iron, calcium, and zinc. After fermentation, the influence on protein and amino acid levels becomes debatable, with some grains, such as maize, millet, and sorghum, showing a rise in protein quality and lysine levels. Food product fermentation enhances and improves the texture, flavor, fragrance, taste, and shelf life of food. During fermentation, a variety of volatile chemicals are produced, including diacetyl acetic acid and butyric acid, which contribute to the production of tastes and aromas. The majority of traditionally fermented drinks and meals contain grains (rice, wheat, maize, or sorghum). The majority of them are used as colorants, spices, drinks, and food. Natural or hybrid cultures of bacteria, fungus, and yeasts carry out fermentation, changing the predominate microbiota. Lactobacillus, Bacillus, Streptococcus, Micrococcus, Leuconostoc, and Pediococcus are common types of bacteria that ferment. Aspergillus, Cladosporium, Paecilomyces, Trichothecium, Penicillium, and Fusarium are among the fungal genera that are often utilized. Lactic acid fermentation is a common method of food fermentation that produces foods with improved nutritional qualities, extended shelf life, widespread acceptability, and safety. The fermentation of cereals is influenced by organic acids, hydrogen peroxide, and antibiotics. The creation of lactic acid creates stress for bacteria that causes spoiling and prevents membrane potential maintenance and active transport (Singh et al. 2015; Sharma et al. 2020; Petrova and Petrov 2020; Papadimitriou et al. 2016).

Biogenic Amine Production by LAB Fermentation in Vegetables and Juices

Biogenic amines (BA) are low molecular weight amines found in food items and beverages. Over-consumption can lead to psychoactive disorders and vasoactive effects. Sauerkraut, a shredded white cabbage known for its sensory attributes and nutritional properties, is preserved through lactic acid fermentation. Studies have found that 121 samples have varying BA concentrations, with histamine content being less than 2 mg/kg in 44% of samples. Pasteurized and bottled German sauerkraut juices contain a significant amount of putrescine, up to 694 mg/dm^3. Researchers have formulated sauerkraut with less BA content under specific conditions to preserve it from bacteria. Factors reducing BA levels in production lead to the primary stoppage of spoilage with monogenic bacteria in shredded cabbage, production machines, transporters, and silos. Inoculants lacking amine are resistant to contamination (Garneau and Moineau 2011; Durak-Dados et al. 2020; Kalač et al. 2000a, b; Karovičová and Kohajdová 2018; Kirschbaum et al. 2000; Hornero-Méndez and Garrido-Fernández 1997).

Biochemical Processes for Enriching Nutritive Properties of Milk

Proteolysis Activity of Biofilm

The process of milk proteolysis converts proteins into amino acids, peptides, and peptones, increasing the levels of vital amino acids such as valine, phenylalanine, tryptophan, leucine, methionine, isoleucine, phenylalanine, and tyrosine. These amino acids are particularly beneficial to people who are physically weak. Proteolysis is carried out by exo- and endo-peptides (Kolesarova 1995). Fermented milk products, such as yogurt, curd, and kefir, have greater protein digestibility rates due to the lactic acid precipitation of the milk, which results in fine curd and an increase in nutritional value. The protein content of these products increases from 85.4% to 90%. Phenylalanine, proline, isoleucine, lysine, arginine, cysteine, and other free amino acids are more abundant throughout the fermentation process and subsequent preservation of the fermented milk (Mehta 2015; Shiby and Mishra 2013; Hao et al. 2019; Savijoki et al. 2006).

Fermented Milk: Key Facts and Alterations (Ye et al. 2017; Yoshii et al. 2019; Vieco-Saiz et al. 2019; García-Burgos et al. 2020)

- Hydrolysis of lactose
  - Lactose hydrolysis by bacteria in milk produces galactose, glucose, and lactose
  - Lactic acid production is essential for lactic acid production and bowel pH regulation
  - Lactic acid is also needed for calcium absorption and organoleptic properties
- Lipolysis
  - Homogenization reduces fat globule size for easy digestion
  - Increased fatty acid content due to LAB causes physiological changes

Vitamin alterations	
	Vitamins such as ascorbic acid, B2, and B1 are reduced due to bacterial culture
	Vitamins such as nicotinamide, riboflavin, and thiamine increase by 2 times
Antibacterial action	
	Antibacterial activity depends on bacteria's antibiotic action
	Other substances such as lactobacilli, lactic acid, hydrogen peroxide, antibiotics, and bacteriocins also have antibacterial activities
Mineral alterations	
	Higher mineral bioavailability due to LAB actions during and post-fermentation

Biochemical Modifications in Fermented Food Products Made of Meat

Anaerobic bacteria *Streptococcus*, *Lactobacillus*, *Leuconostoc*, *Pediococcus*, *Enterococcus*, and *Lactococcus* are responsible for breaking down saccharides into lactic acid, alcohols, amino acids, lipids, and aliphatic compounds. Together, these bacteria in fermented sausage do three tasks: they lower pH, release nitric oxide, and change simple carbohydrates into lactic acid. Sausage should be incubated under ideal, obligatorily anaerobic conditions to promote fast lactic acid development, which converts simple carbohydrates to lactic acid and lowers pH.

The primary process of lactic acid metabolism entails the fermentation of carbohydrates together with the required level of phosphorylation. The three primary variables in lactic acid fermentation are time, temperature, and relative humidity. Smoke coagulates surface proteins and inhibits bacteria by contributing to fragrance and antimicrobial compounds such as phenols, carbonyls, and organic acids. Muscle tissues include enzymes that catalyze the breakdown of fats into aldehydes, alkanes, alcohols, and ketones. Autocatalytic oxidative and hydrolytic changes are often associated with unsaturated lipids. Taste is also influenced by the presence of alcohols, acids, branching aromatic aldehydes, and short-chain nonvolatile fatty acids. Bacteria living in biofilms can adapt to changing environmental circumstances through a process called fermentation based on biofilms. Biofilms allow for the steady growth of cells with increased activity, which creates ideal conditions for industrial fermentation. Both high temperatures and rapid agitation are tolerated by biofilm cells. In conclusion, a variety of bacteria, such as *Streptococcus*, *Lactobacillus*, *Leuconostoc*, *Pediococcus*, *Enterococcus*, and *Lactococcus*, collaborate throughout the fermentation process in sausages to make lactic acid, alcohols, amino acids, lipids, as well as aliphatic compounds. The distinct flavor profile of fermented sausages is a result of the oxidation of lipids, the existence of branched chains of amino acids, and the presence of volatile chemicals (Bintsis 2018; Mastanjević et al. 2017).

3.12 Conclusion

The complex characteristics of biofilms contributes to prominent areas in the field of medicine, environmental management and food safety.

The persistence and adaptability of biofilms are enhanced by their distinct features, which are complex communities of microorganisms embedded in an extracellular matrix that they build on their own. In addition to offering defense against external pressures, this structural arrangement makes it easier for microbial cells to communicate and exchange nutrients.

Biofilms present possibilities as well as challenges for the food industry. Although they are susceptible to contamination and spoiling, their resilience to cleaning agents and longevity call for creative management techniques.

Biofilms are also essential for a number of biotechnological uses, such as bioremediation and wastewater treatment. Its capacity to break down contaminants and promote the cycle of nutrients emphasize its ecological significance. The study of biofilms in natural settings highlights the importance of these organisms in ecosystem functioning and nutrient dynamics, highlighting the need for more research to fully realize biofilms' promise for environmental sustainability.

Biofilms in medical settings are linked to long-term infections and problems with medical devices. Since biofilms on medical devices are resistant to many conventional antibiotics, treating them can be difficult, and other therapeutic strategies such as using bacteriophages, antimicrobial peptides, and biofilm-disrupting agents must be investigated. Targeted interventions to prevent and treat biofilm-associated infections must be developed with a thorough understanding of the genetic and molecular mechanisms underlying biofilm formation and maintenance.

In conclusion, biofilms are a two-edged sword that may be used for both good and bad ends for natural ecosystems and human activity. To maximize their benefits and reduce their hazards, more research must be done on their properties, interactions, and uses. In the end, biofilm research could lead to a greater knowledge of microbial life and its consequences for civilization as well as improvements in food fortification, food safety, environmental health, and the development of innovative medicinal treatments.

References

Abubakr MAS, Hassan Z, Imdakim M, Sharifah NRSA (2012) Antioxidant activity of lactic acid bacteria (LAB) fermented skim milk as determined by 1,1-diphenyl-2-picrylhydrazyl (DPPH) and ferrous chelating activity (FCA). African J Microbiol Res 6:6358–6364. https://doi.org/10.5897/AJMR12.702

Ángeles Pozo-Bayón M, Victoria Moreno-Arribas M (2011) Sherry wines. Adv Food Nutr Res 63:17–40. https://doi.org/10.1016/B978-0-12-384927-4.00002-6

Anukam KC, Reid G (2009) African traditional fermented foods and probiotics. J Med Food 12:1177–1184. https://doi.org/10.1089/jmf.2008.0163

Arroyo-López FN, Bautista-Gallego J, Domínguez-Manzano J, Romero-Gil V, Rodriguez-Gómez F, García-García P et al (2012) Formation of lactic acid bacteria-yeasts communities on the olive surface during Spanish-style Manzanilla fermentations. Food Microbiol 32:295–301. https://doi.org/10.1016/j.fm.2012.07.003

Aslam SN, Cresswell-Maynard T, Thomas DN, Underwood GJ (2012) Production and characterization of the Intra- and Extracellular Carbohydrates and Polymeric Substances (EPS) of three sea-ice diatom species, and evidence for a cryoprotective role for EPS. J Phycol 48(6):1494–1509

Bamidele OP, Fasogbon MB, Oladiran DA, Akande EO (2015) Nutritional composition of fufu analog flour produced from Cassava root (Manihot esculenta) and Cocoyam (Colocasia esculenta) tuber. Food Sci Nutr 3:597–603. https://doi.org/10.1002/fsn3.250

Bintsis T (2018) Lactic acid bacteria as starter cultures: an update in their metabolism and genetics. AIMS Microbiol 4:665–684. https://doi.org/10.3934/microbiol.2018.4.665

Braga RM, Dourado MN, Araújo WL (2016) Microbial interactions: ecology in a molecular perspective. Brazilian J Microbiol 47:86–98. https://doi.org/10.1016/j.bjm.2016.10.005

Coda R, Rizzello CG, Pinto D, Gobbetti M (2012) Selected lactic acid bacteria synthesize antioxidant peptides during sourdough fermentation of cereal flours. Appl Environ Microbiol 78:1087–1096. https://doi.org/10.1128/AEM.06837-11

Dang H, Lovell CR (2015) Microbial surface colonization and biofilm development in marine environments. Microbiol Mol Biol Rev 80:91–138. https://doi.org/10.1128/MMBR.00037-15

Donlan RM (2002) Biofilms: microbial life on surfaces. Emerg Infect Dis 8(9):881–890

Durak-Dados A, Michalski M, Osek J (2020) Histamine and other biogenic amines in food. J Vet Res 64:281–288. https://doi.org/10.2478/jvetres-2020-0029

Fancello F, Multineddu C, Santona M, Deiana P, Zara G, Mannazzu I et al (2020) Bacterial biodiversity of extra virgin olive oils and their potential biotechnological exploitation. Microorganisms 8:97. https://doi.org/10.3390/microorganisms8010097

Ferreira V, Wiedmann M, Teixeira P, Stasiewicz MJ (2014) *Listeria monocytogenes* persistence in food-associated environments: epidemiology, strain characteristics, and implications for public health. J Food Prot 77:150–170. https://doi.org/10.4315/0362-028X.JFP-13-150

Flemming HC, Wingender J, Szewzyk U, Steinberg P, Rice SA, Kjelleberg S (2016a) Biofilms: an emergent form of bacterial life. Nat Rev Microbiol 14(9):563–575

Flemming H-C, Wingender J, Szewzyk U, Steinberg P, Rice SA, Kjelleberg S (2016b) Biofilms: an emergent form of bacterial life. Nat Rev Microbiol 14:563–575. https://doi.org/10.1038/nrmicro.2016.94

Fong JNC, Yildiz FH (2015) Biofilm matrix proteins. Microbiol Spectr 3:10

García-Burgos M, Moreno-Fernández J, Alférez MJM, Díaz-Castro J, López-Aliaga I (2020) New perspectives in fermented dairy products and their health relevance. J Funct Foods 72:104059. https://doi.org/10.1016/j.jff.2020.104059

Garneau JE, Moineau S (2011) Bacteriophages of lactic acid bacteria and their impact on milk fermentations. Microb Cell Factories 10:S20. https://doi.org/10.1186/1475-2859-10-S1-S20

González-Cancho F, Rejano-Navarro L, Rodríguez de la Borbolla, Alcalá JM (1980) Formation of propionic acid during the conservation of table green olives. Grasas Aceites 31:245–250

Gupta RK, Gangoliya SS, Singh NK (2015) Reduction of phytic acid and enhancement of bioavailable micronutrients in food grains. J Food Sci Technol 52:676–684. https://doi.org/10.1007/s13197-013-0978-y

Hall-Stoodley L, Costerton JW, Stoodley P (2004) Bacterial biofilms: from the natural environment to infectious diseases. Nat Rev Microbiol 2(2):95–108

Hao W, Tian P, Zheng M, Wang H, Xu C (2019) Characteristics of proteolytic microorganisms and their effects on proteolysis in total mixed ration silages of soybean curd residue. Asian Australas J Anim Sci 33:100–110. https://doi.org/10.5713/ajas.18.0933

Hasan M, Sultan MZ, Mar-E-Um M (2014) Significance of fermented food in nutrition food science. J Sci Res 6:373–386. https://doi.org/10.3329/jsr.v6i2.16530

Hornero-Méndez D, Garrido-Fernández A (1997) Rapid high-performance liquid chromatography analysis of biogenic amines in fermented vegetable brines. J Food Prot 60:414–419. https://doi.org/10.4315/0362-028X-60.4.414

Jesch ED, Carr TP (2017) Food ingredients that inhibit cholesterol absorption. Prev Nutr food Sci 22:67–80

Jurášková D, Ribeiro SC, Silva CCG (2022) Exopolysaccharides produced by lactic acid bacteria: from biosynthesis to health-promoting properties. Food Secur 11:156

Kalač P, Špička J, KriŽek M, Pelikánová T (2000a) Changes in biogenic amine concentrations during sauerkraut storage. Food Chem 69:309–314. https://doi.org/10.1016/S0308-8146(99)00273-3

Kalač P, Špička J, KriŽek M, Pelikánová T (2000b) The effects of lactic acid bacteria inoculants on biogenic amines formation in sauerkraut. Food Chem 70:355–359. https://doi.org/10.1016/S0308-8146(00)00103-5

Karamoko D, Théodore Deni N, Aboya Moroh J-L, Jean-Paul Bouatenin KM, Dje KM (2016) Biochemical and microbial properties of palm wine: effect of tapping length and varietal differences. Food Nutr Sci 7:763–771. https://doi.org/10.4236/fns.2016.79077

Karovičová J, Kohajdová Z (2018) Lactic acid fermented vegetable juices. Hortic Sci 30:152–158. https://doi.org/10.17221/3878-HORTSCI

Kirschbaum J, Rebscher K, Brückner H (2000) Liquid chromatographic determination of biogenic amines in fermented foods after derivatization with 3,5-dinitrobenzoyl chloride. J Chromatogr A 881:517–530. https://doi.org/10.1016/S0021-9673(00)00257-0

Kolesarova E (1995) Vyskyt a vznik biogennych aminov v potravinach. Bull Potrav Inárskeho Výskumu 34:109–122

Korcz E, Varga L (2021) Exopolysaccharides from lactic acid bacteria: techno-functional application in the food industry. Trends Food Sci Technol 110:375–384

Lear G, Lewis GD (eds) (2012) Microbial biofilms: current research and applications. Caister Academic Press. ISBN 978-1-904455-96-7

Legras J-L, Moreno-Garcia J, Zara S, Zara G, Garcia-Martinez T, Mauricio JC et al (2016) Flor yeast: new perspectives beyond wine aging. Front Microbiol 7:503. https://doi.org/10.3389/fmicb.2016.00503

Li S, Liu SY, Chan SY, Chua SL (2022) Biofilm matrix cloaks bacterial quorum sensing chemoattractants from predator detection. ISME J 16(5):1388–1396

López D, Vlamakis H, Kolter R (2010) Biofilms. Cold Spring Harb Perspect Biol 2(7):398

Mastanjević K, Kovačević D, Frece J, Markov K, Pleadin J (2017) The effect of autochthonous starter culture, sugars and temperature on the fermentation of slavonian kulen. Food Technol Biotechnol 55:67–76. https://doi.org/10.17113/ftb.55.01.17.4688

Mehta BM (2015) Chemical composition of milk and milk products. In: Cheung PCK, Mehta BM (eds) Handbook of food chemistry. Springer, Berlin/Heidelberg, pp 511–553

Melini F, Melini V, Luziatelli F, Ficca AG, Ruzzi M (2019) Health-promoting components in fermented foods: an up-to-date systematic review. Nutrients 11:1189. https://doi.org/10.3390/nu11051189

Meyer RL (2015) Intra- and inter-species interactions within biofilms of important foodborne bacterial pathogens. Front Microbiol 6:841. https://doi.org/10.3389/fmicb.2015.00841

Mine Y, Kwan Wong AH, Jiang B (2005) Fibrinolytic enzymes in Asian traditional fermented foods. Food Res Int 38:243–250. https://doi.org/10.1016/j.foodres.2004.04.008

Minh NP (2014) Investigation of pickled water spinach (Ipomoea aquatic) fermentation by Lactobacillus sp. Int J Multidiscip Res Dev 1:71–80

Miyamoto M, Ueno HM, Watanabe M, Tatsuma Y, Seto Y, Miyamoto T et al (2015) Distinctive proteolytic activity of cell envelope proteinase of Lactobacillus helveticus isolated from airag, a traditional Mongolian fermented mare's milk. Int J Food Microbiol 197:65–71. https://doi.org/10.1016/j.ijfoodmicro.2014.12.012

Mokoena MP (2017) Lactic acid bacteria and their bacteriocins: classification, biosynthesis and applications against uropathogens: a mini-review. Molecules 22:1255. https://doi.org/10.3390/molecules22081255

Mokoena MP, Chelule PK, Gqaleni N (2005) Reduction of fumonisin B1 and zearalenone by lactic acid bacteria in fermented maize meal. J Food Prot 68:2095–2099. https://doi.org/10.4315/0362-028X-68.10.2095

Montriwong A, Rodtong S, Yongsawatdigul J (2015) Detergent-stable salt-activated proteinases from Virgibacillus halodenitrificans SK1-3-7 isolated from fish sauce fermentation. Appl Biochem Biotechnol 176:505–517. https://doi.org/10.1007/s12010-015-1591-5

Moslehishad M, Ehsani MR, Salami M, Mirdamadi S, Ezzatpanah H, Naslaji AN et al (2013) The comparative assessment of ACE-inhibitory and antioxidant activities of peptide fractions obtained from fermented camel and bovine milk by Lactobacillus rhamnosus PTCC 1637. Int Dairy J 29:82–87. https://doi.org/10.1016/j.idairyj.2012.10.015

Nkhata SG, Ayua E, Kamau EH, Shingiro J-B (2018) Fermentation and germination improve nutritional value of cereals and legumes through activation of endogenous enzymes. Food Sci Nutr 6:2446–2458

Nout MJR, Kiers JL (2005) Tempe fermentation, innovation and functionality: update into the third millenium. J Appl Microbiol 98:789–805. https://doi.org/10.1111/j.1365-2672.2004.02471.x

Pannella G, Lombardi SJ, Coppola F, Vergalito F, Iorizzo M, Succi M et al (2020) Effect of biofilm formation by Lactobacillus plantarum on the malolactic fermentation in model wine. Food Secur 9:797

Papadimitriou K, Alegría Á, Bron PA, de Angelis M, Gobbetti M, Kleerebezem M et al (2016) Stress physiology of lactic acid bacteria. Microbiol Mol Biol Rev 80:837–890. https://doi.org/10.1128/MMBR.00076-15

Patel A, Shah N, Prajapati JB (2013) Biosynthesis of vitamins and enzymes in fermented foods by lactic acid bacteria and related genera – a promising approach. Croat J Food Sci Technol 5:85–91

Perpetuini G, Tittarelli F, Mattarelli P, Modesto M, Cilli E, Suzzi G et al (2018) Intraspecies polymorphisms of Kluyveromyces marxianus strains from Yaghnob valley. FEMS Microbiol Lett 365:fny028. https://doi.org/10.1093/femsle/fny028

Petrova P, Petrov K (2020) Lactic acid fermentation of cereals and pseudocereals: ancient nutritional biotechnologies with modern applications. Nutrients 12:1118. https://doi.org/10.3390/nu12041118

Prete R, Alam MK, Perpetuini G, Perla C, Pittia P, Corsetti A (2021) Lactic acid bacteria exopolysaccharides producers: a sustainable tool for functional foods. Food Secur 10:1653

Rollán GC, Gerez CL, LeBlanc JG (2019) Lactic fermentation as a strategy to improve the nutritional and functional values of pseudocereals. Front Nutr 6:98. https://doi.org/10.3389/fnut.2019.00098

Sakuragi Y, Kolter R (2007) Quorum-sensing regulation of the biofilm matrix genes (pel) of Pseudomonas aeruginosa. J Bacteriol 189(14):5383–5386

Samtiya M, Aluko RE, Dhewa T (2020) Plant food anti-nutritional factors and their reduction strategies: an overview. Food Prod Process Nutr 2:6. https://doi.org/10.1186/s43014-020-0020-5

Santona M, Sanna ML, Multineddu C, Fancello F, de la Fuente SA, Dettori S et al (2018) Microbial biodiversity of Sardinian oleic ecosystems. Food Microbiol 70:65–75. https://doi.org/10.1016/j.fm.2017.09.004

Sarkar T, Salauddin M, Pati S, Sheikh HI, Chakraborty R (2021) Application of raw and differently dried pineapple (Ananas comosus) pulp on Rasgulla (sweetened casein ball) to enhance its phenolic profile, shelf life, and in-vitro digestibility characteristics. J Food Process Preserv 45:e15233. https://doi.org/10.1111/jfpp.15233

Savijoki K, Ingmer H, Varmanen P (2006) Proteolytic systems of lactic acid bacteria. Appl Microbiol Biotechnol 71:394–406. https://doi.org/10.1007/s00253-006-0427-1

Sharma R, Garg P, Kumar P, Bhatia SK, Kulshrestha S (2020) Microbial fermentation and its role in quality improvement of fermented foods. Ferment 6:106. https://doi.org/10.3390/fermentation6040106

Shiby VK, Mishra HN (2013) Fermented milks and milk products as functional foods–a review. Crit Rev Food Sci Nutr 53:482–496. https://doi.org/10.1080/10408398.2010.547398

Shrout JD et al (2006) The impact of quorum sensing and swarming motility on Pseudomonas aeruginosa biofilm formation is nutritionally conditional. Mol Microbiol 62:1264–1277

Singh AK, Rehal J, Kaur A, Jyot G (2015) Enhancement of attributes of cereals by germination and fermentation: a review. Crit Rev Food Sci Nutr 55:1575–1589. https://doi.org/10.1080/10408398.2012.706661

Solange A, Georgette KA, Gilbert F, Marcellin DK, Bassirou B (2014) Review on African traditional cereal beverages. Am J Res Commun 2:103–153
Stanley NR, Lazazzera BA (2005) Defining the genetic differences between wild and domestic strains of Bacillus subtilis that affect poly-gamma-dl-glutamic acid production and biofilm formation. Mol Microbiol 57:1143–1158. https://doi.org/10.1111/j.1365-2958.2005.04746.x
Tamang JP, Dewan S, Tamang B, Rai A, Schillinger U, Holzapfel WH (2007) Lactic acid bacteria in Hamei and Marcha of North East India. Indian J Microbiol 47:119–125. https://doi.org/10.1007/s12088-007-0024-8
Vert M, Doi Y, Hellwich KH, Hess M, Hodge P, Kubisa P et al (2012) Terminology for biorelated polymers and applications (IUPAC Recommendations 2012). Pure Appl Chem 84(2):377–410
Vieco-Saiz N, Belguesmia Y, Raspoet R, Auclair E, Gancel F, Kempf I et al (2019) Benefits and inputs from lactic acid bacteria and their bacteriocins as alternatives to antibiotic growth promoters during food-animal production. Front Microbiol 10:57. https://doi.org/10.3389/fmicb.2019.00057
Wang C, Cui Y, Qu X (2018) Mechanisms and improvement of acid resistance in lactic acid bacteria. Arch Microbiol 200:195–201
Wolfe BE, Dutton RJ (2015) Fermented foods as experimentally tractable microbial ecosystems. Cell 161:49–55. https://doi.org/10.1016/j.cell.2015.02.034
Xiang H, Sun-Waterhouse D, Waterhouse GIN, Cui C, Ruan Z (2019) Fermentation-enabled wellness foods: a fresh perspective. Food Sci Human Wellness 8:203–243
Ye A, Cui J, Dalgleish D, Singh H (2017) Effect of homogenization and heat treatment on the behavior of protein and fat globules during gastric digestion of milk. J Dairy Sci 100:36–47. https://doi.org/10.3168/jds.2016-11764
Yoshii K, Hosomi K, Sawane K, Kunisawa J (2019) Metabolism of dietary and microbial vitamin B family in the regulation of host immunity. Front Nutr 6:48. https://doi.org/10.3389/fnut.2019.00048
Zara G, Budroni M, Mannazzu I, Fancello F, Zara S (2020) Yeast biofilm in food realms: occurrence and control. World J Microbiol Biotechnol 36:134. https://doi.org/10.1007/s11274-020-02911-5
Zeng W, Li W, Shu L, Yi J, Chen G, Liang Z (2013) Non-sterilized fermentative co-production of poly(γ-glutamic acid) and fibrinolytic enzyme by a thermophilic Bacillus subtilis GXA-28. Bioresour Technol 142:697–700. https://doi.org/10.1016/j.biortech.2013.05.020
Zhang Y, Hu P, Lou L, Zhan J, Fan M, Li D et al (2017) Antioxidant activities of lactic acid bacteria for quality improvement of fermented sausage. J Food Sci 82:2960–2967. https://doi.org/10.1111/1750-3841.13975

Biofilm Dynamics in Food Environment

4

Saikat Mazumder, Dibyajit Lahiri, and Moupriya Nag

Abstract

The food system and the surrounding environment support the growth of an extensive spectrum of microorganisms. Most bacteria are capable of adhering and forming biofilms, where they can endure and thrive for several days, weeks, or even longer, contingent upon the type of bacterium and the surrounding circumstances. Biofilms undergo several phases of development in their biological cycle, including preliminary attachment, maturity, maintenance, and dispersal. To be more precise, biofilms serve as a continuous breeding ground for bacteria including *Salmonella enterica*, *Bacillus cereus*, *Escherichia coli*, *Pseudomonas aeruginosa*, *Staphylococcus aureus,* and *Listeria* spp. These bacteria can cause serious infections such as nosocomial infections and foodborne illnesses, as well as contaminate food pipelines. Consequently, biofilm-forming pathogenic microbes in the dairy, meat, poultry, and aquaculture food sectors are substantial issues due to these types of biofilms. Accordingly, the food industry places a high priority on preventing the formation of biofilms and eliminating those that have already developed. The production of biofilms has been reported to be inhibited by natural resources such as biopeptides, essential oils, plant extracts, and enzymes such as glycosidases, proteases, and DNases. Cold plasma treatment, pulsed electric field, power ultrasound, and pulsed light technology are some of the cutting-edge methods that have been used recently to identify and assess

S. Mazumder
Department of Biotechnology, University of Engineering and Management, Kolkata, West Bengal, India

Department of Food Technology, Guru Nanak Institute of Technology, Kolkata, West Bengal, India

D. Lahiri (✉) · M. Nag (✉)
Department of Biotechnology, University of Engineering and Management, Kolkata, West Bengal, India
e-mail: dibyajit.lahiri@uem.edu.in; moupriya.nag@uem.edu.in

T. Sarkar et al. (eds.), *Biofilm Applications to Revolutionize Food Technology*,
https://doi.org/10.1007/978-3-031-85205-3_4

microorganisms adhered to surfaces. An improved understanding of the physiology, architecture, and molecular signalling of biofilms can help prevent and manage food-related spoilage and harmful microorganisms. Therefore, the study addresses the development of biofilm in the food environment, possible health risks related to it, environmental variables that play a role, and possible approaches for controlling it.

Keywords

Biofilm formation · Food industry · Food spoilage · Environmental factors · Biofilm inhibition

4.1 Introduction

Food processing environments are more likely to experience spoilage of processed products due to the establishment of biofilms, which may include food-spoiling bacteria and foodborne pathogens. Due to technological malfunctions in the food industry, biofilms not only pose a significant risk to hygiene but also result in significant financial losses. Biofilms have a critical role in the survival and spread of infectious illnesses since they are present practically everywhere in food ecosystems, including industrial settings. Biofilm-forming microbes are responsible for between 65 and 80% of human acute and chronic infections, respectively, in accordance with the National Institutes of Health (NIH). In European hospitals, healthcare-associated infections (HAIs) cause up to 37,000 deaths yearly on average, according to a 2008 estimate from the European Centre for Disease Prevention and Control (ECDC). About 80,000 patients had at least one HAI on an individual day, according to (European Food Safety Authority 2009) estimates. This is approximately one out of every 18 individuals treated in European hospitals (Haque et al. 2018; Marschang and Bernardo 2015; Murphy et al. 2020). Similarly, the US Centre for Disease Control and Prevention calculated that over 1.7 million hospitalised patients get HAIs each year, which results in over 98,000 patient deaths (1 out of 17) from these instances. In Italy, hospitalised patients report 0.45–0.70 million HAIs annually (Antonioli et al. 2020).

Microbes living in adherent colonies encased in extracellular polymeric compounds form biofilms (Costerton et al. 1995). Generally, biofilms are composed of 85% extracellular matrix of polymers and 15% cells, often arranged in microcolonies. Extracellular DNA, water, and polysaccharides, the latter of which vary according to the species contained in the biofilm are the elements that make up the structure known as the matrix (Flemming and Wingender 2010). Biofilms can consist of one kind of microbes or, more frequently in the natural environment, combinations of several species (Manuzon and Wang 2006). Planktonic cells are more susceptible to antimicrobials than biofilm-forming bacteria, which often exhibit a modified phenotype and a 10,009 increase in resistance. Biofilms can develop on humid solid substrates, on living things' soft tissue surfaces, or at liquid–air interfaces. When examining the surfaces of enamel with a crude microscope in 1684, Antonie van Leeuwenhoek first articulated the idea of a biofilm (Donlan and Costerton 2002). The mid-1990s witnessed the first recognition of biofilms as a potential problem in food processing mechanisms by Zottola and Sasahara (1994), and this was further investigated by Hood and Zottola (1995), Kumar and Anand (1998), Poulsen (1999), and subsequently by Shi and Zhu (2009).

Microbes may live stably in biofilms, which are composed of water, polysaccharides, extracellular proteins, and nucleic acid. The components are created by the microbes themselves. Pathogenic microorganisms are severely affecting the health and lives of organisms as a consequence of this adaptation method (Rather et al. 2021b).

The food industry faces significant challenges due to biofilms, which enable bacteria to attach themselves to various surfaces such as rubber products, polypropylene, glass, acrylic, metals such as stainless steel, as well as food items in a matter of minutes. Subsequently, mature biofilms form in a matter of days or even hours (Hall-Stoodley et al. 2004). Consequently, biofilms are thought to be the cause of equipment damage, higher energy expenses, outbreaks, and enhanced resistance to disinfectants in a variety of food sectors, including the processing of seafood, beer, dairy products, and poultry and meat products (Srey et al. 2013). Compared to planktonic cells, there is strong evidence that the biofilm lifestyle of these kinds of cells makes them more susceptible to antibiotic therapy (Figs. 4.1 and 4.2) (Carrascosa et al. 2021). This implies that they must be removed from food processing establishment surfaces, which is an enormous operation (Anwar et al. 1992; Simões et al. 2010). It is commonly known that microbiological assessments frequently count, although a visual assessment cannot immediately eliminate evident

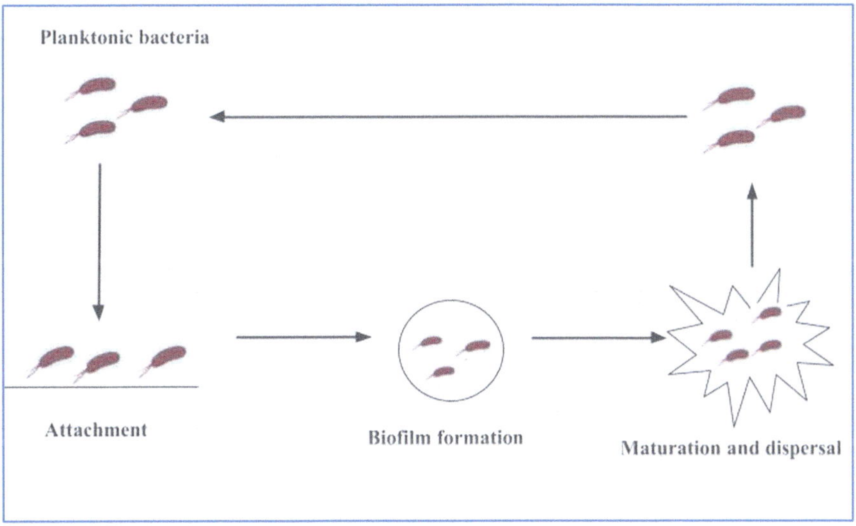

Fig. 4.1 Biofilm formation and development stages

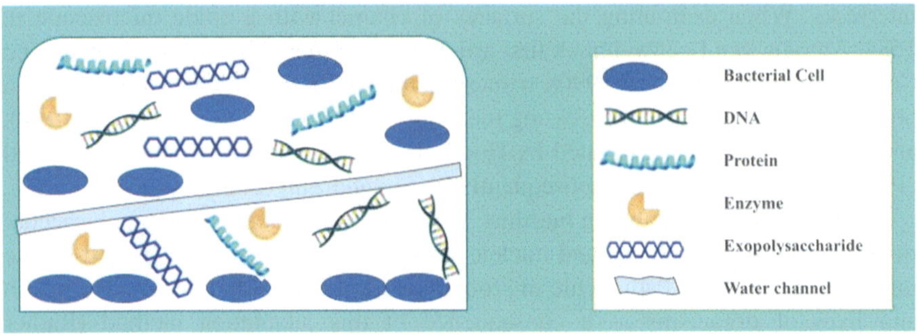

Fig. 4.2 Biofilm structure in the growth and maturation stages

sanitation deficiencies due to bacterial growth (Carrascosa et al. 2021). For the reasons mentioned above, it is necessary to thoroughly assess the sanitation state of food-contact surfaces in order to ensure compliance with microbiological requirements and to make necessary modifications and disparities in hygienic practices (Holah et al. 2002). The food sector finds it increasingly challenging to identify the most effective surfaces due to the sanitary circumstances surrounding food-contact surfaces (González-Rivas et al. 2018).

4.2 Biofilm and the Food Environment: A Crosstalk

The main concern with the food environment is the hygiene of foods. It is reasonable for people to expect to eat healthful, safe food. However, a number of studies have revealed that harmful bacteria are present in a wide range of dietary items. Foodborne illnesses caused by consuming contaminated food, typically by organisms such as bacteria, viruses, toxins, or parasites, are collectively referred to as food poisoning (EFSA 2016). Water, raw foods, dust, machinery, animals, and other things are some of the sources of contamination in the food system. Furthermore, it is acknowledged that food processors are an important conduit of contaminants in the food environment (Abdallah et al. 2014). Furthermore, it has been shown that biofilms are the primary source of contamination when it comes to food goods. Indeed, the enduring growth of biofilm on surfaces and equipment that come into touch with food might potentially serve as an ongoing source of contamination. In addition, some writers highlighted that even with the use of cleaning techniques, biofilms remain on surfaces that come into contact with food. Although research testing the hydrophobicity impact shows contradicting findings, hydrophobic surfaces often attract more microorganisms (Gomes et al. 2015; Veluz et al. 2012), and further studies show that bacterial adhesion is increased on hydrophilic surfaces compared to hydrophobic counterparts (Dhowlaghar et al. 2018; Jindal et al. 2018).

A major societal issue is the presence of harmful bacteria in drinking water (S. Sharma et al. 2003). In actuality, a number of epidemics were connected to the bacterial contamination of water for consumption (Abdallah et al. 2014). The propensity of these bacteria to develop biofilms in the distribution systems of potable water is the most dangerous issue (Chaves Simões and Simões 2013). These types of biofilms have a negative impact on the quality of water, increasing the risk of bio-corrosion, waterborne illnesses, decreased heat exchange efficiency, cross-contamination of surfaces in the food and medical industries, and increased distribution system maintenance expenses (Anaissie et al. 2002; Chaves Simões and Simões 2013).

Methicillin-resistant *S. aureus* (MRSA) and *S. aureus* were discovered to be present in commercial raw chicken specimens in China, according to Wang et al. (2013). *Escherichia coli* and *S. aureus* were also found to be very prevalent in ready-to-eat meals, as demonstrated by Manguiat and Fang (2013). Additional research revealed potentially hazardous bacteria like *S. aureus* and *Escherichia coli* were present in meat products (Gousia et al. 2011), dairy-based food items (Sasidharan et al. 2011), and seafood-based goods (Zarei et al. 2012). *S. aureus* biofilms were shown to have colonised food-related contact surfaces in the dairy, meat, and seafood industries, according to Gutiérrez et al. (2012). Milk lactose stimulates the development of intracellular polysaccharide adhesions in *S. aureus* and activates the LuxS-mediated quorum-sensing system in *Bacillus subtilis* (Duanis-Assaf et al. 2016), both of which produce more biofilms than they would otherwise (Xue et al. 2014). Elevated levels of free Ca2+ and Mg2+ are produced in milk due to enhanced biofilm formation by *Geobacillus* spp. (Somerton et al. 2015).

Food debris, such as meat and milk exudates enriched in proteins, fats, and carbs, facilitates microorganism expansion and proliferation and favours the formation of dual-species biofilms by *E. coli* and *Staphylococcus aureus*. These food matrix components also have an impact on bacterial attachment in situations involving food processing (Dutra et al. 2018; Iñiguez-Moreno et al. 2019). Using the technique of scanning electron microscopy Latorre et al. (2010) discovered the existence of biofilms being formed on dairy farm machinery. The persistence of biofilms on various pasteurisation line segments in industrial plants, even after daily cleaning, was additionally observed by M. Sharma and Anand (2002). A wide range of harmful bacteria, including *S. aureus* and *E. coli*, were also reported to be prevalent by these authors. Moreover, biofilms with a count of 104–106 CFU/cm^2 were found to have formed in the prawn and fish factories when stainless coupons were placed next to food contact surfaces (Gutiérrez et al. 2012). Abdallah et al. (2014) reported that biofilms of *Salmonella* spp., *S. aureus*, and *Listeria monocytogenes* contaminated covered the vast majority of surfaces where food came into contact in small-scale processing plants.

The development of mixed-species biofilms is highly dynamic and is dependent on various factors such as the characteristics of the attachment surface (Tang et al. 2011), the components of the matrix that make up the food (Van Houdt and Michiels 2010), the conditions in the environment (Govaert et al. 2018), and the presence of bacterial cells (Makovcova et al. 2017; Yuan et al. 2018). Nevertheless, other important factors that promote the formation of biofilms include the characteristics of microbial cells, including their hydrophobicity, the components of their cellular membrane (such as proteins and lipopolysaccharides), their appendages (such as pili, flagella, and fimbriae), and the EPS that bacteria exude (Tang et al. 2011). Food safety is thereby compromised, and the risk to the public's health is increased since the food industry creates an environment conducive to the creation of biofilms (Abdallah et al. 2014). Table 4.1 shows the various foodborne microorganisms creating biofilms in the food environment along with their characteristics, growth growth-supporting environmental conditions, and different identification techniques.

4.3 Environmental Influences vs Biofilm Development

Researchers are still at variance about a consensus definition of biofilm, which has undergone substantial evolution since its discovery. Donlan and Costerton (2002) definition, however, is still the most widely accepted. By definition, a biofilm is an organised group of microbial cells that adhere to surfaces, interfaces, and one another. It is encased in a polymeric matrix that the authors have developed. Four typical steps are involved in biofilm development. First, bacteria attach themselves to the surface (1). Next, microcolonies grow (2), and finally, the biofilm matures (3). Dispersion is the last step of the biofilm life cycle, during which the cells exit the biofilm structure to infect additional surfaces (4), and adherence of bacteria to the surface. The most important stage is the creation of a biofilm (Fig. 4.1) (Renner and Weibel 2011). Numerous chemical, physical, and biological interactions appear to assist this phase, which is thought to be reversible. Nonspecific physicochemical interactions including Lewis acid–base, Lifshitz–van der Waals, and electrostatic interactions will be experienced by the entire microbial cell when it gets closer to a substance of interest (Bos et al. 1999; Renner and Weibel 2011). A bacterial adherence that is reversible to the surface will be made possible by the force that results. The cells divide into microcolonies as soon as they connect. Exopolymeric substances (EPS), which are secreted and constitute the biofilm matrix, are the primary cause of irreversible adherence. Proteins, nucleic acids, lipids, and polysaccharides are among the polymeric substances that make up the extracellular matrix (Flemming and Wingender 2010). These materials make it possible for germs to adhere to one another and surfaces (Ghafoor et al. 2011). Biofilm maturation starts at this point, resulting in a mature biofilm where the cells are covered in a matrix that is extracellular with intricate architecture and water channels. In order to stabilise the three-dimensional biofilm structure, such a matrix serves as a scaffold (Dunne 2002).

The regulation of biofilm development by many environmental cues, including temperature and nutrition availability, is becoming more and more evident. The production of biofilms may be modulated by bacteria by sensing certain environmental cues and activating regulatory networks (Karatan and Watnick 2009). Dispersion, the final stage of the biofilm lifecycle, gives the sessile cells the chance to depart, infect new surfaces, and then start the cycle again. A number of environmental factors, including changes in food supply, depletion of oxygen, and other stressful circumstances, cause the biofilm to disperse and encourage the expression of genes responsible for its dispersion (McDougald et al. 2012). As an instance, it has been observed that the change in a carbon supply stimulated the development of flagella,

Table 4.1 Variability in foodborne microorganisms creating biofilms in the food sector

Microorganism	Contaminated food	Surface/ material	Characteristics	Environmental parameters	Identification technique	Harmful spoilage effects	Reference
Listeria monocytogenes	Dairy products, meat, ready-to-eat products, fruit, soft cheeses, ice cream, unpasteurised milk, candied apples, frozen vegetables, poultry	Glass slides, PVC slides, microtitre plates	Gram-positive, rod-shaped, facultative anaerobic, microtitre plate wells	Temperature ranges between 25 °C–37 °C	Microtitre plate assay, plate counting, epifluorescence microscopy,	Listeriosis in the elderly, pregnant women and immune-compromised patients	Chae and Schraft (2000), Costa et al. (2016), Di Bonaventura et al. (2008), Djordjevic et al. (2002), Doijad et al. (2015), Harvey et al. (2007), Lianou and Koutsoumanis (2012), Piercey et al. (2017), Rothrock et al. (2017), and In Lee et al. (2017)
Escherichia coli	Raw milk, fresh meat, fruits and vegetables	Microtitre plates, glass slides, SS coupons, microfluidic device	Gram-negative, rod-shaped	Temperature between 20 °C and37°C	Acridine orange staining and epifluorescence microscopy, microtitre plate assay, confocal laser scanning microscopy	Diarrhoea outbreaks and haemolytic uremic syndrome	Álvarez-Ordóñez et al. (2013), Bridier et al. (2010), Galié et al. (2018), Lianou and Koutsoumanis (2012), Rivas et al. (2007), Vogeleer et al. (2016), and R. Wang et al. (2014)
Salmonella Enterica	Poultry meat, bovine, ovine, porcine, fish	Microtitre plates, glass slides, nitrile butyl-rubber coupons, granite, marble, silestone	Gram-negative, rod-shaped, flagellate, facultative aerobic	Temperature between 25 °C and 37 °C	Microtitre plate assay, confocal laser scanning microscopy, crystal violet staining, scraping of cells and plate counting, epifluorescence microscopy	Can cause gastroenteritis or septicaemia	Chia et al. (2009), Díez-García et al. (2012), Duguid et al. (1966), Lianou and Koutsoumanis (2012), Nguyen et al. (2014), Oliveira et al. (2007), and Vestby et al. (2009)

(continued)

Table 4.1 (continued)

Microorganism	Contaminated food	Surface/ material	Characteristics	Environmental parameters	Identification technique	Harmful spoilage effects	Reference
Bacillus cereus	Dairy products, rice, vegetables, meat	Microtitre plate Wells, Microtitre plates	Gram-positive, spore-forming, anaerobic, facultative anaerobic	Temperature 30 °C	Microtitre plate assay, plate counting and scanning electron microscopy	Diarrhoea and vomiting symptoms	Bottone (2010), Grigore-Gurgu et al. (2020), Hayrapetyan et al. (2015, 2016) and Lianou and Koutsoumanis (2012)
Staphylococcus aureus	Meat products, poultry, egg products, dairy products, salads, bakery products, especially cream-filled pastries and cakes, and sandwich fillings	Microtitre plates	Gram-positive, non-spore forming, non-motile, facultative anaerobic	Temperature between 20 °C and 48 °C	Microtitre plate assay, confocal laser scanning microscopy, staining with safranin O and absorbance measurement, epifluorescence microscopy and scanning electron microscopy	Methicillin resistance, can cause vomiting and diarrhoea	Avila-Novoa et al. (2018), Giaouris et al. (2015), Hammer and Bassler (2003), Oniciuc et al. (2016), Rode et al. (2007) and Vázquez-Sánchez et al. (2013)
Campylobacter jejuni	Animals, poultry, unpasteurised milk	Microtitre plates, epithelial intestinal cells	Gram-negative, aerobic and anaerobic	Temperature between 37 °C and 42 °C	Microtitre plate assay, confocal laser scanning microscopy, confocal laser scanning microscopy	Bloody diarrhoea, fever, stomach cramp, nausea and vomiting	Bronnec et al. (2016), Chlebicz and Śliżewska (2018), Klančnik et al. (2021) and Turonova et al. (2015)

which downregulated the pilus genes (twitching motility) and encouraged *P. aeruginosa*'s biofilm dispersal (Sauer et al. 2004). The physicochemical characteristics of the bacterium and its surrounding surface are crucial for the bacteria's preliminary adhesion (Goulter et al. 2009). Propagation temperature and pH are two factors influencing these attributes (Chavant et al. 2002; Nguyen et al. 2014). The operating temperature and binding surface may alter the biofilm's ultimate architecture (Whitehead and Verran 2015). In the case of stainless steel, *Pseudomonas aeruginosa* biofilms at 20 °C are composed of mushroom-shaped structures linked by thin layers of cells. On polycarbonate, on the other hand, identical mushroom-shaped structures are linked together by thick cell layers, and the resulting biofilms are flat and dense on both surfaces at 30 °C and 37 °C. At 20 and 30 °C, *Staphylococcus aureus* biofilms are flat and uniform on polycarbonate, and uneven with spires and valleys creating the structure on stainless steel at 37 °C (Abdallah et al. 2014).

Numerous investigations have shown how environmental factors affect *Salmonella* spp. biofilm development (Nguyen et al. 2014; Speranza et al. 2011). Biofilm production is accelerated by rising temperatures, pH levels, and water activity (aw) (Dimakopoulou-Papazoglou et al. 2016). The attachment surface's hydrophobic or hydrophilic properties are crucial for the first cell adhesion and the development of biofilms that follow. The attachment surface's hydrophobic or hydrophilic properties are crucial for the first cell adhesion and the development of biofilms that follow. It has been shown in a recent investigation on the wettability of stainless steel (Type 304) that this surface changes over time from a hydrophilic to a hydrophobic condition (Kim et al. 2016).

It has been reported that *Salmonella* spp. (Chia et al. 2009), *Bacillus cereus* (Hayrapetyan et al. 2015), *L. mono cytogenes* (Bonsaglia et al. 2014; Chavant et al. 2002; Di Bonaventura et al. 2008), and *S. typhimurium* (Nguyen et al. 2014) adhere more effectively to surfaces that are hydrophilic than to polystyrene/acrylic (hydrophobic). In contrast, hydrophobic stainless steel with an angle of contact greater than 65° has been observed by other research. Comparing how *L. monocytogenes* and *Salmonella* spp. adhere to PTFE, rubber, and stainless steel at a contact angle of 105.6°, Sinde and Carballo (2000) found that bacteria adhered least to stainless steel, despite it being the least hydrophobic surface in their sample. In their investigation, Faille et al. (2001) hypothesised that stainless steel's moderate hydrophobicity (contact angle of 75°) explained why Bacillus spores formed biofilms on the material. The conditioning of the surface as well as its roughness might affect how well cells adhere to the substratum. Conditioning entails applying a layer of organic molecules, such as bacterial EPS or food-derived proteins, to the attachment surface. This facilitates cell attachment since conditioning modifies the surface's characteristics, such as its hydrophilic/hydrophobic nature (Shi and Zhu 2009). Surface imperfections, or the condition of the surface, have a significant impact on biofilm development and should be taken into account in food processing environments (Rodriguez et al. 2008; Tang et al. 2011). In an investigation on *S. typhimurium* biofilm development on stainless steel, for instance, the biofilms were identical in thickness at first, but the electropolished surface had the least amount of biofilm after one day (23%) despite the fact that preliminary adhesion was as much as four times lower on polished stainless steel surfaces. The mechanically sanded surface had the lowest coverage after 5 days (43%) and the biofilm depth for both mechanically sanded and electropolished surfaces was 30 μm, whereas that of bright alum polished and untreated surfaces was 47 lm. Removing the oxide layer on a stainless steel surface reduces bacterial adhesion although within a very short period of time, this effect is nullified because of the exposure to air (Flint et al. 2000). The study conducted by Bonsaglia et al. (2014) revealed that incubation time played a significant role in the biofilm formation process, with temperature appearing to be unimportant. The authors concluded that *L. monocytogenes* can form biofilm on various surfaces regardless of temperature, but surface composition may be a crucial factor for quicker biofilm development. The rate of *S. typhimurium* biofilm production, however, increased with temperature, despite the fact that the number of connected cells in the biofilm after 10 days reduced with temperature, according to Nguyen et al. (2014).

The physiological heterogeneity that exists inside the biofilm is one example of an internal element that influences the biofilm in addition to external environmental variables (Stewart and Franklin 2008). In some areas of the biofilm and throughout distinct stages of its growth, for instance, there are variations in the availability of nutrients and/or oxygen tension. Cellular oxidative stress brought on by low oxygen levels might raise resistance to oxidising chemicals (biocides) in biofilm cells because of the universal stress response (O'Leary et al. 2015; Yang et al. 2016). It is quite likely that multispecies biofilms, including spore formers like *Bacillus* spp. rather than mono species, biofilms exist in food-processing environments (Faille et al. 2001). Cross-contamination may arise from sporulation within the biofilm (Faille et al. 2014), and the interactions among species may alter the proportion of each within the biofilm, thereby modifying its internal environment and possibly its capability to resist biocides (Kostaki et al. 2012) (Fig. 4.3).

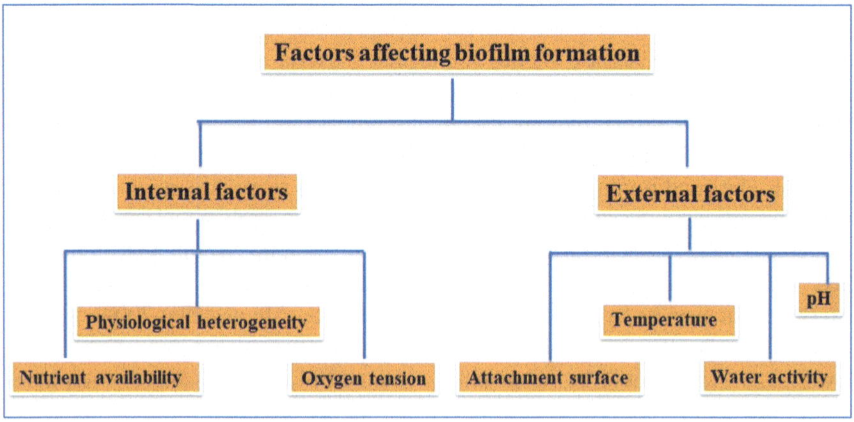

Fig. 4.3 Factors affecting biofilm formation

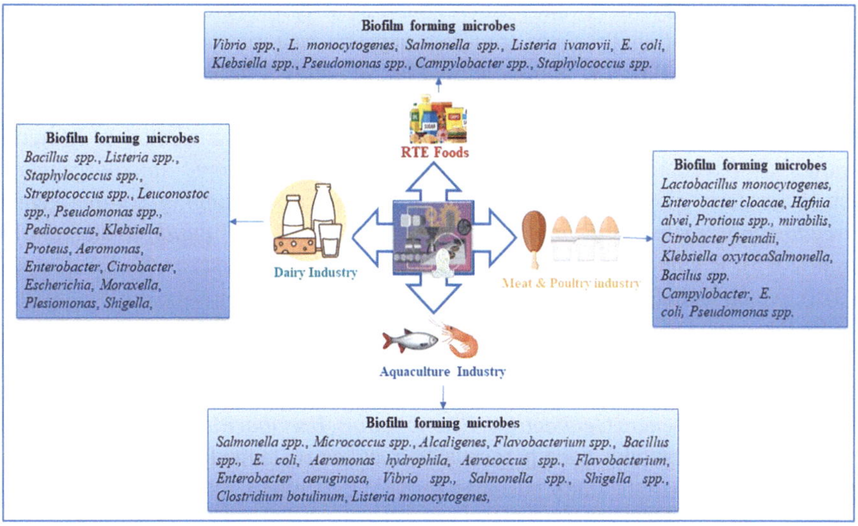

Fig. 4.4 Different types of biofilm-forming microbes in the food sector

4.4 Biofilm Across Different Food Industries: Categories and Variances

Foodborne illnesses have posed a risk to human health and have been linked to several global health crises. According to reports, biofilm-forming bacteria are the source of the majority of foodborne illnesses (Rather et al. 2021b). Food items become contaminated mostly by microbes during the production and preparation stages. The many microbial species that lead to the formation of biofilms in various food industries are illustrated in Fig. 4.4.

Biofilm in the Dairy Sector

Dairy products are susceptible to deterioration in quality and short shelf life due to biofilm growth on dairy processing machines. Delays in transportation and distribution, intricate processing systems, longer processing periods, and other factors have all contributed to the problem of biofilm development, which has raised the likelihood of disease outbreaks and degraded the quality of milk products (Teh et al. 2014). The bacteria that create biofilms impede surfaces such as those found in ultrafiltration and reversal osmosis membranes, regenerative portions of pasteurizers, milking systems, refrigerated tanks, conveyor belts on floors, walls, steps, drains, and floors, and alternatively utilise them as a substrate for the biofilms to grow (Anand et al. 2014). Microorganisms that have been found to form biofilms in dairy products include *Bacillus* spp., *Listeria*, *Staphylococcus*, and lactic acid-producing bacteria (Gram-positive) such as *Streptococcus*, *Pediococcus*, and *Leuconostoc*; additionally, there are Gram-negative bacteria such as *Pseudomonas* spp., *Klebsiella*, *Proteus*, *Aeromonas*, *Enterobacter*, *Citrobacter*, *Escherichia*, *Moraxella*, *Plesiomonas*, *Shigella*, and *Alcali genes* (Varatharaj et al. 2020; Weber et al. 2019). Food decomposing in refrigerators is mostly attributed to bacteria known as psychotrophic microorganisms. Most psychotropic microorganisms found in dairy products are Gram-negative bacteria belonging to the genus *Pseudomonas*, including *P. brenneri* (Machado et al. 2017), *P. putida* (Scatamburlo et al. 2015), *P. koreensis* (Andreani et al. 2016), and *P. fluorescens* (C. Rossi et al. 2018), which is the most prevalent microbe.

B. licheniformis, *B. subtilis*, *B. cereus*, *B. mycoides*, and *B. megaterium* are among the common *Bacillus* spp. discovered in dairy factories and processing facilities (Ostrov et al. 2015). Gopal et al. (2015) explained the spore-producing aerobic bacteria from the genera *Geobacillus*, *Paenibacillus*, *Brevibacillus*, and *Sporosarcina* that are impervious to heat and pasteurisation and cause tainted milk powder. These bacteria are found in the dairy business.

Biofilm in the Processing Sector of Aquaculture

In various water systems such as rivers, ponds, lakes, seas, and artificial water systems, aquaculture refers to the raising, reproduction, and catching of fish, crustaceans, algal cells, and other aquatic plants and animals. In developing nations, the employment of various aquaculture products has improved food security, healthcare, nutrition, and economics (Béné et al. 2016). Aquaculture is now particularly vulnerable to bacterial diseases because bacteria may build biofilms and survive in water without attaching themselves to a host (Pandy et al. 2014). Water is considered the main repository and transmitter of pathogenic organisms that produce biofilms in the aquaculture firm. Kariyawasam and Jayasooriya (2010) revealed that the food contact surfaces of the fish, as well as prawn processing sectors, contained the following microorganisms: *Salmonella*, *Micrococcus* spp., *Alcaligenes*, *Flavobacterium* spp., *Bacillus circulans, E. coli*, *Aeromonas hydrophila*, *Aerococcus* spp., *Flavobacterium*, and *Enterobacter aeroginosa*. The fish manufacturing sector also uses *Vibrio cholerae* (Faruque et al. 2006), *V. vulnificus* (Joseph and Wright 2004), *V. parahaemolyticus* (Enos-Berlage et al. 2005), and *V. alginolyticus* (Yamada et al. 2020). Worldwide, freshwater fish are susceptible to columnaris illness caused by *Flavobacterium columnare*, which also forms noticeable biofilms on the outermost layers of aquaculture systems (Cai et al. 2013). *Vibrio* spp., *Salmonella* spp., *Shigella* spp., *Clostridium botulinum*, *Listeria monocytogenes*, *A. hydrophila*, and other biofilm-forming microbes can cause epidemics of diseases such as *Penaeus monodon* prawn vibriosis caused by *V. parahaemolyticus* and other microbes that are susceptible to the seafood industry. Seafood's shorter shelf life, the emergence of foodborne illnesses, and the deaths that result from them are all caused by bacteria. Therefore, seafood preservation and sterilisation, together with biological preservation, phage lysis, radiation treatment, salting, and high-pressure processing, are prerequisites to minimise seafood rotting and safeguard public health (Kontominas et al. 2021).

Biofilm Within the Ready-to-Eat (RTE) Food Sector

Food and Drug Administration (FDA) defines an RTE food as any plant- or animal-based product that is cleaned, cooked, cooled, frozen, and processed in accordance with US Department of Agriculture guidelines for cleaning, cooling down, drying, processing, preservation in the freezer, and wrapping, and then consumed immediately or after heating. In terms of processing and consumption, it is a rapidly growing sector that is projected to attain an estimated worth of US$195,300 million by the end of 2026 (Popkin et al. 2012). The industry of RTE foods has been correlated with the presence of powerful microorganisms that form biofilms and cause disease. These include *Vibrio* spp., *L. monocytogenes*, *Salmonella* spp., *Listeria ivanovii*, *E. coli*, *Klebsiella* spp., *Pseudomonas* spp., and *Campylobacter* spp. (Beshiru and Igbinosa 2018; Li et al. 2018; Weber et al. 2019). The degradation of food that is ready to eat and human foodborne illnesses are triggered by enterotoxins released by *Staphylococcus* biofilm.

Biofilm: Meat and Poultry Industry

Animal stomachs from slaughter and cross-contamination from processing, shipping, and storing are among the foremost sources of pathogenic bacteria that cause meat to deteriorate (Yang et al. 2017). Of 263 million tonnes of meat products worldwide, 20% are lost because of spoiling, according to information released by the Food and Agriculture Organisation in 2011 (Vilariño et al. 2017). An estimated $150 million is lost annually in the meat business due to biofilm formation by 90% of the bacteria. The most hazardous bacterium for the meat business, according to Danish meat industry standards, is *Lactobacillus monocytogenes*, due to its capacity to thrive and survive at chilly temperatures (Jessen and Lammert 2003). Wang et al. (2018) identified five meat-borne bacteria that form biofilms: *Enterobacter cloacae*, *Hafnia alvei*, *P. mirabilis*, *Citrobacter freundii*, and *Klebsiella oxytoca*. Wang et al. (2018) examined *P. fluorescens's* capacity to build biofilms on surfaces in contact with food, meat degradation, and acute resistance against disinfectants, demonstrating the significance of these bacteria in meat decomposition.

Poultry meat as well as eggs are fundamental food for millions of people due to them being abundant in proteins and other necessary elements (Rather et al. 2021a). Poultry meat has become more abundant and more consumed during the decade preceding; in the decade that follows, it will account for the greatest portion of meat consumption (Merino et al. 2019). As important harmful substances of the chicken industry, *Salmonella*, *Bacilus* spp. *Campylobacter*, *E. coli,* and *Pseudomonas* spp. are well-known biofilm-forming zoonotic bacteria that represent a significant risk to human health (D. A. Rossi et al. 2017). The rising intake of chicken contaminated with *Campylobacter jejuni*, a potent biofilm-forming bacterium found in the poultry industry, is the cause of increased human risks of campylo bacteriosis. In a similar vein, *Salmonella* is considered one of the most powerful pollutants of chicken and is accountable for around 4.5 million infections, 1.2 million hospital admissions, and 1.2 million foodborne illnesses annually in the United States of America alone (Cadena et al. 2019). Risk factors linked to the contamination of chicken products by pathogenic bacteria that produce biofilms include handling, transportation, dust, faeces, and poultry feed (Park et al. 2011).

4.5 Elimination of Biofilm: An Optimistic Trajectory Ahead

The food and health industries are greatly threatened by biofilms. Control over them is therefore crucial. Although they are less effective, traditional methods of eradicating or controlling biofilm development include the application of chemicals that are harsh and antibiotics (Sharma et al. 2019). Even more, microorganisms linked to biofilm-associated chronic infections are difficult to treat since they frequently become resistant to medications (Hall and Mah 2017). Consequently, this motivates researchers to find novel approaches with increased effectiveness. Using natural ingredients, preferably from plants or bacteria themselves, has demonstrated remarkable effects on the regulation of biofilms. Natural ingredients have been employed as effective biofilm-inhibiting agents in clinical settings. Examples of these include biopeptides (Shahid et al. 2020), essential oils (Sahal et al. 2020), biosurfactants (Satpute et al. 2019), and extracts of plants (Gupta et al. 2019; Rather et al. 2021b) based on plants' nanoparticles (Gupta et al. 2014), and so forth.

Karlapudi et al. (2020) improved the removal of biofilms by focusing on quorum sensing while reducing the possibility that biofilm-forming bacteria would acquire multiple drug resistance. Additionally, it has been determined that enzymes such as DNases, proteases, glycosidases, cellulases, and amylases are environmentally friendly substitutes for controlling biofilms (Grigore-Gurgu et al. 2020). Nanotechnology-based antimicrobial surface growth has been demonstrated to be an effective biofilm management method (Olmo et al. 2020; Swar et al. 2020). Phage-based therapies have been proposed recently to eliminate biofilms, and understanding the interactions between phages and biofilms can help develop efficient phage-based techniques (Ferriol-González and Domingo-Calap 2020). In conjunction with traditional methods, innovative technologies such as cold plasma therapy, power ultrasound, pulsed electric field, and pulsed light technology have shown promise in the food industry for managing biofilms. While other methods are better suited for direct food matrix cleansing, power ultrasound and cold plasma techniques are useful in combating biofilms that have developed on food contact surfaces such as pipes or other mechanical parts (Grigore-Gurgu et al. 2020; Zhu et al. 2020). The antibacterial properties of cold plasma therapies have demonstrated great promise in rendering foodborne pathogenic microorganisms inactive. By influencing their metabolic activity and culturability, these treatments have effectively decreased the production of biofilms in *L. monocytogenes*, *S. aureus*, *E. coli, and S. thyphimurium* (Mahnot et al. 2019; Ziuzina et al. 2015).

While the approaches now in use are innovative and successful in combating biofilm-forming bacteria, total elimination or control necessitates an integrated strategy with traditional methodologies. In-depth investigation and enhanced comprehension of the molecular mechanisms behind stress-related genes will facilitate the creation of novel and improved methods to lessen infections' propensity to build biofilms (Rather et al. 2021a). Nevertheless, there are a number of ways to get rid of biofilm, including hydrostatic pressure, non-thermal plasma techniques, enzymatic disruption, chemical treatments, biosurfactants, essential oils, bacteriocins, and steel coatings. The authors have thoroughly examined several approaches and mechanisms of action to regulate the production of biofilms (Galié et al. 2018).

The safety and quality of food can be controlled by following the guidelines set forth by Hazard Analysis and Critical Control Points (HACCP) and Good Manufacturing Practices (GMP) (M. Sharma and Anand 2002). One of the first steps in eliminating and regulating biofilms is identifying critical points where they may multiply and develop (Zottola and Sasahara 1994).

Compared to their planktonic counterparts, microorganisms within biofilms can withstand disinfectants and biocides up to a thousand times better. This property of biofilms may be linked to features of their architecture and physiology, including decreased diffusion, anaerobic growth, physiological alterations brought on by slower growth rates, and the generation of enzymes that break down antimicrobial substances (Høiby et al. 2010; Kumar and Anand 1998; Smith 2005; Sofos and Geornaras 2010). Isothiazolinones, ozone, iodine, hydrogen peroxide, biguanides, surfactants, halogens, and quaternary ammonium compounds are among the most frequently used disinfecting agents in the food industry, but they are insufficient to eradicate biofilms due to these factors (Chaves Simões and Simões 2013; Nilsson et al. 2011; Wirtanen et al. 2001; Zottola and Sasahara 1994). Consequently, it is occasionally necessary to combine sanitizers with other techniques (Chaves Simões and Simões 2013; Meyer 2003). The capacity of steam vapour technologies and aerosolized sanitizer to penetrate and reach hard-to-reach locations makes them potentially useful for disinfecting biofilms on environmental surfaces (S.-H. Park et al. 2012). When treating biofilms containing *Salmonella typhimurium*, *E. coli O157:H7*, and *L. monocytogenes*, aerosolized peracetic acid at a concentration of 100 ppm proved to be more successful than sodium hypochlorite at the same dosage. Cleaning and sanitising using a mix of thermal, mechanical, and chemical techniques are known as Clean-In-Place (CIP), and it is typically used in dairy manufacturing facilities (DMPs) to manage biofilm (Bremer et al. 2001; Nancharaiah et al. 2005). Restricting cell adherence by altering the chemical makeup of surfaces is an additional method of managing biofilms. Surfactants can enhance physical cleaning activities by penetration, emulsification, spreading, wetting, and foaming. They possess a wide range of structural characteristics, including both hydrophilic and hydrophobic structures (Smith 2005). This objective has led researchers to investigate the potential utility of biosurfactants generated by a wide range of bacteria, actinobacteria, and fungi in preventing the formation of biofilms (Sriram et al. 2011; Zezzi do Valle Gomes and Nitschke 2012). Research has demonstrated that applying surfactin (0.25%) as a precondition to polystyrene surfaces decreased the adherence of *L. monocytogenes* and *S. enteritidis* by 42.0%, while applying rhamnolipids (1.0%) inhibited the adherence of *L. monocytogenes* by 578.8% and that of *S. aureus* by 678.8% (Gómez et al. 2012).

In order to facilitate the creation of novel tactics to obstruct pertinent routes for biofilm growth, research has recently focused on the molecular and genetic underpinnings of pathogenic biofilm formation (Hartmann et al. 2010; Lauderdale et al. 2009; Lebeer et al. 2007; Lemon et al. 2007). The two primary matrix components of *Salmonella* and *Stepanović* spp. biofilms are known to be cellulose and thin aggregative fimbriae (Tafi); however, the molecular processes governing biofilm formation by a significant number of foodborne pathogens remain poorly understood (Steenackers et al. 2012).

Bacteriocins are antimicrobial compounds whose ability to regulate biofilms has been extensively studied (García-Almendárez et al. 2008). It is unknown precisely how bacteriocins work in biofilms; however, they are known to cause perforations in the membranes of bacteria. Additionally, they demonstrated that the production of *L. monocytogenes* biofilms could be controlled by utilising a culture medium supernatant that included bacteriocin generated by *Lactobacillus sakei* (Winkelströter et al. 2011). Planktonic and sessile cells of *Lactococcus monocytogenes* adhered to stainless steel chips were decreased by more than 5 log using spray-dried crude bacteriocin fermentate (CBF) of *Lactococcus lactis* UQ2 or *Lactococcus lactis* UQ2 (García-Almendárez et al. 2008).

> The food business may potentially benefit from the possible use of enzymatic cleaning as an adjuvant technique for biofilm removal. Augustin et al. (2004) examined how to employ proteinase disinfecting agents (chymotrypsin, cryotin, alkalase, and krill trypsin) to break down *Lactobacillus lactis*, *Lactobacillus bulgaricus*, and *Streptococcus thermophilus* biofilms. The applications of enzymes are still restricted despite the authors' positive results as it might be challenging to determine which enzymes are efficient against different kinds of biofilms because they are quite expensive (Meyer 2003).

Furukawa et al. (2010) reported that sugar fatty acid esters inhibited *S. aureus*, *Streptococcus mutans*, and *L. monocytogenes* from adhering to the abiotic surface. Bodur and Cagri-Mehmetoglu (2012) found that *L. monocytogenes*, *E. coli O157:H7, S. aureus*, and biofilms on stainless steel surfaces formed less biofilms when exposed to scallop shell powder (SSP). On the other hand, Arevalos-Sánchez et al. (2012) showed that acidic and neutral electrolysed waters exhibited potent antimicrobial activity towards *listerial* biofilms at 65 ppm or greater concentrations of substance, following a 10-min contact time.

Phage removal is another way to get rid of biofilm. The "viruses" of microbial cells known as bacteriophages are obligatory parasites that have a genetic parasitism that is unique to the many strains or serotypes of microbial species (Iacumin et al. 2016). Injecting their DNA, bacteriophage genomes and structures (such as capsids and tails) are produced by the cell under the control of the viruses. Bacteriophage infection has the ability to wipe out the entire colony since, once finished, phages lyse cells (Greer 2005; Soni and Nannapaneni 2010). The FDA/USA allowed the preparation of bacteriophages (LISTEX P100) a few years ago to combat the explicit presence of *L. monocytogenes* in food (EFSA 2016). Using pulverisation, various disinfectants are sprayed to work areas using the aerosolisation technique of disinfection. Using hydrogen peroxide, sodium hypochlorite, and peracetic acid, several authors have demonstrated its effectiveness as a biofilm management strategy in the food industry and hospitals (Choi et al. 2012; Park et al. 2012). Notwithstanding significant advancements in our understanding of the mechanisms governing undesired biofilms, no approach is thought to be entirely effective, and further research is still required in this field.

4.6 Conclusion

An important cause of contamination of food and clinical infections are pathogenic bacteria and spoilage found in biofilms generated in various food industry settings. The bacterial cells and substratum characteristics, as well as the temperature, pH, and nutritional components of the environment, all have a role in the creation of biofilms, as was previously mentioned. Improved comprehension of the principles behind the adhesion of cells and the formation of biofilm should aid in the creation of novel cleaning techniques and prevent the production of biofilms, thereby averting disease outbreaks and monetary losses. Although traditional management tactics are still being developed and employed today, new methods of controlling biofilms are needed since they are becoming more resistant to these conventional procedures. In order to overcome the biofilm resistance, it seems that novel biofilm control tactics must be found. As a result of the unique character of each biofilm, no one technique has been found to be able to eradicate them all. Instead, each biofilm issue must be thoroughly examined to identify its cause before a cleaning and disinfection procedure can be developed and put into place. Furthermore, more research needs to be done to elucidate a few key areas that are still unclear, like how antimicrobial products affect microbial biofilms, how they recover from damage, how microbes can become resistant to one kind of control and go on to survive other effective ones, and other puzzling details surrounding this phenomenon.

References

Abdallah M, Benoliel C, Drider D, Dhulster P, Chihib N-E (2014) Biofilm formation and persistence on abiotic surfaces in the context of food and medical environments. Arch Microbiol 196(7):453–472. https://doi.org/10.1007/s00203-014-0983-1

Álvarez-Ordóñez A, Alvseike O, Omer MK, Heir E, Axelsson L, Holck A, Prieto M (2013) Heterogeneity in resistance to food-related stresses and biofilm formation ability among verocytotoxigenic Escherichia coli strains. Int J Food Microbiol 161(3):220–230. https://doi.org/10.1016/j.ijfoodmicro.2012.12.008

Anaissie EJ, Penzak SR, Dignani MC (2002) The hospital water supply as a source of nosocomial infections. Arch Intern Med 162(13):1483. https://doi.org/10.1001/archinte.162.13.1483

Anand S, Singh D, Avadhanula M, Marka S (2014) Development and control of bacterial biofilms on dairy processing membranes. Compr Rev Food Sci Food Saf 13(1):18–33. https://doi.org/10.1111/1541-4337.12048

Andreani NA, Carraro L, Fasolato L, Balzan S, Lucchini R, Novelli E, Cardazzo B (2016) Characterisation of the thermostable protease AprX in strains of Pseudomonas fluorescens and impact on the shelf-life of dairy products: preliminary results. Italian J Food Saf 5(4):6175. https://doi.org/10.4081/ijfs.2016.6175

Antonioli P, Bolognesi N, Valpiani G, Morotti C, Bernardini D, Bravi F, Di Ruscio E, Stefanati A, Gabutti G (2020) A 2-year point-prevalence surveillance of healthcare-associated infections and antimicrobial use in Ferrara university hospital, Italy. BMC Infect Dis 20(1):75. https://doi.org/10.1186/s12879-020-4791-8

Anwar H, Strap JL, Costerton JW (1992) Establishment of aging biofilms: possible mechanism of bacterial resistance to antimicrobial therapy. Antimicrob Agents Chemother 36(7):1347–1351. https://doi.org/10.1128/AAC.36.7.1347

Arevalos-Sánchez M, Regalado C, Martin SE, Domínguez-Domínguez J, García-Almendárez BE (2012) Effect of neutral electrolyzed water and nisin on listeria monocytogenes biofilms, and on listeriolysin O activity. Food Control 24(1–2):116–122. https://doi.org/10.1016/j.foodcont.2011.09.012

Augustin M, Ali-Vehmas T, Atroshi F (2004) Assessment of enzymatic cleaning agents and disinfectants against bacterial biofilms. J Pharm Pharm Sci 7(1):55–64

Avila-Novoa M-G, Iñíguez-Moreno M, Solís-Velázquez O-A, González-Gómez J-P, Guerrero-Medina P-J, Gutiérrez-Lomelí M (2018) Biofilm formation by *Staphylococcus aureus* isolated from food contact surfaces in the dairy industry of Jalisco, Mexico. J Food Qual 2018:1–8. https://doi.org/10.1155/2018/1746139

Béné C, Arthur R, Norbury H, Allison EH, Beveridge M, Bush S, Campling L, Leschen W, Little D, Squires D, Thilsted SH, Troell M, Williams M (2016) Contribution of fisheries and aquaculture to food security and poverty reduction: assessing the current evidence. World Dev 79:177–196. https://doi.org/10.1016/j.worlddev.2015.11.007

Beshiru A, Igbinosa EO (2018) Characterization of extracellular virulence properties and biofilm-formation capacity of vibrio species recovered from ready-to-eat (RTE) shrimps. Microb Pathog 119:93–102. https://doi.org/10.1016/j.micpath.2018.04.015

Bodur T, Cagri-Mehmetoglu A (2012) Removal of listeria monocytogenes, Staphylococcus aureus and Escherichia coli O157:H7 biofilms on stainless steel using scallop shell powder. Food Control 25(1):1–9. https://doi.org/10.1016/j.foodcont.2011.09.032

Bonsaglia ECR, Silva NCC, Fernades Júnior A, Araújo Júnior JP, Tsunemi MH, Rall VLM (2014) Production of biofilm by listeria monocytogenes in different materials and temperatures. Food Control 35(1):386–391. https://doi.org/10.1016/j.foodcont.2013.07.023

Bos R, van der Mei HC, Busscher HJ (1999) Physico-chemistry of initial microbial adhesive interactions—its mechanisms and methods for study. FEMS Microbiol Rev 23(2):179–230. https://doi.org/10.1111/j.1574-6976.1999.tb00396.x

Bottone EJ (2010) *Bacillus cereus*, a Volatile Human Pathogen. Clin Microbiol Rev 23(2):382–398. https://doi.org/10.1128/CMR.00073-09

Bremer PJ, Monk I, Osborne CM (2001) Survival of listeria monocytogenes attached to stainless steel surfaces in the presence or absence of Flavobacterium spp. J Food Prot 64(9):1369–1376. https://doi.org/10.4315/0362-028X-64.9.1369

Bridier A, Dubois-Brissonnet F, Boubetra A, Thomas V, Briandet R (2010) The biofilm architecture of sixty opportunistic pathogens deciphered using a high throughput CLSM method. J Microbiol Methods 82(1):64–70. https://doi.org/10.1016/j.mimet.2010.04.006

Bronnec V, Turoňová H, Bouju A, Cruveiller S, Rodrigues R, Demnerova K, Tresse O, Haddad N, Zagorec M (2016) Adhesion, biofilm formation, and genomic features of campylobacter jejuni bf, an atypical strain able to grow under aerobic conditions. Front Microbiol 7:1002. https://doi.org/10.3389/fmicb.2016.01002

Cadena M, Kelman T, Marco ML, Pitesky M (2019) Understanding antimicrobial resistance (AMR) profiles of salmonella biofilm and planktonic bacteria challenged with disinfectants commonly used during poultry processing. Food Secur 8(7):275. https://doi.org/10.3390/foods8070275

Cai W, De La Fuente L, Arias CR (2013) Biofilm formation by the fish pathogen Flavobacterium columnare: development and parameters affecting surface attachment. Appl Environ Microbiol 79(18):5633–5642. https://doi.org/10.1128/AEM.01192-13

Carrascosa C, Raheem D, Ramos F, Saraiva A, Raposo A (2021) Microbial biofilms in the food industry—A comprehensive review. Int J Environ Res Public Health 18(4):2014. https://doi.org/10.3390/ijerph18042014

Chae MS, Schraft H (2000) Comparative evaluation of adhesion and biofilm formation of different listeria monocytogenes strains. Int J Food Microbiol 62(1–2):103–111. https://doi.org/10.1016/S0168-1605(00)00406-2

Chavant P, Martinie B, Meylheuc T, Bellon-Fontaine M-N, Hebraud M (2002) *Listeria monocytogenes* LO28: surface physicochemical properties and ability to form biofilms at different temperatures and growth phases. Appl Environ Microbiol 68(2):728–737. https://doi.org/10.1128/AEM.68.2.728-737.2002

Chaves Simões L, Simões M (2013) Biofilms in drinking water: problems and solutions. RSC Adv 3(8):2520–2533. https://doi.org/10.1039/C2RA22243D

Chia TWR, Goulter RM, McMeekin T, Dykes GA, Fegan N (2009) Attachment of different salmonella serovars to materials commonly used in a poultry processing plant. Food Microbiol 26(8):853–859. https://doi.org/10.1016/j.fm.2009.05.012

Chlebicz A, Śliżewska K (2018) Campylobacteriosis, salmonellosis, Yersiniosis, and Listeriosis as zoonotic foodborne diseases: a review. Int J Environ Res Public Health 15(5):863. https://doi.org/10.3390/ijerph15050863

Choi N-Y, Baek S-Y, Yoon J-H, Choi M-R, Kang D-H, Lee S-Y (2012) Efficacy of aerosolized hydrogen peroxide-based sanitizer on the reduction of pathogenic bacteria on a stainless steel surface. Food Control 27(1):57–63. https://doi.org/10.1016/j.foodcont.2012.02.027

Costa A, Bertolotti L, Brito L, Civera T (2016) Biofilm formation and disinfectant susceptibility of persistent and nonpersistent *Listeria monocytogenes* isolates from Gorgonzola cheese processing plants. Foodborne Pathog Dis 13(11):602–609. https://doi.org/10.1089/fpd.2016.2154

Costerton JW, Lewandowski Z, Caldwell DE, Korber DR, Lappin-Scott HM (1995) Microbial biofilms. Ann Rev Microbiol 49(1):711–745. https://doi.org/10.1146/annurev.mi.49.100195.003431

Dhowlaghar N, Bansal M, Schilling MW, Nannapaneni R (2018) Scanning electron microscopy of salmonella biofilms on various food-contact surfaces in catfish mucus. Food Microbiol 74:143–150. https://doi.org/10.1016/j.fm.2018.03.013

Di Bonaventura G, Piccolomini R, Paludi D, D'Orio V, Vergara A, Conter M, Ianieri A (2008) Influence of temperature on biofilm formation by listeria monocytogenes on various food-contact surfaces: relationship with motility and cell surface hydrophobicity. J Appl Microbiol 104(6):1552–1561. https://doi.org/10.1111/j.1365-2672.2007.03688.x

Díez-García M, Capita R, Alonso-Calleja C (2012) Influence of serotype on the growth kinetics and the ability to form biofilms of Salmonella isolates from poultry. Food Microbiol 31(2):173–180. https://doi.org/10.1016/j.fm.2012.03.012

Dimakopoulou-Papazoglou D, Lianou A, Koutsoumanis KP (2016) Modelling biofilm formation of Salmonella enterica ser. Newport as a function of pH and water activity. Food Microbiol 53:76–81. https://doi.org/10.1016/j.fm.2015.09.002

Djordjevic D, Wiedmann M, McLandsborough LA (2002) Microtiter plate assay for assessment of *Listeria monocytogenes* biofilm formation. Appl Environ Microbiol 68(6):2950–2958. https://doi.org/10.1128/AEM.68.6.2950-2958.2002

Doijad SP, Barbuddhe SB, Garg S, Poharkar KV, Kalorey DR, Kurkure NV, Rawool DB, Chakraborty T (2015) Biofilm-forming abilities of listeria monocytogenes serotypes isolated from different sources. PLoS One 10(9):e0137046. https://doi.org/10.1371/journal.pone.0137046

Donlan RM, Costerton JW (2002) Biofilms: survival mechanisms of clinically relevant microorganisms. Clin Microbiol Rev 15(2):167–193. https://doi.org/10.1128/CMR.15.2.167-193.2002

Duanis-Assaf D, Steinberg D, Chai Y, Shemesh M (2016) The LuxS based quorum sensing governs lactose induced biofilm formation by Bacillus subtilis. Front Microbiol 6:1517. https://doi.org/10.3389/fmicb.2015.01517

Duguid JP, Anderson ES, Campbell I (1966) Fimbriae and adhesive properties in salmonellae. J Pathol Bacteriol 92(1):107–137. https://doi.org/10.1002/path.1700920113

Dunne WM (2002) Bacterial adhesion: seen any good biofilms lately? Clin Microbiol Rev 15(2):155–166. https://doi.org/10.1128/CMR.15.2.155-166.2002

Dutra TV, da Fernandes MS, Perdoncini MRFG, dos Anjos MM, de Abreu Filho BA (2018) Capacity of *Escherichia coli* and *Staphylococcus aureus* to produce biofilm on stainless steel surfaces in the presence of food residues. J Food Process Preserv 42(4):e13574. https://doi.org/10.1111/jfpp.13574

EFSA (2016) Evaluation of the safety and efficacy of Listex™ P100 for reduction of pathogens on different ready-to-eat (RTE) food products. EFSA J 14(8):e04565. https://doi.org/10.2903/j.efsa.2016.4565

Enos-Berlage JL, Guvener ZT, Keenan CE, McCarter LL (2005) Genetic determinants of biofilm development of opaque and translucent *Vibrio parahaemolyticus*. Mol Microbiol 55(4):1160–1182. https://doi.org/10.1111/j.1365-2958.2004.04453.x

European Food Safety Authority (2009, May) The community summary report on food-borne outbreaks in the European Union in 2007. EFSA J 7:271r. https://doi.org/10.2903/j.efsa.2009.271r

Faille C, Fontaine F, Bénézech T (2001) Potential occurrence of adhering living *Bacillus* spores in milk product processing lines. J Appl Microbiol 90(6):892–900. https://doi.org/10.1046/j.1365-2672.2001.01321.x

Faille C, Bénézech T, Midelet-Bourdin G, Lequette Y, Clarisse M, Ronse G, Ronse A, Slomianny C (2014) Sporulation of bacillus spp. within biofilms: a potential source of contamination in food processing environments. Food Microbiol 40:64–74. https://doi.org/10.1016/j.fm.2013.12.004

Faruque SM, Biswas K, Udden SMN, Ahmad QS, Sack DA, Nair GB, Mekalanos JJ (2006) Transmissibility of cholera: *in vivo* -formed biofilms and their relationship to infectivity and persistence in the environment. Proc Natl Acad Sci 103(16):6350–6355. https://doi.org/10.1073/pnas.0601277103

Ferriol-González C, Domingo-Calap P (2020) Phages for Biofilm Removal. *Antibiotics* 9(5):268. https://doi.org/10.3390/antibiotics9050268

Flemming H-C, Wingender J (2010) The biofilm matrix. Nat Rev Microbiol 8(9):623–633. https://doi.org/10.1038/nrmicro2415

Flint SH, Brooks JD, Bremer PJ (2000) Properties of the stainless steel substrate, influencing the adhesion of thermo-resistant streptococci. J Food Eng 43(4):235–242. https://doi.org/10.1016/S0260-8774(99)00157-0

Furukawa S, Akiyoshi Y, O'Toole GA, Ogihara H, Morinaga Y (2010) Sugar fatty acid esters inhibit biofilm formation by food-borne pathogenic bacteria. Int J Food Microbiol 138(1–2):176–180. https://doi.org/10.1016/j.ijfoodmicro.2009.12.026

Galié S, García-Gutiérrez C, Miguélez EM, Villar CJ, Lombó F (2018) Biofilms in the food industry: health aspects and control methods. Front Microbiol 9:898. https://doi.org/10.3389/fmicb.2018.00898

García-Almendárez BE, Cann IKO, Martin SE, Guerrero-Legarreta I, Regalado C (2008) Effect of Lactococcus lactis UQ2 and its bacteriocin on listeria monocytogenes biofilms. Food Control 19(7):670–680. https://doi.org/10.1016/j.foodcont.2007.07.015

Ghafoor A, Hay ID, Rehm BHA (2011) Role of exopolysaccharides in Pseudomonas aeruginosa biofilm formation and architecture. Appl Environ Microbiol 77(15):5238–5246. https://doi.org/10.1128/AEM.00637-11

Giaouris E, Heir E, Desvaux M, Hébraud M, Møretrø T, Langsrud S, Doulgeraki A, Nychas G-J, Kačániová M, Czaczyk K, Ölmez H, Simões M (2015) Intra- and inter-species interactions within biofilms of important foodborne bacterial pathogens. Front Microbiol 6:841. https://doi.org/10.3389/fmicb.2015.00841

Gomes LC, Silva LN, Simões M, Melo LF, Mergulhão FJ (2015) *Escherichia coli* adhesion, biofilm development and antibiotic susceptibility on biomedical materials. J Biomed Mater Res A 103(4):1414–1423. https://doi.org/10.1002/jbm.a.35277

Gómez D, Ariño A, Carramiñana JJ, Rota C, Yangüela J (2012) Sponge versus mini-roller for the surface microbiological control of listeria monocytogenes, total aerobic mesophiles and Enterobacteriaceae in the meat industry. Food Control 27(1):242–247. https://doi.org/10.1016/j.foodcont.2012.03.031

González-Rivas F, Ripolles-Avila C, Fontecha-Umaña F, Ríos-Castillo AG, Rodríguez-Jerez JJ (2018) Biofilms in the spotlight: detection, quantification, and removal methods. Compr Rev Food Sci Food Saf 17(5):1261–1276. https://doi.org/10.1111/1541-4337.12378

Gopal N, Hill C, Ross PR, Beresford TP, Fenelon MA, Cotter PD (2015) The prevalence and control of bacillus and related spore-forming bacteria in the dairy industry. Front Microbiol 6:1418. https://doi.org/10.3389/fmicb.2015.01418

Goulter RM, Gentle IR, Dykes GA (2009) Issues in determining factors influencing bacterial attachment: a review using the attachment of *Escherichia coli* to abiotic surfaces as an example. Lett Appl Microbiol 49(1):1–7. https://doi.org/10.1111/j.1472-765X.2009.02591.x

Gousia P, Economou V, Sakkas H, Leveidiotou S, Papadopoulou C (2011) Antimicrobial resistance of major foodborne pathogens from major meat products. Foodborne Pathog Dis 8(1):27–38. https://doi.org/10.1089/fpd.2010.0577

Govaert M, Smet C, Baka M, Janssens T, Van Impe J (2018) Influence of incubation conditions on the formation of model biofilms by *Listeria monocytogenes* and *Salmonella* Typhimurium on abiotic surfaces. J Appl Microbiol 125(6):1890–1900. https://doi.org/10.1111/jam.14071

Greer GG (2005) Bacteriophage control of foodborne bacteria. J Food Prot 68(5):1102–1111. https://doi.org/10.4315/0362-028X-68.5.1102

Grigore-Gurgu L, Ionela Bucur F, Borda D, Alexa E-A, Neagu C, Ioana Nicolau A (2020) Biofilms formed by pathogens in food and food processing environments. In: *Bacterial Biofilms*. IntechOpen. https://doi.org/10.5772/intechopen.90176

Gupta K, Barua S, Hazarika SN, Manhar AK, Nath D, Karak N, Namsa ND, Mukhopadhyay R, Chandra Kalia V, Mandal M (2014) Green silver nanoparticles: enhanced antimicrobial and antibiofilm activity with effects on DNA replication and cell cytotoxicity. RSC Adv 4(95):52845–52855. https://doi.org/10.1039/C4RA08791G

Gupta K, Singh SP, Manhar AK, Saikia D, Namsa ND, Konwar BK, Mandal M (2019) Inhibition of Staphylococcus aureus and Pseudomonas aeruginosa biofilm and virulence by active fraction of Syzygium cumini (L.) Skeels leaf extract: in-vitro and in Silico studies. Indian J Microbiol 59(1):13–21. https://doi.org/10.1007/s12088-018-0770-9

Gutiérrez D, Delgado S, Vázquez-Sánchez D, Martínez B, Cabo ML, Rodríguez A, Herrera JJ, García P (2012) Incidence of Staphylococcus aureus and analysis of associated bacterial communities on food industry surfaces. Appl Environ Microbiol 78(24):8547–8554. https://doi.org/10.1128/AEM.02045-12

Hall CW, Mah T-F (2017) Molecular mechanisms of biofilm-based antibiotic resistance and tolerance in pathogenic bacteria. FEMS Microbiol Rev 41(3):276–301. https://doi.org/10.1093/femsre/fux010

Hall-Stoodley L, Costerton JW, Stoodley P (2004) Bacterial biofilms: from the natural environment to infectious diseases. Nat Rev Microbiol 2(2):95–108. https://doi.org/10.1038/nrmicro821

Hammer BK, Bassler BL (2003) Quorum sensing controls biofilm formation in *Vibrio cholerae*. Mol Microbiol 50(1):101–104. https://doi.org/10.1046/j.1365-2958.2003.03688.x

Haque M, Sartelli M, McKimm J, Abu Bakar M, Bin. (2018) Health care-associated infections – an overview. Infect Drug Resist 11:2321–2333. https://doi.org/10.2147/IDR.S177247

Hartmann I, Carranza P, Lehner A, Stephan R, Eberl L, Riedel K (2010) Genes involved in *Cronobacter sakazakii* biofilm formation. Appl Environ Microbiol 76(7):2251–2261. https://doi.org/10.1128/AEM.00930-09

Harvey J, Keenan KP, Gilmour A (2007) Assessing biofilm formation by listeria monocytogenes strains. Food Microbiol 24(4):380–392. https://doi.org/10.1016/j.fm.2006.06.006

Hayrapetyan H, Muller L, Tempelaars M, Abee T, Nierop Groot M (2015) Comparative analysis of biofilm formation by Bacillus cereus reference strains and undomesticated food isolates and the effect of free iron. Int J Food Microbiol 200:72–79. https://doi.org/10.1016/j.ijfoodmicro.2015.02.005

Hayrapetyan H, Siezen R, Abee T, Nierop Groot M (2016) Comparative genomics of iron-transporting Systems in Bacillus cereus strains and impact of iron sources on growth and biofilm formation. Front Microbiol 7:842. https://doi.org/10.3389/fmicb.2016.00842

Høiby N, Bjarnsholt T, Givskov M, Molin S, Ciofu O (2010) Antibiotic resistance of bacterial biofilms. Int J Antimicrob Agents 35(4):322–332. https://doi.org/10.1016/j.ijantimicag.2009.12.011

Holah JT, Taylor JH, Dawson DJ, Hall KE (2002) Biocide use in the food industry and the disinfectant resistance of persistent strains of *Listeria monocytogenes* and *Escherichia coli*. J Appl Microbiol 92:111S–120S. https://doi.org/10.1046/j.1365-2672.92.5s1.18.x

Hood SK, Zottola EA (1995) Biofilms in food processing. Food Control 6(1):9–18. https://doi.org/10.1016/0956-7135(95)91449-U

Iacumin L, Manzano M, Comi G (2016) Phage inactivation of listeria monocytogenes on san Daniele dry-cured ham and elimination of biofilms from equipment and working environments. Microorganisms 4(1):4. https://doi.org/10.3390/microorganisms4010004

In Lee SH, Barancelli GV, de Camargo TM, Corassin CH, Rosim RE, da Cruz AG, Cappato LP, de Oliveira CAF (2017) Biofilm-producing ability of listeria monocytogenes isolates from Brazilian cheese processing plants. Food Res Int 91:88–91. https://doi.org/10.1016/j.foodres.2016.11.039

Iñiguez-Moreno M, Gutiérrez-Lomelí M, Avila-Novoa MG (2019) Kinetics of biofilm formation by pathogenic and spoilage microorganisms under conditions that mimic the poultry, meat, and egg processing industries. Int J Food Microbiol 303:32–41. https://doi.org/10.1016/j.ijfoodmicro.2019.04.012

Jessen B, Lammert L (2003) Biofilm and disinfection in meat processing plants. Int Biodeterior Biodegradation 51(4):265–269. https://doi.org/10.1016/S0964-8305(03)00046-5

Jindal S, Anand S, Metzger L, Amamcharla J (2018) Short communication: a comparison of biofilm development on stainless steel and modified-surface plate heat exchangers during a 17-h milk pasteurization run. J Dairy Sci 101(4):2921–2926. https://doi.org/10.3168/jds.2017-14028

Joseph LA, Wright AC (2004) Expression of *Vibrio vulnificus* capsular polysaccharide inhibits biofilm formation. J Bacteriol 186(3):889–893. https://doi.org/10.1128/JB.186.3.889-893.2004

Karatan E, Watnick P (2009) Signals, regulatory networks, and materials that build and break bacterial biofilms. Microbiol Mol Biol Rev 73(2):310–347. https://doi.org/10.1128/MMBR.00041-08

Kariyawasam M, Jayasooriya S (2010) A preliminary study on the presence of biofilms on food contact surfaces in selected Sri Lankan fish & shrimp processing factories: setbacks in current cleaning process. Sri Lanka J Aquat Sci 11:75–83. https://doi.org/10.4038/sljas.v11i0.2224

Karlapudi AP, Venkateswarulu TC, Srirama K, Kota RK, Mikkili I, Kodali VP (2020) Evaluation of anti-cancer, anti-microbial and anti-biofilm potential of biosurfactant extracted from an Acinetobacter M6 strain. J King Saud Uni Sci 32(1):223–227. https://doi.org/10.1016/j.jksus.2018.04.007

Kim D, Kim JG, Chu CN (2016) Aging effect on the wettability of stainless steel. Mater Lett 170:18–20. https://doi.org/10.1016/j.matlet.2016.01.107

Klančnik A, Šimunović K, Sterniša M, Ramić D, Smole Možina S, Bucar F (2021) Anti-adhesion activity of phytochemicals to prevent campylobacter jejuni biofilm formation on abiotic surfaces. Phytochem Rev 20(1):55–84. https://doi.org/10.1007/s11101-020-09669-6

Kontominas MG, Badeka AV, Kosma IS, Nathanailides CI (2021) Innovative seafood preservation technologies: recent developments. Animals 11(1):92. https://doi.org/10.3390/ani11010092

Kostaki M, Chorianopoulos N, Braxou E, Nychas G-J, Giaouris E (2012) Differential biofilm formation and chemical disinfection resistance of sessile cells of listeria monocytogenes strains under Monospecies and dual-species (with salmonella enterica) conditions. Appl Environ Microbiol 78(8):2586–2595. https://doi.org/10.1128/AEM.07099-11

Kumar CG, Anand SK (1998) Significance of microbial biofilms in food industry: a review. Int J Food Microbiol 42(1–2):9–27. https://doi.org/10.1016/S0168-1605(98)00060-9

Latorre AA, Van Kessel JS, Karns JS, Zurakowski MJ, Pradhan AK, Boor KJ, Jayarao BM, Houser BA, Daugherty CS, Schukken YH (2010) Biofilm in milking equipment on a dairy farm as a potential source of bulk tank milk contamination with listeria monocytogenes. J Dairy Sci 93(6):2792–2802. https://doi.org/10.3168/jds.2009-2717

Lauderdale KJ, Boles BR, Cheung AL, Horswill AR (2009) Interconnections between Sigma B, *agr*, and Proteolytic Activity in *Staphylococcus aureus* Biofilm Maturation. Infect Immun 77(4):1623–1635. https://doi.org/10.1128/IAI.01036-08
Lebeer S, Verhoeven TLA, Perea Vélez M, Vanderleyden J, De Keersmaecker SCJ (2007) Impact of environmental and genetic factors on biofilm formation by the probiotic strain *Lactobacillus rhamnosus* GG. Appl Environ Microbiol 73(21):6768–6775. https://doi.org/10.1128/AEM.01393-07
Lemon KP, Higgins DE, Kolter R (2007) Flagellar motility is critical for *Listeria monocytogenes* biofilm formation. J Bacteriol 189(12):4418–4424. https://doi.org/10.1128/JB.01967-06
Li X, Kim M-J, Bang W-S, Yuk H-G (2018) Anti-biofilm effect of 405-nm LEDs against listeria monocytogenes in simulated ready-to-eat fresh salmon storage conditions. Food Control 84:513–521. https://doi.org/10.1016/j.foodcont.2017.09.006
Lianou A, Koutsoumanis KP (2012) Strain variability of the biofilm-forming ability of salmonella enterica under various environmental conditions. Int J Food Microbiol 160(2):171–178. https://doi.org/10.1016/j.ijfoodmicro.2012.10.002
Machado SG, Baglinière F, Marchand S, Van Coillie E, Vanetti MCD, De Block J, Heyndrickx M (2017) The biodiversity of the microbiota producing heat-resistant enzymes responsible for spoilage in processed bovine milk and dairy products. Front Microbiol 8:302. https://doi.org/10.3389/fmicb.2017.00302
Mahnot NK, Mahanta CL, Keener KM, Misra NN (2019) Strategy to achieve a 5-log salmonella inactivation in tender coconut water using high voltage atmospheric cold plasma (HVACP). Food Chem 284:303–311. https://doi.org/10.1016/j.foodchem.2019.01.084
Makovcova J, Babak V, Kulich P, Masek J, Slany M, Cincarova L (2017) Dynamics of mono- and dual-species biofilm formation and interactions between *Staphylococcus aureus* and gram-negative bacteria. Microb Biotechnol 10(4):819–832. https://doi.org/10.1111/1751-7915.12705
Manguiat LS, Fang TJ (2013) Microbiological quality of chicken- and pork-based street-vended foods from Taichung, Taiwan, and Laguna, Philippines. Food Microbiol 36(1):57–62. https://doi.org/10.1016/j.fm.2013.04.005
Manuzon MY, Wang HH (2006) Mixed culture biofilms. In: Biofilms in the food environment. Wiley, pp 105–126. https://doi.org/10.1002/9780470277782.ch5
Marschang S, Bernardo G (2015) Prevention and control of healthcare-associated infection in Europe: a review of patients' perspectives and existing differences. J Hosp Infect 89(4):357–362. https://doi.org/10.1016/j.jhin.2015.01.017
McDougald D, Rice SA, Barraud N, Steinberg PD, Kjelleberg S (2012) Should we stay or should we go: mechanisms and ecological consequences for biofilm dispersal. Nat Rev Microbiol 10(1):39–50. https://doi.org/10.1038/nrmicro2695
Merino L, Procura F, Trejo FM, Bueno DJ, Golowczyc MA (2019) Biofilm formation by salmonella sp. in the poultry industry: detection, control and eradication strategies. Food Res Int 119:530–540. https://doi.org/10.1016/j.foodres.2017.11.024
Meyer B (2003) Approaches to prevention, removal and killing of biofilms. Int Biodeterior Biodegradation 51(4):249–253. https://doi.org/10.1016/S0964-8305(03)00047-7
Murphy F, Tchetchik A, Furxhi I (2020) Reduction of health care-associated infections (HAIs) with antimicrobial inorganic nanoparticles incorporated in medical textiles: an economic assessment. Nano 10(5):999. https://doi.org/10.3390/nano10050999
Nancharaiah YV, Venugopalan VP, Wuertz S, Wilderer PA, Hausner M (2005) Compatibility of the green fluorescent protein and a general nucleic acid stain for quantitative description of a pseudomonas putida biofilm. J Microbiol Methods 60(2):179–187. https://doi.org/10.1016/j.mimet.2004.09.016
Nguyen HDN, Yang YS, Yuk HG (2014) Biofilm formation of salmonella Typhimurium on stainless steel and acrylic surfaces as affected by temperature and pH level. LWT Food Sci Technol 55(1):383–388. https://doi.org/10.1016/j.lwt.2013.09.022
Nilsson RE, Ross T, Bowman JP (2011) Variability in biofilm production by listeria monocytogenes correlated to strain origin and growth conditions. Int J Food Microbiol 150(1):14–24. https://doi.org/10.1016/j.ijfoodmicro.2011.07.012

O'Leary D, McCabe EM, McCusker MP, Martins M, Fanning S, Duffy G (2015) Acid environments affect biofilm formation and gene expression in isolates of salmonella enterica Typhimurium DT104. Int J Food Microbiol 206:7–16. https://doi.org/10.1016/j.ijfoodmicro.2015.03.030

Oliveira K, Oliveira T, Teixeira P, Azeredo J, Oliveira R (2007) Adhesion of salmonella Enteritidis to stainless steel surfaces. Braz J Microbiol 38(2):318–323. https://doi.org/10.1590/S1517-83822007000200026

Olmo JA-D, Ruiz-Rubio L, Pérez-Alvarez L, Sáez-Martínez V, Vilas-Vilela JL (2020) Antibacterial coatings for improving the performance of biomaterials. Coatings 10(2):139. https://doi.org/10.3390/coatings10020139

Oniciuc E-A, Cerca N, Nicolau AI (2016) Compositional analysis of biofilms formed by Staphylococcus aureus isolated from food sources. Front Microbiol 7:390. https://doi.org/10.3389/fmicb.2016.00390

Ostrov I, Sela N, Freed M, Khateb N, Kott-Gutkowski M, Inbar D, Shemesh M (2015) Draft genome sequence of bacillus licheniformis S127, isolated from a sheep udder clinical infection. Genome Announc 3(5):e00971-15. https://doi.org/10.1128/genomeA.00971-15

Pandy PK, Vivekanand B, Kundan K (2014) Biofilm in aquaculture production. Afr J Microbiol Res 8(13):1434–1443. https://doi.org/10.5897/AJMR2013.6445

Park SH, Jarquin R, Hanning I, Almeida G, Ricke SC (2011) Detection of salmonella spp. survival and virulence in poultry feed by targeting the hilA gene. J Appl Microbiol 111(2):426–432. https://doi.org/10.1111/j.1365-2672.2011.05054.x

Park S-H, Cheon H-L, Park K-H, Chung M-S, Choi SH, Ryu S, Kang D-H (2012) Inactivation of biofilm cells of foodborne pathogen by aerosolized sanitizers. Int J Food Microbiol 154(3):130–134. https://doi.org/10.1016/j.ijfoodmicro.2011.12.018

Piercey MJ, Ells TC, Macintosh AJ, Truelstrup Hansen L (2017) Variations in biofilm formation, desiccation resistance and Benzalkonium chloride susceptibility among listeria monocytogenes strains isolated in Canada. Int J Food Microbiol 257:254–261. https://doi.org/10.1016/j.ijfoodmicro.2017.06.025

Popkin BM, Adair LS, Ng SW (2012) Global nutrition transition and the pandemic of obesity in developing countries. Nutr Rev 70(1):3–21. https://doi.org/10.1111/j.1753-4887.2011.00456.x

Poulsen LV (1999) Microbial biofilm in food processing. LWT Food Sci Technol 32(6):321–326. https://doi.org/10.1006/fstl.1999.0561

Rather MA, Gupta K, Bardhan P, Borah M, Sarkar A, Eldiehy KSH, Bhuyan S, Mandal M (2021a) Microbial biofilm: a matter of grave concern for human health and food industry. J Basic Microbiol 61(5):380–395. https://doi.org/10.1002/jobm.202000678

Rather MA, Gupta K, Mandal M (2021b) Inhibition of biofilm and quorum sensing-regulated virulence factors in Pseudomonas aeruginosa by Cuphea carthagenensis (Jacq.) J. F. Macbr. Leaf extract: an in vitro study. J Ethnopharmacol 269:113699. https://doi.org/10.1016/j.jep.2020.113699

Renner LD, Weibel DB (2011) Physicochemical regulation of biofilm formation. MRS Bull 36(5):347–355. https://doi.org/10.1557/mrs.2011.65

Rivas L, Fegan N, Dykes GA (2007) Attachment of Shiga toxigenic Escherichia coli to stainless steel. Int J Food Microbiol 115(1):89–94. https://doi.org/10.1016/j.ijfoodmicro.2006.10.027

Rode TM, Langsrud S, Holck A, Møretrø T (2007) Different patterns of biofilm formation in Staphylococcus aureus under food-related stress conditions. Int J Food Microbiol 116(3):372–383. https://doi.org/10.1016/j.ijfoodmicro.2007.02.017

Rodriguez A, Autio WR, Mclandsborough LA (2008) Effect of surface roughness and stainless steel finish on listeria monocytogenes attachment and biofilm formation. J Food Prot 71(1):170–175. https://doi.org/10.4315/0362-028X-71.1.170

Rossi DA, Melo RT, Mendonça EP, Monteiro GP (2017) Biofilms of salmonella and campylobacter in the poultry industry. In: *Poultry science*. InTech. https://doi.org/10.5772/65254

Rossi C, Serio A, Chaves-López C, Anniballi F, Auricchio B, Goffredo E, Cenci-Goga BT, Lista F, Fillo S, Paparella A (2018) Biofilm formation, pigment production and motility in pseudomonas spp. isolated from the dairy industry. Food Control 86:241–248. https://doi.org/10.1016/j.foodcont.2017.11.018

Rothrock MJ, Davis ML, Locatelli A, Bodie A, McIntosh TG, Donaldson JR, Ricke SC (2017) Listeria occurrence in poultry flocks: detection and potential implications. Front Vet Sci 4:125. https://doi.org/10.3389/fvets.2017.00125

Sahal G, Woerdenbag HJ, Hinrichs WLJ, Visser A, Tepper PG, Quax WJ, van der Mei HC, Bilkay IS (2020) Antifungal and biofilm inhibitory effect of Cymbopogon citratus (lemongrass) essential oil on biofilm forming by Candida tropicalis isolates; an in vitro study. J Ethnopharmacol 246:112188. https://doi.org/10.1016/j.jep.2019.112188

Sasidharan S, Prema B, Yoga Latha L (2011) Antimicrobial drug resistance of Staphylococcus aureus in dairy products. Asian Pac J Trop Biomed 1(2):130–132. https://doi.org/10.1016/S2221-1691(11)60010-5

Satpute SK, Mone NS, Das P, Banat IM, Banpurkar AG (2019) Inhibition of pathogenic bacterial biofilms on PDMS based implants by L. acidophilus derived biosurfactant. BMC Microbiol 19(1):39. https://doi.org/10.1186/s12866-019-1412-z

Sauer K, Cullen MC, Rickard AH, Zeef LAH, Davies DG, Gilbert P (2004) Characterization of nutrient-induced dispersion in *Pseudomonas aeruginosa* PAO1 biofilm. J Bacteriol 186(21):7312–7326. https://doi.org/10.1128/JB.186.21.7312-7326.2004

Scatamburlo TM, Yamazi AK, Cavicchioli VQ, Pieri FA, Nero LA (2015) Spoilage potential of pseudomonas species isolated from goat milk. J Dairy Sci 98(2):759–764. https://doi.org/10.3168/jds.2014-8747

Shahid A, Rasool M, Akhter N, Aslam B, Hassan A, Sana S, Hidayat Rasool M, Khurshid M (2020) Innovative strategies for the control of biofilm formation in clinical settings. In: *Bacterial Biofilms*. IntechOpen. https://doi.org/10.5772/intechopen.89310

Sharma M, Anand SK (2002) Biofilms evaluation as an essential component of HACCP for food/dairy processing industry–a case. Food Control 13(6–7):469–477. https://doi.org/10.1016/S0956-7135(01)00068-8

Sharma S, Sachdeva P, Virdi JS (2003) Emerging water-borne pathogens. Appl Microbiol Biotechnol 61(5–6):424–428. https://doi.org/10.1007/s00253-003-1302-y

Sharma D, Misba L, Khan AU (2019) Antibiotics versus biofilm: an emerging battleground in microbial communities. Antimicrob Resist Infect Control 8(1):76. https://doi.org/10.1186/s13756-019-0533-3

Shi X, Zhu X (2009) Biofilm formation and food safety in food industries. Trends Food Sci Technol 20(9):407–413. https://doi.org/10.1016/j.tifs.2009.01.054

Simões M, Simões LC, Vieira MJ (2010) A review of current and emergent biofilm control strategies. LWT Food Sci Technol 43(4):573–583. https://doi.org/10.1016/j.lwt.2009.12.008

Sinde E, Carballo J (2000) Attachment of salmonella spp. and listeria monocytogenes to stainless steel, rubber and polytetrafluorethylene: the influence of free energy and the effect of commercial sanitizers. Food Microbiol 17(4):439–447. https://doi.org/10.1006/fmic.2000.0339

Smith A (2005) Biofilms and antibiotic therapy: is there a role for combating bacterial resistance by the use of novel drug delivery systems? Adv Drug Deliv Rev 57(10):1539–1550. https://doi.org/10.1016/j.addr.2005.04.007

Sofos JN, Geornaras I (2010) Overview of current meat hygiene and safety risks and summary of recent studies on biofilms, and control of Escherichia coli O157:H7 in nonintact, and listeria monocytogenes in ready-to-eat, meat products. Meat Sci 86(1):2–14. https://doi.org/10.1016/j.meatsci.2010.04.015

Somerton B, Lindsay D, Palmer J, Brooks J, Flint S (2015) Changes in sodium, calcium, and magnesium ion concentrations that inhibit Geobacillus biofilms have no effect on Anoxybacillus flavithermus biofilms. Appl Environ Microbiol 81(15):5115–5122. https://doi.org/10.1128/AEM.01037-15

Soni KA, Nannapaneni R (2010) Bacteriophage significantly reduces listeria monocytogenes on raw Salmon fillet tissue. J Food Prot 73(1):32–38. https://doi.org/10.4315/0362-028X-73.1.32

Speranza B, Corbo MR, Sinigaglia M (2011) Effects of nutritional and environmental conditions on *Salmonella* sp. Biofilm formation. J Food Sci 76(1):M12-6. https://doi.org/10.1111/j.1750-3841.2010.01936.x

Srey S, Jahid IK, Ha S-D (2013) Biofilm formation in food industries: a food safety concern. Food Control 31(2):572–585. https://doi.org/10.1016/j.foodcont.2012.12.001

Sriram MI, Kalishwaralal K, Deepak V, Gracerosepat R, Srisakthi K, Gurunathan S (2011) Biofilm inhibition and antimicrobial action of lipopeptide biosurfactant produced by heavy metal tolerant strain Bacillus cereus NK1. Colloids Surf B: Biointerfaces 85(2):174–181. https://doi.org/10.1016/j.colsurfb.2011.02.026

Steenackers H, Hermans K, Vanderleyden J, De Keersmaecker SCJ (2012) Salmonella biofilms: an overview on occurrence, structure, regulation and eradication. Food Res Int 45(2):502–531. https://doi.org/10.1016/j.foodres.2011.01.038

Stewart PS, Franklin MJ (2008) Physiological heterogeneity in biofilms. Nat Rev Microbiol 6(3):199–210. https://doi.org/10.1038/nrmicro1838

Swar S, Máková V, Stibor I (2020) The covalent tethering of Poly(ethylene glycol) to Nylon 6 surface via N,N′-Disuccinimidyl carbonate conjugation: a new approach in the fight against pathogenic bacteria. Polymers 12(10):2181. https://doi.org/10.3390/polym12102181

Tang L, Pillai S, Revsbech NP, Schramm A, Bischoff C, Meyer RL (2011) Biofilm retention on surfaces with variable roughness and hydrophobicity. Biofouling 27(1):111–121. https://doi.org/10.1080/08927014.2010.544848

Teh KH, Flint S, Palmer J, Andrewes P, Bremer P, Lindsay D (2014) Biofilm – an unrecognised source of spoilage enzymes in dairy products? Int Dairy J 34(1):32–40. https://doi.org/10.1016/j.idairyj.2013.07.002

Turonova H, Briandet R, Rodrigues R, Hernould M, Hayek N, Stintzi A, Pazlarova J, Tresse O (2015) Biofilm spatial organization by the emerging pathogen campylobacter jejuni: comparison between NCTC 11168 and 81-176 strains under microaerobic and oxygen-enriched conditions. Front Microbiol 6:709. https://doi.org/10.3389/fmicb.2015.00709

Van Houdt R, Michiels CW (2010) Biofilm formation and the food industry, a focus on the bacterial outer surface. J Appl Microbiol 109(4):1117–1131. https://doi.org/10.1111/j.1365-2672.2010.04756.x

Varatharaj A, Thomas N, Ellul MA, Davies NWS, Pollak TA, Tenorio EL, Sultan M, Easton A, Breen G, Zandi M, Coles JP, Manji H, Al-Shahi Salman R, Menon DK, Nicholson TR, Benjamin LA, Carson A, Smith C, Turner MR et al (2020) Neurological and neuropsychiatric complications of COVID-19 in 153 patients: a UK-wide surveillance study. Lancet Psychiatry 7(10):875–882. https://doi.org/10.1016/S2215-0366(20)30287-X

Vázquez-Sánchez D, Habimana O, Holck A (2013) Impact of food-related environmental factors on the adherence and biofilm formation of natural Staphylococcus aureus isolates. Curr Microbiol 66(2):110–121. https://doi.org/10.1007/s00284-012-0247-8

Veluz GA, Pitchiah S, Alvarado CZ (2012) Attachment ofSalmonella serovars andListeria monocytogenes to stainless steel and plastic conveyor belts. Poult Sci 91(8):2004–2010. https://doi.org/10.3382/ps.2011-01689

Vestby LK, Møretrø T, Langsrud S, Heir E, Nesse LL (2009) Biofilm forming abilities of Salmonellaare correlated with persistence in fish meal- and feed factories. BMC Vet Res 5(1):20. https://doi.org/10.1186/1746-6148-5-20

Vilariño MV, Franco C, Quarrington C (2017) Food loss and waste reduction as an integral part of a circular economy. Front Environ Sci 5:21. https://doi.org/10.3389/fenvs.2017.00021

Vogeleer P, Tremblay YDN, Jubelin G, Jacques M, Harel J (2016) Biofilm-forming abilities of Shiga toxin-producing Escherichia coli isolates associated with human infections. Appl Environ Microbiol 82(5):1448–1458. https://doi.org/10.1128/AEM.02983-15

Wang X, Tao X, Xia X, Yang B, Xi M, Meng J, Zhang J, Xu B (2013) Staphylococcus aureus and methicillin-resistant Staphylococcus aureus in retail raw chicken in China. Food Control 29(1):103–106. https://doi.org/10.1016/j.foodcont.2012.06.002

Wang R, Kalchayanand N, King DA, Luedtke B e, Bosilevac JM, Arthur TM (2014) Biofilm formation and sanitizer resistance of Escherichia coli O157:H7 strains isolated from "high event period" meat contamination. J Food Prot 77(11):1982–1987. https://doi.org/10.4315/0362-028X.JFP-14-253

Wang H, Cai L, Li Y, Xu X, Zhou G (2018) Biofilm formation by meat-borne Pseudomonas fluorescens on stainless steel and its resistance to disinfectants. Food Control 91:397–403. https://doi.org/10.1016/j.foodcont.2018.04.035

Weber M, Liedtke J, Plattes S, Lipski A (2019) Bacterial community composition of biofilms in milking machines of two dairy farms assessed by a combination of culture-dependent and –independent methods. PLoS One 14(9):e0222238. https://doi.org/10.1371/journal.pone.0222238

Whitehead KA, Verran J (2015) Formation, architecture and functionality of microbial biofilms in the food industry. Curr Opin Food Sci 2:84–91. https://doi.org/10.1016/j.cofs.2015.02.003

Winkelströter LK, Gomes BC, Thomaz MRS, Souza VM, De Martinis ECP (2011) Lactobacillus sakei 1 and its bacteriocin influence adhesion of listeria monocytogenes on stainless steel surface. Food Control 22(8):1404–1407. https://doi.org/10.1016/j.foodcont.2011.02.021

Wirtanen G, Salo S, Helander IM, Mattila-Sandholm T (2001) Microbiological methods for testing disinfectant efficiency on pseudomonas biofilm. Colloids Surf B: Biointerfaces 20(1):37–50. https://doi.org/10.1016/S0927-7765(00)00173-9

Xue T, Chen X, Shang F (2014) Short communication: effects of lactose and milk on the expression of biofilm-associated genes in Staphylococcus aureus strains isolated from a dairy cow with mastitis. J Dairy Sci 97(10):6129–6134. https://doi.org/10.3168/jds.2014-8344

Yamada Y, Guillemette R, Baudoux A-C, Patel N, Azam F (2020) Viral attachment to biotic and abiotic surfaces in seawater. Appl Environ Microbiol 86(3):e01687-19. https://doi.org/10.1128/AEM.01687-19

Yang Y, Mikš-Krajnik M, Zheng Q, Lee S-B, Lee S-C, Yuk H-G (2016) Biofilm formation of Salmonella Enteritidis under food-related environmental stress conditions and its subsequent resistance to chlorine treatment. Food Microbiol 54:98–105. https://doi.org/10.1016/j.fm.2015.10.010

Yang Y, Hoe YW, Zheng Q, Chung H-J, Yuk H-G (2017) Biofilm formation by salmonella Enteritidis in a simulated liquid egg processing environment and its sensitivity to chlorine and hot water treatment. Food Control 73:595–600. https://doi.org/10.1016/j.foodcont.2016.09.008

Yuan L, Sadiq FA, Burmølle M, Liu T, He G (2018) Insights into bacterial Milk spoilage with particular emphasis on the roles of heat-stable enzymes, biofilms, and quorum sensing. J Food Prot 81(10):1651–1660. https://doi.org/10.4315/0362-028X.JFP-18-094

Zarei M, Maktabi S, Ghorbanpour M (2012) Prevalence of *Listeria monocytogenes* , *Vibrio parahaemolyticus* , *Staphylococcus aureus* , and *Salmonella* spp. in seafood products using multiplex polymerase chain reaction. Foodborne Pathog Dis 9(2):108–112. https://doi.org/10.1089/fpd.2011.0989

Zezzi do Valle Gomes M, Nitschke M (2012) Evaluation of rhamnolipid and surfactin to reduce the adhesion and remove biofilms of individual and mixed cultures of food pathogenic bacteria. Food Control 25(2):441–447. https://doi.org/10.1016/j.foodcont.2011.11.025

Zhu Y, Li C, Cui H, Lin L (2020) Feasibility of cold plasma for the control of biofilms in food industry. Trends Food Sci Technol 99:142–151. https://doi.org/10.1016/j.tifs.2020.03.001

Ziuzina D, Boehm D, Patil S, Cullen PJ, Bourke P (2015) Cold plasma inactivation of bacterial biofilms and reduction of quorum sensing regulated virulence factors. PLoS One 10(9):e0138209. https://doi.org/10.1371/journal.pone.0138209

Zottola EA, Sasahara KC (1994) Microbial biofilms in the food processing industry—should they be a concern? Int J Food Microbiol 23(2):125–148. https://doi.org/10.1016/0168-1605(94)90047-7

Resilience and Adaptability of Biofilm

5

Mohan Das, Sayantan Santra, Shibasis Mukherjee, Devalina Khamaru, Sarobi Das, and Rintu Banerjee

Abstract

Antimicrobial resistance is a major concern for medical and agro-food industry, as frequent exposure to antimicrobial drugs/chemicals makes pathogens resistant. Bacteria in biofilms are especially resilient, as they are embedded in a protective extracellular matrix that shields them from antimicrobials. Quorum Sensing and other chemical messengers play a key role in density-dependent decision-making within large groups of microbial cells. A range of innovative experimental methodologies and procedures have been developed to explore metabolic interactions, phylogenetic connections, and competition among biofilm members. This chapter discusses the biochemistry and molecular biology of microbial biofilm formation followed by understanding how these biopolymeric microenvironment provide resistance against antimicrobial agents. Moreover, the classical and novel strategies for controlling this biofilm have been discussed. Hence, understanding this intricate relationship among similar or cross microbial species across space and time may direct novel strategies to investigate and influence microbial communities.

M. Das (✉) · S. Santra · D. Khamaru · R. Banerjee
Microbial Biotechnology and Downstream Processing Laboratory, Agricultural and Food Engineering Department, Indian Institute of Technology Kharagpur, Kharagpur, West Bengal, India
e-mail: mohan@agfe.iitkgp.ac.in

S. Mukherjee
The Agricultural Biotechnology Division, School of Agriculture and Rural Development, Ramakrishna Mission Vivekananda Educational and Research Institute, Kolkata, India

S. Das
Department of Biology, University of Texas, Arlington, TX, USA

T. Sarkar et al. (eds.), *Biofilm Applications to Revolutionize Food Technology*,
https://doi.org/10.1007/978-3-031-85205-3_5

Keywords

Biofilm · Biofilm adaptability · Cell-to-cell communication · Extracellular polysaccharides · Sociomicrobiology

5.1 Introduction

The concept that microorganisms can coexist and form biofilms is a long-established one, tracing its origins back to the early foundational studies by Antonie van Leeuwenhoek. However, the existence of this microbial colony in self-secreted matrix composed of diverse materials, such as glycol-proteins, glycol-lipids, and mono/polysaccharides, was ignored for long time (Santos et al. 2018). Until recently, microbes were considered unsocial beings and molecular machines too simplistic to exhibit the complex social behaviors typically associated with larger, more charismatic species. However, studies from the last few decades have demonstrated that microorganisms form extensive, interactive communities with the ability to regulate a wide range of behavioral patterns which is similar to interaction and cooperation as observed in animal and plant kingdom. Sociomicrobiology is an evolving research area exploring the complex behavioral pattern of microbes and its underlying mechanism of socializing. In general, a community refers to a social group, regardless of its size, where members share a common locality, characteristics, or interests. They establish communication platforms and often have a shared heritage. In the microbial world, this concept promotes the integration, fulfillment, and promotion of beneficial features that are superior to living alone (Sonawane et al. 2022). The microbial interactions between different species can have positive, neutral, or negative effects on the individuals within the group. The cumulative effect of these interactions determines the structure and role of the microbial community within an ecosystem. Microorganisms in nature may be found in two distinct forms: planktonic, which refers to bacterial populations that are freely floating, and biofilm, which describes the clustering and adherence of microbial cells to surfaces (Gupta et al. 2016). The transition from free planktonic to community structure offers resistance to environmental stress factors, improved nutrient acquisition, protection from antimicrobial agents, and distribution of labor for metabolite production. Prior research has demonstrated that bacteria have the ability to adjust to varying environmental circumstances by enhancing their fitness through the formation of biofilm structures. For instance, the presence of microbial biofilm in the rhizosphere demonstrated superior soil fertility and plant development in comparison to the planktonic form. The Allee effect posits that there is an inherent advantage to clustering, as indicated by a positive association between the size or density of a community and the average fitness of individuals. However, cooperation often incurs costs, making it challenging to understand its evolution in Darwinian terms, especially when the burden significantly impacts the individual reproductive fitness of those who cooperate. Interestingly, during cooperative interactions, there can be individual known as "cheaters" who gain the advantages of collaboration such as acquisition of "public goods" without having to pay the costs. These cheaters may gradually become more dominant in the community because they have better individual fitness relative to those who cooperate (Rendueles and Ghigo 2015). This microbial biofilm consists of 3D structures that contain water channels, which aid in the absorption of nutrients and oxygen and the elimination of waste products. In addition, the biofilm's varied microenvironments facilitate intricate microbial interactions, including chemical and electrochemical cell-to-cell communication, and facilitate increased horizontal gene transfer. Biofilms are abundant in diverse settings, spanning from living creatures to the food and medical sectors, and may even be observed in severe environmental conditions. According to National Institutes of Health (NIH) report, around 65% of acute infections in people are caused by bacteria that create biofilms (Rather et al. 2021a). Bacteria within biofilms exhibit heightened resistance to both microbicidal agents and the host's immune system. This chapter describes biochemistry and molecular basis of microbial biofilm formation and its importance in agro-food industry. Additionally, diverse treatments for controlling biofilm growth as well as microbial defense mechanism for combating these chemical/biological agents have been discussed.

5.2 Composition, Architecture and Cell-to-Cell Communication in Biofilm Formation

Biofilms are elaborate groups of bacteria which attach to surfaces and interfaces and are covered in an extracellular matrix (ECM) made of proteins, polysaccharides, and nucleic acids that they generate on their own (Huang et al. 2014). These common place patterns are important in many different contexts, including industrial settings, natural ecosystems, and even in human bodies. The constantly changing procedure of biofilm formation, which involves complex cell-to-cell communication, is what enables bacteria to transform from single, free-floating entities into multicellular organisms. Initial multiple stages of biofilm growth are orchestrated by this communication. It is started with the initial attachment and continues through surface colonization, maturation, and distribution (Rumbaugh and Sauer 2020).

As cell-to-cell communication is one of the major architectural causes of biofilm formation, Quorum sensing (QS) is one of the most researched methods of cell-to-cell communication in biofilm production. Using this method, bacteria are able to assess the density of their population by identifying particular signaling molecules identified as autoinducers (AIs). The quantity of AIs in the environment around them grows with the bacterial population. These populations' densities are coordinated how to respond through combinations of gene expression when the concentration of AIs reaches a crucial level. This behavior may have a consequence on a number of biofilm development factors, such as:

Physical Attachment

Bacterial biofilms can be viewed as "composites of colloids embedded in a cross-linked polymer gel," whereas EPS secreted by bacterial can be considered as "molecular glue" for guiding the initial attachment to the surface (Karimi et al. 2015). QS can regulate the production of adhesins, molecules that allow bacteria to adhere to surfaces with each other. This facilitates the initial steps of biofilm formation, and adhesins play a dominant role in quorum sensing, a sophisticated communication mechanism among bacteria. There are several types of adhesins in the context of quorum sensing found in all the groups of bacterial species based on their functions and environmental utilities. This adaptability allows bacteria to thrive in diverse conditions. Several factors influence how strongly bacteria attach to surfaces and form biofilms. Rougher surfaces offer more area and protection for microbes, while hydrophobic surfaces are generally more favorable for bacterial attachment compared to hydrophilic ones. Bacterial characteristics like surface properties, presence of appendages (fimbriae and flagella), and EPS production also play a significant role. The attachment process itself involves various forces like electrostatic interactions, hydrogen bonds, and London dispersion forces. These forces contribute to the overall stability and strength of the biofilm matrix. Importantly, EPS allows for irreversible attachment to various surfaces, solidifying the biofilm's hold. Till now, eight types of physical attachment structures mostly related to cytoskeleton and membrane were found in bacteria which were responsible for cell-to-cell communication and biofilm formation, that is, curli fibers, fibrils, fimbriae, flagella, membrane vesicles, tubular spinae, and tunneling nanotubes.

Among all the structures, Pilli is the major structure by which bacteria directly interact, distinct from flagella. These interactions, mediated by specific protein-protein interactions between pilin subunits and host receptors, facilitate biofilm formation, enhance communication by potentially concentrating signaling molecules, and enable horizontal gene transfer of genetic material among cells of the same or different species (Chang 2018). Pili can also play a role in adhesion to host tissues during infection (Abby and Rocha 2017). Examples include type I pili in *E. coli*, which interact with host receptors through dedicated usher proteins, and type IV pili in *Neisseria* sp. and *Pseudomonas* sp., known for their role in DNA uptake during conjugation. On the other hand, flagella are crucial for bacterial motility and virulence. They propel bacteria toward favorable environments and facilitate attachment to host cells, initiating colonization. Flagella come in various arrangements (monotrichous, amphitrichous, and peritrichous), affecting motility patterns. Studies with non-motile mutants confirm the strong link between flagella and disease progression (Yuan et al. 2024). Beyond motility, flagella also act as surface sensors, allowing bacteria to respond to environmental changes and initiate surface-dependent functions through biofilm formation. This type grants swimming motility, like a boat propeller. Examples include *Vibrio* spp. (cholera) and *Pseudomonas* spp. during planktonic growth. *Campylobacter jejuni* (causes foodborne illness) is an example of amphitrichous flagellary action. "run and tumble" motility, is an example of peritrichous flagella actions include *Salmonella enterica* (food poisoning), *E. coli* (intestinal bacteria, can cause foodborne illness), and *Bacillus subtilis* (soil bacteria) (Khan and Scholey 2018; Rahman et al. 2022). Another important structure involved in communication and biofilm production is tunneling nanotubes (TNTs), which act as bridges between bacteria, facilitating the exchange of genetic material and cytoplasmic contents. Studies on species like *E. coli*, *Staphylococcus aureus*, and *Bacillus subtilis* highlight their role in bacterial communication. The nanotubes were mostly function in the exchange of nuclear materials (plasmids) and responsible for gene transfer actions also. Membrane vesicles are another important structure involved in the connection to recipient cells and performing the supplying of signal molecules preferably, which was observed in the species of *Vibrio* and *Pseudomonas* genus (Pérez-Cruz et al. 2015; Brameyer et al. 2018). Due to the different membrane architecture of gram-positive and gram-negative bacteria, differences were observed in membrane vesicle structure in both groups (Toyofuku et al. 2019). A unique attribute of membrane vesicles These MVs showed various signaling strategies like—in *Pseudomonas* sp. quinolone signal (PQS), it is a good example of ligand-mediated signaling in pathogenic *Pseudomonas aeuroginosa* for managing several unmanageable conditions (Häussler and Becker 2008). Similarly, CAI-1, a long-chain amino ketone, is secreted by the pathogen *Vibrio harveyi* through outer membrane vesicles (Brameyer et al. 2018). Additionally, N-acyl-homoserine lactones, as well as lipase, chitinase, and alkaline phosphatase, are also delivered by membrane vesicles to the marine pathogen *Vibrio shilonii* (Li et al. 2016). These vesicles are the principal component for the release of long-chain C16-N-(hexadecanoyl)-L-homoserine lactone, which is fundamental for cell-to-cell communication. Even its presence was observed in soil bacteria such as *Paracoccus denitrificans* PD1222. Many gram-negative bacteria, such as *Spinomonas maritime,* contain an appendage called spinae. These appendages are mainly formed at low salt, neutral pH, and slightly raised temperature (Easterbrook and Sperker 1982). These external structures act as cementing material with sediments as well as pathogens. It is captivating that *S. maritime* cells are cultured in low salinity, signifying the potential for function in extended cell-to-cell communication (Bayer and Easterbrook 1991). This type of structure shows the early phases of multicellular concepts in addition to facilitating the overall findings. In bacteria such as *Myxococcus xanthus*, these bigger and more solid forms of pili serve as necessary for adhesion to surfaces and cell-to-cell contact, hence fostering social behaviors. Fibrils can perform unique duties in addition to this main function. For instance, *Agrobacterium tumefaciens* produces cellulose fibrils that aid in their connection to host cells and contribute to the development of tumors. Likewise, to the way Bordetella's hemagglutinin filament contributes to its pathogenicity, *Salmonella typhimurium*'s fibrillar tendrils are required for internalization into host cells (Matthysse 1983). These diverse instances illustrate the versatility fibrils are in supporting bacterial interactions. The protein extensions on bacteria called fimbriae, which resemble hairs, particularly vital to the early phases of infection (Miller and Bassler 2001). Fimbriae, which imitate fibrils but differ in size and structure (0.5–10 micrometers long, 2–8 nanometers broad), function as adhesins and assist bacteria in binding to host cells (Miller and Bassler 2001). For bacterial accumulation, settlement, and biofilm formation—all of which contribute to pathogenesis—this attachment is a prerequisite. Originally, hemagglutination patterns—the capacity of the mannose monosaccharide to prevent adhesion—and shape led to the splitting of fimbriae into seven distinct groups (categories I to VI and type F), whereas Enterobacteriaceae type of bacteria has been the subject of extensive study due to the prevalence of these seven kinds, such as *Salmonella* spp., *E. Coli*, and *Klebsiella pneumoniae* (Clegg and Gerlach 1987). Subsequent classifications of fimbriae, however, revealed three primary assembly pathways: chaperone-usher (CU), nucleation-precipitation (N/P), and type IV fimbriae. These categories were based on genetic similarity and serological examinations. Comprehensive instances showed of type I fimbriae in *Salmonella enteritidis* include SEF 14 and SEF 21 (Naughton et al. 2001). Therefore, fimbriae's fall out in tendencies for latching to host cell receptors aids in host-pathogen proximity and has likelihood to cause serious infections and diseases like salmonellosis.

Matrix Production and Maturation

Liking to QS, another promising approaches which contribute in bioflim formation is the matrix production, QS can control the production of exopolysaccharides (EPS), the primary components of the biofilm matrix. EPS production contributes to cell-to-cell aggregation, nutrient acquisition, and protection from environmental stresses. A collection of bacteria, algae, fungus, and protozoa encased in a matrix made of a blend of polymeric substances, mostly polysaccharides, known as extracellular polymeric material (EPS) is called a biofilm. These biopolymers are habitat to more than 99% of all Earth's microbes. All microbial gathering necessitates the establishment of biofilms for them to exist at all (Sutherland 2001). Biofilms are complex structures formed by bacteria in various environments. The slimy layer composed of sugar and protein moieties is called extra polymeric substances or EPS (Vu et al. 2009). The exact physical, chemical, and molecular characteristics of this polymer varies based on the composition of bacterial species (Mayer et al. 1999). Its width can vary significantly, ranging from a mere 10–30 nanometers in some species to a more substantial 0.2–1.0 micrometers in others. This obscure structure is composed of various biomolecules like polysaccharides, proteins, nucleic acids, glycoproteins, and even several types of lipids. Predominantly, five carbon or six carbon sugars were involved in the skeleton of EPS. Several organic acids like acetate, fumarate, or malate were found in this structure (Harimawan and Ting 2016). The enthralling feature of EPS is the distinct variation in the linkages and combination of polysaccharides and proteins in species to species. Homopolysaccharides and heteropolysaccharides are the major groups of polysaccharides found in EPS (Monsan et al. 2001). D-glucose, D-galactose, L-fructose, L-rhamnose, D-glucuronic acid, L-guluronic acid, and D-mannuronic acid were the major examples of polysaccharides and its derivatives found in EPS. The precise measures of these sugar units, including the type of linkage and branch patterns, ultimately settle on the structure of the resulting polysaccharide. Moreover, heteropolysaccharides like pyruvates, succinates, and formates were found in EPS functioning several ways. Alginate, a well-known example of heteropolysaccharide with an asymmetrical structure composed of D-mannurosyl and L-guluronosyl substrates. *Pseudomonas aeruginosa* and *Azotobacter vinelandii* were the major example which produce or take up these types of molecules from environment (De Vuyst and Degeest 1999). The proteins found in EPS are high in molecular weight (>10,000 daltons). Interestingly, the amino acid compositions of proteins in EPS vary, for example, in the extracellular protein layer of *Bacillus subtilis* contains both L- and D-amino acids. Conversely, nutrient concentration should be sufficient enough to generate enhanced EPS. *Escherichia* spp. and *Pseudomonas* spp. are the typical bacterial examples that have been shown to convert a significant portion of available glucose into EPS under low ammonium conditions. On the other hand, high nitrogen content seems to favor the production of extracellular proteins. *Pseudomonas* spp. and *Rhodococcus* spp. have been observed to increase their production of these proteins when grown in media rich in ammonium salts (Gandhi et al. 1997).

The intricate mechanisms governing EPS production in microbes remain an area of active research. While the enzymes responsible for creating EPS precursors appear to be under separate control from the genes involved in EPS molecule biosynthesis itself (with some exceptions). Recent studies are shedding light on the regulatory landscape. *Xanthomonas campestris* offers a unique case where Tang et al. (1990) identified a single genomic region (rpf) that controls both EPS production and the enzymes needed to convert precursor molecules. This rpf cluster regulates various extracellular enzymes, including cellulase, polygalacturonate lyase, amylase, and protease. Interestingly, under harsh environmental conditions, taxonomically diverse microorganisms can produce similar or identical EPS. Pseudomonas spp. and *Azotobacter* spp. exemplify this phenomenon. Another research suggests that in both *Pseudomonas aeruginosa* and *Azotobacter vinelandii*, a similar gene cluster controls the biosynthesis and secretion of EPS. This cluster includes genes like algA (encoding GDP-mannose pyrophosphorylase), algD (coding for GDP-mannose dehydrogenase), and algE (potentially involved in alginate export). However, Pseudomonas spp. possesses additional chromosomal genes (algC and algK) that are not found in *Azotobacter* spp. These genes, encoding phosphomannomutase (algC) and a protein with unknown function (algK), are believed to contribute to the specific characteristics of alginate produced by *Pseudomonas aeruginosa*.

In contrast to chromosomal control, certain microbes (*Escherichia* spp., *Streptococcus* spp., and *Lactobacillus* spp.) have been found to exhibit positive EPS regulation by megaplasmid. Megaplasmids show the expressions of gene clusters having immense assembly of EPS, and these clusters can even be exchanged between different bacterial strains. An interaction existed between genetic and environmental variables in influencing bacterial extracellular polymeric substance synthesis. The earlier studies on the regulation system of EPS indicated that *Pseudomonas aeruginosa*, a well-known pathogen, exemplifies the importance of QS. It utilizes two key QS systems, LasI/LasR and RhlI/RhlR, similar to the LuxI/

LuxR (Miller and Bassler 2001; Waters and Bassler 2005). LsrR/LsrK system is a key player involved in *Escherichia coli*, controlling various cellular functions. Mutant strains of *E. coli* pointed out distorted cell fimbriae, matrix structure, and biofilm thickness compared to the wild type. Recent research highlighted the presence of a QS system (AfeI/AfeR) in the *Acidithiobacillus ferrooxidans* (Ruiz et al. 2008). Nutrient starvation showed a positive impact on enhancing lipopolysaccharide production and afeI gene transcription in *A. ferrooxidans* (Farah et al. 2005).

Dispersion

Dispersion is typically considered the final stage of biofilm formation, indicating the moment when previously connected cells detach from the biofilm and transition to a planktonic growth state. If the dispersion makes the biofilm residents vulnerable to environmental stress, then the question is why they leave. The answer is simple—to avoid overcrowding and degraded biofilm microcolony. As the biofilm grows larger, the steepening in chemical gradient makes it harder for all members of the community to access resources equally. This results in the development of subgroups within the community. This dispersion can be native dispersion, that is, triggered by self-synthesized molecules or chemical gradient and environmentally triggered dispersion. In case of native dispersion, it has been observed that cells residing near the center of the biofilm microenvironment initiate this process and leave the matrix by a small opening. The diameter and height of biofilm have a strong correlation with biofilm dispersion, for example, microscopic studies of biofilms grown in flow cells under continuous flow conditions revealed that dispersion events within the biofilm's interior only took place when the biofilms reached a minimum diameter of 40 μm and a height of 10 μm (Rumbaugh and Sauer 2020). There are several inducers playing pivotal roles in biofilm dispersion, which are secreted by microbes itself and removed (either by transporters or diffusion) from biofilm when size is small, but with increasing size of biofilm these inducers accumulate in biofilm. One example of such autoinducers is *cis*-2-decenoic acid which is a fatty acid signal molecule first identified in *P. aeruginosa*, is responsible for controlling more than 666 genes related to motility, chemotaxis, cell attachment, EPS synthesis and secretion. It has been demonstrated that, *cis*-2-decenoic acid (concentration > 2.5 nM) are able to induce biofilm dispersion in *P. aeruginosa* PAO1 biofilms as well as different groups of gram-positive and gram-negative bacteria (Davies and Marques 2009). In addition to this, pyruvate depletion is another triggering factor for biofilm dispersion; for example, the depletion of pyruvate led to a decrease in biofilm biomass in vivo and increased the effectiveness of tobramycin antibiotic in case of *P. aeruginosa* biofilm (Goodwine et al. 2019). Anaerobic respiration in the interior of the biofilm releases nitric oxide, may react with superoxide, and generate toxic $ONOO^-$ radical, leading to cellular damage and ultimately resulting loosening of biofilm matrix (Rumbaugh and Sauer 2020). In contrast to native dispersion, environmental triggered dispersion is triggered by environmental factors. In both cases, the c-di-GMP level plays a pivotal role as a messenger for transforming sessile form of life to planktonic form. These complex metabolic cascades finally lead to expression of some endonuclease genes such as *endA* (degrades matrix-entrapped DNA) and glycoside hydrolases (*pslG* and *pelA*). These PslG and PelA are able to degrade two vital components of biofilm matrix, that is, Psl (neutral pentasaccharide) and Pel (glucose-rich polysaccharide), respectively (Cherny and Sauer 2020). As a result of this, the viscoelastic structure of biofilm eventually degrades, and cells residing in the biofilm return to their planktonic state.

5.3 Beyond Quorum Sensing: Alternative Communication Pathways in Biofilm Microenvironment

While QS plays a crucial role in biofilm formation, it is not the only key player in this process. Bacteria utilize a diverse set of communication pathways to coordinate their activities within a biofilm. Here are some key molecules recently studied in the signaling, chemical, and other pathways having important contributions to biofilm formation.

Cyclic Di-guanosine Monophosphate (c-di-GMP)

It is one of the widely studied second messenger molecules that plays a critical role in regulating biofilm formation in various bacterial species. High c-di-GMP levels promote biofilm formation by activating EPS production and motility inhibition (Valentini and Filloux 2016). This molecule mostly interacts with the outer environment and manages to build intra-cellular connections in some of the species (Römling et al. 2013; Jenal et al. 2017). It acts as a switching system inside the cell by regulating two key enzymes, that is, diguanylate cyclase and phosphodiesterase (Simm et al. 2004; Ryjenkov et al. 2005; Chan et al. 2004). It should be noted that the breakdown of GTP molecules to c-di-GMP and the energy released during this reaction stimulates biofilm formation (Römling and Amikam 2006; Hengge 2009). Recent advancements in biofilm research clearly demonstrate a proportional relation between the biofilm formation and c-di-GMP concentration through a protein intervention, *n*amely, HD-GYP (Dow et al. 2006). Understanding this organization offers precious insights into bacterial functions and potentially paves the colonization behavior of bacteria (Gupta et al. 2014). In essence, c-di-GMP acts as a master switch, regulating various aspects of bacterial behavior by controlling its concentration within the cell. Understanding this system offers valuable insights into how bacteria function and potentially paves the way for developing new strategies to control bacterial growth and behavior.

Small Regulatory RNAs (sRNAs)

These non-coding RNAs can regulate gene expression and influence biofilm development. sRNAs can act by modulating the expression of genes involved in EPS production, motility, and other biofilm-related processes. An extensive study reported a group of non-coding RNA in both gram-positive and gram-negative bacteria. In the case of gram-negative bacteria, four different sRNAs were reported from *E. coli*, that is, RydC, McaS, CsrB, and CsrC; on the other hand, *Pseudomonas aeuroginosa* reported two non-coding RNAs, that is, SrbA and ErsA. One non-coding RNA was detected on *Salmonella typhi*, AsfD. *Staphylococcus aureus* and *Bacillus subtilis* are two representative model gram-positive bacteria that reported RsaA and RNAIII sRNAs, respectively. Conversely, *Enterococcus faecalis* is also reported to be AswalR non-coding RNA (Mitra and Mukhopadhyay 2023). A case study investigated a small regulatory molecule (sRNA0426) in the bacterium *Streptococcus mutans*. The results suggest that sRNA0426 plays a significant role in biofilm formation, the sticky community's bacteria form on surfaces. Higher levels of sRNA0426 were linked to increased biofilm formation and exopolysaccharide (EPS) production, a key component of biofilms. Computational analysis suggested that sRNA0426 might influence metabolic pathways, particularly those related to carbon utilization, which is essential for bacterial growth. Five specific genes potentially targeted by sRNA0426 were identified. Three of these genes showed increased expression along with sRNA0426, while one showed a decrease. Overall, this study highlights the potential importance of sRNA0426 in regulating biofilm formation in *S. mutans* (Yin et al. 2020). Understanding this regulatory mechanism can provide valuable insights into controlling biofilm development in this bacterium, which is associated with dental caries.

Intercellular Signaling Molecules

Bacteria can communicate using a variety of other signaling molecules besides AIs. These molecules can be peptides, nucleotides, or even gases like nitric oxide. Their specific roles in biofilm formation are still under investigation. Second, messenger signaling is not a linear pathway. Often, multiple-second messengers work together in a network, with their interactions determining the final cellular response. Additionally, the activity of enzymes that synthesize and degrade these molecules plays a crucial role in maintaining the appropriate signal strength within the cell. Some common second messenger signaling molecules are cAMP: In some bacteria, cAMP signaling plays a complex role. It might promote or inhibit biofilm formation depending on the specific species and environmental conditions. c-di-GMP signaling is a well-charted territory compared to the relatively unexplored landscape of cAMP-CRP regulation (Liu et al. 2020). The cAMP-CRP pathway's influence on biofilm formation varies greatly across bacteria. Studies show it can be a promoter, like in *E. coli* and *Pseudomonas aeruginosa* (Müller et al. 2009; Sutrina et al. 2015), or an inhibitor, as seen in *Serratia marcescens* and *Vibrio cholerae* (Kalivoda et al.,

2013). Additionally, cAMP-CRP regulates diverse cellular functions, creating a link between these processes and biofilm development (Sutrina et al. 2015). Calcium ions act as versatile second messengers, influencing various cellular processes. In biofilm formation, they can trigger changes in cell surface properties, promoting adhesion and biofilm development. Calcium ions (Ca^{2+}) act as molecular glue and play a key role in maintaining biofilm integrity by preventing its dispersal in *Bacillus subtilis* (Nishikawa and Kobayashi 2021). In a case study, three unique strains of *Pseudomonas fluorescens* were used in this study to explore the effects of calcium (Ca^{2+}) on biofilm formation. For instance, there were notable differences in the strains' reactions to Ca^{2+}. The consequences of Ca^{2+} on biofilm formation and EPS creation were strain-specific, although it consistently decreased flocking mobility in all strains. Surprisingly all strains produced more EPS and formed biofilms at moderate Ca^{2+} concentrations (Yuan et al. 2024).

5.4 The Threat of Biofilms in Agro-Food Industry

Under the favorable condition, a wide variety of pathogenic microorganisms can create biofilms on food and their surfaces. One of the earliest reports on biofilms found in food was the adherence of *Salmonella* to various food-related surfaces (Duguid et al. 1966). Numerous studies lately detailed the bacterial assemblages of biofilms that develop on the outer layers of machines for processing used in different food enterprises (Guobjoernsdottir et al. 2005.; Gunduz and Tuncel 2006). A broad variety of ailments can be produced by chemicals which were ingested with food, a condition known as foodborne infections. According to the World Health Organization (WHO), environmental contaminants, such as those in the water, soil, or air, can contaminate food at any point from manufacturing to ingestion. Furthermore, biofilms may influence heat transfer devices by decreasing heat transmission and performance. Even in systems with elevated residual chlorine levels, biofilms may develop, which in turn increases energy use, mechanical obstruction, and metallic surface deterioration (Kumar and Anand 1998). The process of biofilm development in food industries is very intricate, involving the initial accumulation of organic compounds of food on tool surfaces and the preparation of a conditioning layer. Furthermore, this layer exhibits an attraction to microbes. In addition, certain microorganisms persist throughout the washing and sanitization process and subsequently commence proliferation. Finally, the development of mature biofilms is facilitated by the activation of genes and the process of quorum sensing. The creation of biofilms involves the interplay of several elements such as the characteristics of the substrate and cells, environmental conditions, and genetic control of microbes. These components determine whether the adherence of microbes to a substratum is reversible or irreversible and influence the establishment of microcolonies that eventually develop into an extensive biofilm (Shi and Zhu 2009).

Agricultural Sector

At now, the tactics for controlling microbes are not sufficiently effective to completely eliminate harmful microbes without negatively impacting the quality of the item. Pruning, chopping, cleaning, scrubbing, drainage, and packing are all common practices in the agricultural sector. These activities have been identified as the main causes of contamination. One of these essential control factors is packaging, since most people eat fruits and vegetables in their fresh form. To prevent further contamination, the final phase needs careful and frequent monitoring. Between 1990 and 2007, 12.3% of food poisoning cases were traced back to fresh products; 10% to inappropriate handling after harvesting, while the remaining cases were linked to the cultivation, packaging, transportation, or processing of fruit that was freshly picked. Intact cantaloupes infected with *Listeria monocytogenes* caused an infection surge in 2011. The occurrences are believed to be caused by the filthy state of the packaging yard. In addition, the bacteria had been identified in various other places such as the processing area, drying area, and basement drainage. Disease-causing organisms are believed to adhere strongly to hard-to-reach areas and create a biofilm before they penetrate into the food. Upon seeding and culture at 37 °C for 24 h, it was observed that undamaged green pepper interfaces did not exhibit any noteworthy development of *E. coli* O157:H7. In contrast, undamaged surfaces exhibited substantial proliferation and growth. According to reports, bacterial biofilm is resistant to disinfectants such as ozone, chlorine, and organic acids (Ölmez and Temur 2010). For example, even after being submerged with 20,000 ppm of chlorine, *E. coli* was found to have successfully grown to 6 log CFU/g of radish sprouts—a number that was far greater than what was originally recorded in those seeds. Research has shown that microbes may be discovered in radish sprouts on their outer layers as well as their roots. They are additionally prevalent in biofilms linked to the sprouts' tissues as they sprout (Fransisca et al. 2011). The presence of native microflora that can create biofilms is common in settings where fresh food is processed.

Dairy Sector

Because of its high perishability, milk is extremely susceptible to contamination by many different kinds of microbes. It is often believed that apparatus that has not been adequately handled and sterilized is the primary cause of tainted milk and dairy products (Jessen and Lammert 2003). *Enterobacter, Lactobacillus,* and *Streptococcus* are extensively reported species of bacteria that is commonly found in dairy products surroundings (Salo et al. 2006). One possible explanation is the fact that the dairy specimens included bacteria capable of forming biofilms. When comparing fresh and pasteurized milk, one may find that the former contains more thermostable *Streptococci* and *Bacillus* species. This might be due to biofilm infestation (Flint et al. 1997). Currently, several researches have been conducted that specifically examine the different types of bacteria found in dairy industries.

Numerous investigations were unable to find any link between rough edges or imperfections and the ability of bacteria to adhere. Having said that, studies have looked into the possibility of poisoning in dairy industries due to biofilms carrying *L. monocytogenes* in milking instruments. It was believed that the existence of single colonies of bacteria, as shown by scanning electron microscope, were predominantly associated with irregular surfaces after beneficial results had been acquired (Latorre et al. 2010). The microbial community in unpasteurized, refrigerated milk may grow biofilms on stainless steel equipment, according to a study. The variety of bacteria in milk makes this process possible, and the biofilms can include more than one species. The existence of such proteolytic cold-adapted microbes in a biofilm is worrisome and has the potential to undermine the purity of unprocessed milk and its products (Oliveira et al. 2019). Although other chemicals showed a significant influence on the adhesion, lactose did not affect the adherence across the two surfaces in comparison with the control poly (butylene succinate). The scientists hypothesized that, reduced adherence of bacteria for biofilm formation is possible because of repulsion of microbial cell surface and negatively charged proteins. Biofilm formed by many species, which can persist during washing and sterilization, are frequently formed by a wide range of bacteria on milk ultrafiltration and reverse osmosis filters. These biofilms can potentially contaminate dairy products (Anand and Singh 2013; Hassan et al. 2010). Researchers have also studied the existence of *L. monocytogenes* bacterial species within unpasteurized milk biofilms. They added specific isolates to that milk to see significant modifications in the biofilm residues, the chemistry, and the harmony of bacterial communities (Weiler et al. 2013). *Staphylococcus aureus* is the primary causative agent of both chronic as well as acute bovine mastitis, which is the greatest economically burdensome ailment for the global dairy sector (Budd et al. 2016). A key virulence feature in the development of mastitis is the ability of this bacterium to form biofilms (Gomes et al. 2016; Salimena et al. 2016). It often produces polysaccharide intercellular adhesin (PIA) and a number of proteins in conjunction with biofilm development (Oniciuc et al. 2016).

Seafood Industry

Despite its importance in a balanced diet, seafood poses a danger of food poisoning due to the many different types of bacteria, viruses, and parasites that can infect it (Iwamoto et al. 2010). Although almost half of these infections are caused by viruses, the majority of infections and fatalities are caused by bacteria. A constant coating of slime surrounds fish, providing anatomical, chemical, and physiological shields against potential external threats. Dermal mucus is a key entrance point for harmful microbes, the primary disease-causing agents in fish, despite the fact that mucus contains several antimicrobial components like immunoglobulins, agglutinins, lectins, lysins, and lysozyme (Benhamed et al. 2014). Several elements contribute to the intrinsic threat of certain marine food products compared to others. These variables include the habitat from where they originate, their dietary habits, the time of harvest, and the preparation and serving methods (Iwamoto et al. 2010). Therefore, every kind of seafood is vulnerable to becoming contaminated on its exterior or within its tissues due to the marine ecosystem. Disease-causing organisms from a human source can infect areas of cultivation when they come into contact with human waste. Bivalve mollusks have the ability to capture and strengthen infectious agents that naturally occur in the marine environment where they are harvested, such as *Vibrios*. They do this by filtering substantial amounts of saltwater while they eat. The primary cause linked to infection is the ingestion of seafood that is either raw or not fully cooked (Giaouris and Simões 2018). A number of investigations have examined the makeup of the bacteria present in seafood equipment used for processing. Researchers at four separate seafood-processing factories examined the microorganisms that clung to the processing apparatus both during and after sterilization and washing (Bagge-Ravn et al. 2003). The findings indicated that a significant number of bacteria frequently identified in seafood were discovered as well in the seafood-processing factories. Nevertheless, the caviar processing resulted in the prevalence of halo- and osmo-tolerant microbes, indicating a selection based on processing conditions. Furthermore, the prevailing microorganisms that stuck to surfaces after being cleaned were pseudomonads and yeasts, regardless of the microorganisms present throughout the processing stage (Bagge-Ravn et al. 2003). A further investigation found the microbes that stick to the materials used in processing industries of fish and shrimp (Guobjoernsdottir et al. 2005). Marine water is utilized in the seafood business rather than freshwater due to its financial benefits, making it equally significant to other constituents. Despite undergoing treatment using a combination chlorination and ultraviolet (UV) radiation system, analysis revealed that the treated seawater had been infected with *Vibrio* spp. This contamination had been attributed to the development of biofilm in the marine water transportation network during the course of processing. Several publications asserted the inability of chlorine to prevent the development as well as the maturation of biofilm was not caused by the pH's influence (Momba and Makala 2004). Studies have confirmed that water quality deteriorates after receiving treatment, mostly owing to the detachment of organisms from the biofilm in the circulation network. This detachment occurs because residual chlorine is not efficient in preventing it. This phenomenon is widespread in water supply networks. Additionally, the production of biofilm by *Vibrio harveyi*, a bacterium found in poisoned seafood, on various substrates, such as plastic, cement slabs, and steel coupons, was also assessed (Karunasagar et al. 1996). The microbial community that managed to survive the cleaning procedure of conveyor chains in salmon processing factories was primarily composed of gram-negative bacteria, with a notable prevalence of *Pseudomonas* spp. (Langsrud et al. 2016). Pathogens that have the ability to produce biofilms often inhabit specific varieties of shellfish, like oysters (Froelich and Oliver 2013). The adherence and resulting colonization of seafood materials by microorganisms might diminish the effectiveness of management strategies employed to manage them. Remarkably, research has demonstrated that the presence of *Listeria* and *Salmonella* on shrimp shells can actually increase their ability to withstand high temperatures, chlorine, and acidic conditions (Norhana et al. 2010).

Meat Industry

Although organic matter left behind in processed food can cause cross-contamination and provide a habitat for the growth of microorganisms and biofilm development, this is a growing issue among scientists (McLandsborough et al. 2006; Simões et al. 2010). With the global rise in the intake of meat, there is a consequent rise in distress and difficulties around food sanitation and security (Sofos and Geornaras 2010). Meat is commonly exposed to microbes after the animal is slaughtered and undergoes additional processing. Research on the adherence of bacteria to meat contact points and its potential to cause food contamination has been done, and in vivo, biofilms were identified in meat-processing configurations (Giaouris et al. 2012). Biofilms may be formed on multiple substrates by *Salmonella*, a major foodborne disease, particularly in poultry (Giaouris 2015). The capacity mentioned is tightly controlled through a well-synchronized and intricate system of gene expression and protein function. It may be affected by factors such as serotype, cell surface features, ambient circumstances, and physical and chemical characteristics associated with these interfaces (Schonewille et al. 2012; Simm et al. 2014). Nevertheless, the shape and chemistry of such fimbriae (curli) may vary greatly across various strains and the environmental circumstances in which they exist (Wang et al. 2013). Biofilms likely have a significant impact upon the ability of salmonellae to survive and thrive under adverse environmental circumstances, for instance, in livestock slaughterhouses as well as processing factories (Giaouris 2015; Vestby et al. 2009). *L. monocytogenes* is a notable foodborne bacterium that induces listeriosis, a very infrequent yet potentially fatal infection that mostly impacts those with weakened immune systems. This organism has the ability to adhere to and establish biofilms on processing instruments. Its resilience in meat-processing facilities makes it challenging to eradicate, hence increasing the risk of food contamination (Valderrama and Cutter 2013). It remains present in meat-processing industries for extended periods, perhaps spanning many years or even decades. This persistence is a significant contributing element to its spreading and has been identified as the cause of several occurrences of listeriosis in humans (Ferreira et al. 2014). While a few scientists have observed that enduring *L. monocytogenes* strains exhibit distinct traits that could aid in their capacity to endure, such as the capacity to stick to solid surfaces, form biofilms, and more effectively adjust to stressful circumstances, other scientists have not discovered significant variations in the physical features between persistent and nonpersistent strains that could contribute to persistence (Ferreira et al. 2014; Szlavik et al. 2012). Therefore, it was proposed that the primary factor contributing to the continued presence of *L. monocytogenes* populations in commercial ready to eat meat is their enhanced ability to adhere and develop biofilms, instead of their resistance to sanitizers (Wang et al. 2015).

5.5 Mechanism of Resistance to Antimicrobial Agents

Biofilm enhances the microorganism's ability to withstand antimicrobial substances and contributes to the development of human infections. While antibiotic resistance has received a lot of attention, resistance to other food preservation chemicals, disinfecting agents, and antimicrobial agents has received far less attention. The ability of bacteria to regulate gene expression is a characteristic strategy of resistance inside biofilms (Anderl et al. 2000).

Glycocalyx

Both gram-types of bacteria have been shown to produce glycocalyx, which is an essential component of biofilms (Costerton 1988.). For the biofilm to adhere to the solid substrate and remain cohesive for maturation, Glycocalyx uses electrostatic, Van Der Waal, and hydrogen bond interactions (Peña et al. 2011). Glycocalyx's chemistry is adaptable and regulated by biofilm development, which helps harmful microbe to endure in extremely unfavorable host circumstances (Anwar et al. 1990). Multiple external variables have an impact on the constituents of biofilm glycocalyx, including polysaccharides and glycoproteins. Not only antibiotics, but also other antimicrobial agents are resisted by glycocalyx. Remarkably, the glycocalyx sheath has the ability to build up antimicrobial molecules that account for around 25% of its total mass. The adsorption areas inside the scaffolding restrict the movement of biocides and provide a surface for extracellular enzymes to attach to (Sugano et al. 2016). The extracellular enzymes serve to safeguard the mobility of certain antimicrobial compounds and serve as a surface for the breakdown of biocide metabolites, which then results in a reduction in the effectiveness of susceptible medications (Arciola et al. 2012).

Extracellular Enzyme

Enzymes facilitate the conversion of bactericide into a harmless state, hence conferring immunity against antimicrobial agents. Only a finite number of microbial species have been found as having the ability to degrade hazardous chemicals, including aromatic and phenolic chemicals, as well as heavy metals including Cd, Ni, Cu, Ag, Co, Zn, and Pb (Brown et al. 1995). Detoxification typically takes place through the biochemical digestion of ions and metal-resistance genes. The inclusion of heavy metals resulted in the development of a wider range of resistance characteristics (Kümmerle et al. 1996).

Persister Cells

Persisters refer to a group of cells that are resistant to antibacterial agents and have a significant role in causing serious and persistent infections. The identification of particular strains of bacteria is a significant problem in clinical environments (Shan et al. 2017). The research investigating the impact on ATP-dependent persister production concludes that reducing ATP levels diminishes the action of antibiotic targets, resulting in the creation of persisters. Bacterial biofilm harbors resilient persister cells which show resistance to many drugs and bactericidal agents (Lewis and Ausubel 2006). Any type of bacteria that grow slowly and show endurance or persistence can also show signs of antibiotic resistance as well as resistance to multiple drugs. The development phases of different bacterial populations, which multiply quickly and endure in exposure of deadly concentrations of antimicrobial drugs, regulate the creation of persister cells (Brooun et al. 2000). Biofilms' resistivity increased in tandem with the number of persister cells generated by stationary phase microorganisms (Del Pozo and Patel 2007). Free-living bacterial populations have their antibiotic activity neutralized by the body's immune response, yet persister cells manage to evade elimination. The glycocalyx matrix aids in the persistence of biofilm persisters by providing protection from the immune system (Lewis 2007). Upon completion of antibiotic treatment, persisters within the irreversibly attached microbial population initiate the re-establishment of bacterial biofilm development. The production of persister cells is determined by the phases of bacterial development (Keren et al. 2004). The development of persister cells is similarly contingent upon the metabolism of bacteria. It has been proposed that persister cells are a latent variation of the natural form rather than mutant cells. Curiously, persister cells do not exhibit a response to encounters with bactericidal agents. Antibiotics kill bacteria by interfering with their normal activity and not by inhibiting them. Persister resistance has been associated with programmed cell death (PCD), which occurs when antimicrobial substances cause cellular harm rather than cell mortality (Gilbert et al. 1989). Autolysis, carried out by autolysin enzymes, is the predominant phenomenon observed inside biofilms.

Genetic Adaptation

Genetic modification is necessary for biofilms to decrease vulnerability and acquire a unique and predominantly protected appearance. The mar operons are a universal regulator that governs the expression of many genes in *E. coli* and facilitates multidrug-resistant characteristics, encompassing immunity to antimicrobial agents, organic solvents, and different disinfecting agents (Williams et al. 2007). In addition, *M. tuberculosis* enters a period of hibernation for long periods in harsh environmental circumstances, which goes undetected by anti-TB medications. The majority of microbes are capable of fermentation, producing enzymes that degrade and fix oxidants, and they also display a response to oxidative stress. Within a few hours of being exposed to low levels of a detrimental substance, cells like these that respond to stress grow more immunity to it (Suntharalingam and Cvitkovitch 2005). Furthermore, other regulatory genes, including oxyR and soxR, were studied and identified. These genes play a role in determining the redox potential inside cells and trigger a stress response whenever organisms are exposed to oxidizing substances (Singh et al. 2017).

Outer Membrane Architecture

Bacterial resistance to antibiotics is attributed to the change of the cell's membrane or adaptation, as antimicrobial drugs need to invade to get to their intended location. Hydrophilic antimicrobial chemicals are mostly hindered from penetrating across the exterior membrane due to the presence of the lipopolysaccharide barrier and the underpinning phospholipids. On the other hand, hydrophobic substances are excluded by exterior membrane-associated proteins. In addition, *Nocardia farcinica*, a kind of gram-positive bacterium, possesses an intricate cell wall that may attach to various lipids and permeabilizing proteins. This allows the formation of a hydrophilic channel through the cell wall, which plays a role in its relationship with antibiotics (Singh et al. 2015). Some bacteria that are resistant to antibiotics either do not have or produce an excessive amount of exterior membrane proteins. As an illustration, the *P. aeroginosa* strain does not possess Opr D, which is a specific porin that selectively allows particular carbon sources to enter (Brözel and Cloete 1994). Changes in the composition of the exterior membrane protein result in the prevention of the bactericidal activity of sodium dimethyl dithiocarbonate.

Efflux Pumps

During harsh environments, such as those caused by antimicrobial drugs, bacteria are able to survive because they have a network of efflux mechanisms. Both innate and acquired resistance to various antimicrobial drugs from identical or other families is exerted by efflux pumps (Liaw et al. 2010; Hogan and Kolter 2002). Excessive synthesis of efflux pumps can result in the formation of multidrug resistance (MDR). Microbial efflux pumps exhibit an MDR pattern in conjunction with additional strategies for resistance, which include site change and antibiotic deactivation (Bolla et al. 2011; Davin-Regli et al. 2008). *Acinetobacter baumannii* poses a significant problem in hospitals because of its high level of tolerance to several drugs, its ability to survive in environment, and its capacity to develop a biofilm on injured surfaces (Richmond et al. 2016). Bacterial biofilms display MDR when exposed to xenobiotics like salicylate and chlorinated phenols, as well as reduced levels of antibiotics like chloramphenicol and tetracycline (Levy 1992; Ma et al. 1993). One important factor in reducing the antibiotic sensitivity of *Pseudomonas aeruginosa* biofilm is the presence of mutants that express more efflux pumps (Ma et al. 1996). In addition to being a key component of gram-negative bacteria's MDR, efflux pumps provide significant challenges to the pharmaceutical industry. New insights into efflux pumps may give a bacterial model to guide drug development (De Kievit et al. 2001).

Horizontal Gene Transfer (HGT)

The capability of bacteria to develop resistance to antibiotics is attributed to the fact that biofilms store antibiotic-resistance genes by HGT (Von Wintersdorff et al. 2016). Bacteria undergo acclimatization through HGT, which promotes biofilm development and antibiotic resistance (Madsen et al. 2012). HGT generates new variations without changing the existing ones (Kaplan 2014). A study group has shown that HGT is essential in preserving the skeletal rigidity, integrity, and resilience of dwelling bacterial communities (Fan et al. 2018). Endodontic infections caused by *E. faecalis*, which are characterized by virulence and antibiotic resistance, are influenced by HGT (Jain et al. 2016).

5.6 Control of Microbial Biofilm

Due to the fact that biofilm is extremely immune to traditional antibiotics, there exists an immediate need to find new and effective treatment approaches to this challenge in order to enhance medical care, food security, and other fields in the industry (Fig. 5.1).

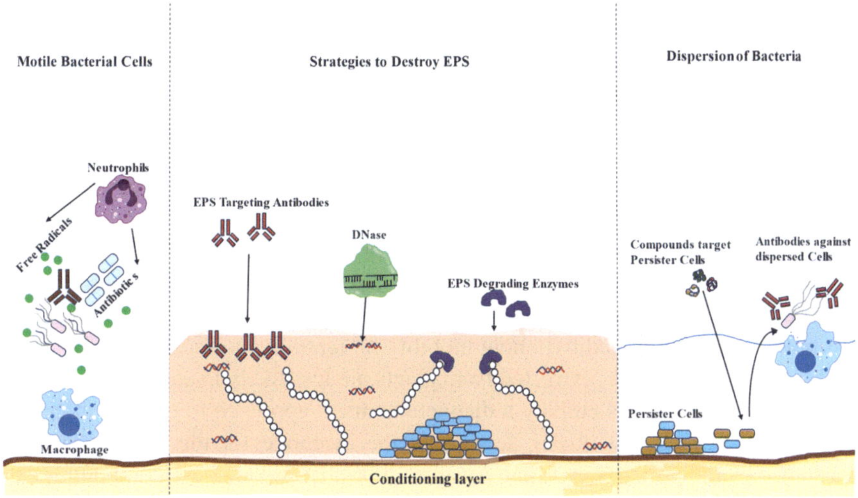

Fig. 5.1 Strategies for controlling microbial biofilm

Plant-Derived Products

The therapeutic properties of plants have been recognized for decades, and people have utilized them to cure a variety of disorders and prevent foodstuffs from becoming imperishable. There has been a rapid increase in the amount of research on the antibacterial properties of many plants. Researchers are increasingly more interested in discovering plant compounds that can inhibit the emergence of biofilms as well as virulence parameters of harmful microbes than in actively eliminating them (Rumbaugh and Ahmad 2014). While numerous plant-based compounds have been investigated for potential anti-biofilm effects, there are still many undiscovered herbal extracts. Just like human beings and animals, plants often encounter harmful microbes; thus, vegetation have evolved a complex biological defense mechanism to ward off these invaders. Phenolics, flavonoids, alkaloids, and terpenoids have been shown in the past to be beneficial against infectious microbes that create biofilms, even at doses below the minimum inhibitory concentration (MIC) (Koh et al. 2013).In the vast majority of harmful bacteria, quorum sensing regulates a number of metabolic processes, including pathogenicity or virulence, the expression of genes critical for biofilm formation, and other related phenomena. The emergence of resistance to antibiotics in microbes is due to changes in bacterial phenotypic as well as other metabolic traits. Instead of affecting the expansion of bacteria, inhibiting QS might be more effective to reduce biofilm development since it inhibits the production of QS chemicals that are necessary for expanding pathogenicity (Borges et al. 2016). According to some recent reports, compounds extracted from plants possess the potential to be employed as QS inhibitors due to their benign nature, efficacy, and stability. These compounds provide a promising alternative to commonly employed antibiotics for biofilm suppression, dispersion, and removal without causing bacteria to acquire resistance (Gupta et al. 2019; Rather et al. 2021b). Some phytochemicals are well known for their QS-inhibiting capability like farnesol, cinnamaldehyde, quercetin, kaempferol, tannic acid, and curcumin (Bouyahya et al. 2017). Plant extracts utilize many ways to hinder the production of biofilms. Several studies have shown that medicinal herbs can have a role in inhibiting or reducing the formation of biofilm. The specific plants that have been found to have this effect include: Extracted leaf juice of Kaffir lime effectively suppresses the activity of virulence-promoting factors and regulator genes in *S. mutans* (Kooltheat et al. 2016). Curcumin, derived from *Curcuma longa*, has been found to hinder protease activity, pyocyanin synthesis, formation of acyl-homoserine lactone (HSL), and downregulate QS genes in *P. aeruginosa* (Rudrappa and Bais 2008). Naringin, a glycosylated flavanone extracted from oranges, has been shown to decrease the swimming, swarming, and twitching mobility of *Yersinia enterocolitica* (Truchado et al. 2012). Compounds, extracted from plants are less likely to cause resistance in microbes because plants employ a distinct chemical approach to combat infection by microorganisms, potentially decreasing the incentive for the development of resistance to antibiotics (Lewis and Ausubel 2006). However, these compounds can also impede growth, resulting in no benefit compared to existing antibiotics. In addition, it is hypothesized that plants have developed the ability to produce

QS-inhibiting compounds in order to combat microbial diseases associated with the development of biofilms.

Advanced Chemical Treatments

Biofilms are able to be treated with a variety of pharmaceutical disinfectants that work depending on dosage and duration. Through a procedure known as sanitization, the purpose is to bring down bacteria communities to human-safe levels. The only certain way to keep different batches of food from being contaminated is to sanitize the instrument used to process them (Bayoumi et al. 2012). The food sector mostly uses sanitizers that include chlorine; however, certain microorganisms have developed tolerance to this type of disinfection. One trait that was shown to be associated with chlorine tolerance in *S. enterica* was its ability to produce cellulose. The food manufacturing industries' stressful environmental settings determined this trait (Yang et al. 2016). In unpasteurized milk microorganisms, *Staphylococcus aureus* and *S. enterica* are prevalent organisms that may readily establish biofilms in milk processing plants. Such biofilms may be effectively removed from metallic and polypropylene materials with sodium hypochlorite. Nonetheless, in identical circumstances, this antimicrobial agent failed to remove the biofilms of the bacterium *Cronobacter sakazakii* (Bayoumi et al. 2012). The hazardous gas ozone also has strong oxidizing properties. By degrading the cell membranes, it kills several microbes, including biofilms (Srey et al. 2013). The usage of it in the dairy sector serves to inhibit the excessive growth of mold on metallic equipment surfaces and powdered formulations, especially on cheeses (Varga and Szigeti 2016). One further instance of the development of tolerance to biocides is observed in *Campylobacter jejuni*, which is frequently encountered in humans. This microbe has the ability to develop a tolerance for three frequently employed chemicals in the food sector: CH_3COOH, NaOCl, and Na_3PO_4. The ability to resist is acquired by its development in progressively higher sublethal doses of these chemicals. Additionally, it is important to mention that this microbe has the ability to create biofilms with distinctive shapes when it comes into contact with various chemicals. This indicates that the bacteria have the capability to release several extracellular matrixes, which is contingent upon the chemical conditions in its surroundings (Techaruvichit et al. 2016). It is worth mentioning that the prolonged use of disinfectants in food processing plants over the years might have been a significant factor in the emergence of antibiotic resistance in microorganisms and its spread to pathogenic organisms. Hence, the advancement of new technologies might serve as an initial measure in mitigating this significant global health issue (Capita and Alonso-Calleja 2013).

Bio Control by Bacteriophages

Bacteriophages have antimicrobial properties that are harmless to other higher organisms because they are very selective in host choice and target prokaryotic cells for destruction. Consequently, phage treatment is presently a compelling substitute for antibiotics. Commercialized bacteriophage treatments are often utilized to attack specific pathogen organisms, such as *Salmonella enterica* or *Escherichia coli* (Gutiérrez et al. 2016). The inability of phage therapies to reach and kill bacteria within biofilms is their main drawback. Because of the complex biofilm architecture and extracellular substance that physically blocks phage transport, this constraint is present. The exopolysaccharide depolymerase found in a few phages, nevertheless, provides a fantastic remedy to this transmission issue (Pires et al. 2016). Enzymes like these make it easier for phages to invade and spread throughout the biofilm. Endolysins evaluated potential treatments for removing biofilms due to their ability to effectively permeate the biofilms (Gutiérrez et al. 2016; Shen et al. 2013). Phage therapies are likely affected by a number of factors, such as the interaction with bacteriophages and traditional disinfectants, as well as the temperatures often employed in the food sector, which might not be ideal for bacteriophages. Yet another pertinent concern is the impact of such bacteriophages on heterogeneous biofilms composed of several species, a frequent occurrence on substrates in the food processing sectors (Gutiérrez et al. 2016). To summarize, bacteriophages and phage-associated proteins have proven to be quite efficient at removing biofilms in laboratory settings; however, there are a few industrial examples available. Nevertheless, further study is required prior to its full integration into the conventional cleaning methods in the food processing sectors (Gutiérrez et al. 2016).

Biofilm Disruption: Enzymatic Approach

Enzymes are regarded as environmentally friendly techniques to combat biofilm development due to their biodegradability and reduced toxicity. These characteristics provide them with an efficient means for managing biofilms, and as a result, they are extensively employed in detergents for usage in the food processing sector (Huang et al. 2014; Torres et al. 2011). A biofilm's primary structural elements are proteins and polysaccharides. Consequently, biofilm eradication with protein-degrading enzymes or proteases and carbohydrate-degrading enzymes or glycosidases is consistently recommended (Boels 2011; Huang et al. 2014; Meireles et al. 2016). Enzymes like pectin methyl esterases can break down biofilms in bioreactors. This step is critical for the food processing plants since it cleanses the pipelines and machinery that are utilized in the sector (Torres et al. 2011). The food sector also frequently uses commercial detergents that contain enzymes including amylases, cellulases, glycosidases, and DNases to eliminate biofilms (Coughlan et al. 2016). Enzymes such as cellulase and alpha-amylase are capable of breaking down biofilms caused by *Staphylococcus aureus* (Wang et al. 2016; Yamaguchi et al. 2013). Because of their less substrate specificity, proteases work better on biofilms composed of organic compounds (Huang et al. 2014). After the proteases partially destroy the framework of the biofilm, it is able to be fully eliminated by mechanical procedures and becomes more susceptible to the effects of disinfectants (Coughlan et al. 2016). Although protease compositions appeared effective in removing

S. aureus biofilms from polystyrene substrates, combinations of protease, amylase, and cellulase were essential for the elimination of biofilms from *P. aeruginosa* (Stiefel et al. 2016). During biofilm development, some bacteria, such as *Escherichia coli*, *Listeria monocytogenes*, or *Staphylococcus aureus*, also produce extracellular DNA. To eliminate these complex complexes made of cells, bovine DNase might be used in such situations (Da Silva and De Martinis 2013). Biofilm clearance in various species, such as *E. coli* or *Bacillus* sp., was boosted by combining enzymes with varying activity with other chemical as well as physical therapies (Meireles et al. 2016). Nevertheless, the prohibitive expense of these therapies, mostly as a result of patent restrictions, continues to act as a barrier to their industrial-scale deployment.

5.7 Conclusion

Biofilms are intricate microbial communities residing within polysaccharides and/or protein matrices, which can be produced by various microorganisms, including different types of bacteria and fungi. This complex viscoelastic biopolymeric colony provides favorable environment against antibiotic resistance. Several pathogenic as well as non-pathogenic bacteria adapt this strategy to bypass harsh nature, which creates problems in agro-food industrial sectors. In recent days, diverse biotechnological techniques have been adapted to control the development of biofilm. Although biofilm is one of the most extensive areas, a complete understanding of biofilm formation remains somewhat unclear to date. Starting from the initial attachment of the cell to biofilm formation and, ultimately, dispersion is a complex mechanism involving numerous metabolic pathways and environmental stress factors. Therefore, the understanding of the nutrient need and nature of social messengers of microbiome members residing inside biofilm could lead to targeted interventions and the development of precise and customized treatment strategies for controlling biofilm growth.

References

Abby SS, Rocha EP (2017) Identification of protein secretion systems in bacterial genomes using MacSyFinder. In: Bacterial protein secretion systems: methods and protocols, pp 1–21. https://doi.org/10.1007/978-1-4939-7033-9_1

Anand S, Singh D (2013) Resistance of the constitutive microflora of biofilms formed on whey reverse-osmosis membranes to individual cleaning steps of a typical clean-in-place protocol. J Dairy Sci 96(10):6213–6222. https://doi.org/10.3168/jds.2013-7012

Anderl JN, Franklin MJ, Stewart PS (2000) Role of antibiotic penetration limitation in Klebsiella pneumoniae biofilm resistance to ampicillin and ciprofloxacin. Antimicrob Agents Chemother 44(7):1818–1824. https://doi.org/10.1128/aac.44.7.1818-1824.2000

Anwar H, Dasgupta MK, Costerton J (1990) Testing the susceptibility of bacteria in biofilms to antibacterial agents. Antimicrob Agents Chemother 34(11):2043–2046. https://journals.asm.org/doi/pdf/10.1128/aac.34.11.2043

Arciola CR, Campoccia D, Speziale P, Montanaro L, Costerton JW (2012) Biofilm formation in staphylococcus implant infections. A review of molecular mechanisms and implications for biofilm-resistant materials. Biomaterials 33(26):5967–5982. https://doi.org/10.1016/j.biomaterials.2012.05.031

Bagge-Ravn D, Ng Y, Hjelm M, Christiansen JN, Johansen C, Gram L (2003) The microbial ecology of processing equipment in different fish industries – analysis of the microflora during processing and following cleaning and disinfection. Int J Food Microbiol 87(3):239–250. https://doi.org/10.1016/S0168-1605(03)00067-9

Bayer ME, Easterbrook K (1991) Tubular spinae are long-distance connectors between bacteria. Microbiology 137(5):1081–1086. https://doi.org/10.1099/00221287-137-5-1081

Bayoumi MA, Kamal RM, Abd El Aal SF, Awad EI (2012) Assessment of a regulatory sanitization process in Egyptian dairy plants in regard to the adherence of some food-borne pathogens and their biofilms. Int J Food Microbiol 158(3):225–231. https://doi.org/10.1016/j.ijfoodmicro.2012.07.021

Benhamed S, Guardiola FA, Mars M, Esteban MÁ (2014) Pathogen bacteria adhesion to skin mucus of fishes. Vet Microbiol 171(1–2):1–12. https://doi.org/10.1016/j.vetmic.2014.03.008

Boels G (2011) Enzymatic removal of biofilms: a report. Virulence 2(5):490–489. https://doi.org/10.4161/viru.2.5.17317

Bolla JM, Alibert-Franco S, Handzlik J, Chevalier J, Mahamoud A, Boyer G, Kieć-Kononowicz K, Pags JM (2011) Strategies for bypassing the membrane barrier in multidrug resistant Gram-negative bacteria. FEBS Lett 585(11):1682–1690. https://doi.org/10.1016/j.febslet.2011.04.054

Borges A, Abreu AC, Dias C, Saavedra MJ, Borges F, Simões M (2016) New perspectives on the use of phytochemicals as an emergent strategy to control bacterial infections including biofilms. Molecules 21(7):877. https://doi.org/10.3390/molecules21070877

Bouyahya A, Dakka N, Et-Touys A, Abrini J, Bakri Y (2017) Medicinal plant products targeting quorum sensing for combating bacterial infections. Asian Pac J Trop Med 10(8):729–743. https://doi.org/10.1016/j.apjtm.2017.07.021

Brameyer S, Plener L, Müller A, Klingl A, Wanner G, Jung K (2018) Outer membrane vesicles facilitate trafficking of the hydrophobic signaling molecule CAI-1 between Vibrio harveyi cells. J Bacteriol 200(15):10–1128. https://doi.org/10.1128/jb.00740-17

Brooun A, Liu S, Lewis KIM (2000) A dose-response study of antibiotic resistance in Pseudomonas aeruginosa biofilms. Antimicrob Agents Chemother 44(3):640–646. https://doi.org/10.1128/aac.44.3.640-646.2000

Brown SM, Howell ML, Vasil ML, Anderson AJ, Hassett DJ (1995) Cloning and characterization of the katB gene of Pseudomonas aeruginosa encoding a hydrogen peroxide-inducible catalase: purification of KatB, cellular localization, and demonstration that it is essential for optimal resistance to hydrogen peroxide. J Bacteriol 177(22):6536–6544. https://doi.org/10.1128/jb.177.22.6536-6544.1995

Brözel VS, Cloete TE (1994) Resistance of Pseudomonas aeruginosa to isothiazolone. J Appl Bacteriol 76(6):576–582. https://doi.org/10.1111/j.1365-2672.1994.tb01655.x

Budd KE, Mitchell J, Keane OM (2016) Lineage associated expression of virulence traits in bovine-adapted Staphylococcus aureus. Vet Microbiol 189:24–31. https://doi.org/10.1016/j.vetmic.2016.04.013

Capita R, Alonso-Calleja C (2013) Antibiotic-resistant bacteria: a challenge for the food industry. Crit Rev Food Sci Nutr 53(1):11–48. https://doi.org/10.1080/10408398.2010.519837

Chan C, Paul R, Samoray D, Amiot NC, Giese B, Jenal U, Schirmer T (2004) Structural basis of activity and allosteric control of diguanylate cyclase. Proc Natl Acad Sci 101(49):17084–17089. https://doi.org/10.1073/pnas.0406134101

Chang CY (2018) Surface sensing for biofilm formation in *Pseudomonas aeruginosa*. Front Microbiol 8:294514. https://doi.org/10.3389/fmicb.2017.02671

Cherny KE, Sauer K (2020) Untethering and degradation of the polysaccharide matrix are essential steps in the dispersion response of Pseudomonas aeruginosa biofilms. J Bacteriol 202(3):10–1128. https://doi.org/10.1128/jb.00575-19

Clegg S, Gerlach GF (1987) Enterobacterial fimbriae. J Bacteriol 169(3):934–938. https://journals.asm.org/doi/pdf/10.1128/jb.169.3.934-938.1987

Costerton JW (1988) Structure and plasticity at various organization levels in the bacterial cell. Can J Microbiol 34(4):513–521. https://doi.org/10.1139/m88-088
Coughlan LM, Cotter PD, Hill C, Alvarez-Ordóñez A (2016) New weapons to fight old enemies: novel strategies for the (bio) control of bacterial biofilms in the food industry. Front Microbiol 7:1641. https://doi.org/10.3389/fmicb.2016.01641
Da Silva EP, De Martinis ECP (2013) Current knowledge and perspectives on biofilm formation: the case of listeria monocytogenes. Appl Microbiol Biotechnol 97:957–968. https://doi.org/10.1007/s00253-012-4611-1
Davies DG, Marques CN (2009) A fatty acid messenger is responsible for inducing dispersion in microbial biofilms. J Bacteriol 191(5):1393–1403. https://doi.org/10.1128/jb.01214-08
Davin-Regli A, Bolla JM, James CE, Lavigne JP, Chevalier J, Garnotel E, Molitor A (2008) Membrane permeability and regulation of drug "influx and efflux" in enterobacterial pathogens. Curr Drug Targets 9(9):750–759. https://www.ingentaconnect.com/content/ben/cdt/2008/00000009/00000009/art00004
De Kievit TR, Parkins MD, Gillis RJ, Srikumar R, Ceri H, Poole K, Iglewski BH, Storey DG (2001) Multidrug efflux pumps: expression patterns and contribution to antibiotic resistance in Pseudomonas aeruginosa biofilms. Antimicrob Agents Chemother 45(6):1761–1770. https://doi.org/10.1128/AAC.45.6.1761-1770.2001
De Vuyst L, Degeest B (1999) Heteropolysaccharides from lactic acid bacteria. FEMS Microbiol Rev 23(2):153–177. https://doi.org/10.1111/j.1574-6976.1999.tb00395.x
Del Pozo JL, Patel R (2007) The challenge of treating biofilm-associated bacterial infections. Clin Pharmacol Ther 82(2):204–209. https://doi.org/10.1038/sj.clpt.6100247
Dow JM, Fouhy Y, Lucey JF, Ryan RP (2006) The HD-GYP domain, cyclic di-GMP signaling, and bacterial virulence to plants. Mol Plant-Microbe Interact 19(12):1378–1384. https://doi.org/10.1094/MPMI-19-1378
Duguid JP, Anderson ES, Campbell I (1966) Fimbriae and adhesive properties in salmonellae. J Pathol Bacteriol 92(1):107–137. https://doi.org/10.1002/path.1700920113
Easterbrook KB, Sperker S (1982) Physiological controls of bacterial spinae production in complex medium and their value as indicators of spina function. Can J Microbiol 28(1):130–136. https://doi.org/10.1139/m82-014
Fan Y, Xiao Y, Momeni B, Liu YY (2018) Horizontal gene transfer can help maintain the equilibrium of microbial communities. J Theor Biol 454:53–59. https://doi.org/10.1016/j.jtbi.2018.05.036
Farah C, Vera M, Morin D, Haras D, Jerez CA, Guiliani N (2005) Evidence for a functional quorum-sensing type AI-1 system in the extremophilic bacterium Acidithiobacillus ferrooxidans. Appl Environ Microbiol 71(11):7033–7040. https://doi.org/10.1128/AEM.71.11.7033-7040.2005
Ferreira V, Wiedmann M, Teixeira P, Stasiewicz MJ (2014) Listeria monocytogenes persistence in food-associated environments: epidemiology, strain characteristics, and implications for public health. J Food Prot 77(1):150–170. https://doi.org/10.4315/0362-028X.JFP-13-150
Flint SH, Bremer PJ, Brooks JD (1997) Biofilms in dairy manufacturing plant-description, current concerns and methods of control. Biofouling 11(1):81–97. https://doi.org/10.1080/08927019709378321
Fransisca L, Zhou B, Park H, Feng H (2011) The effect of calcinated calcium and chlorine treatments on Escherichia coli O157: H7 87–23 population reduction in radish sprouts. J Food Sci 76(6):M404–M412. https://doi.org/10.1111/j.1750-3841.2011.02270.x
Froelich B, Oliver JD (2013) The interactions of Vibrio vulnificus and the oyster Crassostrea virginica. Microb Ecol 65(4):807–816. https://doi.org/10.1007/s00248-012-0162-3
Gandhi HP, Ray RM, Patel RM (1997) Exopolymer production by bacillus species. Carbohydr Polym 34(4):323–327. https://doi.org/10.1016/S0144-8617(97)00132-X
Giaouris E (2015) Ability of foodborne bacterial pathogens to attach to meat and meat contact surfaces. In: Biofilms in the food environment. Wiley, pp 145–175. https://doi.org/10.1002/9781118864036.ch6
Giaouris EE, Simões MV (2018) Pathogenic biofilm formation in the food industry and alternative control strategies. In: Foodborne diseases. Elsevier, pp 309–377. https://doi.org/10.1016/B978-0-12-811444-5.00011-7

Giaouris E, Chorianopoulos N, Skandamis P, Nychas GJ (2012) Attachment and biofilm formation by salmonella in food processing environments. In: Salmonella—a dangerous foodborne pathogen. Intechopen, pp 157–180. https://doi.org/10.5772/28107
Gilbert P, Allison DG, Evans DJ, Handley PS, Brown MR (1989) Growth rate control of adherent bacterial populations. Appl Environ Microbiol 55(5):1308–1311. https://doi.org/10.1128/aem.55.5.1308-1311.1989
Gomes F, Saavedra MJ, Henriques M (2016) Bovine mastitis disease/pathogenicity: evidence of the potential role of microbial biofilms. FEMS Pathog Dis 74(3):ftw006. https://doi.org/10.1093/femspd/ftw006
Goodwine J, Gil J, Doiron A, Valdes J, Solis M, Higa A, Davis S, Sauer K (2019) Pyruvate-depleting conditions induce biofilm dispersion and enhance the efficacy of antibiotics in killing biofilms in vitro and in vivo. Sci Rep 9(1):3763. https://doi.org/10.1038/s41598-019-40378-z
Gunduz GT, Tuncel G (2006) Biofilm formation in an ice cream plant. Anton Leeuw Int J Gen Mol Microbiol 89(3–4):329–336. https://doi.org/10.1007/s10482-005-9035-9
Guobjoernsdottir B, Einarsson H, Thorkelsson G (2005) Microbial adhesion to processing lines for fish fillets and cooked shrimp: influence of stainless steel surface finish and presence of gram-negative bacteria on the attachment of listeria monocytogenes. Food Technol Biotechnol 43(1):55–61. https://hrcak.srce.hr/file/162696
Gupta K, Liao J, Petrova OE, Cherny KE, Sauer K (2014) Elevated levels of the second messenger c-di-GMP contribute to antimicrobial resistance of Pseudomonas aeruginosa. Mol Microbiol 92(3):488–506. https://doi.org/10.1111/mmi.12587
Gupta P, Sarkar S, Das B, Bhattacharjee S, Tribedi P (2016) Biofilm, pathogenesis and prevention—a journey to break the wall: a review. Arch Microbiol 198:1–15. https://doi.org/10.1007/s00203-015-1148-6
Gupta K, Singh SP, Manhar AK, Saikia D, Namsa ND, Konwar BK, Mandal M (2019) Inhibition of Staphylococcus aureus and Pseudomonas aeruginosa biofilm and virulence by active fraction of Syzygium cumini (L.) Skeels leaf extract: in-vitro and in Silico studies. Indian J Microbiol 59(1):13–21. https://doi.org/10.1007/s12088-018-0770-9
Gutiérrez D, Rodríguez-Rubio L, Martínez B, Rodríguez A, García P (2016) Bacteriophages as weapons against bacterial biofilms in the food industry. Front Microbiol 7:825. https://doi.org/10.3389/fmicb.2016.00825
Harimawan A, Ting YP (2016) Investigation of extracellular polymeric substances (EPS) properties of P. Aeruginosa and B. Subtilis and their role in bacterial adhesion. Colloids Surf B: Biointerfaces 146:459–467. https://doi.org/10.1016/j.colsurfb.2016.06.039
Hassan GM, Al-Ashmawy MAM, Meshref AMS, Afify SI (2010) Studies on enterotoxigenic bacillus cereus in raw milk and some dairy products. J Food Saf 30(3):569–583. https://doi.org/10.1111/j.1745-4565.2010.00226.x
Häussler S, Becker T (2008) The pseudomonas quinolone signal (PQS) balances life and death in Pseudomonas aeruginosa populations. PLoS Pathog 4(9):e1000166. https://doi.org/10.1371/journal.ppat.1000166
Hengge R (2009) Principles of c-di-GMP signalling in bacteria. Nat Rev Microbiol 7(4):263–273. https://doi.org/10.1038/nrmicro2109
Hogan D, Kolter R (2002) Why are bacteria refractory to antimicrobials? Curr Opin Microbiol 5(5):472–477. https://doi.org/10.1016/S1369-5274(02)00357-0
Huang H, Ren H, Ding L, Geng J, Xu K, Zhang Y (2014) Aging biofilm from a full-scale moving bed biofilm reactor: characterization and enzymatic treatment study. Bioresour Technol 154:122–130. https://doi.org/10.1016/j.biortech.2013.12.031
Iwamoto M, Ayers T, Mahon BE, Swerdlow DL (2010) Epidemiology of seafood-associated infections in the United States. Clin Microbiol Rev 23(2):399–411. https://doi.org/10.1128/CMR.00059-09
Jain H, Mulay S, Mullany P (2016) Persistence of endodontic infection and enterococcus faecalis: role of horizontal gene transfer. Gene Rep 5:112–116. https://doi.org/10.1016/j.genrep.2016.09.010

Jenal U, Reinders A, Lori C (2017) Cyclic di-GMP: second messenger extraordinaire. Nat Rev Microbiol 15(5):271–284. https://doi.org/10.1038/nrmicro.2016.190

Jessen B, Lammert L (2003) Biofilm and disinfection in meat processing plants. Int Biodeterior Biodegradation 51(4):265–269. https://doi.org/10.1016/S0964-8305(03)00046-5

Kalivoda EJ, Brothers KM, Stella NA, Schmitt MJ, Shanks RM (2013) Bacterial cyclic AMP-phosphodiesterase activity coordinates biofilm formation. PLoS One 8(7):e71267. https://doi.org/10.1371/journal.pone.0071267

Kaplan T (2014) The role of horizontal gene transfer in antibiotic resistance. Eukaryon 10:80–81

Karimi A, Karig D, Kumar A, Ardekani AM (2015) Interplay of physical mechanisms and biofilm processes: review of microfluidic methods. Lab Chip 15(1):23–42. https://doi.org/10.1039/C4LC01095G

Karunasagar I, Otta SK, Karunasagar I (1996) Biofilm formation by Vibrio harveyi on surfaces. Aquaculture 140(3):241–245. https://doi.org/10.1016/0044-8486(95)01180-3

Keren I, Shah D, Spoering A, Kaldalu N, Lewis K (2004) Specialized persister cells and the mechanism of multidrug tolerance in Escherichia coli. J Bacteriol 186(24):8172–8180. https://doi.org/10.1128/JB.186.24.8172-8180.2004

Khan S, Scholey JM (2018) Assembly, functions and evolution of archaella, flagella and cilia. Curr Biol 28(6):R278–R292. https://doi.org/10.1016/j.cub.2018.01.085

Koh CL, Sam CK, Yin WF, Tan LY, Krishnan T, Chong YM, Chan KG (2013) Plant-derived natural products as sources of anti-quorum sensing compounds. Sensors 13(5):6217–6228. https://doi.org/10.3390/s130506217

Kooltheat N, Kamuthachad L, Anthapanya M, Samakchan N, Sranujit RP, Potup P, Ferrante A, Usuwanthim K (2016) Kaffir lime leaves extract inhibits biofilm formation by streptococcus mutants. Nutrition 32(4):486–490. https://doi.org/10.1016/j.nut.2015.10.010

Kumar CG, Anand SK (1998) Significance of microbial biofilms in food industry: a review. Int J Food Microbiol 42(1–2):9–27. https://doi.org/10.1016/S0168-1605(98)00060-9

Kümmerle N, Feucht HH, Kaulfers PM (1996) Plasmid-mediated formaldehyde resistance in Escherichia coli: characterization of resistance gene. Antimicrob Agents Chemother 40(10):2276–2279. https://doi.org/10.1128/aac.40.10.2276

Langsrud S, Moen B, Møretrø T, Løype M, Heir E (2016) Microbial dynamics in mixed culture biofilms of bacteria surviving sanitation of conveyor belts in salmon-processing plants. J Appl Microbiol 120(2):366–378. https://doi.org/10.1111/jam.13013

Latorre AA, Van Kessel JS, Karns JS, Zurakowski MJ, Pradhan AK, Boor KJ et al (2010) Biofilm in milking equipment on a dairy farm as a potential source of bulk tank milk contamination with listeria monocytogenes. J Dairy Sci 93(6):2792–2802. https://doi.org/10.3168/jds.2009-2717

Levy SB (1992) Active efflux mechanisms for antimicrobial resistance. Antimicrob Agents Chemother 36(4):695–703. https://journals.asm.org/doi/pdf/10.1128/aac.36.4.695

Lewis K (2007) Persister cells, dormancy and infectious disease. Nat Rev Microbiol 5(1):48–56. https://doi.org/10.1038/nrmicro1557

Lewis K, Ausubel FM (2006) Prospects for plant-derived antibacterials. Nat Biotechnol 24(12):1504–1507. https://doi.org/10.1038/nbt1206-1504

Li J, Azam F, Zhang S (2016) Outer membrane vesicles containing signalling molecules and active hydrolytic enzymes released by a coral pathogen *Vibrio shilonii* AK1. Environ Microbiol 18(11):3850–3866. https://doi.org/10.1111/1462-2920.13344

Liaw SJ, Lee YL, Hsueh PR (2010) Multidrug resistance in clinical isolates of Stenotrophomonas maltophilia: roles of integrons, efflux pumps, phosphoglucomutase (SpgM), and melanin and biofilm formation. Int J Antimicrob Agents 35(2):126–130. https://doi.org/10.1016/j.ijantimicag.2009.09.015

Liu C, Sun D, Zhu J, Liu J, Liu W (2020) The regulation of bacterial biofilm formation by cAMP-CRP: a mini-review. Front Microbiol 11:802. https://doi.org/10.3389/fmicb.2020.00802

Ma D, Cook DN, Alberti M, Pon NG, Nikaido H, Hearst JE (1993) Molecular cloning and characterization of acrA and acrE genes of Escherichia coli. J Bacteriol 175(19):6299–6313. https://doi.org/10.1128/jb.175.19.6299-6313.1993

Ma D, Alberti M, Lynch C, Nikaido H, Hearst JE (1996) The local repressor AcrR plays a modulating role in the regulation of acrAB genes of Escherichia coli by global stress signals. Mol Microbiol 19(1):101–112. https://doi.org/10.1046/j.1365-2958.1996.357881.x

Madsen JS, Burmølle M, Hansen LH, Sørensen SJ (2012) The interconnection between biofilm formation and horizontal gene transfer. FEMS Immunol Med Microbiol 65(2):183–195. https://doi.org/10.1111/j.1574-695X.2012.00960.x

Matthysse AG (1983) Role of bacterial cellulose fibrils in agrobacterium tumefaciens infection. J Bacteriol 154(2):906–915. https://doi.org/10.1128/jb.154.2.906-915.1983

Mayer C, Moritz R, Kirschner C, Borchard W, Maibaum R, Wingender J, Flemming HC (1999) The role of intermolecular interactions: studies on model systems for bacterial biofilms. Int J Biol Macromol 26(1):3–16. https://doi.org/10.1016/S0141-8130(99)00057-4

McLandsborough L, Rodriguez A, Pérez-Conesa D, Weiss J (2006) Biofilms: at the interface between biophysics and microbiology. Food Biophys 1(2):94–114. https://doi.org/10.1007/s11483-005-9004-x

Meireles A, Borges A, Giaouris E, Simões M (2016) The current knowledge on the application of anti-biofilm enzymes in the food industry. Food Res Int 86:140–146. https://doi.org/10.1016/j.foodres.2016.06.006

Miller MB, Bassler BL (2001) Quorum sensing in bacteria. Annu Rev Microbiol 55(1):165–199. https://doi.org/10.1146/annurev.micro.55.1.165

Mitra A, Mukhopadhyay S (2023) Regulation of biofilm formation by non-coding RNA in prokaryotes. Curr Res Pharmacol Drug Discov 4:100151. https://doi.org/10.1016/j.crphar.2022.100151

Momba MN, Makala N (2004) Comparing the effect of various pipe materials on biofilm formation in chlorinated and combined chlorine-chloraminated water systems. Water SA 30(2):175–182. https://doi.org/10.4314/wsa.v30i2.5061

Monsan P, Bozonnet S, Albenne C, Joucla G, Willemot RM, Remaud-Siméon M (2001) Homopolysaccharides from lactic acid bacteria. Int Dairy J 11(9):675–685. https://doi.org/10.1016/S0958-6946(01)00113-3

Müller CM, Åberg A, Straseviçiene J, Emődy L, Uhlin BE, Balsalobre C (2009) Type 1 fimbriae, a colonization factor of uropathogenic Escherichia coli, are controlled by the metabolic sensor CRP-cAMP. PLoS Pathog 5(2):e1000303. https://doi.org/10.1371/journal.ppat.1000303

Naughton PJ, Grant G, Bardocz S, Allen-Vercoe E, Woodward MJ, Pusztai A (2001) Expression of type 1 fimbriae (SEF 21) of salmonella enterica serotype enteritidis in the early colonisation of the rat intestine. J Med Microbiol 50(2):191–197. https://doi.org/10.1099/0022-1317-50-2-191

Nishikawa M, Kobayashi K (2021) Calcium prevents biofilm dispersion in Bacillus subtilis. J Bacteriol 203(14):10–1128. https://doi.org/10.1128/jb.00114-21

Norhana MW, Poole SE, Deeth HC, Dykes GA (2010) Prevalence, persistence and control of salmonella and listeria in shrimp and shrimp products: a review. Food Control 21(4):343–361. https://doi.org/10.1016/j.foodcont.2009.06.020

Oliveira GS, Lopes DRG, Andre C, Silva CC, Baglinière F, Vanetti MCD (2019) Multispecies biofilm formation by the contaminating microbiota in raw milk. Biofouling 35(8):819–831. https://doi.org/10.1080/08927014.2019.1666267

Ölmez H, Temur SD (2010) Effects of different sanitizing treatments on biofilms and attachment of Escherichia coli and listeria monocytogenes on green leaf lettuce. LWT-Food Sci Technol 43(6):964–970. https://doi.org/10.1016/j.lwt.2010.02.005

Oniciuc EA, Cerca N, Nicolau AI (2016) Compositional analysis of biofilms formed by Staphylococcus aureus isolated from food sources. Front Microbiol 7:390. https://doi.org/10.3389/fmicb.2016.00390

Peña J, Bargar JR, Sposito G (2011) Role of bacterial biomass in the sorption of Ni by biomass-birnessite assemblages. Environ Sci Technol 45(17):7338–7344. https://doi.org/10.1021/es201446r

Pérez-Cruz C, Delgado L, López-Iglesias C, Mercade E (2015) Outer-inner membrane vesicles naturally secreted by gram-negative pathogenic bacteria. PLoS One 10(1):e0116896. https://doi.org/10.1371/journal.pone.0116896

Pires DP, Oliveira H, Melo LDR, Sillankorva S, Azeredo J (2016) Bacteriophage-encoded depolymerases: their diversity and biotechnological applications. Appl Microbiol Biotechnol 100(5):2141–2151. https://doi.org/10.1007/s00253-015-7247-0

Rahman MU, Fleming DF, Wang L, Rumbaugh KP, Gordon VD, Christopher GF (2022) Microrheology of Pseudomonas aeruginosa biofilms grown in wound beds. NPJ Biofilms microbi 8(1):49. https://doi.org/10.1038/s41522-022-00311-1

Rather MA, Gupta K, Bardhan P, Borah M, Sarkar A, Eldiehy KS, Bhuyan S, Mandal M (2021a) Microbial biofilm: a matter of grave concern for human health and food industry. J Basic Microbiol 61(5):380–395. https://doi.org/10.1002/jobm.202000678

Rather MA, Gupta K, Mandal M (2021b) Inhibition of biofilm and quorum sensing-regulated virulence factors in Pseudomonas aeruginosa by Cuphea carthagenensis (Jacq.) J. F. Macbr. Leaf extract: an in vitro study. J Ethnopharmacol 269:113699. https://doi.org/10.1016/j.jep.2020.113699

Rendueles O, Ghigo JM (2015) Mechanisms of competition in biofilm communities. In: Microbial biofilms, pp 319–342. https://doi.org/10.1128/9781555817466.ch16

Richmond GE, Evans LP, Anderson MJ, Wand ME, Bonney LC, Ivens A, Chua KL, Webber MA, Mark Sutton J, Peterson ML, Piddock LJV (2016) The Acinetobacter baumannii two-component system aders regulates genes required for multidrug efflux, biofilm formation, and virulence in a strain-specific manner. MBio 7(2):e00430-16. https://doi.org/10.1128/mBio.00430-16

Römling U, Amikam D (2006) Cyclic di-GMP as a second messenger. Curr Opin Microbiol 9(2):218–228. https://doi.org/10.1016/j.mib.2006.02.010

Römling U, Galperin MY, Gomelsky M (2013) Cyclic di-GMP: the first 25 years of a universal bacterial second messenger. Microbiol Mol Biol Rev 77(1):1–52. https://doi.org/10.1128/mmbr.00043-12

Rudrappa T, Bais HP (2008) Curcumin, a known phenolic from Curcuma longa, attenuates the virulence of Pseudomonas aeruginosa PAO1 in whole plant and animal pathogenicity models. J Agric Food Chem 56(6):1955–1962. https://doi.org/10.1021/jf072591j

Ruiz LM, Valenzuela S, Castro M, Gonzalez A, Frezza M, Soulère L, Rohwerder T, Queneau T, Doutheau A, Sand W, Jerez CA, Guiliani N (2008) AHL communication is a widespread phenomenon in biomining bacteria and seems to be involved in mineral-adhesion efficiency. Hydrometallurgy 94(1–4):133–137. https://doi.org/10.1016/j.hydromet.2008.05.028

Rumbaugh KP, Ahmad I (2014) Springer series on biofi lms Antibiofi lm agents from diagnosis to treatment and prevention. http://www.springer.com/series/7142

Rumbaugh KP, Sauer K (2020) Biofilm dispersion. Nat Rev Microbiol 18(10):571–586. https://doi.org/10.1038/s41579-020-0385-0

Ryjenkov DA, Tarutina M, Moskvin OV, Gomelsky M (2005) Cyclic diguanylate is a ubiquitous signaling molecule in bacteria: insights into biochemistry of the GGDEF protein domain. J Bacteriol 187(5):1792–1798. https://doi.org/10.1128/jb.187.5.1792-1798.2005

Salimena APS, Lange CC, Camussone C, Signorini M, Calvinho LF, Brito MAVP, Borges CAV, Guimarães AS, Ribeiro JB, Mendonça LC, Piccoli RH (2016) Genotypic and phenotypic detection of capsular polysaccharide and biofilm formation in Staphylococcus aureus isolated from bovine milk collected from Brazilian dairy farms. Vet Res Commun 40(3–4):97–106. https://doi.org/10.1007/s11259-016-9658-5

Salo S, Ehavald H, Raaska L, Vokk R, Wirtanen G (2006) Microbial surveys in Estonian dairies. LWT-Food Sci Technol 39(5):460–471. https://doi.org/10.1016/j.lwt.2005.03.008

Santos AD, Galdino ACM, Mello TD, Ramos LDS, Branquinha MH, Bolognese AM, Neto JC, Roudbary M (2018) What are the advantages of living in a community. A Microbial Biofilm perspective. Mem Inst Oswaldo Cruz 113(9):e180212. https://doi.org/10.1590/0074-02760180212

Schonewille E, Nesse LL, Hauck R, Windhorst D, Hafez HM, Vestby LK (2012) Biofilm building capacity of salmonella enterica strains from the poultry farm environment. FEMS Immunol Med Microbiol 65(2):360–365. https://doi.org/10.1111/j.1574-695X.2012.00966.x

Shan Y, Brown Gandt A, Rowe SE, Deisinger JP, Conlon BP, Lewis K (2017) ATP-dependent persister formation in *Escherichia coli*. MBio 8(1):10–1128. https://doi.org/10.1128/mBio.02267-16

Shen Y, Köller T, Kreikemeyer B, Nelson DC (2013) Rapid degradation of streptococcus pyogenes biofilms by PlyC, a bacteriophage-encoded endolysin. J Antimicrob Chemother 68(8):1818–1824. https://doi.org/10.1093/jac/dkt104

Shi X, Zhu X (2009) Biofilm formation and food safety in food industries. Trends Food Sci Technol 20(9):407–413. https://doi.org/10.1016/j.tifs.2009.01.054

Simm R, Morr M, Kader A, Nimtz M, Römling U (2004) GGDEF and EAL domains inversely regulate cyclic di-GMP levels and transition from sessility to motility. Mol Microbiol 53(4):1123–1134. https://doi.org/10.1111/j.1365-2958.2004.04206.x

Simm R, Ahmad I, Rhen M, Le Guyon S, Römling U (2014) Regulation of biofilm formation in salmonella enterica serovar Typhimurium. Future Microbiol 9(11):1261–1282. https://doi.org/10.2217/fmb.14.88

Simões M, Simões LC, Vieira MJ (2010) A review of current and emergent biofilm control strategies. LWT-Food Sci Technol 43(4):573–583. https://doi.org/10.1016/j.lwt.2009.12.008

Singh PR, Bajaj H, Benz R, Winterhalter M, Mahendran KR (2015) Transport across the outer membrane porin of mycolic acid containing actinomycetales: Nocardia farcinica. Biochim Biophys Acta Biomembr 1848(2):654–661. https://doi.org/10.1016/j.bbamem.2014.11.020

Singh S, Singh SK, Chowdhury I, Singh R (2017) Understanding the mechanism of bacterial biofilms resistance to antimicrobial agents. Open Microbiol J 11(1):53–62. https://doi.org/10.2174/1874285801711010053

Sofos JN, Geornaras I (2010) Overview of current meat hygiene and safety risks and summary of recent studies on biofilms, and control of Escherichia coli O157:H7 in nonintact, and listeria monocytogenes in ready-to-eat, meat products. Meat Sci 86(1):2–14. https://doi.org/10.1016/j.meatsci.2010.04.015

Sonawane JM, Rai AK, Sharma M, Tripathi M, Prasad R (2022) Microbial biofilms: recent advances and progress in environmental bioremediation. Sci Total Environ 824:153843. https://doi.org/10.1016/j.scitotenv.2022.153843

Srey S, Jahid IK, Ha SD (2013) Biofilm formation in food industries: a food safety concern. Food Control 31(2):572–585. https://doi.org/10.1016/j.foodcont.2012.12.001

Stiefel P, Mauerhofer S, Schneider J, Maniura-Weber K, Rosenberg U, Ren Q (2016) Enzymes enhance biofilm removal efficiency of cleaners. Antimicrob Agents Chemother 60(6):3647–3652. https://doi.org/10.1128/AAC.00400-16

Sugano M, Morisaki H, Negishi Y, Endo-Takahashi Y, Kuwata H, Miyazaki T, Yamamoto M (2016) Potential effect of cationic liposomes on interactions with oral bacterial cells and biofilms. J Liposome Res 26(2):156–162. https://doi.org/10.3109/08982104.2015.1063648

Suntharalingam P, Cvitkovitch DG (2005) Quorum sensing in streptococcal biofilm formation. Trends Microbiol 13(1):3–6. https://doi.org/10.1016/j.tim.2004.11.009

Sutherland IW (2001) Biofilm exopolysaccharides: a strong and sticky framework. Microbiology 147(1):3–9. https://doi.org/10.1099/00221287-147-1-3

Sutrina SL, Daniel K, Lewis M, Charles NT, Anselm CK, Thomas NA, Holder N (2015) Biofilm growth of Escherichia coli is subject to cAMP-dependent and cAMP-independent inhibition. J Mol Microbiol Biotechnol 25(2–3):209–225. https://doi.org/10.1159/000375498

Szlavik J, Paiva DS, Mørk N, van den Berg F, Verran J, Whitehead K, Knøchel S, Nielsen DS (2012) Initial adhesion of listeria monocytogenes to solid surfaces under liquid flow. Int J Food Microbiol 152(3):181–188. https://doi.org/10.1016/j.ijfoodmicro.2011.09.006

Tang JL, Gough CL, Daniels MJ (1990) Cloning of genes involved in negative regulation of production of extracellular enzymes and polysaccharide of Xanthomonas campestris pathovar campestris. Mol Gen Genet MGG 222:157–160. https://doi.org/10.1007/BF00283038

Techaruvichit P, Takahashi H, Kuda T, Miya S, Keeratipibul S, Kimura B (2016) Adaptation of campylobacter jejuni to biocides used in the food industry affects biofilm structure, adhesion strength, and cross-resistance to clinical antimicrobial compounds. Biofouling 32(7):827–839. https://doi.org/10.1080/08927014.2016.1198476

Torres CE, Lenon G, Craperi D, Wilting R, Blanco Á (2011) Enzymatic treatment for preventing biofilm formation in the paper industry. Appl Microbiol Biotechnol 92(1):95–103. https://doi.org/10.1007/s00253-011-3305-4

Toyofuku M, Nomura N, Eberl L (2019) Types and origins of bacterial membrane vesicles. Nat Rev Microbiol 17(1):13–24. https://doi.org/10.1038/s41579-018-0112-2
Truchado P, Giménez-Bastida JA, Larrosa M, Castro-Ibáñez I, Espín JC, Tomás-Barberán FA, García-Conesa MT, Allende A (2012) Inhibition of quorum sensing (QS) in yersinia enterocolitica by an Orange extract rich in glycosylated flavanones. J Agric Food Chem 60(36):8885–8894. https://doi.org/10.1021/jf301365a
Valderrama WB, Cutter CN (2013) An ecological perspective of listeria monocytogenes biofilms in food processing facilities. Crit Rev Food Sci Nutr 53(8):801–817. https://doi.org/10.1080/10408398.2011.561378
Valentini M, Filloux A (2016) Biofilms and cyclic di-GMP (c-di-GMP) signaling: lessons from Pseudomonas aeruginosa and other bacteria. J Biol Chem 291(24):12547–12555. https://doi.org/10.1074/jbc.R115.711507
Varga L, Szigeti J (2016) Use of ozone in the dairy industry: a review. Int J Dairy Technol 69(2):157–168. https://doi.org/10.1111/1471-0307.12302
Vestby LK, Møretrø T, Langsrud S, Heir E, Nesse LL (2009) Biofilm forming abilities of salmonella are correlated with persistence in fish meal- and feed factories. BMC Vet Res 5:20. https://doi.org/10.1186/1746-6148-5-20
Von Wintersdorff CJH, Penders J, Van Niekerk JM, Mills ND, Majumder S, Van Alphen LB, Savelkoul PHM, Wolffs PFG (2016) Dissemination of antimicrobial resistance in microbial ecosystems through horizontal gene transfer. Front Microbiol 7:173. https://doi.org/10.3389/fmicb.2016.00173
Vu B, Chen M, Crawford R, Ivanova E (2009) Bacterial extracellular polysaccharides involved in biofilm formation. Molecules 14:2535–2554. https://doi.org/10.3390/molecules14072535
Wang R, Kalchayanand N, Schmidt JW, Harhay DM (2013) Mixed biofilm formation by Shiga toxin-producing Escherichia coli and salmonella enterica serovar typhimurium enhanced bacterial resistance to sanitization due to extracellular polymeric substances. J Food Prot 76(9):1513–1522. https://doi.org/10.4315/0362-028X.JFP-13-077
Wang J, Ray AJ, Hammons SR, Oliver HF (2015) Persistent and transient listeria monocytogenes strains from retail deli environments vary in their ability to adhere and form biofilms and rarely have inlA premature stop codons. Foodborne Pathog Dis 12(2):151–158. https://doi.org/10.1089/fpd.2014.1837
Wang H, Wang H, Xing T, Wu N, Xu X, Zhou G (2016) Removal of salmonella biofilm formed under meat processing environment by surfactant in combination with bio-enzyme. LWT 66:298–304. https://doi.org/10.1016/j.lwt.2015.10.049
Waters CM, Bassler BL (2005) Quorum sensing: cell-to-cell communication in bacteria. Annu Rev Cell Dev Biol 21(1):319–346. https://doi.org/10.1146/annurev.cellbio.21.012704.131001
Weiler C, Ifland A, Naumann A, Kleta S, Noll M (2013) Incorporation of listeria monocytogenes strains in raw milk biofilms. Int J Food Microbiol 161(2):61–68. https://doi.org/10.1016/j.ijfoodmicro.2012.11.027
Williams P, Winzer K, Chan WC, Camara M (2007) Look who's talking: communication and quorum sensing in the bacterial world. Philos Trans R Soc B: Biol Sci 362(1483):1119–1134. https://doi.org/10.1098/rstb.2007.2039
Yamaguchi S, Yamamoto E, Mannen T, Nagamune T, Nagamune T (2013) Protein refolding using chemical refolding additives. Biotechnol J 8(1):17–31. https://doi.org/10.1002/biot.201200025
Yang Y, Mikš-Krajnik M, Zheng Q, Lee SB, Lee SC, Yuk HG (2016) Biofilm formation of salmonella Enteritidis under food-related environmental stress conditions and its subsequent resistance to chlorine treatment. Food Microbiol 54:98–105. https://doi.org/10.1016/j.fm.2015.10.010
Yin L, Zhu W, Chen D, Zhou Y, Lin H (2020) Small noncoding RNA sRNA0426 is involved in regulating biofilm formation in *Streptococcus mutans*. Microbiology 9(9):e1096. https://doi.org/10.1002/mbo3.1096
Yuan L, Zhang Y, Mi Z, Zheng X, Wang S, Li H, Yang Z (2024) Calcium-mediated modulation of Pseudomonas fluorescens biofilm formation. J Dairy Sci 107(4):1950–1966. https://doi.org/10.3168/jds.2023-23860

Textural Symphony: Biofilm's Impact on Food

6

Swapnali Patil, Pranali Pangam, Shreyasi Kadam, Poournima Sankpal, Akshay Nalawade, and Shubham Yadav

Abstract

Biofilms, complex communities of microorganisms adhering to surfaces, play a pivotal yet often overlooked role in food ecosystems. This abstract delves into the intricate interplay between biofilms and food, elucidating their profound impact on texture. Through a synthesis of scientific literature, it explores how biofilms alter food texture, from enhancing mouthfeel to compromising freshness. Additionally, it examines the multifaceted implications of biofilm formation in food processing, storage, and safety. Furthermore, this abstract discusses emerging strategies to mitigate biofilm-related textural alterations, ranging from novel antimicrobial agents to innovative packaging technologies. Ultimately, understanding the textural symphony orchestrated by biofilms in food systems is crucial for optimizing product quality, safety, and consumer satisfaction.

Keywords

Biofilm · Food texture · Shelf-life · Microorganisms · Antimicrobial agents · Control strategies · Spoilage · Consumer satisfaction

S. Patil (✉) · S. Kadam
Department of Pharmaceutics, Vijayrao Naik college of pharmacy, Tal-Kankavli, India

P. Pangam (✉)
Department of Pharmaceutical Quality Assurance, Vijayrao Naik college of pharmacy, Tal-Kankavli, India

P. Sankpal (✉)
Department of Pharmaceutical Chemistry, Bharati Vidyapeeth College of Pharmacy, Palus, India

A. Nalawade · S. Yadav
Department of Pharmaceutics, Bharati Vidyapeeth College of Pharmacy, Palus, India

T. Sarkar et al. (eds.), *Biofilm Applications to Revolutionize Food Technology*,
https://doi.org/10.1007/978-3-031-85205-3_6

6.1 Introduction

In the dynamic landscape of food production and preservation, biofilms represent complex microbial communities that adhere to surfaces and form intricate structures embedded within a matrix of extracellular polymeric substances (EPS). These biofilms, often imperceptible to the naked eye, exert a significant influence on various aspects of the food industry. At their core, biofilms serve as architects of texture and flavor, contributing to the development of desirable sensory attributes in fermented foods like cheese, yogurt, and sourdough bread. Their presence can enhance the richness and complexity of flavors through enzymatic activities and metabolic processes, thereby enriching the gastronomic experience. In the vibrant tapestry of food production, biofilms weave a complex narrative that shapes the very essence of what we eat. Defined as intricate communities of microorganisms ensconced within a matrix of extracellular polymeric substances, biofilms form an integral part of the microbial ecosystems found in food processing environments. Their significance transcends mere presence, as biofilms wield a profound influence on food quality, safety, and shelf-life, leaving an indelible mark on every stage of the food supply chain (Gurtler 2023; Landis 2021).

At the heart of biofilms lies their dual nature: they are both creators and disruptors, architects of flavor and texture while also harboring potential threats to food safety. On one hand, biofilms contribute to the development of distinctive flavors and textures in fermented foods, where they participate in complex metabolic processes and enzymatic reactions that impart richness and complexity to the final product. From the tangy tang of aged cheese to the airy crumb of artisanal bread, biofilms play a silent yet crucial role in shaping our culinary experiences (Mlambo et al. 2022).

However, the impact of biofilms extends beyond culinary enhancement, as they also wield considerable influence over food safety and shelf-life. Within food processing environments, biofilms can colonize surfaces of equipment and infrastructure, serving as reservoirs for pathogenic microorganisms such as *Salmonella*, *Listeria*, and *Escherichia coli*. These pathogens, shielded within the protective matrix of biofilms, pose significant challenges to sanitation practices and contamination control efforts, increasing the risk of foodborne illnesses and necessitating stringent regulatory measures (Mota and Vilela 2024).

Moreover, biofilms contribute to the persistence of spoilage microorganisms, accelerating the degradation of food quality and shortening shelf-life. Through mechanisms such as nutrient scavenging and metabolic interactions, biofilms facilitate the survival and proliferation of spoilage bacteria and fungi, leading to undesirable changes in the texture, appearance, and flavor of food products. Additionally, biofilms can impede the effectiveness of antimicrobial treatments and preservatives, rendering conventional preservation methods less efficacious in combating microbial contamination and spoilage (Panneerselvam et al. 2024).

Thus, the intricate interplay between biofilms and food presents a multifaceted challenge for the food industry, requiring a comprehensive understanding of biofilm dynamics and the development of innovative strategies for their control and management (Dunkley and Hekmat 2020). By elucidating the mechanisms underlying biofilm formation, persistence, and impact on food quality and safety, this chapter seeks to provide insights into the complex relationship between biofilms and food, empowering food producers and regulators to navigate the challenges posed by biofilm-associated issues and ensure the integrity of food products from farm to fork (Munteanu-Ichim et al. 2024).

6.2 Mechanisms of Biofilm Formation

A. **Initial Adhesion and Surface Conditioning.**
   1. **Role of Extracellular Polymeric Substances (EPS).**
   Extracellular polymeric substances (EPS) are a fundamental component of biofilms, providing structural support and contributing to the biofilm's resilience. EPS consists of a complex mixture of polysaccharides, proteins, lipids, and extracellular DNA. The primary functions of EPS include the following:

- **Adhesion**: EPS facilitates the initial attachment of microorganisms to surfaces by creating a sticky matrix that holds cells in place. The polysaccharides in EPS, such as exopolysaccharides (EPSs), play a crucial role in anchoring microbial cells to surfaces and to each other (Alotaibi and Bukhari 2021; Andryukov et al. 2020).
- **Protection**: EPS provides a protective barrier against environmental stresses such as desiccation, antimicrobial agents, and immune responses. This protection is crucial for the survival of microorganisms within biofilms, particularly in hostile environments such as food processing surfaces.
- **Nutrient Retention**: The EPS matrix helps retain nutrients and provides a microenvironment that supports microbial growth and metabolism. The presence of nutrients within the EPS matrix can enhance the proliferation of microorganisms within the biofilm (Flemming et al. 2023; Saha et al. 2020).

The production of EPS is a dynamic process regulated by quorum sensing, a cell-to-cell communication system that allows microorganisms to coordinate their activities based on population density. This coordination ensures the formation of a cohesive and functional biofilm (Saini et al. 2023).

6.3 Physicochemical Interactions with Food Contact Surfaces

The initial adhesion of microorganisms to food contact surfaces is influenced by various physicochemical interactions, including the following:

- **Surface Energy**: The surface energy of food contact materials affects microbial adhesion. Surfaces with high energy, such as stainless steel, tend to attract microorganisms more readily compared to low-energy surfaces. Surface energy influences the wetting properties and the formation of initial microbial attachments (Aguirre-Ramírez et al. 2021; DeFlorio et al. 2021).
- **Hydrophobicity and Hydrophilicity**: The hydrophobic or hydrophilic nature of a surface affects microbial adhesion. Hydrophobic surfaces, such as certain plastics and coatings, tend to attract hydrophobic microorganisms, while hydrophilic surfaces are more likely to attract hydrophilic microorganisms.
- **Surface Roughness**: The texture and roughness of food contact surfaces can influence biofilm formation. Rough or irregular surfaces provide more sites for microbial attachment and biofilm development compared to smooth surfaces. Surface roughness can also affect the efficacy of cleaning and sanitization procedures (Ganesan et al. 2022).
- **Surface Conditioning Films**: Before microorganisms can adhere to a surface, conditioning films made of organic and inorganic substances, such as proteins and lipids from food or environmental sources, may form. These films alter the surface properties and facilitate microbial adhesion by providing initial attachment sites.

B. **Microbial Composition and Succession.**

1. **Dominant Bacterial and Fungal Species.**

The microbial composition of biofilms is highly variable and depends on several factors, including the environment and the type of food product. Common bacterial and fungal species found in food-related biofilms include the following:

- **Bacteria**:
  - *Listeria monocytogenes*: A pathogen frequently found in biofilms in dairy and meat processing environments. Its ability to form robust biofilms makes it a significant concern for food safety.
  - *Salmonella* spp.: Known for its role in foodborne illnesses, Salmonella can persist in biofilms in food processing environments, contributing to contamination (Jiang et al. 2021).
  - *Escherichia coli*: Including pathogenic strains like *E. coli* O157, which can form biofilms in various food processing settings and pose serious health risks (Wang et al. 2023).
- **Fungi**:
  - *Aspergillus* spp.: Commonly found in biofilms on grains and nuts, contributing to spoilage and mycotoxin production.
  - *Candida* spp.: Yeasts like *Candida albicans* can form biofilms on various food surfaces and in food processing environments, affecting food quality and safety (Deng et al. 2020).

The dominance of specific microbial species in a biofilm depends on factors such as nutrient availability, surface properties, and environmental conditions. For instance, nutrient-rich environments may favor the growth of specific bacteria, while low-nutrient conditions might select for more resilient species (Shen et al. 2021).

6.4 Environmental Influences on Biofilm Communities

The composition and structure of biofilm communities are influenced by several environmental factors:

- **Nutrient Availability**: The presence and concentration of nutrients in the environment affect microbial growth and biofilm development. Nutrient-rich environments, such as those found in food processing areas, can support diverse microbial populations and enhance biofilm formation (Zhou et al. 2024).
- **Temperature**: Temperature variations can influence the growth rates and metabolic activities of microorganisms. Optimal temperatures for biofilm formation vary depending on the microbial species, but extreme temperatures can impact biofilm structure and function.
- **pH**: The pH of the environment affects microbial growth and biofilm formation. Certain microorganisms thrive in acidic or alkaline conditions, and changes in pH can influence the microbial composition and the stability of the biofilm (Exton et al. 2024).
- **Hydrodynamics**: Flow conditions, such as water or liquid flow rates, impact biofilm formation and development. High shear forces can dislodge biofilm cells, while low shear conditions can promote the accumulation and growth of biofilms.
- **Disinfectants and Antimicrobials**: The presence of disinfectants and antimicrobial agents in the environment can affect biofilm composition by selecting for resistant strains. Biofilms often exhibit increased resistance to such agents, making their control more challenging (Exton et al. 2024; Lin et al. 2024).

6.5 Biofilms in Food Production Systems

A. Beneficial Effects on Food Products

1. Biochemical Contributions to Flavor Development

Biofilms formed by beneficial microorganisms play a crucial role in the development of unique flavors and textures in various food products. The biochemical activities of these microbial communities contribute significantly to the sensory qualities of the food (Uddin Mahamud et al. 2024).

a. **Fermented Dairy Products**

In fermented dairy products, such as cheese and yogurt, biofilms formed by lactic acid bacteria (LAB) are essential for flavor development. LAB, including strains of *Lactobacillus*, *Streptococcus*, and *Lactococcus*, form biofilms on the surfaces of cheese curds and in the fermentation vats. These bacteria produce lactic acid, which lowers the pH and contributes to the tangy flavor of the products. Additionally, LAB enzymes break down proteins and lipids in milk, resulting in the development of complex flavors and textures. For example, in cheeses like Brie and Camembert, the biofilm-associated microorganisms contribute to the soft, creamy texture and characteristic flavor through proteolysis and lipolysis (Emiliano et al. 2024).

b. **Fermented Vegetables**

Similarly, in fermented vegetables such as sauerkraut and kimchi, microbial biofilms play a role in flavor development. Lactic acid bacteria in these biofilms ferment sugars in vegetables, producing lactic acid, acetic acid, and other metabolites that contribute to the characteristic tangy flavor and preserve the vegetables. The biofilm matrix also protects the microorganisms from environmental stresses, allowing them to thrive and produce the desired sensory attributes. The fermentation process enhances the nutritional value of the vegetables by increasing the levels of vitamins and bioactive compounds (Huang 2024; Perpetuini et al. 2024).

2. Protective Functions Against Pathogenic Microorganisms

In some cases, biofilms can have protective effects against pathogenic microorganisms, contributing to the safety and quality of food products.

- **Competitive Exclusion**: Beneficial microbial biofilms can inhibit the growth of pathogenic microorganisms through competitive exclusion. By occupying available niches and utilizing nutrients, these beneficial microbes prevent pathogenic organisms from establishing themselves. For instance, in fermented dairy products, the presence of LAB biofilms can prevent the growth of spoilage organisms and pathogens (Li et al. 2024).
- **Production of Antimicrobial Substances**: Certain microorganisms within biofilms produce antimicrobial substances, such as bacteriocins and organic acids, which inhibit the growth of harmful pathogens. For example, LAB in fermented products produces lactic acid and other antimicrobial compounds that help suppress the growth of pathogenic bacteria (Mgomi et al. 2023).

B. **Detrimental Effects on Food Quality and Safety**

1. **Biofilm-Mediated Food Spoilage**

While biofilms can have beneficial effects, they can also contribute to food spoilage and quality deterioration.

a. **Mechanisms of Spoilage**

Biofilms formed by spoilage microorganisms can lead to significant quality issues in food products. Spoilage bacteria and fungi produce enzymes and metabolic by-products that cause off-flavors, odors, and textural changes. For example, in dairy products, biofilms formed by spoilage bacteria can produce proteolytic and lipolytic enzymes that break down proteins and fats, resulting in rancid flavors and textures. In meat products, biofilms formed by microorganisms can lead to sliminess and off-odors due to the production of polysaccharides and other by-products (Korber et al. 2009).

b. **Economic Losses**

The economic impact of biofilm-mediated spoilage is considerable. The presence of spoilage microorganisms in biofilms can lead to product losses, reduced shelf-life, and increased costs for cleaning and maintenance. Spoiled food products may need to be discarded, leading to financial losses for producers and distributors. Additionally, the need for enhanced cleaning protocols and equipment maintenance to manage biofilms increases operational costs (Chitlapilly Dass and Wang 2022).

2. Biofilms as Reservoirs for Foodborne Pathogens

Biofilms are also significant concerns for food safety due to their role as reservoirs for pathogenic microorganisms.

a. **Persistence and Resistance Mechanisms**

Biofilms provide a protective environment for pathogenic microorganisms, making them more resistant to cleaning and disinfection procedures. The EPS matrix and the physical structure of biofilms enhance the survival of pathogens in harsh conditions, such as high temperatures, low pH, and exposure to disinfectants. For instance, pathogens like *Listeria monocytogenes* and *Salmonella* spp. can persist in biofilms on food contact surfaces, leading to ongoing contamination risks (Galié et al. 2018).

b. **Public Health Risks**

The presence of pathogenic microorganisms in biofilms poses significant public health risks. Contaminated food products can lead to outbreaks of foodborne illnesses, with potential consequences for consumer health and safety. Biofilm-associated pathogens are often difficult to detect and control, increasing the likelihood of contamination and subsequent health issues. Effective biofilm management and control measures are essential to mitigate these risks and ensure the safety of food products (Lu et al. 2022; Muhammad et al. 2020).

6.6 Biofilm-Induced Textural Modifications in Food Products

Biofilms, complex communities of microorganisms embedded in a self-produced matrix, play a significant role in the textural characteristics of various food products. Depending on the type of microorganisms involved and the specific food matrix, biofilms can either enhance or deteriorate the texture of food. Beneficial biofilms, often formed by lactic acid bacteria (LAB) and yeasts, contribute to desirable textural attributes in fermented foods by producing acids and enzymes that modify food structure (Lin et al. 2017, 2022). Conversely, spoilage biofilms, composed of bacteria and fungi, can lead to undesirable changes such as sliminess, softening, and off-textures due to their metabolic activities. The following table provides a detailed overview of how biofilms influence the texture of different food products, highlighting both beneficial and detrimental effects (Alvarez-Ordóñez et al. 2019; Carrascosa et al. 2021; Galié et al. 2018; Giaouris et al. 2014) (Table 6.1).

6.7 Strategies for Biofilm Control in the Food Industry

A. **Sanitation Protocols and Hygiene Practices**
   1. **Efficacy of Chemical Cleaners and Disinfectants**
   Chemical cleaners and disinfectants are essential tools in controlling biofilms in the food industry. Their efficacy depends on the type of chemical used, its concentration, and the contact time with the biofilm (Bai et al. 2021).

- **Chlorine-Based Disinfectants**: Chlorine compounds, such as sodium hypochlorite, are commonly used due to their broad-spectrum antimicrobial activity. They work by oxidizing cellular components, leading to microbial death. However, biofilms can exhibit increased resistance to chlorine, necessitating higher concentrations or prolonged exposure times.
- **Quaternary Ammonium Compounds (QACs)**: QACs are cationic surfactants that disrupt cell membranes, leading to cell lysis. They are effective against a wide range of microorganisms but may be less effective against biofilms due to the protective extracellular polymeric substances (EPS) (Lu et al. 2022; Simões et al. 2010).
- **Hydrogen Peroxide and Peracetic Acid**: These oxidizing agents are effective in breaking down biofilms by generating reactive oxygen species that damage cellular components. They are also effective against a wide range of microorganisms and can be used in combination with other disinfectants for enhanced efficacy.
- **Enzyme-Based Cleaners**: Enzymatic cleaners contain specific enzymes, such as proteases, lipases, and amylases, which target and degrade components of the biofilm matrix. These cleaners can enhance the effectiveness of traditional disinfectants by disrupting the biofilm structure (Galié et al. 2018; Giaouris and Simões 2018).

Table 6.1 Biofilm-driven texture alterations in food

Food product	Type of biofilm	Textural modification	Mechanism
Cheese	Lactic acid bacteria (LAB)	Softening and creaminess	Proteolysis of casein proteins and lipolysis of fats by LAB
Fermented dairy (e.g., yogurt)	Lactic acid bacteria (LAB)	Smooth and creamy texture	Fermentation reduces pH, causing milk proteins to gel
Sauerkraut	Lactic acid bacteria (LAB)	Crunchiness with slight softening	Fermentation produces lactic acid, which alters texture
Kimchi	Lactic acid bacteria (LAB)	Crunchy texture with slight softening	Similar to sauerkraut, lactic acid production affects texture
Meat	Spoilage bacteria (e.g., *Pseudomonas*)	Sliminess and off-texture	Production of polysaccharides and proteolytic enzymes
Poultry	Spoilage bacteria (e.g., *Enterobacter*)	Slimy, sticky texture	Biofilm-associated enzymes break down proteins, causing spoilage
Vegetables (e.g., carrots)	Spoilage bacteria (e.g., *Bacillus*)	Softening and mushiness	Microbial enzymes break down structural polysaccharides
Grains (e.g., rice)	Fungi (e.g., *Aspergillus*)	Softening and loss of firmness	Enzymatic activity by fungi affects grain texture
Bread and baked goods	Yeasts (e.g., *Saccharomyces*)	Spongy and airy texture	Yeast fermentation produces CO_2, causing dough to rise
Pickles	Lactic acid bacteria (LAB)	Crispness and firmness	Lactic acid fermentation maintains firmness by preserving cell walls
Wine	Yeasts (e.g., *Saccharomyces*)	Smooth mouthfeel	Yeast fermentation produces glycerol, contributing to mouthfeel
Beer	Yeasts (e.g., *Saccharomyces*)	Effervescence and body	Yeast fermentation produces CO_2 and alcohol, affecting texture
Vinegar	Acetic acid bacteria (e.g., *Acetobacter*)	Firmness and integrity in vegetables	Acetic acid production preserves texture
Tofu	Spoilage bacteria (e.g., *Pseudomonas*)	Sliminess and off-texture	Production of exopolysaccharides and proteolytic enzymes
Fish	Spoilage bacteria (e.g., *Shewanella*)	Slimy texture and off-flavors	Production of trimethylamine and proteolytic activity
Juices (e.g., apple juice)	Spoilage yeasts (e.g., *Zygosaccharomyces*)	Cloudiness and off-flavors	Yeast fermentation causes cloudiness and flavor changes
Cured meats (e.g., salami)	Lactic acid bacteria (LAB)	Firm texture and flavor development	LAB fermentation produces lactic acid and lowers pH, firming meat

(continued)

Table 6.1 (continued)

Food product	Type of biofilm	Textural modification	Mechanism
Olives	Lactic acid bacteria (LAB)	Firmness and texture improvement	Fermentation preserves cell structure and enhances firmness
Soy sauce	Molds (e.g., *Aspergillus oryzae*)	Smooth texture and flavor complexity	Enzymatic breakdown of soybeans and wheat, producing umami flavor
Sourdough bread	Lactic acid bacteria (LAB) and Yeasts	Chewy texture and complex flavors	LAB and yeast fermentation produces organic acids and CO_2
Dried fruits	Spoilage fungi (e.g., *Penicillium*)	Softening and texture degradation	Enzymatic degradation of cell walls and polysaccharides
Nuts	Spoilage fungi (e.g., *Aspergillus*)	Softening and off-flavors	Enzymatic activity and mycotoxin production by fungi
Chocolate	Spoilage bacteria (e.g., *Leuconostoc*)	Graininess and textural defects	Polysaccharide production affecting texture

2. **Mechanical and Physical Cleaning Techniques**

Mechanical and physical cleaning techniques are crucial for physically removing biofilms from surfaces in food processing environments. These methods include the following:

- **High-Pressure Washing**: High-pressure water jets can effectively remove biofilms from surfaces by dislodging and flushing away the biofilm matrix and associated microorganisms. This method is particularly useful for cleaning large equipment and surfaces (Chmielewski and Frank 2003).
- **Brushing and Scrubbing**: Manual brushing and scrubbing can effectively remove biofilms from smaller or irregular surfaces where high-pressure washing may not be feasible. This technique can be combined with chemical cleaners for improved results.
- **Ultrasonic Cleaning**: Ultrasonic waves can disrupt biofilms by causing cavitation, which generates localized high-energy shock waves that detach biofilms from surfaces. This method is effective for cleaning complex equipment and hard-to-reach areas.
- **Thermal Treatments**: Hot water or steam cleaning can kill microorganisms and disrupt biofilms. High temperatures denature proteins and other biofilm components, leading to biofilm degradation. This method is effective but may not be suitable for all surfaces and equipment due to potential heat damage (Cappitelli et al. 2014; Dallagi et al. 2023).

B. Innovative Approaches for Biofilm Management

1. Use of Biofilm-Disrupting Agents

a. Enzymatic Treatments

Enzymatic treatments involve the application of specific enzymes that target and degrade biofilm components. Enzymes such as proteases, polysaccharide-degrading enzymes (e.g., alginate lyase), and DNases can break down proteins, polysaccharides, and extracellular DNA in the biofilm matrix, respectively. This targeted approach weakens the biofilm structure, making it more susceptible to removal by mechanical or chemical means.

- **Proteases**: These enzymes break down proteins in the biofilm matrix, reducing its structural integrity.
- **Polysaccharide-Degrading Enzymes**: Enzymes like alginate lyase target polysaccharides, which are key components of the EPS, thereby disrupting the biofilm matrix.
- **DNases**: These enzymes degrade extracellular DNA, which contributes to the stability and adhesion of the biofilm (Holah 2014).

b. Bacteriophage Applications

Bacteriophages, or phages, are viruses that specifically infect and kill bacteria. Phage therapy for biofilm control involves using phages that can penetrate biofilms and lyse bacterial cells.

- **Specificity**: Phages can be highly specific to target bacteria, reducing the impact on beneficial microorganisms.
- **Efficacy**: Phages can penetrate the biofilm matrix and infect bacteria, leading to biofilm disruption.
- **Adaptability**: Phages can co-evolve with bacteria, potentially reducing the development of bacterial resistance (Chan and Abedon 2015; Liu et al. 2022).

2. Development of Anti-Biofilm Surfaces and Coatings

a. Material Science Innovations

Advances in material science have led to the development of surfaces and coatings that resist biofilm formation. These materials can be designed with various properties to prevent microbial adhesion and biofilm development.

- **Hydrophobic Surfaces**: Hydrophobic materials repel water and reduce microbial adhesion, making it difficult for biofilms to establish.
- **Anti-Adhesive Coatings**: Coatings with anti-adhesive properties, such as Teflon or silicone, can prevent microorganisms from adhering to surfaces.
- **Antimicrobial Surfaces**: Incorporating antimicrobial agents, such as silver nanoparticles or copper alloys, into surfaces can provide continuous antimicrobial activity, preventing biofilm formation (Motlagh et al. 2016).

b. **Implementation of Food Processing Equipment**

Applying anti-biofilm surfaces and coatings to food processing equipment can significantly reduce biofilm-related issues.

- **Conveyor Belts**: Coating conveyor belts with anti-adhesive or antimicrobial materials can prevent biofilm formation and facilitate easier cleaning.
- **Pipes and Tanks**: Treating the internal surfaces of pipes and tanks with anti-biofilm coatings can reduce biofilm buildup and improve sanitation.
- **Cutting and Slicing Equipment**: Applying antimicrobial coatings to cutting and slicing equipment can minimize the risk of biofilm formation and cross-contamination (Shineh et al. 2023).

6.8 Conclusion

Biofilms have a significant impact on food texture and safety, with both beneficial and detrimental effects. While beneficial biofilms enhance flavors and textures in fermented foods, detrimental biofilms cause spoilage and pose public health risks. Effective biofilm control in the food industry requires integrated strategies, including sanitation protocols, innovative treatments, and advanced material science solutions. Combining mechanical, chemical, and enzymatic cleaning methods, along with the use of biofilm-disrupting agents and anti-biofilm surfaces, is crucial. Future research should focus on improving biofilm detection and monitoring, as well as developing novel interventions and preventive measures. By adopting comprehensive biofilm management practices, the food industry can ensure higher food quality and safety, ultimately protecting consumer health and reducing economic losses.

References

Aguirre-Ramírez M, Silva-Jiménez H, Banat IM, Díaz De Rienzo M (2021) Surfactants: physicochemical interactions with biological macromolecules. Biotechnol Lett 43(3):523–535
Alotaibi GF, Bukhari MA (2021) Factors influencing bacterial biofilm formation and development. Am J Biomed Sci Res 12(6):617–626

Alvarez-Ordóñez A, Coughlan LM, Briandet R, Cotter PD (2019) Biofilms in food processing environments: challenges and opportunities. Annu Rev Food Sci Technol 10(1):173–195
Andryukov B, Romashko R, Efimov T, Lyapun I, Bynina M, Matosova E (2020) Mechanisms of adhesive–cohesive interaction of bacteria in the formation of biofilm. Mol Genet Microbiol Virol 35(4):195–201
Bai X, Nakatsu CH, Bhunia AK (2021) Bacterial biofilms and their implications in pathogenesis and food safety. Food Secur 10(9):2117
Cappitelli F, Polo A, Villa F (2014) Biofilm formation in food processing environments is still poorly understood and controlled. Food Eng Rev 6:29–42
Carrascosa C, Raheem D, Ramos F, Saraiva A, Raposo A (2021) Microbial biofilms in the food industry—a comprehensive review. Int J Environ Res Public Health 18(04):2014
Chan BK, Abedon ST (2015) Bacteriophages and their enzymes in biofilm control. Curr Pharm Des 21(1):85–99
Chitlapilly Dass S, Wang R (2022) Biofilm through the looking glass: A microbial food safety perspective. Pathogens 11(3):346
Chmielewski R, Frank J (2003) Biofilm formation and control in food processing facilities. Compr Rev Food Sci Food Saf 2(1):22–32
Dallagi H, Jha PK, Faille C, Le-Bail A, Rawson A, Benezech T (2023) Removal of biocontamination in the food industry using physical methods; an overview. Food Control 148:109645
DeFlorio W, Liu S, White AR, Taylor TM, Cisneros-Zevallos L, Min Y, Scholar EM (2021) Recent developments in antimicrobial and antifouling coatings to reduce or prevent contamination and cross-contamination of food contact surfaces by bacteria. Compr Rev Food Sci Food Saf 20(3):3093–3134
Deng Y, Wang R, Wang Y, Sun L, Tao S, Li X, Gooneratne R, Zhao J (2020) Diversity and succession of microbial communities and chemical analysis in dried Lutianus erythropterus during storage. Int J Food Microbiol 314:108416
Dunkley KE, Hekmat S (2020) Development of probiotic vegetable juice using Lactobacillus Rhamnosus GR-1. Nutr Food Sci 50(5):955–968
Emiliano JVDS, Fusieger A, Camargo AC, Rodrigues FFDC, Nero LA, Perrone ÍT, Carvalho AFD (2024) Staphylococcus aureus in dairy industry: enterotoxin production, biofilm formation, and use of lactic acid bacteria for its biocontrol. Foodborne Pathog Dis 21:601
Exton B, Hassard F, Medina-Vaya A, Grabowski RC (2024) Undesirable river biofilms: the composition, environmental drivers, and occurrence of sewage fungus. Ecol Indic 161:111949
Flemming H-C, van Hullebusch ED, Neu TR, Nielsen PH, Seviour T, Stoodley P, Wingender J, Wuertz S (2023) The biofilm matrix: multitasking in a shared space. Nat Rev Microbiol 21(2):70–86
Galié S, García-Gutiérrez C, Miguélez EM, Villar CJ, Lombó F (2018) Biofilms in the food industry: health aspects and control methods. Front Microbiol 9:898
Ganesan S, Ruendee T, Kimura SY, Chawengkijwanich C, Janjaroen D (2022) Effect of biofilm formation on different types of plastic shopping bags: Structural and physicochemical properties. Environ Res 206:112542
Giaouris EE, Simões MV (2018) Pathogenic biofilm formation in the food industry and alternative control strategies. In: Foodborne diseases. Elsevier, pp 309–377
Giaouris E, Heir E, Hébraud M, Chorianopoulos N, Langsrud S, Møretrø T, Habimana O, Desvaux M, Renier S, Nychas G-J (2014) Attachment and biofilm formation by foodborne bacteria in meat processing environments: causes, implications, role of bacterial interactions and control by alternative novel methods. Meat Sci 97(3):298–309
Gurtler V (2023) Biofilms. Elsevier
Holah J (2014) Cleaning and disinfection practices in food processing. In: Lelieveld HLM, Holah JT, Napper D (eds) Hygiene in food processing, pp 259–304
Huang R (2024) Exploring kombucha: production, microbiota biotransformation, flavor, health benefits and potential risks. ACS Food Sci Technol 4:1610

Jiang S, Xing Y, Liu G, Hu C, Wang X, Yan G, Wang Q (2021) Changes in soil bacterial and fungal community composition and functional groups during the succession of boreal forests. Soil Biol Biochem 161:108393

Korber D, Mangalappalli-Illathu A, Vidović S (2009) Biofilm formation by food spoilage microorganisms in food processing environments. In: Biofilms in the food and beverage industries. Elsevier, pp 169–199

Landis EA (2021) Deconstructing and reconstructing fermented food microbiomes: revealing microbial interactions and developing new educational models. Tufts University

Li Y, Liang X, Chen N, Yuan X, Wang J, Wu Q, Ding Y (2024) The promotion of biofilm dispersion: a new strategy for eliminating foodborne pathogens in the food industry. Crit Rev Food Sci Nutr:1–25

Lin S, Yang L, Chen G, Li B, Chen D, Li L, Xu Z (2017) Pathogenic features and characteristics of food borne pathogens biofilm: biomass, viability and matrix. Microb Pathog 111:285–291

Lin Y-T, Wang Y-C, Xue Y-M, Tong Z, Jiang G-Y, Hu X-R, Crittenden JC, Wang C (2024) Decoding the influence of low temperature on biofilm development: The hidden roles of c-di-GMP. Sci Total Environ 927:172376

Liu D, Huang Q, Gu W, Zeng X-A (2022) A review of bacterial biofilm control by physical strategies. Crit Rev Food Sci Nutr 62(13):3453–3470

Lu J, Hu X, Ren L (2022) Biofilm control strategies in food industry: Inhibition and utilization. Trends Food Sci Technol 123:103–113

Mgomi FC, Yang Y-R, Cheng G, Yang Z-Q (2023) Lactic acid bacteria biofilms and their antimicrobial potential against pathogenic microorganisms. Biofilms 5:100118

Mlambo V, Mnisi CM, Matshogo TB, Mhlongo G (2022) Prospects of dietary seaweeds and their bioactive compounds in sustainable poultry production systems: a symphony of good things? Front Anim Sci 3:998042

Mota J, Vilela A (2024) Aged to perfection: the scientific symphony behind port wine, vinegar, and acetic acid bacteria. Fermentation 10(4):200

Motlagh AM, Bhattacharjee AS, Goel R (2016) Biofilm control with natural and genetically-modified phages. World J Microbiol Biotechnol 32:1–10

Muhammad MH, Idris AL, Fan X, Guo Y, Yu Y, Jin X, Qiu J, Guan X, Huang T (2020) Beyond risk: bacterial biofilms and their regulating approaches. Front Microbiol 11:928

Munteanu-Ichim R-A, Canja C-M, Lupu M, Bădărău C-L, Matei F (2024) Tradition and innovation in yoghurt from a functional perspective—a review. Fermentation 10(7):357

Panneerselvam D, Murugesan P, Moses J (2024) Silk fibroin and prospective applications in the food sector. Eur Polym J 113058

Perpetuini G, Rossetti AP, Rapagnetta A, Arfelli G, Prete R, Tofalo R (2024) Wine barrel biofilm as a source of yeasts with non-conventional properties. Microorganisms 12(5):880

Saha I, Datta S, Biswas D (2020) Exploring the role of bacterial extracellular polymeric substances for sustainable development in agriculture. Curr Microbiol 77(11):3224–3239

Saini S, Tewari S, Dwivedi J, Sharma V (2023) Biofilm-mediated wastewater treatment: a comprehensive review. Mater Adv 4(6):1415–1443

Shen D, Shen H, Yang Q, Chen S, Dun Y, Liang Y, Zheng J, Zhao S (2021) Deciphering succession and assembly patterns of microbial communities in a two-stage solid-state fermentation system. Microbiol Spectr 9(2):e00718–e00721

Shineh G, Mobaraki M, Perves Bappy MJ, Mills DK (2023) Biofilm formation, and related impacts on healthcare, food processing and packaging, industrial manufacturing, marine industries, and sanitation–a review. Appl Microbiol 3(3):629–665

Simões M, Simões LC, Vieira MJ (2010) A review of current and emergent biofilm control strategies. LWT-Food Sci and Technol 43(4):573–583

Uddin Mahamud AS, Nahar S, Ashrafudoulla M, Park SH, Ha S-D (2024) Insights into antibiofilm mechanisms of phytochemicals: prospects in the food industry. Crit Rev Food Sci Nutr 64(6):1736–1763

Wang Y, Zhang C, Liu F, Jin Z, Xia X (2023) Ecological succession and functional characteristics of lactic acid bacteria in traditional fermented foods. Crit Rev Food Sci Nutr 63(22):5841–5855
Zhou Y, Meng F, Ochieng B, Xu J, Zhang L, Kimirei IA, Feng M, Zhu L, Wang J (2024) Climate and environmental variables drive stream biofilm bacterial and fungal diversity on tropical mountainsides. Microb Ecol 87(1):28

Aroma Amplification by Biofilm

7

Mohan Das, Sayantan Santra, Priyadharshini Jayaseelan, Devalina Khamaru, Sarobi Das, and Rintu Banerjee

Abstract

The process of biofilm development is a dynamic phenomenon that encompasses several processes involved in the adhesion and proliferation of bacteria. Extracellular polymeric compounds play a vital role in the adherence and establishment of bacteria on surfaces that come into contact with food. Nevertheless, the presence of these naturally established biofilms in the environment might result in detrimental outcomes, including contamination of equipment and issues with food safety. But recent reports demonstrated that microbial fermentation facilitated by biofilms offers a promising method for efficient biochemical production due to their ability to self-immobilize, resist toxic substances, and sustain long-term cellular activity. This expands a new horizon for application of biofilm in food industry, specifically for improving the aroma and texture of the food components. Therefore, this chapter discusses the mechanism and signalling of biofilm formation followed by application of biofilm for improving the aroma and texture of different types of food and beverages. Additionally, it also includes the chemistry of aromatic compounds produced by microbial biofilm with health beneficial activities. Hence, the use of biofilm-mediated fermentation is expected to be promising prospects for the future development of biorefinery platforms.

M. Das (✉) · S. Santra · P. Jayaseelan · D. Khamaru · R. Banerjee
Microbial Biotechnology and Downstream Processing Laboratory,
Agricultural and Food Engineering Department, Indian Institute of Technology Kharagpur,
Kharagpur, West Bengal, India
e-mail: mohan@agfe.iitkgp.ac.in

S. Das
Department of Biology, University of Texas at Arlington, Arlington, TX, USA

T. Sarkar et al. (eds.), *Biofilm Applications to Revolutionize Food Technology*,
https://doi.org/10.1007/978-3-031-85205-3_7

Keywords

Aroma chemistry · Biofilm · Biofilm safety · Microbial fermentation

7.1 Introduction

A biofilm can be defined as the collection of surface-associated cells of different microbial communities encased within an extracellular polymeric material in a matrix. Biofilms are dynamic and complex systems with a heavy cell density of 10^8 to 10^{11} cells g^{-1} wet weight range (Flemming et al. 2016). Biofilms can contain only one type of species or exist with multiple microbial species, forming a heterogeneous community composed of different types of microorganisms depending upon the specific environment where they are found. Some commonly seen microbes include bacteria, archaea, fungi, protozoa, algae, actinomycetes, and viruses like bacteriophages (Sharma et al. 2023). Each microbial community has a specific role in the biofilm, as they interact with each other to coexist. The biofilms are found in biotic as well as abiotic surfaces, providing favourable conditions for the survival of microbial communities despite the harsh ecology and other crises (O'Toole et al. 2000). The formation of biofilms is regulated by both the genetics of the microorganism and the niche or the environmental conditions they are surrounded with. The molecular mechanism in which the microbes regulate the formation of the biofilm will vary according to the genetic makeup of the organism. However, we can generalize the attributes of biofilm to compose an extracellular matrix made up of polysaccharide biopolymer with lipids, protein, and extracellular DNA that can hold the microbial cells together (López et al. 2010). The characteristics of the matrix will differ based on the medium in which it is growing, the growth conditions, and the substrates. Although biofilms are generally regarded as on-beneficial or harmful to human health, recent findings suggest that microbial biofilm can be helpful for several industrial aspects such as slowing metal corrosion, treatment of effluents and removal of toxic substances. Interestingly, biofilm-mediated fermentation is an age-old traditional treatment for processing foods with improved health benefits, texture, and aroma. This chapter discusses the formation and biochemistry of microbial biofilm, its application in food fermentation, the aroma chemistry of biofilm-mediated fermented foods and its safety concerns.

7.2 Formation and Characteristics of Biofilms

The formation of the biofilm, that is, conversion of planktonic cells to community structure, can be divided into different stages: firstly, the attachment of the free-floating microorganism to a surface, then cell-to-cell adhesion, followed by the secretion of EPS material, along with the multiplication and accumulation of microorganisms within the EPS matrix and finally dispersion. The mobility of the bacteria can be accounted for by fimbriae and flagella. The longitudinal attachment of the bacterial cell pole or the flagellum with the surface of the substratum occurs during the reversible attachment (Sauer et al. 2022). The forces responsible for such attachment are van der Waals, electrostatic, and other non-bonded attractive forces between the organism and the surface. After the establishment of the initial connection, the stable, irreversible connection is made and maintained through specific ligand-receptor interaction between the bacterial protein and the surface. The hydrophobicity, the surface charge of the cell, and the surface adhesion of fimbriae or glycocalyx play an important role here (Nandakumar et al. 2013).

Once the microorganisms attach to the surface of the biofilm, maturation begins, in which the cells multiply and divide to form microcolonies along with the secretion of EPS substrate. The maturation phase is divided into two phases. The maturation phase-I lasts up to 3 days, in which the microbes produce EPS, and the microcolonies increase to a cell density range of 10 μm of multi-layer thickness. The second phase of maturation stays for nearly 6 days, during which the microbial colonies expand to a layer thickness of 100 μm (Marić and Vraneš 2007). It has been observed that the microorganism inside the EPS changes its protein structure to cope with the environmental adaptations. A significant phenotypic change in the protein structure of the microbial cells in phase II of maturation was in comparison with the free-living/planktonic bacteria (Funari and Shen 2022).

Finally, the last stage of biofilm formation involves dispersion, where the planktonic bacteria are dispersed to go and further colonize. The biofilm has two layers: one is the foundational layer, and the other is the superficial or surface layer; this second layer acts as the dispersal zone from where the microbes spread to the new niche. This happens as the biofilm grows old, the metabolic by-products accumulate, and the toxicity level increases as well as the resources for survival to the biofilm community deplete. In such instances, autophosphorylation occurs, triggering the release of short-chain fatty acids such as cis-11-methy-2-dodecenoic acids, which in turn activates the breakdown of cyclic diguanylate Guanosine monophosphate phosphodiesterase (c-di-GMP) (Veerachamy et al. 2014). The breakdown of c-di-GMP causes sheer pressure on the biofilm surface layers, causing the dispersion and release of microbes. Apart from this mechanism, the enzymes produced by the native microbes facilitate the breakdown by lysis of saccharides of the surface layer of EPS, resulting in dispersion (Gupta et al. 2016; Grande et al. 2014). The pictorial representation of the stages involved in the formation of the biofilm is presented in Fig. 7.1.

7.3 Architecture of Biofilms Formed

The biofilm is a heterogeneous structure that comprises about 2–5% of microbial cells, polysaccharides of 1–2%, with proteins less than 1–2% and external DNA and RNA accounting for less than a percentage; the rest of the composition is water up to 94–97% (Balducci et al. 2023). Considering the EPS fraction of the biofilm, the polysaccharides are a major component, as they are involved in many useful functions, from the formation of the matrix to the maintenance of the biofilm, facilitating

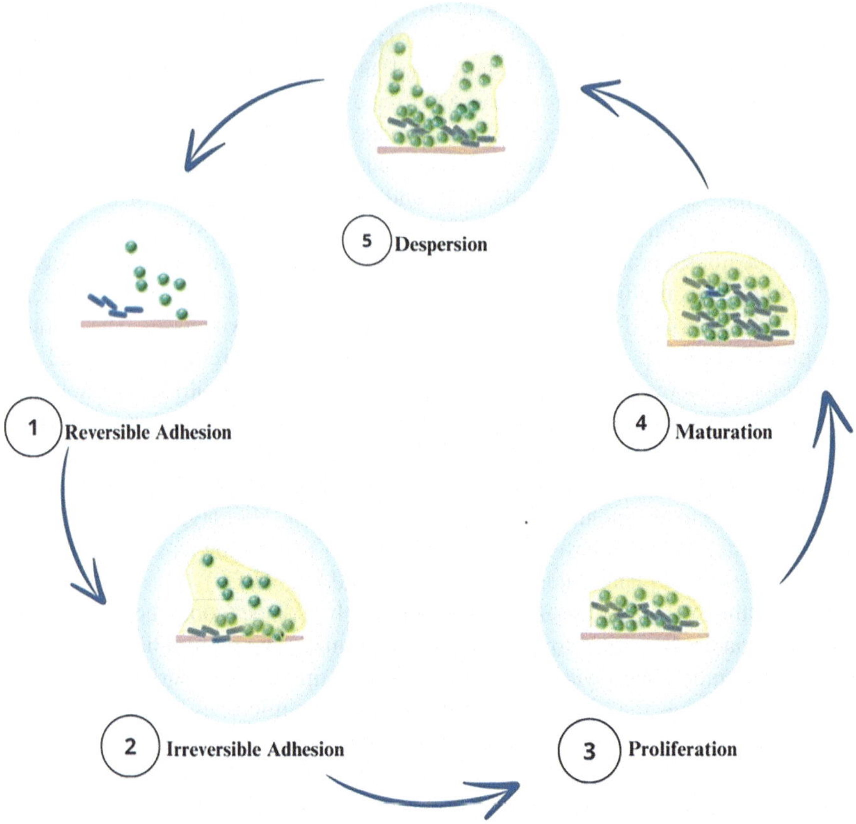

Fig. 7.1 Representation of different stages of biofilm formation by microorganism

the survival of the microbial colonies even in hostile environments (Flemming and Wingender 2010). The polysaccharides involved here are exopolysaccharides of different charges like polyanionic xanthan, alginate, colonic acids, and polycationic PNAG (Poly-β-(1 $\rightarrow$ 6)-*N*-acetylglucosamine) and Pel. Also, some homopolysaccharides like glucans, fructans, and cellulose are also involved. In fungi, the presence of galactosaminogalactans and galactomannans group of heteropolysaccharides was observed.

The PNAG was first discovered in the biofilms formed by *S. aureus* and other such pathogenic bacterial species (Nguyen et al. 2020). PNAG in the biofilm aids in the adherence to polymeric substrates and biomaterials and also helps in bacterial intracellular linking. Despite the benefits, it provides to the bacterial biofilm it is also used as an indicator in determining infections caused by such pathogens in disease detection (Olson et al. 2006). The polysaccharides such as alginate, Psl, and Pel are found in biofilm developed by *P.aeruginosa* in its different stages. The heteropolymer Psl is a combination of mannose, glucose, and rhamnose units; this polysaccharide facilitates the adhesion of cell-to-surface and cell-to-cell in both mucoid and non-mucoid strains. Apart from that, Psl also stimulates the production of c-di-GMP, which helps in producing a firm biofilm surface. In contrast, the Pel heteropolymer is found responsible for the surface attachment of the substrate and the cell, and it is only found in the non-mucoid strains (Jennings et al. 2015). Alginate offers various defence benefits to the biofilm; it is only seen when the biofilm is subjected to stress as a response (Pezzoni et al. 2022).

7.4 Signalling and Ecology of Biofilm Formation

The intracellular communication within a biofilm to regulate and synchronize the gene expression when the population of cells increases is called quorum sensing. The biofilm formation, maturation, and dispersion are interdependent with the quorum sensing between the cell population inside a biofilm. The cells release autoinducers, which act as chemical signals between the cells; they are of three different types depending upon the type of bacteria involved in the biofilm formation. The gram-negative bacteria have the acyl-homoserine lactone QS system (AHL), while the gram-positive bacteria use autoinducing peptide (AIP), and autoinducer-2 (AI-2) is present in both types of bacteria (Nyanasegran et al. 2023). The AHL molecule has a homoserine lactone ring with variation in its substitutes and length depending upon the strain. It has been observed in various models that the increase in AHL molecule has positively correlated with bacterial growth within the biofilm (Sharma et al. 2023). The AIP secretion increases as the cell density increases, and they bind with the histidine kinase sensor; upon binding, phosphorylation occurs, which results in the alteration of the target gene expression. For example, in the case of *S. aureus*, the target gene is *agr*, which is responsible for the production of various toxins and exozymes. The LuxS synthase is catalysed by the AI-2 molecule seen in specific strains of gram-positive and gram-negative bacteria. The LuxS gene activation corresponds to the activation of the methylation cycle, which in turn controls numerous genes specific to the adhesion, toxin production, and dispersion of the biofilm (Preda and Săndulescu 2019).

The various environmental factors in the ecosystem have a substantial influence on the formation of the biofilm. The chemical and physical variables that are involved include temperature, pH, the presence of toxins, nutrients, and light availability (Alotaibi and Bukhari 2021). In addition, biological factors, such as the existence of other microbial communities, have an impact. The first stage of bacterial biofilms involves adhesion, which is specific to the optimum temperature and pH of a particular strain. It has been observed that some of the *Staphylococcus* sp. (Nostro et al. 2012) could not withstand alkaline pH, resulting in impairment of biofilm formation. Observations revealed that many aquatic *Pseudomonas* sp. exhibited a decline in surface adherence as the temperature declined (Garrett et al. 2008). Conversely, in the case of *S. aureus*, elevating the temperature by 15–17 °C beyond the optimal level resulted in increased hydrophobicity and enhanced surface adhesion (Khelissa et al. 2017). Oxygen availability is a crucial component that affects biofilm formation. A drop in oxygen levels can cause the biofilm to disperse in certain microbes, whereas others, such as *P. aeruginosa*, thrive in low oxygen level.

7.5 Aroma and Texture Amplification of Fermented Food by Biofilm

Biofilm production is quite likely in food processing lines, particularly in the food business, because of the continuous supply of nutrients and the undisturbed nature of pipes and trays in mass manufacturing units, which allows for the adherence and maturation of the biofilm. The predominant substrate for biofilm growth in the food industry is stainless-steel plates or pipelines (Lindsay and Von Holy 2006). While these materials are corrosion-resistant and do not offer a conducive rough surface for bacterial attachment, there is a potential for the development of fractures or fissures over time. These fractures provide sufficient space for airborne or food-borne bacteria to build biofilms inside, as physical cleaning and other sanitation techniques are challenging to carry out in these crevices and fissures. The food waste or spillage also acts as a matrix for the attachment of microbes, as these contain needed nutrients that would foster their growth and multiplication (Gomes et al. 2015). Biofilm's impact on the food sector is both advantageous and disadvantageous. Undesirable or unplanned biofilms can cause significant issues such as food-borne illnesses, cross-contamination, and the development of off-flavours and disagreeable aromas in food. Conversely, biofilms can provide several advantages when seen in appropriate circumstances. Biofilm exerts a beneficial influence on fermentation processes such as cheese, yoghurt, and plant-based fermented meals. Additionally, it improves the quality of food and promotes the production of aroma. It also plays a role in food packaging through the use of edible biofilms. Though traditionally, biofilm holds the stigma of spoilage, controlled growth of biofilms confers the best results. It has been proven that biofilms can provide a protective environment and improve the metabolic activity of microorganisms, which can be exploited for fermenting microbes.

Beverage Fermentation

Microbes play an essential part in the unique and complicated combination of chemistry, biology, and culture that is wine. The intricacy of the volatile elements that contribute to wine scent has prompted much research into the chemical components of vine and wine flavour. Of the three components that make up wine flavour—taste, fragrance, and visual attributes—the most important is aroma, or scent (Polášková et al. 2008). According to research, the microbiome of grapes may provide clues about the amount and the presence of different kinds of metabolic compounds that affect wine quality (Bokulich et al. 2016). The fact that wine fermentation involves more than one type of microbes is already common knowledge, and scientists all over the globe are studying numerous microbes that contribute to the winemaking operation. While various microbes, including fermentative and prominent grape epiphytes, may contribute to wine flavour, little is known about their precise function in the winemaking process (Belda et al. 2016). The majority of a wine's fragrance is often composed of volatile chemicals that are produced during fermentation. Several alcohols and esters are generated as a result of alcoholic fermentation, which is mostly accomplished by *Saccharomyces cerevisiae* (Álvarez-Pérez et al. 2012). Lactic acid bacteria (LAB) perform a crucial role in wine making process where it transform malic acid into lactic acid, which is called malolactic fermentation (MLF). This typically follows alcoholic fermentation as a characteristic of wine. Numerous genera of LAB, such as *Lactobacillus*, *Leuconostoc*, and *Oenococcus*, have the ability to decarboxylate L-malate. *Oenococcus oeni*, among the above options, demonstrates the highest capability to sustain its metabolic processes in conditions characterized by mild acidity (about pH 3.5) and an abundance of sulphur dioxide (Lonvaud-Funel 1999). The utilization of *O. oeni* in MLF enhances the sensory characteristics and microbial equilibrium of wine by consuming the remaining sugar and facilitating bacterial fermentation of other products, resulting in the generation of lactic acid (Lonvaud-Funel 1995; Nedovic et al. 2000). MLF is an essential procedure for the majority of red wines and certain white wines. It enhances the flavour of wines by diminishing the sharpness caused by malic acid, while also offering benefits such as microbiological stability and enhanced aromatic richness. *O. oeni* has complex defences that allow it to thrive in difficult circumstances like wine. One of these techniques is modifying the proportion of saturated to unsaturated fatty acids to increase or decrease the integrity of the membrane (Grandvalet et al. 2008; Maitre et al. 2014). *Oenococcus oeni* is a common microorganism that is often found in natural MLF. Additionally, it is often utilized as a starter culture for this particular reason (Bastard et al. 2016). It inhibits the growth of microorganisms that might cause spoiling and enhances the perceptual qualities of wine. Previous investigations have demonstrated that *O. oeni* exhibits the ability to endure for several months under challenging winemaking circumstances when attached to oak barrels. Biofilm-forming strain of *O. oeni* provided (1) enhanced resistance under adverse conditions in wine and (2) demonstrated excellent malolactic activity, contributing to its functionality.

The *Saccharomyces* yeast species play a crucial role in large-scale food fermentations in the industry. Nevertheless, several yeast species have great potential for enhancing flavour profiles, although research into them remains limited. The process of winemaking involves the production of fusel or higher alcohols, which are aliphatic and aromatic in nature. The phenomenon of increased alcohol production by yeasts is highly significant in the realm of fermented drinks. Medium-chain alcohols have a significant influence on the overall taste of alcoholic drinks, even when present in small amounts. Although high amounts of fusel alcohols can cause undesirable flavours, modest levels of these molecules and their esters perform a key function in creating the fundamental profile of flavours and aroma found in wine. In the method of producing wine, the presence of *Lactobacillus plantarum* Lpls22 facilitated transformation of L-malate into lactate by the formation of biofilm. This biofilm exhibited beneficial characteristics for the production of additional aromatic compounds (Pannella et al. 2020).

The evolution of biofilm indicates that the biological state of similar bacterial cells at various times does not constitute a fundamental distinction between biofilm and free-floating cells (McDougald et al. 2012). Many earlier researches using metabolomics, proteomics, and transcriptomics techniques described the variations in metabolites and gene expression in both sessile state of biofilm free-floating state (Favre et al. 2018; Jia et al. 2017; Pannella et al. 2020). Research has shown that the biofilm state has the ability to produce a greater variety of ester compounds but a lower overall ester concentration, both in pure culture and in combined-culture with the *B. velezensis*. Interestingly, the biofilm form of *S. fibuligera* had a better capacity for synthesizing ethyl caproate, with a level double as much as the planktonic form *of S. fibuligera*. How does the existence of biofilm impact the process of taste generation in beverages? Exopolysaccharides in a biofilm have been proposed to provide protection to cells under challenging conditions and enhance their ability to survive. Conversely, exopolysaccharides consist of carbohydrates and proteins. Their monomers are produced during hydrolysis and can serve as substrates for the production of esters and other aroma-enhancing compounds (Zeng et al. 2023).

Plant-Derived Food Fermentation

Growing food in a way that does not harm the environment has made plant-based diets very popular (Kinnunen et al. 2020; Willett et al. 2019). Foods that come from plants have a limited shelf life, which makes them harder to get. However, it is acknowledged that plant-based meals are important for a nutritious as well as sustainable diet. To boost the chain of distribution of food, fermentation seems to be a good way to keep food fresh longer and make shipping easier. Historically, biofilms were originally seen as an indication of deterioration and considerable endeavours have been focused on managing and eradicating biofilms. For example, the presence of biofilms may trigger a subsequent fermentation process, which can then result in overoxidation, turbidity, textural disintegration, unwanted flavours as well as taste (Perpetuini et al. 2018; Yun et al. 2019). While biofilms have traditionally been connected to negative outcomes, recent research indicates that numerous biofilms actually have protective properties that enhance and regulate the metabolic processes of fermenting microbes. Vinegar is a widely used condiment made by fermenting cereals with *A. pasteurianus* bacteria. This fermentation process produces acetic acid, which gives vinegar its flavour. To withstand the extremely high levels of acetic acid (8%) throughout fermentation, acetic acid bacteria, LAB, and yeasts create a delicate, transparent, yellow-orange to reddish brown-coloured biofilm around junction between the fluid and air (Yun et al. 2019). This biofilm primarily consists of diverse microorganisms, such as *A. pasteurianus* and *Gluconacetobacter xylinus*. Cellulose acetate that is excreted by these microorganisms. Biofilm development and the synthesis of lactate and alcohol are aided by the myxoid protein released by *Lactobacillus plantarum* and mannan released by *S. cerevisiae*, respectively, as vinegar ages. As fermentation advances, the alcohol concentration rises, causing the biofilm layer, which consists of koji, yeasts, and LAB, to slide towards the centre of fermenting vessel. Additionally, acetic acid bacteria contribute to the development of the biofilm (Furukawa 2015; Furukawa et al. 2013).

Cereal, Grains, and Soy Fermentation

Cereal and soybean are frequently utilized as primary ingredients in the production of fermented food products, including vinegar, sufu, and beer. The process of making soybean curd includes firstly coagulating it using calcium ions (Ca^{2+}) and magnesium ions (Mg^{2+}). Then, fungi or bacteria are added to the curd and the mixture is cultured for a period of 3–4 days. Ultimately, the process of maturation takes place throughout a span of 3 to 6 months in a solution that consists of saline liquid as well as alcohol. Throughout the entire procedure, the mucor's filaments gradually spreads and creates a compact biofilm around the exterior of the soybean curd that has been inoculated (Wei et al. 2019b). The development of a biofilm, which may look like a ropey web made of spiders or an irregular coating with 'bubbles', during the fermentation process of wild or sour beer, might be interpreted as an indication of the aerobic metabolism of yeasts like *Lactobacillus* or *Pediococcus* sp. and the generation of flavourful (Ashtavinayak and Elizabeth 2016; Gong et al. 2021). Unique functional beers having rich flavours and a reduced alcohol concentration can be made by using non-traditional yeasts like *Cyberlindnera saturnus* and *Wickerhamomyces anomalus* (Basso et al. 2016; Hernández et al. 2018). The Fukuyama pot vinegar is made by the application of biofilms contacting *Aspergillus* sp. and lactic acid producing bacteria (LAB). In this instance, the microorganisms initially ferment rice into reducing sugars by utilizing the available oxygen. Subsequently, the LAB, such as *Lactobacillus plantarum*, is employed to produce myxoid protein, while *S. cerevisiae* produces mannan (Furukawa 2015). Lactic acid and ethanol are generated during anaerobic fermentation. As the concentration of ethanol in the medium increases, the biofilm migrates to the centre of the broth, where the acetic acid bacteria (AAB) undergo fermentation, resulting in the production of vinegar and a white biofilm (Wei et al. 2019a, b). The moulds/biofilms of mucor are employed to ferment soy curd for a duration of 3–6 months in the production of alcohol and saline in the case of soybean products. In the same vein, the utilization of biofilms of *Wickerhamomyces* sp. in the production of beer results in the production of a complex, flavourful beverage (Hernández et al. 2018). Fermented items made from grains and soybeans might have enhanced in quality and flavour by adjusting fermentation parameters to promote the generation of advantageous biofilms.

Vegetable Fermentation

Paocai serves as a notable illustration of biofilm development in the presence of periodic oxygen exposure. However, biofilm development is commonly considered a characteristic indication of aerobic degradation in paocai. The process of biofilm development in paocai is characterized by the sequential colonization of microorganisms and the occurrence of intricate metabolic processes. Paocai is made by immersing fresh vegetables in a jar filled with a 4–8% brine solution. This brine solution contains LAB and is used to facilitate anaerobic fermentation. They break down carbohydrates present in the raw materials and convert them into lactic acid. The pH value of the environment drops as the lactic acid content increases. Whenever the expansion of LAB is limited due to the rising concentration of lactic acid, acid-tolerant microorganisms including *Pediococcus* sp. and *Lactococcus* sp. have the ability to convert lactic acid to acetic acid in aerobic environment (Rao et al. 2018, 2019).

Pickled red pepper (PRP) is a popular fermented vegetable commonly eaten in Southeast China. A comparison was made between the physicochemical and microbiological characteristics, as well as sensory qualities, of the PRP with no biofilm and the PRP containing biofilm in order to demonstrate the impact of the biofilm. Both the biofilm-free and biofilm-containing PRP samples had low nitrite levels, and none of the prevalent pathogens were identified. Titratable acidity showed up in decreased pH and an increased microbial community in PRP-containing biofilm compared to PRP without biofilm. When comparing PRP with and without biofilm, the former exhibited more volatile and aroma-active chemical content, a wider range of volatile and aromatic molecules, and more distinctive flavours. Nevertheless, PRP with biofilm included certain undesirable fragrances, including 4-ethylguaiacol and 4-vinylguaiacol (Zhang et al. 2020).

Cocoa and coffee beans are widely consumed in South America. It is common practice to ferment these beans with the fruit pulp, which contains a diverse array of microbes, after peeling the beans. The pulp encircling the densely packed seeds creates a luscious layer that stops air from getting in during the anaerobic fermentation process. The biofilm-dwelling *S. cerevisiae* can release pectinase, which breaks down the cocoa beans' pectin into simple sugars and, eventually, ethanol—the first product of fermentation. Lactic acid, ethanol, and mannitol are produced via micro-aerobic fermentation by LAB in the biofilm as fermentation advances. The breakdown of the flesh results in the release of excess liquid and the introduction of air, initiating the aerobic stage. When oxygen is present in the biofilm, AAB (acetic acid bacteria) transforms alcohol and acids into acetic acid. This acetic acid continues to be metabolized into CO_2 and water. At elevated temperatures, the combined effect of ethanol and acetic acid causes the cocoa beans to become biologically inactive. The intricate biological interactions are responsible for the distinct colour, flavour, and fragrance of cocoa beans (Almeida et al. 2020).

Dairy Products Fermentation

LAB performs a key function in enhancing the aroma and texture of numerous kinds of cheeses. This is achieved via the activity of living cells as well as the discharge of intracellular components, notably several enzymes, as these cells break down (Garde et al. 2007). The diversification of native LAB is recognized as a crucial element in ascertaining the organoleptic properties of artisanal cheese. The production of Ragusano cheese with Protected Designation of Origin (PDO) excludes employment of starter cultures. Instead, the procedure of fermentation is initiated by naturally occurring LAB that originates from unpasteurized milk, the biofilm found on the interior of the containers, and the immediate surroundings (Licitra et al. 2007; Lortal et al. 2009). These containers might be covered by biofilm. It has just been verified that using the wooden vessel is an excellent method for adding thermophilic LAB to unpasteurized milk, which helps with the acidification process. It is strongly implied that LAB colonization via containers' biofilms is the underlying mechanism behind the acceleration of raw milk acidity in 80% of instances after only a couple of minutes upon contact with these vessels (Licitra et al. 2007). By manipulating the conventional Caciocavallo Palermitano cheese in a wooden vessel, researchers were able to determine its bacterial makeup and biodiversity, proving that the wood structure is an abundant reservoir of *S. thermophilus* strains that contributes to lowering the pH of the curd (Settanni et al. 2012). A myriad of metabolic events, including the intricate interactions of microbial communities from milk to cheese ripening, aid in aroma development components in cheese (Steele et al. 2013). The microbial community present in PDO cheese, which originates from elements like milk, livestock feed, the surroundings, wooden vessels and equipment, contributes significantly to the development of particular fragrance-producing volatile organic compounds (VOCs) seen in unpasteurized milk cheeses (Delgado et al. 2011; Gatti et al. 2014; Neviani et al. 2013; Ordiales et al. 2013). The interactions between microbes that take place inside the biofilm and later in the cultured milk are an intricate and not yet fully understood issue. They are influenced by various factors including pH, process settings, water activity, substrate unavailability, and microbial metabolites. These factors collectively contribute to the prevention of pathogen growth throughout the cheese-making procedure. A study has found that the high number of LAB species present in cultured milk may be responsible for the wide range of volatile chemicals found in cheese preparations. The existence of lactis species, which is capable of producing alcohol substances ranging from valine, isoleucine, and leucine—the main VOCs found in artisanal cheeses like Pecorino, La Serena, and Ibores—may have some relationship to these chemicals (Carbonell et al. 2002; Delgado et al. 2011; Ferreira et al. 2009; Randazzo et al. 2007).

The existence of *E. hirae* was favourably associated with capryl alcohol and non-2-en-1-ol, according to statistical interactions between microbiome and VOCs. Conversely, the synthesis of aldehydes like tridecanal and 3-methylbutanal was favourably connected with *Lactobacillus lactis*, and that was likewise connected with *Lactobacillus plantarum* species. When found in cheese, aldehydes are temporary chemicals since they quickly transform into main alcohols or even acids. They might have come from bacterial breakdown, Strecker breakdown, or lipid oxidation, giving off smells of herbs and greens (Moio et al. 1993). The existence of straight-chain aldehydes, namely caprylic aldehyde and non-aldehyde, was not shown to be correlated with the presence of *L. fermentum* and *L. helveticus*. However, according to a study, these aldehydes only became apparent in the control population, as clearly seen in the principal component analysis chart (Carpino et al. 2017). Ketones are frequently found in milk and milk products and are responsible for the distinct aromas attributed to methyl ketones. These compounds are formed of free fatty acids by a different pathway than β-oxidation, resulting in the development of sweet, flowery, and fruity overtones (Engels et al. 1997; McSweeney and Sousa 2000). Also, the study of volatile chemicals shows that LABs residing inside biofilm could have a key role in the development of cheese VOCs, which helps make Ragusano cheese unique.

Chemistry of Aromatic Compounds and Biofilm

Microorganisms are utilized in the preservation, transformation, and enhancement of several components, including wine, beer, fermented fruits and vegetables, and milk. Various bacteria adapt to distinct metabolic pathways to generate distinct flavour compounds. Biofilm development on the exterior of solid fermented items serves to preserve the food's appealing appearance, limit the loss of water, and mitigate oxidative browning caused by prolonged exposure to air. The synergistic impact of all of these factors yields favourable results in terms of the unique appearance and the overall quality of fermented foods. Once sufu is coated with a biofilm, microbes break down the sustenance in sufu to facilitate the process of ripening and ageing. Meanwhile, *Monascus* is capable of preserving the vibrant red hue of sufu without causing noticeable water loss (He and Chung 2019a, b). In addition, the method of fermentation might result in the formation of significant-sized pores due to the metabolic activities of microbes within the biofilm. During the process of maturation, the protein component of sufu undergoes homogenization, resulting in a uniform texture, while the quantity of pores gradually decreases (Wei et al. 2019a, b). The results of many studies on the production of volatile flavouring compounds (VFCs) through the biosynthetic process using various substrates and specific organisms are documented. Biotechnological manufacturing of VFCs has received considerable attention in the last decade for a number of reasons, including the widely held notion that it offers a sustainable alternative to artificial flavourings in food processing. The primary pathways for the generation of VFC in cheese are attributed to the metabolic processes of lactate and lactose. The lactate undergoes many transformations that contribute to the flavour of the cheese. These transformations are influenced by factors like as the kind of dairy products, the microorganisms used, and its maturation environments (Ianni et al. 2020).The alternative pathway generates compounds that are that originate from fat, including esters, free fatty acids (FFAs), ketones, and lactones, which are formed by lipolysis and lipid oxidation processes, resulting in modest concentrations of aroma (Thierry et al. 2017). The distinctive smell of cheese is a result of VFCs that are generated through enzymatic activity throughout the period of ripening. The analysis of volatile components in Pecorino Abruzzese and Ewes' milk cheese showed the presence of diacetyl, ethanol, and acetoin following 15 days of preservation at 10 °C. There were significant variations in the levels of these molecules among the different strains of *Enterococcus* (Calasso et al. 2017; Serio et al. 2010).

The increased amount of fatty acids, that contribute fruity flavours to the taste, is typically attributed to the lipolytic action of LAB. Moreover, enterococci possess the ability to generate volatile chemicals and promote the proliferation of LAB, suggesting that they possess the metabolic properties necessary for the formation of fragrance and flavour. Butanoic acid is a prominent flavour component that contributes to the distinct strong and cheesy smell found in several types of cheese, such as Camembert, Pecorino, and Ragusano (Molimard and Spinnler 1996). Hexanoic acid is a distinct aromatic compound found in Grana Padano and Roncal cheese. It is known for its light to intense, strong, goat-like fragrance. Similarly, pentanoic acid is also observed in several varieties of cheese, contributing to a light to intense savoury cheesy aroma (Frank et al. 2004; Randazzo et al. 2008). Another research demonstrated that, 39 volatile organic components are present which were not fully quantified, produced from kefir milk. The chemicals present in the mixture was diverse compounds including alcohol, aldehyde, ketone, and ester. All VOCs concentration was increased over time during preservation (Walsh et al. 2016).

In dairy items, the lactones are responsible for a plethora of different flavours. Butter, especially freshly prepared butter, possesses a wonderful fruity aroma and is a source of free lactones and their precursors. δ-Lactone is detected in both animal-derived fat as well as vegetable oils which have undergone auto-oxidation. Some microbes, including *Aspergillus niger*, *Cladosporium suaveolens*, and *Pichia etchelisii*, were discovered to create decalactone from *R. communis* oil (Prabakaran et al. 2020). The majority of esters found in cheese are characterized by their fruity, sweet, and flowery aromas. Specifically, ethyl esters have a significant role in generating a fruity flavour in cheese. Specifically, this molecule demonstrated the ability to generate an orange-like aroma in Pecorino cheese (Curioni and Bosset 2002). Thus, the synthesis of ester in cheeses can be attributed to a concurrent rise in the percentage of short- and medium-chain fatty acids, which occurs as a result of the breakdown of fats by enterococci and lactobacilli (Dahl et al. 2000). The particular fragrances of cheese matured under distinctive influence of *Penicillium* could be attributed to 2-alkanones, secondary alcohols, and FFAs (Caron et al. 2021). LAB is one of the microbes that create diacetyl in milk and dairy products (Wang et al. 2019). Acetaldehyde is the primary flavouring ingredient in fermented dairy goods such as yoghurt, and it is produced by a number of Lactobacillus species. Several other forms of yeast have been reported to produce acetaldehyde (Dan et al. 2019) (Fig. 7.2).

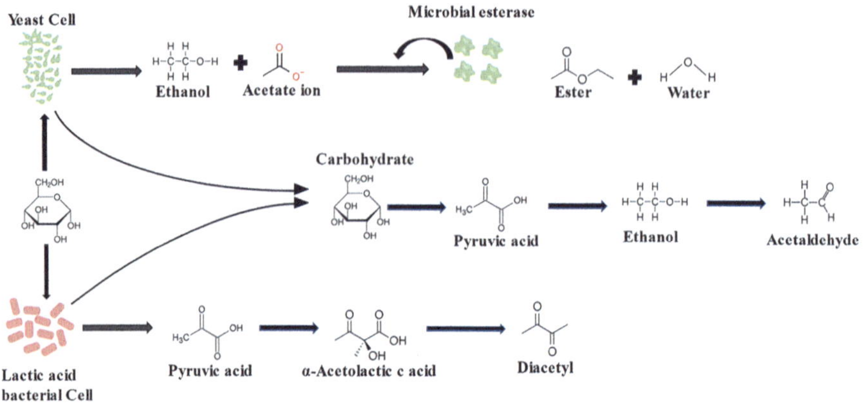

Fig. 7.2 Production of various microbial volatile flavour compounds by yeasts and lactic acid bacteria

Fermented Liquor

The biofilm developed on oak modulates the taste components found in wine. The enzyme activity of *O. oeni* glycosidase was demonstrated to liberate aromatic compound like vanillin from oak (Bloem et al. 2008). The white wine, which undergoes malolactic fermentation facilitated by biofilm on wood, likewise displayed these distinct variations in its aromatic attributes, characterized by a reduction in oak-derived aromatic components. Curiously, the same wine contained greater quantities of trans-whisky lactone, indicating the chemical relationship between wood and wine, facilitated by the *O. oeni* biofilm, may influence the flavour richness of the wine. This phenomenon can be elucidated with the presence of an intricate framework that envelops the oak exterior and functions as a filter (Dunne Jr 2002). These chemicals may either remain attached to the EPS or undergo conversion by biofilm enzymes. Given the little organoleptic impact of trans-whisky lactone, aroma evaluations strongly indicate the prospective value of *O. oeni* biofilms for analysing the oak maturation process of wines to achieve precise extraction of wood smells. Similarly, the colour of wine acquired throughout the ageing process is influenced by the existence of biofilm on wood (Bastard et al. 2016). Acetaldehyde, a crucial flavouring molecule, is produced by different *Lactobacillus* species in dairy products such as yoghurt. Additionally, certain forms of yeast have been discovered to also generate acetaldehyde (Dan et al. 2019; Rosca et al. 2016).

There was speculation suggesting the presence of EPS in a biofilm might offer protection to cells in difficult circumstances and enhance their chances of survivability. Conversely, EPS is comprised of carbohydrates and proteins. The hydrolysis products of this material, which consist of glucose and amino acids, might serve as substrates for the production of esters and additional taste compounds in baijiu (Zeng et al. 2023). It is worth mentioning that acetaldehyde influences both the speed at which fermentation occurs and the overall excellence of wine. Recently conducted research revealed that wines implanted with Saccharomyces cerevisiae, subjected to elevated amounts of sulphur dioxide, and fermented at elevated temperatures had increased acetaldehyde concentrations. A clear relationship existed between its overall activity and the overall volume of acetaldehyde produced by *S. cerevisiae* yeasts (October 2020).

Beverage

There are numerous specific biochemical alterations that occur while fermentation. Yeasts break down the only resource for carbon, sucrose, producing glucose and fructose at the beginning of fermentation. These sugars are then utilized by other microorganism drivers, leading to several metabolic modifications. Although yeasts undertook alcoholic fermentation, the ethanol concentrations were quite low, indicating that the ethanol was rapidly oxidized into acetic acid by AAB. A significant and quick rise in the yeast and AAB counts throughout the first 11 days, suggesting a correlation with these results. Indeed, AAB appears in considerable quantities in biofilm specimens near the liquid-air interface, creating optimal circumstances for the effective synthesis of acetic acid. Additionally, many other kinds of organic acids underwent a rise throughout the fermentation process, maintained rather consistent levels, or only existed in little quantities (Savary et al. 2021). Among the various organic acids found in kombucha, glucuronic acid stands out as a key player due to its possible beneficial effects on health. True, this acid's detoxifying effects in humans are due to the glucuronidation procedure, which increases the clearance of xenobiotics and natural metabolites like bilirubin, oxidized fatty acids, and unnecessary steroid hormones. In addition, the polyphenols included in the tea utilized to make kombucha are more easily transported and absorbed, which means that the beverage's antioxidant capabilities may be enhanced (Leal et al. 2018; Vīna et al. 2014).

Meat Products

Fermented meat products have produced almost 200 or more distinct volatile chemicals (Sharma et al. 2020a, b). Preservation induces modifications in the biological and organoleptic characteristics. Several studies have been undertaken to investigate the impact of VFCs on the development of matured aroma. These investigations reported that the sausages had brief periods of vacuum preservation and a moderate decrease in nitrates. This decrease was associated with the presence of chemicals that produce odours, such as a buttery or cheesy aroma (Perea-Sanz et al. 2019, 2020; Silva et al. 2020). Furthermore, the influence of introducing *Debaryomyces hansenii* on aroma generation was investigated in order to decrease the concentrations of nitrite and nitrate in dried fermented sausages. Various drying durations were investigated to analyse changes in biological and organoleptic characteristics, aroma, and VFCs (Perea-Sanz et al. 2020). The diameter exerted a substantial influence on the generation of VOCs such as aldehydes (particularly hexanal) and ketones (dimethyl ketone, 2,3-butandione, methyl ethyl ketone, and acetoin), that exhibited their greatest fluctuations. The presence of native *Lactobacillus sakei* inside the saucer that were injected with pediococci had a less pronounced impact on the starter cultures (Montanari et al. 2018). Aldehydes are regarded as essential elements for the flavour formation in fermented meat due to their ability to emit green grass, nutty, sweet, and cheesy odours, as well as their low tolerance levels (Dajanta et al. 2011).

Sea Food Products

Research into the method of fermentation and its effects on fish item flavours has mostly concentrated on VFCs and their generation, extraction, and characterization. A particular flavour is supposedly created during the natural fermentation of fish by breaking down proteins with the help of enzymes derived from both microorganisms and fish (Marti-Quijal et al. 2020). The predominant VFCs found in fermented Mandarin fish were anise camphor, indole, linalool, piperitone, 2-Methyloctan-3-one, and vinyl amyl carbinol (Yang et al. 2020). While the period of volatile microbial fermentation of Suan-zuo-yu, LAB reduced the levels of long-chain proteins, myosin, and actin while increasing the levels of aldehydes, alcohols, and esters—the latter three of which mostly contributed to the flavour (Wang et al. 2020). Primary VFCs generated during seafood fermenting process included acids, alcohols, aldehydes, esters, furans, hydrocarbons, ketones, and compounds rich in nitrogen (Wang et al. 2017). A study examined the flavour of fermented fish and demonstrated that S. cerevisiae 152 broke down aromatic amino acids such as phenylalanine, leucine, and valine to generate phenylethanol, active amyl alcohol, isobutanol, and isoamyl alcohol. In addition, the breakdown of leucine and phenylalanine was observed, leading to the production of phenylethanol and isoamyl alcohol by *Lactobacillus pentococcus* 22 and *Lactobacillus plantarum* 120, correspondingly (Wang et al. 2017). Esters play a crucial role in Suanyu fermentation by indirectly affecting FFA metabolic processes while also plays a role in the production of esters through esterification and alcoholic reaction. The presence of LAB was seen to stimulate the formation of acetate molecules, while *Staphylococcus* and yeast were discovered to enhance the creation of ethyl molecules (Sidira et al. 2016). The metabolic processes of microorganisms can generate many volatile compounds, which ultimately influence the flavour and texture of amino acids (AAs) and fatty acids (FAs) digestion. The inclusion of metabolic versatility and diversity of microbes may potentially create the possibility for novel and improved materials. A study found that the rise in *Staphylococcus* bacteria was correlated with the rise in esters, while *Macrococcus* bacteria, which have the ability to break down proteins and lipids, were also observed during the course of fermentation (Patel et al. 2013).

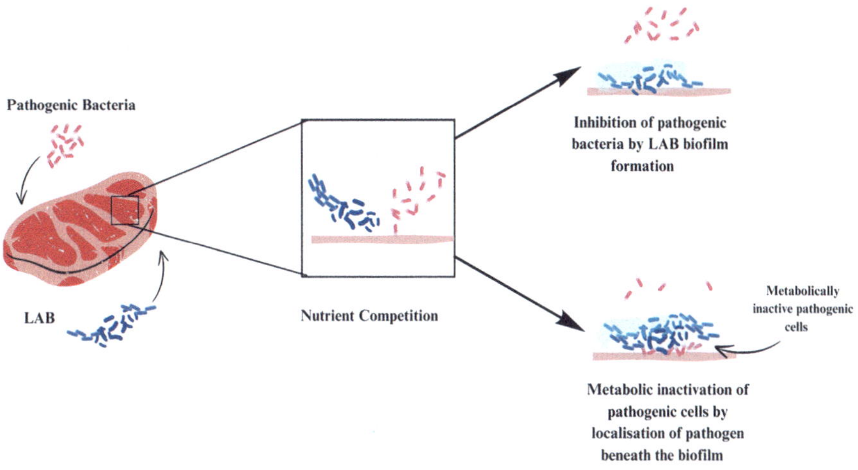

Fig. 7.3 Illustration of competitive exclusion of pathogenic bacteria by LAB

7.6 Importance and Negative Impact of Biofilm in Food Industry

It has been observed that the LAB bacterial biofilm confers excellent benefits in enhancing nutrient availability and resistance to other pathogenic microorganisms. They confer antibacterial properties to the substrate they are attached to by either competitive exclusion or by producing antibacterial agents (Mgomi et al. 2023). It has been observed that between LAB and any pathogenic microbes, LAB precedes in establishing the biofilm over the substrate, thereby depriving the other microbial colony access to nutrients and space to grow (Fig. 7.3). In the cheese ripening, the presence of natural LAB outruled the contamination of *L. monocytogenes* by up to 2 $\log_{10}$(CFU/cm^2), acting as biocontrol of the pathogen (Mariani et al. 2011).

The important locations in which biofilm formation in the food industry may arise as a potential threat are pipelines, pasteurizing vessels, membrane filters, food kept in storage, contact surfaces of storage vessels, packaging material, etc. Some microbial biofilms are resistant to sanitization procedures by forming endospores. The *Geobacilllus* sp. is capable of growing at a temperature of around 65 °C and is capable of producing spores that can survive heat; these cause problems in milk powder factories. It has been observed that *Bacillus cereus* is capable of surviving in a wide range of temperature gradients, and it can also endure multiple sterilization procedures in the food processing industry by forming endospores. This has a serious impact on the dairy industry, as they are capable of forming biofilm persistently when we relax the sterilization procedures. The enterotoxins produced by it are capable of inflicting abdominal pain, diarrhoea, and gastric disturbance with vomiting (Christison et al. 2007). In the poultry and meat industry, the biofilm of *Salmonella* sp. is a huge problem; they produce a biofilm that contains endotoxins

Table 7.1 The harmful effects of bacterial biofilm in food and its impact of human health

Microbial biofilm	Contaminated food	Impact on human health	Reference
Bacillus cereus	Milk products, cereals, meat	Diarrhoea and vomiting symptoms	Carrascosa et al. (2021)
Escherichia coli	Fruits and vegetables, chilled pork	Haemolytic uremic syndrome	Li et al. (2018)
Listeria monocytogenes	Soft cheese, raw milk, vegetables, poultry, canned foods	Listeriosis	Saldivar et al. (2018)
Staphylococcus aureus	Bakery products, cream, milk, egg, poultry, stored vegetables	Diarrhoea and vomiting symptoms	Kadariya et al. (2014)
Vibrio parahemolyticus	Salmonella, Vannamei	Gastroenteritis, wound infections, sepsis	Su and Liu (2007)
Psychrophilic Clostridium sp.	Meat and poultry	Tetanus, botulism, and gas gangrene	Minton and Clarke (2013)
Pseudomonas sp.	Raw milk	Acute and chronic life-threatening infections	Li et al. (2023)
Shewanellan putrefaciens	Cold-stored marine fishes	Abdominal infections in the soft tissues, biliary tract infections, and peritonitis	Müller et al. (2023)
Campylobacter jejuni	Raw milk, uncooked poultry, contaminated water	Stomach cramps, fever, bloody diarrhoea	Klančnik et al. (2021)

that are virulent enough to cause a decline in myocardial function and affect lymphocyte function. *Staphylococcus aureus* is another major food-borne microbe whose enterotoxins bind with the T cells to trigger acute response with vomiting and gastrointestinal disturbance (Schillaci et al. 2013). Some of the commonly seen food-borne pathogens are listed in Table 7.1.

Apart from the health threat, biofilm formation also hampers the mechanical efficiency of the functioning of the huge system in the food processing industry. Untreated growth of biofilm, when intensified, causes blockage in the values and disturbs the fluid transfer in the pipelines. The prolonged stay of biofilms also leads to corrosion of the metal surfaces (Bryers 1987).

7.7 Conclusion

In recent decades, there has been a profound expansion in biofilm research, which has shifted its main emphasis from prevention and eradication, especially in the medical domain, to investigating methods to exploit biofilms for advantageous objectives. The impact of biofilms on the quality, safety, and processability of fermented foods is a double-edged sword, offering both advantages and disadvantages depending on the particular food, the biofilms involved, fermentation conditions, and other factors. This discipline is simply in its early stages of adopting novel concepts, and the possibilities for utilizing the biofilm-forming capabilities of microbes seem almost boundless. The application of biofilm for food fermentation is an age-old process which has been practiced to improve the aroma and texture of the food. In recent days, sophisticated technologies have been utilized to check the quality and safety of biofilm-mediated fermented food. It is expected that the combined progress of synthetic biology and materials science will augment the process of biofilm development, providing significant knowledge for their broader use in agro-food industrial bio-processing.

References

Almeida OGGD, Pinto UM, Matos CB, Frazilio DA, Braga VF, von Zeska-Kress MR, De Martinis ECP (2020) Does quorum sensing play a role in microbial shifts along spontaneous fermentation of cocoa beans? An in silico perspective. Food Res Int 131:109034. https://doi.org/10.1016/j.foodres.2020.109034

Alotaibi GF, Bukhari MA (2021) Factors influencing bacterial biofilm formation and development. Am J Biomed Sci Res 12(6):617–626. https://doi.org/10.34297/AJBSR.2021.12.001820

Álvarez-Pérez JM, Campo E, San-Juan F, Coque JJR, Ferreira V, Hernández-Orte P (2012) Sensory and chemical characterisation of the aroma of Prieto Picudo rosé wines: the differential role of autochthonous yeast strains on aroma profiles. Food Chem 133(2):284–292. https://doi.org/10.1016/j.foodchem.2012.01.024

Ashtavinayak P, Elizabeth HA (2016) Review: gram negative bacteria in brewing. Adv Microbiol 06(03):195–209. https://doi.org/10.4236/aim.2016.63020

Balducci E, Papi F, Capialbi DE, Del Bino L (2023) Polysaccharides' structures and functions in biofilm architecture of antimicrobial-resistant (AMR) pathogens. Int J Mol Sci 24(4):4030. https://doi.org/10.3390/ijms24044030

Basso RF, Alcarde AR, Portugal CB (2016) Could non-Saccharomyces yeasts contribute on innovative brewing fermentations? Food Res Int 86:112–120. https://doi.org/10.1016/j.foodres.2016.06.002

Bastard A, Coelho C, Briandet R, Canette A, Gougeon R, Alexandre H et al (2016) Effect of biofilm formation by Oenococcus oeni on malolactic fermentation and the release of aromatic compounds in wine. Front Microbiol 7:613. https://doi.org/10.3389/fmicb.2016.00613

Belda I, Ruiz J, Alastruey-Izquierdo A, Navascués E, Marquina D, Santos A (2016) Unraveling the enzymatic basis of wine "flavorome": a phylo-functional study of wine related yeast species. Front Microbiol 7:12. https://doi.org/10.3389/fmicb.2016.00012

Bloem A, Lonvaud-Funel A, de Revel G (2008) Hydrolysis of glycosidically bound flavour compounds from oak wood by Oenococcus oeni. Food Microbiol 25(1):99–104. https://doi.org/10.1016/j.fm.2007.07.009

Bokulich NA, Collins TS, Masarweh C, Allen G, Heymann H, Ebeler SE, Mills DA (2016) Associations among wine grape microbiome, metabolome, and fermentation behavior suggest microbial contribution to regional wine characteristics. MBio 7(3):10–1128. https://doi.org/10.1128/mBio.00631-16

Bryers JD (1987) Biologically active surfaces: processes governing the formation and persistence of biofilms. Biotechnol Prog 3(2):57–68. https://doi.org/10.1002/btpr.5420030202

Calasso M, Mancini L, De Angelis M, Conte A, Costa C, Del Nobile MA, Gobbetti M (2017) Multiple microbial cell-free extracts improve the microbiological, biochemical and sensory features of ewes' milk cheese. Food Microbiol 66:129–140. https://doi.org/10.1016/j.fm.2017.04.011

Carbonell M, Nuñez M, Fernández-García E (2002) Seasonal variation of volatile compounds in ewe raw milk La Serena cheese. Lait 82(6):699–711. https://doi.org/10.1051/lait:2002043

Caron T, Le Piver M, Péron AC, Lieben P, Lavigne R, Brunel S, Roueyre D, Place M, Bonnarme P, Giraud T, Branca A, Landaud S, Chassard C (2021) Strong effect of Penicillium roqueforti populations on volatile and metabolic compounds responsible for aromas, flavor and texture in blue cheeses. Int J Food Microbiol 354:109174. https://doi.org/10.1016/j.ijfoodmicro.2021.109174

Carpino S, Randazzo CL, Pino A, Russo N, Rapisarda T, Belvedere G, Caggia C (2017) Influence of PDO Ragusano cheese biofilm microbiota on flavour compounds formation. Food Microbiol 61:126–135. https://doi.org/10.1016/j.fm.2016.09.006

Carrascosa C, Raheem D, Ramos F, Saraiva A, Raposo A (2021) Microbial biofilms in the food industry – a comprehensive review. Int J Environ Res Public Health 18(04):2014. https://doi.org/10.3390/ijerph18042014

Christison CA, Lindsay D, Von Holy A (2007) Cleaning and handling implements as potential reservoirs for bacterial contamination of some ready-to-eat foods in retail delicatessen environments. J Food Prot 70(12):2878–2883. https://doi.org/10.4315/0362-028X-70.12.2878

Curioni PMG, Bosset JO (2002) Key odorants in various cheese types as determined by gas chromatography-olfactometry. Int Dairy J 12(12):959–984. https://doi.org/10.1016/S0958-6946(02)00124-3

Dahl S, Tavaria FK, Malcata FX (2000) Relationships between flavour and microbiological profiles in Serra da Estrela cheese throughout ripening. Int Dairy J 10(4):255–262. https://doi.org/10.1016/S0958-6946(00)00042-X

Dajanta K, Apichartsrangkoon A, Chukeatirote E, Frazier RA (2011) Free-amino acid profiles of thua nao, a Thai fermented soybean. Food Chem 125(2):342–347. https://doi.org/10.1016/j.foodchem.2010.09.002

Dan T, Ren W, Liu Y, Tian J, Chen H, Li T, Liu W (2019) Volatile flavor compounds profile and fermentation characteristics of milk fermented by Lactobacillus delbrueckii subsp. bulgaricus. Front Microbiol 10:2183. https://doi.org/10.3389/fmicb.2019.02183

Delgado FJ, González-Crespo J, Cava R, Ramírez R (2011) Formation of the aroma of a raw goat milk cheese during maturation analysed by SPME-GC-MS. Food Chem 129(3):1156–1163. https://doi.org/10.1016/j.foodchem.2011.05.096

Dunne WM Jr (2002) Bacterial adhesion: seen any good biofilms lately? Clin Microbiol Rev 15(2):155–166. https://doi.org/10.1128/CMR.15.2.155-166.2002

Engels WJM, Dekker R, De Jong C, Neeter R, Visser S (1997) A comparative study of volatile compounds in the water-soluble fraction of various types of ripened cheese. Int Dairy J 7(4):255–263. https://doi.org/10.1016/S0958-6946(97)00003-4

Favre L, Ortalo-Magné A, Pichereaux C, Gargaros A, Burlet-Schiltz O, Cotelle V, Culioli G (2018) Metabolome and proteome changes between biofilm and planktonic phenotypes of the marine bacterium Pseudoalteromonas lipolytica TC8. Biofouling 34(2):132–148. https://doi.org/10.1080/08927014.2017.1413551

Ferreira IMPLVO, Pinho O, Sampaio P (2009) Volatile fraction of DOP "Castelo Branco" cheese: influence of breed. Food Chem 112(4):1053–1059. https://doi.org/10.1016/j.foodchem.2008.06.048

Flemming HC, Wingender J (2010) The biofilm matrix. Nat Rev Microbiol 8(9):623–633. https://doi.org/10.1038/nrmicro2415

Flemming HC, Wingender J, Szewzyk U, Steinberg P, Rice SA, Kjelleberg S (2016) Biofilms: an emergent form of bacterial life. Nat Rev Microbiol 14(9):563–575. https://doi.org/10.1038/nrmicro.2016.94

Frank DC, Owen CM, Patterson J (2004) Solid phase microextraction (SPME) combined with gas-chromatography and olfactometry-mass spectrometry for characterization of cheese aroma compounds. LWT 37(2):139–154. https://doi.org/10.1016/S0023-6438(03)00144-0

Funari R, Shen AQ (2022) Detection and characterization of bacterial biofilms and biofilm-based sensors. ACS Sens 7(2):347–357. https://doi.org/10.1021/acssensors.1c02722

Furukawa S (2015) Studies on formation, control and application of biofilm formed by food related microorganisms. Biosci Biotechnol Biochem 79(7):1050–1056. https://doi.org/10.1080/09168451.2015.1018126

Furukawa S, Watanabe T, Toyama H, Morinaga Y (2013) Significance of microbial symbiotic coexistence in traditional fermentation. J Biosci Bioeng 116(5):533–539. https://doi.org/10.1016/j.jbiosc.2013.05.017
Garde S, Ávila M, Fernández-García E, Medina M, Nuñez M (2007) Volatile compounds and aroma of Hispánico cheese manufactured using lacticin 481-producing Lactococcus lactis subsp. lactis INIA 639 as an adjunct culture. Int Dairy J 17(6):717–726. https://doi.org/10.1016/j.idairyj.2006.07.005
Garrett TR, Bhakoo M, Zhang Z (2008) Bacterial adhesion and biofilms on surfaces. Prog Nat Sci 18(9):1049–1056. https://doi.org/10.1016/j.pnsc.2008.04.001
Gatti M, Bottari B, Lazzi C, Neviani E, Mucchetti G (2014) Invited review: microbial evolution in raw-milk, long-ripened cheeses produced using undefined natural whey starters. J Dairy Sci 97(2):573–591. https://doi.org/10.3168/jds.2013-7187
Gomes LC, Silva LN, Simões M, Melo LF, Mergulhão FJ (2015) Escherichia coli adhesion, biofilm development and antibiotic susceptibility on biomedical materials. J Biomed Mater Res A 103(4):1414–1423. https://doi.org/10.1002/jbm.a.35277
Gong C, He Y, Tang Y, Hu R, Lv Y, Zhang Q et al (2021) Biofilms in plant-based fermented foods: formation mechanisms, benefits and drawbacks on quality and safety, and functionalization strategies. Trends Food Sci Technol 116:940–953. https://doi.org/10.1016/j.tifs.2021.08.026
Grande R, Nistico L, Sambanthamoorthy K, Longwell M, Iannitelli A, Cellini L et al (2014) Temporal expression of agrB, cidA, and alsS in the early development of Staphylococcus aureus UAMS-1 biofilm formation and the structural role of extracellular DNA and carbohydrates. Pathog Dis 70(3):414–422. https://doi.org/10.1111/2049-632X.12158
Grandvalet C, Assad-García JS, Chu-Ky S, Tollot M, Guzzo J, Gresti J, Tourdot-Maréchal R (2008) Changes in membrane lipid composition in ethanol- and acid-adapted Oenococcus oeni cells: characterization of the cfa gene by heterologous complementation. Microbiology 154(9):2611–2619. https://doi.org/10.1099/mic.0.2007/016238-0
Gupta P, Sarkar S, Das B, Bhattacharjee S, Tribedi P (2016) Biofilm, pathogenesis and prevention – a journey to break the wall: a review. Arch Microbiol 198:1–15. https://doi.org/10.1007/s00203-015-1148-6
He W, Chung HY (2019a) Comparison between quantitative descriptive analysis and flash profile in profiling the sensory properties of commercial red sufu (Chinese fermented soybean curd). J Sci Food Agric 99(6):3024–3033. https://doi.org/10.1002/jsfa.9516
He W, Chung HY (2019b) Multivariate relationships among sensory, physicochemical parameters, and targeted volatile compounds in commercial red sufus (Chinese fermented soybean curd): comparison of QDA® and Flash Profile methods. Food Res Int 125:108548. https://doi.org/10.1016/j.foodres.2019.108548
Hernández A, Pérez-Nevado F, Ruiz-Moyano S, Serradilla MJ, Villalobos MC, Martín A, Córdoba MG (2018) Spoilage yeasts: what are the sources of contamination of foods and beverages? Int J Food Microbiol 286:98–110. https://doi.org/10.1016/j.ijfoodmicro.2018.07.031
Ianni A, Bennato F, Martino C, Grotta L, Martino G (2020) Volatile flavor compounds in cheese as affected by ruminant diet. Molecules 25(3):461. https://doi.org/10.3390/molecules25030461
Jennings LK, Storek KM, Ledvina HE, Coulon C, Marmont LS, Sadovskaya I et al (2015) Pel is a cationic exopolysaccharide that cross-links extracellular DNA in the Pseudomonas aeruginosa biofilm matrix. Proc Natl Acad Sci 112(36):11353–11358. https://doi.org/10.1073/pnas.1503058112
Jia K, Wang G, Liang L, Wang M, Wang H, Xu X (2017) Preliminary transcriptome analysis of mature biofilm and planktonic cells of Salmonella Enteritidis exposure to acid stress. Front Microbiol 8:1861. https://doi.org/10.3389/fmicb.2017.01861
Kadariya J, Smith TC, Thapaliya D (2014) Staphylococcus aureus and staphylococcal food-borne disease: an ongoing challenge in public health. Biomed Res Int 2014(1):827965. https://doi.org/10.1155/2014/827965
Khelissa SO, Jama C, Abdallah M, Boukherroub R, Faille C, Chihib NE (2017) Effect of incubation duration, growth temperature, and abiotic surface type on cell surface properties, adhesion

and pathogenicity of biofilm-detached Staphylococcus aureus cells. AMB Express 7:1–13. https://doi.org/10.1186/s13568-017-0492-0

Kinnunen P, Guillaume JHA, Taka M, D'Odorico P, Siebert S, Puma MJ, Jalava M, Kummu M (2020) Local food crop production can fulfil demand for less than one-third of the population. Nat Food 1(4):229–237. https://doi.org/10.1038/s43016-020-0060-7

Klančnik A, Šimunović K, Sterniša M, Ramić D, Smole Možina S, Bucar F (2021) Anti-adhesion activity of phytochemicals to prevent Campylobacter jejuni biofilm formation on abiotic surfaces. Phytochem Rev 20:55–84. https://doi.org/10.1007/s11101-020-09669-6

Leal JM, Suárez LV, Jayabalan R, Oros JH, Escalante-Aburto A (2018) A review on health benefits of kombucha nutritional compounds and metabolites. CYTA J Food 16(1):390–399. https://doi.org/10.1080/19476337.2017.1410499

Li Y, Zhou D, Hu S, Xiao X, Yu Y, Li X (2018) Transcriptomic analysis by RNA-seq of Escherichia coli O157: H7 response to prolonged cold stress. LWT 97:17–24. https://doi.org/10.1016/j.lwt.2018.06.025

Li X, Gu N, Huang TY, Zhong F, Peng G (2023) Pseudomonas aeruginosa: a typical biofilm forming pathogen and an emerging but underestimated pathogen in food processing. Front Microbiol 13:1114199. https://doi.org/10.3389/fmicb.2022.1114199

Licitra G, Ogier JC, Parayre S, Pediliggieri C, Carnemolla TM, Falentin H, Madec MN, Carpino S, Lortal S (2007) Variability of bacterial biofilms of the "tina" wood vats used in the Ragusano cheese-making process. Appl Environ Microbiol 73(21):6980–6987. https://doi.org/10.1128/AEM.00835-07

Lindsay D, von Holy A (2006) What food safety professionals should know about bacterial biofilms. Br Food J 108(1):27–37. https://doi.org/10.1108/00070700610637616

Lonvaud-Funel A (1995) Microbiology of the malolactic fermentation: molecular aspects. FEMS Microbiol Lett 126(3):209–214. https://doi.org/10.1111/j.1574-6968.1995.tb07420.x

Lonvaud-Funel A (1999) Lactic acid bacteria in the quality improvement and depreciation of wine. Anton Leeuw Int J Gen Mol Microbiol 76(1–4):317–331. https://doi.org/10.1023/A:1002088931106

López D, Vlamakis H, Kolter R (2010) Biofilms. Cold Spring Harb Perspect Biol 2(7):a000398. https://cshperspectives.cshlp.org/content/2/7/a000398.short

Lortal S, Di Blasi A, Madec MN, Pediliggieri C, Tuminello L, Tanguy G, Fauquant J, Lecuona Y, Campo P, Carpino S, Licitra G (2009) Tina wooden vat biofilm: a safe and highly efficient lactic acid bacteria delivering system in PDO Ragusano cheese making. Int J Food Microbiol 132(1):1–8. https://doi.org/10.1016/j.ijfoodmicro.2009.02.026

Maitre M, Weidmann S, Dubois-Brissonnet F, David V, Covés J, Guzzoa J (2014) Adaptation of the wine bacterium oenococcus oeni to ethanol stress: role of the small heat shock protein lo18 in membrane integrity. Appl Environ Microbiol 80(10):2973–2980. https://doi.org/10.1128/AEM.04178-13

Mariani C, Oulahal N, Chamba JF, Dubois-Brissonnet F, Notz E, Briandet R (2011) Inhibition of Listeria monocytogenes by resident biofilms present on wooden shelves used for cheese ripening. Food Control 22(8):1357–1362. https://doi.org/10.1016/j.foodcont.2011.02.012

Marić S, Vraneš J (2007) Characteristics and significance of microbial biofilm formation. Period Biol 109:115–121. https://www.croris.hr/crosbi/publikacija/prilog-casopis/134967

Marti-Quijal FJ, Remize F, Meca G, Ferrer E, Ruiz MJ, Barba FJ (2020) Fermentation in fish and by-products processing: an overview of current research and future prospects. Curr Opin Food Sci 31:9–16. https://doi.org/10.1016/j.cofs.2019.08.001

McDougald D, Rice SA, Barraud N, Steinberg PD, Kjelleberg S (2012) Should we stay or should we go: mechanisms and ecological consequences for biofilm dispersal. Nat Rev Microbiol 10(1):39–50. https://doi.org/10.1038/nrmicro2695

McSweeney PL, Sousa MJ (2000) Biochemical pathways for the production of flavour compounds in cheeses during ripening: a review. Lait 80(3):293–324. https://doi.org/10.1051/lait:2000127

Mgomi FC, Yang YR, Cheng G, Yang ZQ (2023) Lactic acid bacteria biofilms and their antimicrobial potential against pathogenic microorganisms. Biofilms 5:100118. https://doi.org/10.1016/j.bioflm.2023.100118

Minton NP, Clarke DJ (eds) (2013) Clostridia, vol 3. Springer

Moio L, Dekimpe J, Etievant PX, Addeo F (1993) Volatile flavour compounds of water buffalo Mozzarella cheese. https://www.cabidigitallibrary.org/doi/full/10.5555/19930458947

Molimard P, Spinnler HE (1996) Compounds involved in the flavor of surface mold-ripened cheeses: origins and properties. J Dairy Sci 79(2):169–184. https://doi.org/10.3168/jds.S0022-0302(96)76348-8

Montanari C, Gatto V, Torriani S, Barbieri F, Bargossi E, Lanciotti R, Grazia L, Magnani R, Tabanelli G, Gardini F (2018) Effects of the diameter on physico-chemical, microbiological and volatile profile in dry fermented sausages produced with two different starter cultures. Food Biosci 22:9–18. https://doi.org/10.1016/j.fbio.2017.12.013

Müller S, von Bonin S, Schneider R, Krüger M, Quick S, Schröttner P (2023) Shewanella putrefaciens, a rare human pathogen: a review from a clinical perspective. Front Cell Infect Microbiol 12:1033639. https://doi.org/10.3389/fcimb.2022.1033639

Nandakumar V, Chittaranjan S, Kurian VM, Doble M (2013) Characteristics of bacterial biofilm associated with implant material in clinical practice. Polym J 45(2):137–152. https://doi.org/10.1038/pj.2012.130

Nedovic VA, Durieuxb A, Van Nedervelde L, Rosseels P, Vandegans J, Plaisant AM, Simon JP (2000) Continuous cider fermentation with co-immobilized yeast and Leuconostoc oenos cells. Enzym Microb Technol 26(9–10):834–839. https://doi.org/10.1016/S0141-0229(00)00179-4

Neviani E, Bottari B, Lazzi C, Gatti M (2013) New developments in the study of the microbiota of raw-milk, long-ripened cheeses by molecular methods: the case of Grana Padano and Parmigiano Reggiano. Front Microbiol 4:36. https://doi.org/10.3389/fmicb.2013.00036

Nguyen HT, Nguyen TH, Otto M (2020) The staphylococcal exopolysaccharide PIA–biosynthesis and role in biofilm formation, colonization, and infection. Comput Struct Biotechnol J 18:3324–3334. https://doi.org/10.1016/j.csbj.2020.10.027

Nostro A, Cellini L, Di Giulio M, D'Arrigo M, Marino A, Blanco AR, Favaloro A, Cutroneo G, Bisignano G (2012) Effect of alkaline pH on staphylococcal biofilm formation. APMIS 120(9):733–742. https://doi.org/10.1111/j.1600-0463.2012.02900.x

Nyanasegran PK, Nathan S, Firdaus-Raih M, Muhammad NAN, Ng CL (2023) Biofilm signaling, composition and regulation in Burkholderia pseudomallei. J Microbiol Biotechnol 33(1):15. https://doi.org/10.4014/jmb.2207.07032

O'Toole G, Kaplan HB, Kolter R (2000) Biofilm formation as microbial development. Annu Rev Microbiol 54(1):49–79. https://doi.org/10.1146/annurev.micro.54.1.49

October FM (2020) Effect of yeasts and oenological parameters on acetaldehyde production during alcoholic fermentation of South African grape musts. https://scholar.sun.ac.za/handle/10019.1/108105

Olson ME, Garvin KL, Fey PD, Rupp ME (2006) Adherence of Staphylococcus epidermidis to biomaterials is augmented by PIA. Clin Orthop Relat Res 451:21–24. https://doi.org/10.1097/01.blo.0000229320.45416.0c

Ordiales E, Martín A, Benito MJ, Hernández A, Ruiz-Moyano S, de Córdoba MG (2013) Role of the microbial population on the flavor of the soft-bodied cheese Torta del Casar. J Dairy Sci 96(9):5477–5486. https://doi.org/10.3168/jds.2013-6587

Pannella G, Lombardi SJ, Coppola F, Vergalito F, Iorizzo M, Succi M, Tremonte P, Iannini C, Sorrentino E, Coppola R (2020) Effect of biofilm formation by lactobacillus plantarum on the malolactic fermentation in model wine. Food Secur 9(6):797. https://doi.org/10.3390/foods9060797

Patel A, Shah N, Prajapati JB (2013) Biosynthesis of vitamins and enzymes in fermented foods by lactic acid bacteria and related genera – a promising approach. Croat J Food Sci Technol 5(2):85–91. https://hrcak.srce.hr/113693

Perea-Sanz L, Montero R, Belloch C, Flores M (2019) Microbial changes and aroma profile of nitrate reduced dry sausages during vacuum storage. Meat Sci 147:100–107. https://doi.org/10.1016/j.meatsci.2018.08.026

Perea-Sanz L, López-Díez JJ, Belloch C, Flores M (2020) Counteracting the effect of reducing nitrate/nitrite levels on dry fermented sausage aroma by Debaryomyces hansenii inoculation. Meat Sci 164:108103. https://doi.org/10.1016/j.meatsci.2020.108103

Perpetuini G, Tittarelli F, Schirone M, Di Gianvito P, Corsetti A, Arfelli G, Suzzi G, Tofalo R (2018) Adhesion properties and surface hydrophobicity of Pichia manshurica strains isolated from organic wines. LWT 87:385–392. https://doi.org/10.1016/j.lwt.2017.09.011

Pezzoni M, Lemos M, Pizarro RA, Costa CS (2022) UVA as environmental signal for alginate production in Pseudomonas aeruginosa: role of this polysaccharide in the protection of planktonic cells and biofilms against lethal UVA doses. Photochem Photobiol Sci 21(8):1459–1472. https://doi.org/10.1007/s43630-022-00236-w

Polášková P, Herszage J, Ebeler SE (2008) Wine flavor: chemistry in a glass. Chem Soc Rev 37(11):2478–2489. https://doi.org/10.1039/b714455p

Prabakaran R, Marie JM, Xavier AJM (2020) Biobased unsaturated polyesters containing castor oil-derived ricinoleic acid and itaconic acid: synthesis, in vitro antibacterial, and cytocompatibility studies. ACS Appl Bio Mater 3(9):5708–5721. https://doi.org/10.1021/acsabm.0c00480

Preda VG, Săndulescu O (2019) Communication is the key: biofilms, quorum sensing, formation and prevention. Discover 7(3):e10. https://doi.org/10.15190/d.2019.13

Randazzo CL, De Luca S, Todaro A, Restuccia C, Lanza CM, Spagna G, Caggia C (2007) Preliminary characterization of wild lactic acid bacteria and their abilities to produce flavour compounds in ripened model cheese system. J Appl Microbiol 103(2):427–435. https://doi.org/10.1111/j.1365-2672.2006.03261.x

Randazzo CL, Pitino I, De Luca S, Scifò GO, Caggia C (2008) Effect of wild strains used as starter cultures and adjunct cultures on the volatile compounds of the Pecorino Siciliano cheese. Int J Food Microbiol 122(3):269–278. https://doi.org/10.1016/j.ijfoodmicro.2007.12.005

Rao Y, Qian Y, She X, Yang J, He P, Jiang Y, Wang M, Xiang W (2018) Pellicle formation, microbial succession and lactic acid utilisation during the aerobic deteriorating process of Sichuan pickle. Int J Food Sci Technol 53(3):767–775. https://doi.org/10.1111/ijfs.13652

Rao Y, Qian Y, Tao Y, She X, Li Y, Che Z, Li H, Liu L (2019) Influence of oxygen exposure on fermentation process and sensory qualities of Sichuan pickle (paocai). RSC Adv 9(66):38520–38530. https://doi.org/10.1039/c9ra05994f

Rosca I, Petrovici AR, Brebu M, Stoica I, Minea B, Marangoci N (2016) An original method for producing acetaldehyde and diacetyl by yeast fermentation. Braz J Microbiol 47(4):949–954. https://doi.org/10.1016/j.bjm.2016.07.005

Saldivar JC, Davis ML, Johnson MG, Ricke SC (2018) Listeria monocytogenes adaptation and growth at low temperatures: mechanisms and implications for foodborne disease. In: Food and feed safety systems and analysis. Academic, pp 227–248. https://doi.org/10.1016/B978-0-12-811835-1.00013-0

Sauer K, Stoodley P, Goeres DM, Hall-Stoodley L, Burmølle M, Stewart PS, Bjarnsholt T (2022) The biofilm life cycle: expanding the conceptual model of biofilm formation. Nat Rev Microbiol 20(10):608–620. https://doi.org/10.1038/s41579-022-00767-0

Savary O, Mounier J, Thierry A, Poirier E, Jourdren J, Maillard MB, Penland M, Decamps C, Coton E, Coton M (2021) Tailor-made microbial consortium for Kombucha fermentation: microbiota-induced biochemical changes and biofilm formation. Food Res Int 147:110549. https://doi.org/10.1016/j.foodres.2021.110549

Schillaci D, Napoli EM, Cusimano MG, Vitale M, Ruberto G (2013) Origanum vulgare subsp. hirtum essential oil prevented biofilm formation and showed antibacterial activity against planktonic and sessile bacterial cells. J Food Prot 76(10):1747–1752. https://doi.org/10.4315/0362-028X.JFP-13-001

Serio A, Chaves-López C, Paparella A, Suzzi G (2010) Evaluation of metabolic activities of enterococci isolated from Pecorino Abruzzese cheese. Int Dairy J 20(7):459–464. https://doi.org/10.1016/j.idairyj.2010.02.005

Settanni L, Di Grigoli A, Tornambé G, Bellina V, Francesca N, Moschetti G, Bonanno A (2012) Persistence of wild Streptococcus thermophilus strains on wooden vat and during the

manufacture of a traditional Caciocavallo type cheese. Int J Food Microbiol 155(1–2):73–81. https://doi.org/10.1016/j.ijfoodmicro.2012.01.022
Sharma R, Garg P, Kumar P, Bhatia SK, Kulshrestha S (2020a) Microbial fermentation and its role in quality improvement of fermented foods. Fermentation 6(4):106. https://doi.org/10.3390/fermentation6040106
Sharma A, Sharma P, Singh J, Singh S, Nain L (2020b) Prospecting the potential of agroresidues as substrate for microbial flavor production. Front Sustain Food Syst 4:18. https://doi.org/10.3389/fsufs.2020.00018
Sharma S, Mohler J, Mahajan SD, Schwartz SA, Bruggemann L, Aalinkeel R (2023) Microbial biofilm: a review on formation, infection, antibiotic resistance, control measures, and innovative treatment. Microorganisms 11(6):1614. https://doi.org/10.3390/microorganisms11061614
Sidira M, Kandylis P, Kanellaki M, Kourkoutas Y (2016) Effect of curing salts and probiotic cultures on the evolution of flavor compounds in dry-fermented sausages during ripening. Food Chem 201:334–338. https://doi.org/10.1016/j.foodchem.2016.01.084
Silva R, Pereira J, Rouxinol M, Patarata L (2020) Sensory changes and listeria monocytogenes behavior in sliced cured pork loins during extended storage. Food Secur 9(5). https://doi.org/10.3390/foods9050621
Steele J, Broadbent J, Kok J (2013) Perspectives on the contribution of lactic acid bacteria to cheese flavor development. Curr Opin Biotechnol 24(2):135–141. https://doi.org/10.1016/j.copbio.2012.12.001
Su YC, Liu C (2007) Vibrio parahaemolyticus: a concern of seafood safety. Food Microbiol 24(6):549–558. https://doi.org/10.1016/j.fm.2007.01.005
Thierry A, Collins YF, Mukdsi MA, McSweeney PL, Wilkinson MG, Spinnler HE (2017) Lipolysis and metabolism of fatty acids in cheese. In: Cheese. Academic Press, pp 423–444. https://doi.org/10.1016/B978-0-12-417012-4.00017-X
Veerachamy S, Yarlagadda T, Manivasagam G, Yarlagadda PK (2014) Bacterial adherence and biofilm formation on medical implants: a review. Proc Inst Mech Eng H J Eng Med 228(10):1083–1099. https://doi.org/10.1177/0954411914556137
Vīna I, Semjonovs P, Linde R, Deniņa I (2014) Current evidence on physiological activity and expected health effects of kombucha fermented beverage. J Med Food 17(2):179–188. https://doi.org/10.1089/jmf.2013.0031
Walsh AM, Crispie F, Kilcawley K, O'Sullivan O, O'Sullivan MG, Claesson MJ, Cotter PD (2016) Microbial succession and flavor production in the fermented dairy beverage kefir. MSystems 1(5). https://doi.org/10.1128/msystems.00052-16
Wang S, Chen P, Dang H (2019) Lactic acid bacteria and γ-aminobutyric acid and diacetyl. Lactic Acid Bacteria. Bioengineering and Industrial Applications 1–19. https://doi.org/10.1007/978-981-13-7283-4_1
Wang W, Xia W, Gao P, Xu Y, Jiang Q (2017) Proteolysis during fermentation of Suanyu as a traditional fermented fish product of China. Int J Food Prop 20:S166–S176. https://doi.org/10.1080/10942912.2017.1293089
Wang Z, Xu Z, Sun L, Dong L, Wang Z, Du M (2020) Dynamics of microbial communities, texture and flavor in Suan zuo yu during fermentation. Food Chem 332:127364. https://doi.org/10.1016/j.foodchem.2020.127364
Wei Z, Chen C, Liu YJ, Dong S, Li J, Qi K, Liu S, Ding X, De Ora LO, Munoz-Gutíerrez I, Li Y, Yao H, Lamed R, Bayer EA, Cui Q, Feng Y (2019a) Alternative σI/anti-σI factors represent a unique form of bacterial σ/anti-σ complex. Nucleic Acids Res 47(11):5988–5997. https://doi.org/10.1093/nar/gkz355
Wei G, Wang K, Liu Y, Regenstein JM, Liu X, Zhou P (2019b) Characteristic of low-salt solid-state fermentation of Yunnan oil furu with Mucor racemosus: microbiological, biochemical, structural, textural and sensory properties. Int J Food Sci Technol 54(4):1342–1354. https://doi.org/10.1111/ijfs.14022
Willett W, Rockström J, Loken B, Springmann M, Lang T, Vermeulen S et al (2019) Food in the Anthropocene: the EAT–Lancet Commission on healthy diets from sustainable food systems. Lancet 393(10170):447–492. https://doi.org/10.1016/S0140-6736(18)31788-4

Yang Z, Liu S, Lv J, Sun Z, Xu W, Ji C, Liang H, Li S, Yu C, Lin X (2020) Microbial succession and the changes of flavor and aroma in Chouguiyu, a traditional Chinese fermented fish. Food Biosci 37:100725. https://doi.org/10.1016/j.fbio.2020.100725

Yun JH, Kim JH, Lee JE (2019) Surface film formation in static-fermented rice vinegar: a case study. Mycobiology 47(2):250–255. https://doi.org/10.1080/12298093.2019.1575585

Zeng X, Mo Z, Zheng J, Wei C, Dai Y, Yan Y, Qiu S (2023) Effects of biofilm and co-culture with Bacillus velezensis on the synthesis of esters in the strong flavor Baijiu. Int J Food Microbiol 394:110166. https://doi.org/10.1016/j.ijfoodmicro.2023.110166

Zhang Q, Zhang F, Gong C, Tan X, Ren Y, Yao K, Chi Y, Zhang Q (2020) Physicochemical, microbial, and aroma characteristics of Chinese pickled red peppers (Capsicum annuum) with and without biofilm. RSC Adv 10(11):6609–6617. https://doi.org/10.1039/d0ra00490a

Fresh Produce Challenges: Biofilm Solutions

8

Subhasis Chakrabarty, Joyeeta Bhattacharya, Asim Halder, Partha Roy, and Soma Das

Abstract

The production of fresh and healthy food has always been a challenge for the agriculture industry. With increasing demand and changing consumer preferences, there is constant pressure to produce food in large quantities while maintaining its quality and freshness. However, traditional farming methods face various challenges such as limited resources, climate change, and the risk of contamination from harmful bacteria. This is where the application of biofilm technology comes in as a perfect solution. Biofilm is a thin layer of microorganisms that form on surfaces, providing protection against external factors and promoting the growth of beneficial bacteria. By incorporating biofilm into farming practices, agriculture persons can prepare a controlled environment that can improve crop yields, reduce water consumption as well as large-scale production, and protect against harmful pathogens. This innovative solution not only addresses the challenges faced by traditional farming methods but also promotes sustainable agriculture practices for producing fresher and healthier food for consumers. Furthermore, biofilm technology can be applied in various stages of the supply chain, ensuring the freshness of produce from farm to table. Overall, the use of biofilm presents a promising solution to overcome the challenges of producing fresh food and promoting sustainable agricultural practices.

S. Chakrabarty · J. Bhattacharya
Department of Pharmaceutical Technology, DmbH Institute of Medical Science, Rasulpur, India

A. Halder · S. Das (✉)
Department of Pharmaceutical Technology, Adamas University, Barasat, India

P. Roy
GITAM School of Pharmacy, GITAM University, Visakhapatnam, India

T. Sarkar et al. (eds.), *Biofilm Applications to Revolutionize Food Technology*,
https://doi.org/10.1007/978-3-031-85205-3_8

Keywords

Food processing · Agriculture industry · Challenges · Application of biofilms · Solution and futuristic approach

8.1 Introduction

Biofilm is a community of microbial that can be denoted through its adhesive property specifically to the superficial area of the solid surface. An extracellular polymeric matrix was fabricated by the biofilm itself where huge numbers of microbes can be encapsulated (Xu et al. 2020). This extracellular polymeric matrix helps the inner microbes to contribute the required amount of nutrients and oxygen, as well as it can also provide shelter to the inner microbes from the external environment (Silva et al. 2021). This extracellular matrix channelizes the intracellular aggregation, retains the adhesion property, maintains the fluidity level in the cell, and also provides the proper protection against some environmental issues. Due to these advantages provided by polymeric extracellular matrix, the biofilm can be produced easily (Kasza et al. 2021). Quorum sensing and cell-to-cell communication are important when the biofilm is initiated to design. Quorum sensing is one of the major steps in the production of biofilms; it can recognize the alteration in cellular density by sensing the tiny signalling molecules. These molecules are called autoinducers (Cámara et al. 2022). These are responsible for the subsequent growth and the formation of biofilms. Initially, the concept was revealed that the microbes are basically floating cells. In 1970, the concept became transparent when a Canadian microbiologist invented the aggregation of microbes and termed as biofilms (Shineh et al. 2023). The extracellular matrix of the biofilms is formed by different polymeric substances such as protein, lipid materials, carbohydrates, polysaccharides, and extracellular DNA. The special three-dimensional structure of biofilms is created by Extra Cellular Matrix (ECM) scaffolds (Zhang et al. 2020). Biofilms are produced by different intermediate steps. This can be elucidated with five important steps (Asma et al. 2022). Primarily, different bacterial aggregations are accumulated and migrated, then attached to the surface successfully. Under positive circumstances, the attached chunks of bacteria are moving to their next goal and developing the first phase of biofilms successfully (Khan et al. 2021). This primarily formed biofilm is coated with a polymeric substance. In the next phase, the extracellular polymeric substances are secreted by different bacteria, and due to this effect, they can adhere to the surface. It can lead to the agglomeration of the bacterial colony and form the matrix (Darvishi et al. 2022). In the third phase of biofilm production, the microcolonies of the bacteria are created in the presence of water with different layers. In the fourth stage of biofilm production, it gets the full maturation and starts the operation as a three-dimensional community. At the last stage of biofilm production, the biofilm allows releasing the bacterial colonies from the layers and escalates the infection (Sen et al. 2021). Different environmental conditions such as the presence of ambient temperature, water, and requirement of oxygen can lead to the exponential growth of the bacteria through quorum sensing (QS). Such produced biofilms are very tough to penetrate through the antibiotics to terminate the growth of the bacteria (Sen et al. 2021).

It is also analysed from the peer-reviewed journals that through the application of quorum sensing the gene expression may occur and initiate the secretion of auto-inducers due to different changes such as the accessibility of various nutrients and also the variation of the density of the cell production (Rather et al. 2021a). The existence of the biofilm is found in different surfaces such as plastics, metals, bio-films, glass, different packaging-related products, and agro (Sharma et al. 2023). In the case of the bacterial life cycle, the formation of the biofilm is an important step. The inside microbes in the shelter of biofilm are comparatively safer than the normal ones. Extracellular polymeric matrix in biofilms also gives protection from the entry of other microbial agents from outside. Here, the extracellular polymeric matrix (EPS) acts as an anion exchanger. The EPS also limits the entry of the other adjuvant in the biofilm through the diffusion mechanism. However, these particular features depend on the extracellular polymeric substance and also on the microbial agents (da Silva et al. 2021). This polymeric matrix is capable of isolating different toxins, metallic ions, and other compounds, and due to this property, it was reported through scientific research report that it can lead to some protective barrier from the UV sensitive light, radiation, different pH sensations, and desiccation (Assefa and Amare 2022). The layers of water also provide nutrient support and maintain the water balance inside the matrix, and due to the presence of aqueous bulk, it can remove unwanted elements such as toxins, impurities, and others (Lv et al. 2023).

Nevertheless, there is also some positive role of biofilm observed in different contagious diseases as it develops some specific tolerance, and due to that, it is partially resistant to various common antibiotics (Su et al. 2022). This phenomenon is truly a vital step for the healthcare segments. Thereby, the production of biofilm is really a challenge for some important industries such as the pharma industry, agro-industry, food industry, beverage industry, and medical equipment (Abebe 2020). As a result, innovation and research are required to know properly the step-by-step process of biofilm production and its resistance mechanism, especially to antibiotics/antimicrobials. Antimicrobial resistance surely affects the treatment of some chronic infections and wounds; as a result, the treatment cost can be increased, and prolonged hospitalization may be required. It is also analysed from some peer-reviewed journals that the biofilm initiates the development of some specific diseases such as meningitis, osteomyelitis, chronic bacterial prostatitis, endocarditis, infection in the kidney, some chronic non-curing diabetic wounds, keratitis as associated with the contact lens, and lastly implantable device-related infections (IDRI) (Singh et al. 2022). These IDRI infections are associated with pacemakers, stents for angioplasty, prosthetic surgery, and knee replacements. One of the major causes of human mortality is due to IDRI infection, where one of the causes is the accumulation of biofilm (Quinn et al. 2020). IDRI deals with the devices which can be implanted in the human body. These devices penetrate through the mucosal layers of the skin. The rate of infections depends upon either the life cycle of the bacteria or the site of implantation into the body. So, there is a scarcity of potent antimicrobials that can manage this biofilm formation and mitigate microbial infection on different surfaces (Xiu et al. 2021). Thereby, the production of biofilm is challenging in some parts of the healthcare segments. A brief research is also required to know the key to eliminate the exponential growth rate of the biofilm and remove the chances of contamination through this. There are a good number of articles published regarding biofilm and its importance and challenges for production. However, there are some important roles also associated with biofilms, such as it can contribute to bacterial persistence in different industries such as food, agro, and packaging. Biofilms also play a definite important role in food processing (Chang et al. 2022). This chapter will focus on different challenges in the production of biofilm and its solution for the betterment of healthcare segments.

Formulation Strategies of Biofilm Production

Biofilm can be designed by different processes, and it has several steps to complete its formation. These steps are farmed with initiation and adsorption into the surface, adhesion with the surface materials, formation of the microcolonial structures, development, maturation, and dispersion (Al et al. 2022). In biofilm formation, the aqueous (water) plays an important role in providing conditional support and the formation of the microbial colony. Water also supplies the nutrients support to the bacterial cells and makes the cell more robust, especially on the surface of different materials (Al et al. 2022). In the biofilm colony, the cells can be attached with a compact aggregate, and due to the close and tight accumulation of the cells, the nutritional support may be high, and the transfer of the gene expression may occur (Li et al. 2021a). Biofilms are suited to formulate in an environment where the turbulent flow is present, and the Reynolds number is >5000 Re. In fluid mechanics, the lower level of Reynolds number suggests the laminar flow, and the higher level of Reynolds number suggests the turbulence flow (Rath et al. 2021). It is established from different research reports that biofilm production may be increased due to the turbulence flow. Biofilm can be formed in both smooth and rough surfaces of different materials. In addition, biofilm is formed by various microorganisms, and they have a better degree of correlation (Kaiser et al. 2021). The twenty-first century reveals how the production of biofilm is controlled by various natural forces and other physiological aspects. It was already stated in that section that how the five stages are important for the formation of biofilms. These stages are described in this section.

(i) *Initial Attachment:* In the initial attachment, the planktonic, free-floating bacteria can be stroked at different surfaces, and that begins the initial attachment for biofilm production (Dostert et al. 2021). The bacteria conjugate to the surface by the application of flagella or pilli or by the specific application of physical forces. This process is an easy and reversible process where the bacterial encounter at the surface material can take place (Yin et al. 2022). The adhesion of the bacteria towards the surface of choice can be governed by the nature of the bacteria, positive environmental conditions, temperature, the structural belongings of the bacterial cells, and the pressure. The forces that are responsible for the bacterial adhesion at different surfaces of the rough or smooth materials are basically electrostatic forces, hydrophobic forces, steric, van der Wall forces, and the protein coupling reaction (Bowler et al. 2020). These forces are quite capable of sticking the bacteria at the surface of the materials and also effective in sustaining the bacteria against the repulsive forces.

(ii) *Aggregation of the Bacteria and Adhesion Mechanism:* This second stage of biofilm production is called the anchoring phase or the latching phase. In this phase, the microbes are attached at the molecular level through the bacterial cling at the outer part of the surface. The loosely connected bacteria can be solidified through the adhesion mechanism through the formation of an extracellular matrix. This extracellular matrix is responsible for interacting with the surface of the receptors with the support of the microbial pilli (Thambirajoo et al. 2021). This phenomenon helps the bacteria to adhere strongly to the surface. In the second part, the aggregation of the bacteria may be stronger and also irreversible without any addition of chemical or physical forces. The microbes are able to accumulate at the surface without any hesitation in the same way that a cocoon can conjugate at the surface of a leaf (Tran et al. 2020). The microbes apply different adhesins to make them strongly adhere to the surface of any material according to the different environmental conditions. In these circumstances, the planktonic bacteria are highly able to attach themselves to one another to form an agglomeration at the site of the substratum corneum (Mishra et al. 2024). The presence of microbes at the outer surface of the stratum corneum can initiate the stimulation of other microbes to adhere to the previous one and make the agglomeration successful. All the microbes are able to produce the adhesin to accumulate the bacteria; it is also noted from the scientific report that some of the adhesins are able to authorize the transcription level. This can lead to the transition of the microbes from an immobile situation to a planktonic structure, according to the environmental factors (Cui et al. 2022).

(iii) *Formation of Microcolony:* The third stage is to fabricate the microcolony from the bacteria itself. In this step, the bacteria attach to each other and initiate bacterial cell division to create the microcolony (Rather et al. 2021b). The microcolonies are sustained in the biofilm, and they consist of various kinds of micro-communities. It is observed from various scientific research reports that the microcolonies are interconnected by the micro-community. This interconnection is really important for transferring different metabolites, nutrients, and aqueous substances with a specific flow; this micro-community is also helpful in discarding metabolic waste (Jiang et al. 2020). At least three different species of bacteria are required to do the anaerobic digestion and to break down the complex organic materials from CH_4 to CO_2 (Jiang et al. 2020). After that the bacteria again start to ferment them to produce different alcohol and acid and again initiate to produce the composite organic compounds. Whereas, accetogenic bacteria treat the organic compounds as their food. Methanogenic bacteria get their energy by transforming hydrogen, acetate, and CO_2 into methane (Zhao et al. 2023). Biofilms proffer an excellent composite environment for the formation of syntrophic collaboration (Li et al. 2022).

(iv) *Maturation of Bacterial Film:* After the microcolonial formation, the next step is the maturation step, where the bacteria get full maturation and complete growth toward the next phase. Maturation of the bacteria is allowed by the stimulation of signalling factors from the attached cell of bacteria thereby, the result reflects through the impression of the biofilm-particulate gene (Mirghani et al. 2022). The signalling factor also cooperates to modify the gene regulation to increase the bacterial virulence factors. The process starts after the ejection of the extracellular matrix from the core of the bacterial cell. These extracellular matrixes provide a shield to the sheltered bacteria inside the biofilm and also protect them from the anti-microbial agents (Jiang et al. 2021). According to the scientific report, the environmental DNA plays an important role in regulating intracellular signalling and the robustness of biofilm. In the time of accumulation or aggregation of the bacteria on the material surface, many layers or channels can be formed. These cell layers then create the microcolonies which are responsible for transferring different substances. In this maturation phase, the bacteria are able to acknowledge the specific measurement of the neighbouring group that can initiate the formation of the clusters, and it can make a bond with the neighbouring cells successfully (Coenye et al. 2020).

(v) *Dispersion Phase:* This phase is assumed to be the last phase to formulate the biofilm structure. It can transmit from one infected person to another healthy. This phenomenon is really important for the fabrication of the biofilm. Biofilms are composed of two completely separate layers (Zea et al. 2020). The first component is the primary layer where the bacteria can be safe and stay primarily. Another part is called as surface layer, and this layer works as a particular place for dispersion; in this area, biofilm initiates to spread normally (Srinivasan et al. 2021). This stage is actually responsible for the chronic infection and other infectious disease. In this phase, a sudden attack in the body due to sudden shock, stroke, or clotting of blood may be observed. If it happens, it needs urgent treatment and special medical attention to cure. One of the most common phenomena in this stage is metastatic seeding (Liu et al. 2021). Sometimes, it may be difficult to control if the biofilm is produced and spread strongly and sustained for a long time. This can enhance the severity of the infection. The bacterial cells then transmit and disperse into another part of the infected one. This can also enhance their activity on the surface of the medical biodevice. There, it can grow because of the nutritional support and oxygen demands. Different cells such as a single cell or an accumulation of cells can be stripped from the biofilm and start the fabrication of dispersion (Fig. 8.1) (Alves et al. 2021).

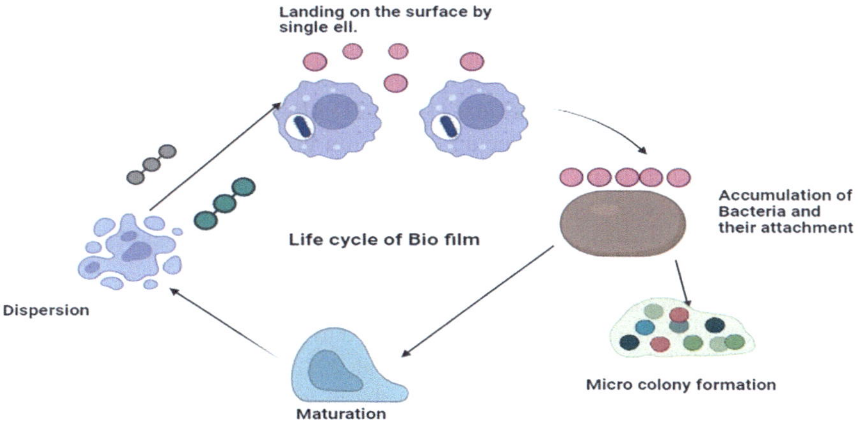

Fig. 8.1 Life cycle of biofilm production

8.2 Challenges to Produce Biofilm

To know the biofilm mechanisms, first understand the basic information and fabrication mechanism of biofilm and its properties and fabricate a specific biofilm model (Schilcher and Horswill 2020). This model system can be responsible for the analysis of the biofilm phenotype successfully. Over the last two decades, scientists have been trying to develop different in vitro and in vivo models to fabricate the system of biofilm production. For growing the adhesive biofilm populace, both the in vitro models (static and dynamic) are responsible. Different flat substances are required for growing the static test systems such as petri plates, various plastic tubes, and small petri dishes. It is also analysed from some peer-reviewed journals that a 96-well plate is considered the most popular static biofilm fabrication system (Funari and Shen 2022). In this system, each and every plate was utilized to add the bacterial suspension and incubated for the prescribed time (Muhammad et al. 2020). However, this static biofilm cannot intimate the natural biofilm production system thoroughly as per the expectation because, in a natural biofilm system, there is some obvious nutritional support through different layers. However, in an in vitro static system, this nutritional support is missing (Cacaci et al. 2020). As a result, there was an obvious need to fabricate a strong in vitro process to create a biofilm system with constant support of nutrients. It is reported from some scientific reports that there are some flow displacement model systems available in the global market such as modified Robbins devices (MRDs) and Centre for Disease Control biofilm reactors (Hu et al. 2022). The modified Robbins device (MRD) is a very common tank where the reaction can take place, and it has many numbers of ports that actually cling the disc for plug holding where the biofilm can be produced (de Barros et al. 2020). The free-flowing bacterial suspension flows accordingly with the device after the adhesion phase of the biofilm production; at this time, a medium flow can be incorporated to copy the in vivo biofilm fabrication process. However, in the in vitro dynamic method, better control can be observed to produce the biofilm successfully. Due to this, the dynamic process can be used to analyse the various resistance issues of the biofilm population. Both the in vitro processes are equally important to characterize the biofilm population (Schulze et al. 2021). However, there are some major limitations also observed to producing the biofilm in this process, such as some environmental effects (e.g., some signalling molecules and gene expression) that can be present in the time of the biofilm population. This limitation can be overcome a little bit through the application of the in vivo model (Barzegari et al. 2020). Several animal models have been approached to analyse the BAI because they have a better affinity towards the physiological behaviour of biofilm as well as the host. The rat model and rabbit model can be approached to know the exact behaviour and the formation of the biofilm production (Luo et al. 2022). These models are not only applicable to study and testing purposes. In the in vivo CVC model, an external jugular vein can be used to insert the catheters and microbial contamination can happen; in this texture, the rat model can be applied (Wang 2021). This model is one of the best examples to study the effect of biofilm after attaching to CVC and also analyse the effect of the antibiotics after being administered through the parenteral route. In the case of the implantable products, it was inserted beneath the skin tactfully to analyse the biofilm production inside the implantable products (Amankwah et al. 2021). It is also noted that the animals can be affected due to bacterial contamination pre- or postoperative conditions. This model is a well-established model to know the adherence strength of biofilm upon different substrates. In addition, the inserted foreign material into the peritoneal cavity can be in focus for the chronic infection study (Amankwah et al. 2021). Some other studies also reveal the biofilm adherence study to insert the plastic tube into the bladder, resulting in the effect of urinary tract infections and others. There are some other ex vivo models that are also applicable to studying the biofilm adhesion mechanism in different organs or tissues in the human body, such as the kidney or teeth model (Azevedo et al. 2020). These models are capable of explaining different infections related to the wound inside the human body. It is also reported from different peer-reviewed reports that the withdrawn tissue from the body can be placed for the in vivo experiments (Sauer et al. 2022).

Application of Biofilm in Different Areas

Biofilm is a self-regenerating system, and it has tremendous outcomes, especially in developing different biological processes. It can also enhance the activity of different biological processes either by economic aspects or to increase the overall activity of the biological process. This concept is more or less true that various products and materials subproducts may incorporate toxicity into the cell; as a result, the cell's viability or other metabolite activity may be affected (Abusrewil et al. 2020). However, the application of the biofilm can protect the cell from the outside environment when it interacts with the environment itself. As a result, the inside good bacteria can be safe. There are different types of acidic materials can be produced through the application of the biofilm, such as acetic acid, succinic acid, butanol, lactic acid, and fumaric acid. Biofilm can also be applied to prepare microbial fuel cells (MFCs) either by the use of monospecies, multispecies, or the use of biofilms (Guzmán-Soto et al. 2021). Reactors for the production of biofilm are already developed to produce different biofilms but that can also be challenging to produce the biofilm through the use of the reactors. Biofilm reactors are often used to prepare different biologically active products and materials to use in laboratories and hospitals, treating different biowaste, wastewater treatment, and bioremediation (Falcone et al. 2021). It is also helpful to prepare the biologically based materials through the applications of the bioreactors. It is already established that biofilms are self-immobilized, resistant to toxic environments, fast in fermentation activity, and sustainable activity; they can percolate cell recycling and make it easy for product delivery (Ferriol-González and Domingo-Calap 2020). These characteristics also decrease the production time as well as reduce the cost of production. Biofilms are considered eco-friendly materials to produce different materials without affecting the environment. This tool is important for preparing industrial products also. Inside the reactor, the biofilm is present to catalyse the reaction and proceed with the reaction for sustainable effect. A higher concentration of biofilm can reduce the production time and thereby the production cost will be less as compared to others (Malone and Schultz 2022).

Fresh Produce Challenges

Biofilm has a negative impact on food processing industries; due to biofilm production, plenty of processed food can be spoiled. It is also noted from a peer-reviewed research report that the QS-modulated, produced biofilms are responsible for impacting the food industry severely, and due to bacterial contamination, the processed food can be contaminated, and as a result, different foodborne diseases can occur (Batoni et al. 2021). Maximum biofilms are actually human pathogens. It is also analysed from different statistical data that 60% of the infected diseases from food are due to biofilm production (Munir et al. 2020). The formation of the biofilm is simple, and it can be configured with any substances such as plastics, wood, metal, polyethylene, jute, medical equipment, and medical devices, thereby, it can cause issues to the machinery. These incidents initiate the challenges for the food processing industry (Ding et al. 2022). The biofilms are especially immobilized and made with different species. Thereby, it is also noted from different scientific reports that primarily the biochemical properties of the biofilm should be known to get the actual nature of the biofilm in a suited manner (Thaarup and Bjarnsholt 2021). These produced biofilms can be the primary cause affecting the fresh-produced materials such as the food industry, beverage industry, dairy products, and meat and fish processes industry. As a result, it is important to control biofilm production to ensure the processed foods and their safety and efficiency towards human life (Murray et al. 2021). In every stage of processed food, especially the storage, production, and distribution, the supply chain is equally important and sensitive to bacterial contaminations (Barman et al. 2024). This area should be improved to enhance the production of fresh produce materials through different industries. There are various species of bacteria that are responsible for creating the biofilm and impacting the fresh produce materials such as *Bacillus cereus*, *Escherichia Coli, Listeria monocytogens,* and some of the species of *Salmonella* and *Coccus* (Rodríguez-Merchán et al. 2021). Various species of bacterial biofilm can also impact on fresh preparation of the fish such as *S. enterica* and *Aeromonas hydrophila* (Sahli et al. 2022). Biofilm also plays a responsible part in the fermentation process. It is already established that different biochemical products guided by the biofilm process are in the market, and they are potent enough for different biochemical work (Wang et al. 2020). This process is more efficient because it is less toxic and it maintains a sustained effect on the cell. In fermented products such as beer, distillates, cheese, meats, fish, and package materials, the microbial groups are present, and they are protected because of the biofilm production and engulfed by the extracellular matrixes. This can give an open space for growing the inside materials under the biofilm umbrella (Nag et al. 2021). Presently, in agricultural sectors, biofilm has a tremendous impact. In the green era of the world, different countries witnessed an important follow through to produce different raw materials of foods; primarily, it was analysed that the green revolution happened due to the application of improved technology seeds (high-yielding varieties seeds), in addition with different use of pesticides and fertilizers. However, the excessive use of chemicals and also herbicides has a negative impact, especially on soil quality as well as has tremendous effect on humans and other animals. As a result, the agricultural sectors have planned to produce the crop again with some novel nontoxic methods; biofilm is one of them (Saccomano et al. 2021). The application of biofilm in the agricultural sector can modify the sustainability of the product without any harm to the soil and human life. The agriculturally important microbes are actually responsible for the huge growth of the green with less harmful effects towards the human race, such as plant growth–promoting rhizobacteria (PGPR) (Saccomano et al. 2021). These bacteria can make a colony inside the root area of the specific plant and create the biofilm; this biofilm supports the plant for nutritional growth and also the growth hormonal effect for the huge production of crops. This PGPR can be helpful to give the shelter of *Rhizobium*, *Pseudomonas*, *Azotobacter*, etc. This can also play an important role in bioremediation (Fasiku et al. 2021). It has been observed from different scientific reports that the natural biofilms are potent and they are single species. In recent eras, the multispecies biofilm has also gained focus because they are combing with fungi and commonly found in mycorrhiza (Li et al. 2021b). Compared to the single species, the multispecies are more advantageous in the bulk production of crops as well as in improving the soil's ecological performance. It is also noted that the application of biofilm is also very important to produce biogas. This biogas is one of the major sources of sustainable energy. Most likely, wastewater treatment is also an important part where the biofilm can incorporate and upgrade the technology on an advanced scale (Dzianach et al. 2019). However, all the production may be hampered due to the excessive biofilm production and also the application of different bacteria from the outside. As a result, it may be challenging to produce fresh materials. To overcome this issue, there may be an incorporation of some novel technology in addition to the biofilm (Dzianach et al. 2019). This section will discuss the biofilm solution over different issues.

Biofilm Solution As per the statistics of WHO, around 4.95 million people announced death due to antibiotic resistance. This report was displayed through antimicrobial resistance (AMR) (Kumar et al. 2019). The multidrug resistance is a common issue in human races as well as different plants and animals. This multidrug resistance can be grown to different bacteria to make it more potent to survive under different negative conditions. One of the common threats of antimicrobial resistance is the indiscriminate use of antibiotics; besides this, the improper application of sanitization, cleanliness, and water distribution systems may also impact the formation of bacteria and bacterial biofilm. However, to overcome this issue, different approaches can be adopted to combat the biofilm such as the use of advanced category antibiotics, antiseptics, disinfectants, various enzymes, surface-modified agents, and at last the QS inhibitors. Some anti-biofilm production processes may be discussed in the below-mentioned sections (Vickery 2019).

Application of Phytoextracts The World is the natural source of different plants. Various plant extracts may be an approach to combat the biofilm-related issue as reported by some peer-reviewed journals. As per the reported study, it was stated that the plant extracts are excellent as a secondary metabolite, and the plant-conjugated extract has a positive impact in combating various kinds of biofilm infections (Gebreyohannes et al. 2019). Several plants can be identified to decline the growth of the biofilm. Various antifungal and antibacterial compounds can be synthesized from the plant, and they have tremendous efficacy in combating biofilm activity. For oral infection, the *Candida* species is very common, and this species is responsible for severe oral infection in humans. These biofilms are tested for evaluation purposes in the presence of some plant extracts. Some essential oil from plants can be extracted and charged upon this fungal biofilm, and lastly, the result was found suitable to decline this fungal infection. The essential oil was extracted from the bulb of *Allium sativum L.* and *Cinnamomum zeylanica* Blume leaves (Jorge et al. 2019). It was reported that both the agents found from the plants have tremendous activity against the biofilm. In another study, it was stated that the combination therapy of quercetin glycolic extracts and curcumin was really effective against some bacterial biofilms (*C. albicans, C. dubliniensis, and C. tropicalis*). The infection was reduced remarkably. *Piper betle* extract has proved itself very potent against *Bacillus gaemokensis* biofilm. In this extraction process, as a solvent, chloroform was applied (Liu et al. 2019) (Table 8.1).

Nanotechnological Intervention to Combat Biofilm Nanotechnology can be incorporated to target the biofilm. This technology is innovative and effective to be tested for the pre-clinical and clinical trials to decline biofilm formation. Nanotechnological intervention can lead to an increase in the antibacterial payload to the affected area and result in a definite bactericidal effect (Jaśkiewicz et al.

Table 8.1 Different phytochemical extracts and their activity against the biofilm formation

Name of the bacteria	Minimum inhibitory concentration	Plant source	References
Staphylococcus aureus	5 mg/ml	*Cinnamomum verum*	Chirca (2019)
Bacillus cereus	0.156 mg/mL	*Allium sativum*	Hu et al. (2019)
Pseudomonas aeruginosa	12–19 mg/mL	*Cinnamomum cassia*	Mandakhalikar (2019)
Escherichia coli	26–35 mg/mL	*Cinnamomum cassia*	Elbourne et al. (2019)
Enterobacter spp.	0.78 mg/mL	*Syzygium aromaticum*	Dostert et al. (2019)
Enterococcus faecalis	9.63 mg/mL	*Piper nigrum*	Alvarez-Ordóñez et al. (2019)
Vibrio parahaemolytics	6.25 mg/mL	Cinnamon	Percival et al. (2019)

2019). The nanosized particle can penetrate through the extracellular matrix and directly impact the microbial. Due to the tiny size, suited shape, excellent surface area, and better aspect ratio, the nanoparticles can easily penetrate through the extracellular matrix and destroy the biofilms. The drugs are encapsulated by the nanosystem; and due to this, the protection of the molecule from various enzymatic reactions can easily percolate (Narayana and Srihari 2019). Due to the severe toxicity and stability issue, the nanoparticles are not applied in full fledge to decline the growth of biofilm. Especially for the antimicrobial effect, different metallic nanoparticles are applied such as gold, silver, zinc, and iron (Kadam et al. 2019). When the nanoparticles somehow move towards the extracellular matrix, the interaction initiates, and at the last stage, due to the ionization mechanism, the EPS membrane can be ruptured, and the metallic nanoparticles enter into the biofilm heart and inhibit further growth (Vaishnavi et al. 2019). Some factors are responsible for initiating this mechanism such as the biochemical properties of the biofilm, the density, the charge, and surface texture (Vaishnavi et al. 2019). The surface modification, particle size, shape, and other morphological characters can impact the nanoparticle and the biofilm interactions. Some common bactericidal nanotechnological method is implemented to inhibit the growth of the biofilm such as DNA leakage, chain break, protein denaturation, and oxidative damage (Fig. 8.2).

Application of Antimicrobial Peptide A minimum 5–50 amino acid chain is required to form AMP. AMP has a positive charge and has a maximum molecular mass of ~10KDa (Melo et al. 2019). This AMA is present naturally in human animals and plants in primary defence to combat pathogens. This AMP actually acts as an antibacterial or antimicrobial agent. Various approaches may be taken to utilize the AMP against the adhesion activity of biofilm on different surfaces (Pinto et al. 2019). A group of scientists claim that the AMP system can disrupt and destroy the membrane surface of the biofilm and interact with the bacterial QS-system, disrupt the extracellular matrix completely, and decline the growth of alarmone. This is responsible for monitoring the bacterial stress level (Mosaddad et al. 2019).

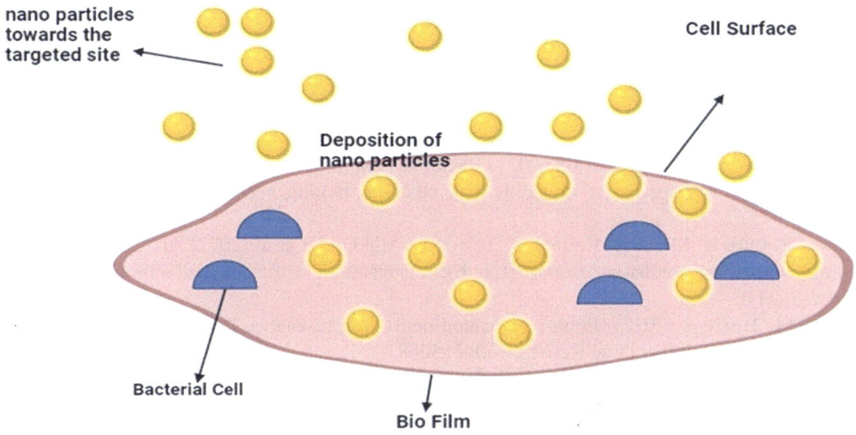

Fig. 8.2 Nanotechnological intervention in targeting biofilms

Conclusion There is no doubt that microbial biofilm has a different impact from environmental science to healthcare segments. Conventional antibiotics may be useful to treat bacterial infections, but due to the indiscriminate use of antibiotics, antibiotic resistance may occur especially to treat the biofilm infection as well as to inhibit the growth of bacterial infections. As a result, there should be an alternative potent approach to combat the biofilm infection. Fresh produce materials are affected regularly due to the regular intervention of biofilm and other colonies of bacteria. Knowing the specific application and the clinical aspects of biofilm, it is somehow important to design effective methods to combat biofilm-related issues. Different alternative strategies such as the nanotechnological intervention, using phytochemicals, and using the AMP system can replace the conventional antibiotic therapy to combat the biofilm, as well as inhibit bacterial growth. However, still, positive research is required to know the exact mechanism of biofilm formation. The safety and toxicity profile of the new regime treatment should be in focus for further pre-clinical and clinical trials. The proper knowledge of the biofilm technology and further revealing of the loopholes can mitigate the gap in research in biofilm technology; in this way, we can generate a smooth transition of combating process from the biofilm-associated issues.

References

Abebe GM (2020) The role of bacterial biofilm in antibiotic resistance and food contamination. Int J Microbiol 2020(1):1705814

Abusrewil S, Alshanta OA, Albashaireh K, Alqahtani S, Nile CJ, Scott JA, McLean W (2020) Detection, treatment and prevention of endodontic biofilm infections: what's new in 2020? Crit Rev Microbiol 46(2):194–212

Alvarez-Ordóñez A, Coughlan LM, Briandet R, Cotter PD (2019) Biofilms in food processing environments: challenges and opportunities. Annu Rev Food Sci Technol 10(1):173–195

Alves PJ, Barreto RT, Barrois BM, Gryson LG, Meaume S, Monstrey SJ (2021) Update on the role of antiseptics in the management of chronic wounds with critical colonisation and/or biofilm. Int Wound J 18(3):342–358

Al-Wrafy FA, Al-Gheethi AA, Ponnusamy SK, Noman EA, Fattah SA (2022) Nanoparticles approach to eradicate bacterial biofilm-related infections: a critical review. Chemosphere 288:132603

Amankwah S, Abdella K, Kassa T (2021) Bacterial biofilm destruction: a focused review on the recent use of phage-based strategies with other antibiofilm agents. Nanotechnol Sci Appl 14:161–177

Asma ST, Imre K, Morar A, Herman V, Acaroz U, Mukhtar H et al (2022) An overview of biofilm formation–combating strategies and mechanisms of action of antibiofilm agents. Life 12(8):1110

Assefa M, Amare A (2022) Biofilm-associated multi-drug resistance in hospital-acquired infections: a review. Infect Drug Resist 15:5061–5068

Azevedo MM, Lisboa C, Cobrado L, Pina-Vaz C, Rodrigues A (2020) Hard-to-heal wounds, biofilm and wound healing: an intricate interrelationship. Br J Nurs 29(5):S6–S13

Barman S, Kurnaz LB, Leighton R, Hossain MW, Decho AW, Tang C (2024) Intrinsic antimicrobial resistance: molecular biomaterials to combat microbial biofilms and bacterial persisters. Biomaterials 311:122690

Barzegari A, Kheyrolahzadeh K, Hosseiniyan Khatibi SM, Sharifi S, Memar MY, Zununi Vahed S (2020) The battle of probiotics and their derivatives against biofilms. Infect Drug Resist 13:659–672

Batoni G, Maisetta G, Esin S (2021) Therapeutic potential of antimicrobial peptides in polymicrobial biofilm-associated infections. Int J Mol Sci 22(2):482

Bowler P, Murphy C, Wolcott R (2020) Biofilm exacerbates antibiotic resistance: is this a current oversight in antimicrobial stewardship? Antimicrob Resist Infect Control 9:1–5

Cacaci M, Martini C, Guarino C, Torelli R, Bugli F, Sanguinetti M (2020) Graphene oxide coatings as tools to prevent microbial biofilm formation on medical device. Adv Microbiol Infect Dis Public Health 14:21–35

Cámara M, Green W, MacPhee CE, Rakowska PD, Raval R, Richardson MC et al (2022) Economic significance of biofilms: a multidisciplinary and cross-sectoral challenge. npj Biofilms Microb 8(1):42

Chang C, Yu X, Guo W, Guo C, Guo X, Li Q, Zhu Y (2022) Bacteriophage-mediated control of biofilm: a promising new dawn for the future. Front Microbiol 13:825828

Chirca I (2019) The hospital environment and its microbial burden: challenges and solutions. Future Microbiol 14(12):1007–1010

Coenye T, Kjellerup B, Stoodley P, Bjarnsholt T (2020) The future of biofilm research–report on the '2019 Biofilm Bash'. Biofilms 2:100012

Cui F, Li T, Wang D, Yi S, Li J, Li X (2022) Recent advances in carbon-based nanomaterials for combating bacterial biofilm-associated infections. J Hazard Mater 431:128597

da Silva RA, Afonina I, Kline KA (2021) Eradicating biofilm infections: an update on current and prospective approaches. Curr Opin Microbiol 63:117–125

Darvishi S, Tavakoli S, Kharaziha M, Girault HH, Kaminski CF, Mela I (2022) Advances in the sensing and treatment of wound biofilms. Angew Chem 134(13):e202112218

de Barros PP, Rossoni RD, de Souza CM, Scorzoni L, Fenley JDC, Junqueira JC (2020) Candida biofilms: an update on developmental mechanisms and therapeutic challenges. Mycopathologia 185(3):415–424

Ding M, Zhao W, Song LJ, Luan SF (2022) Stimuli-responsive nanocarriers for bacterial biofilm treatment. Rare Metals 41(2):482–498

Dostert M, Belanger CR, Hancock RE (2019) Design and assessment of anti-biofilm peptides: steps toward clinical application. J Innate Immun 11(3):193–204

Dostert M, Trimble MJ, Hancock RE (2021) Antibiofilm peptides: overcoming biofilm-related treatment failure. RSC Adv 11(5):2718–2728

Dzianach PA, Dykes GA, Strachan NJ, Forbes KJ, Pérez-Reche FJ (2019) Challenges of biofilm control and utilization: lessons from mathematical modelling. J R Soc Interface 16(155):20190042
Elbourne A, Truong VK, Cheeseman S, Rajapaksha P, Gangadoo S, Chapman J, Crawford RJ (2019) The use of nanomaterials for the mitigation of pathogenic biofilm formation. In: Methods in microbiology, vol 46. Academic Press, pp 61–92
Falcone M, De Angelis B, Pea F, Scalise A, Stefani S, Tasinato R et al (2021) Challenges in the management of chronic wound infections. J Glob Antimicrob Resist 26:140–147
Fasiku VO, Omolo CA, Devnarain N, Ibrahim UH, Rambharose S, Faya M et al (2021) Chitosan-based hydrogel for the dual delivery of antimicrobial agents against bacterial methicillin-resistant Staphylococcus aureus biofilm-infected wounds. ACS Omega 6(34):21994–22010
Ferriol-González C, Domingo-Calap P (2020) Phages for biofilm removal. Antibiotics 9(5):268
Funari R, Shen AQ (2022) Detection and characterization of bacterial biofilms and biofilm-based sensors. ACS Sensors 7(2):347–357
Gebreyohannes G, Nyerere A, Bii C, Sbhatu DB (2019) Challenges of intervention, treatment, and antibiotic resistance of biofilm-forming microorganisms. Heliyon 5(8):e02192
Guzmán-Soto I, McTiernan C, Gonzalez-Gomez M, Ross A, Gupta K, Suuronen EJ et al (2021) Mimicking biofilm formation and development: recent progress in in vitro and in vivo biofilm models. Iscience 24(5):102443
Hu C, Wang LL, Lin YQ, Liang HM, Zhou SY, Zheng F et al (2019) Nanoparticles for the treatment of oral biofilms: current state, mechanisms, influencing factors, and prospects. Adv Healthc Mater 8(24):1901301
Hu Y, Ruan X, Lv X, Xu Y, Wang W, Cai Y et al (2022) Biofilm microenvironment-responsive nanoparticles for the treatment of bacterial infection. Nano Today 46:101602
Jaśkiewicz M, Janczura A, Nowicka J, Kamysz W (2019) Methods used for the eradication of staphylococcal biofilms. Antibiotics 8(4):174
Jiang Y, Geng M, Bai L (2020) Targeting biofilms therapy: current research strategies and development hurdles. Microorganisms 8(8):1222
Jiang Y, Liu Y, Zhang X, Gao H, Mou L, Wu M et al (2021) Biofilm application in the microbial biochemicals production process. Biotechnol Adv 48:107724
Jorge P, Magalhães AP, Grainha T, Alves D, Sousa AM, Lopes SP, Pereira MO (2019) Antimicrobial resistance three ways: healthcare crisis, major concepts and the relevance of biofilms. FEMS Microbiol Ecol 95(8):fiz115
Kadam S, Shai S, Shahane A, Kaushik KS (2019) Recent advances in non-conventional antimicrobial approaches for chronic wound biofilms: have we found the 'chink in the armor'? Biomedicines 7(2):35
Kaiser P, Wächter J, Windbergs M (2021) Therapy of infected wounds: overcoming clinical challenges by advanced drug delivery systems. Drug Deliv Transl Res 11:1545–1567
Kasza K, Gurnani P, Hardie KR, Camara M, Alexander C (2021) Challenges and solutions in polymer drug delivery for bacterial biofilm treatment: a tissue-by-tissue account. Adv Drug Deliv Rev 178:113973
Khan J, Tarar SM, Gul I, Nawaz U, Arshad M (2021) Challenges of antibiotic resistance biofilms and potential combating strategies: a review. 3 Biotech 11:1–15
Kumar S, Chandra N, Singh L, Hashmi MZ, Varma A (eds) (2019) Biofilms in human diseases: treatment and control. Springer, Cham
Li X, Sun L, Zhang P, Wang Y (2021a) Novel approaches to combat medical device-associated biofilms. Coatings 11(3):294
Li C, Cornel EJ, Du J (2021b) Advances and prospects of polymeric particles for the treatment of bacterial biofilms. ACS Appl Polym Mater 3(5):2218–2232
Li Z, Wang X, Wang J, Yuan X, Jiang X, Wang Y et al (2022) Bacterial biofilms as platforms engineered for diverse applications. Biotechnol Adv 57:107932
Liu Y, Shi L, Su L, van der Mei HC, Jutte PC, Ren Y, Busscher HJ (2019) Nanotechnology-based antimicrobials and delivery systems for biofilm-infection control. Chem Soc Rev 48(2):428–446

Liu Y, Li Y, Shi L (2021) Controlled drug delivery systems in eradicating bacterial biofilm-associated infections. J Control Release 329:1102–1116
Luo X, Zhang B, Lu Y, Mei Y, Shen L (2022) Advances in application of ultraviolet irradiation for biofilm control in water and wastewater infrastructure. J Hazard Mater 421:126682
Lv X, Wang L, Mei A, Xu Y, Ruan X, Wang W et al (2023) Recent nanotechnologies to overcome the bacterial biofilm matrix barriers. Small 19(6):2206220
Malone M, Schultz G (2022) Challenges in the diagnosis and management of wound infection. Br J Dermatol 187(2):159–166
Mandakhalikar KD (2019) Medical biofilms. In: Introduction to biofilm engineering. American Chemical Society, pp 83–99
Melo LD, Pires DP, Monteiro R, Azeredo J (2019) Phage therapy of infectious biofilms: challenges and strategies. In: Phage therapy: a practical approach, pp 295–313
Mirghani R, Saba T, Khaliq H, Mitchell J, Do L, Chambi L et al (2022) Biofilms: formation, drug resistance and alternatives to conventional approaches. AIMS Microbiol 8(3):239
Mishra A, Aggarwal A, Khan F (2024) Medical device-associated infections caused by biofilm-forming microbial pathogens and controlling strategies. Antibiotics 13(7):623
Mosaddad SA, Tahmasebi E, Yazdanian A, Rezvani MB, Seifalian A, Yazdanian M, Tebyanian H (2019) Oral microbial biofilms: an update. Eur J Clin Microbiol Infect Dis 38:2005–2019
Muhammad MH, Idris AL, Fan X, Guo Y, Yu Y, Jin X et al (2020) Beyond risk: bacterial biofilms and their regulating approaches. Front Microbiol 11:928
Munir S, Shah AA, Shahid M, Manzoor I, Aslam B, Rasool MH et al (2020) Quorum sensing interfering strategies and their implications in the management of biofilm-associated bacterial infections. Braz Arch Biol Technol 63:e20190555
Murray E, Draper LA, Ross RP, Hill C (2021) The advantages and challenges of using endolysins in a clinical setting. Viruses 13(4):680
Nag M, Lahiri D, Sarkar T, Ghosh S, Dey A, Edinur HA et al (2021) Microbial fabrication of nanomaterial and its role in disintegration of exopolymeric matrices of biofilm. Front Chem 9:690590
Narayana PSVVS, Srihari PSVV (2019) Biofilm resistant surfaces and coatings on implants: a review. Mater Today Proc 18:4847–4853
Percival SL, Mayer D, Kirsner RS, Schultz G, Weir D, Roy S et al (2019) Surfactants: role in biofilm management and cellular behaviour. Int Wound J 16(3):753–760
Pinto RM, Lopes-de-Campos D, Martins MCL, Van Dijck P, Nunes C, Reis S (2019) Impact of nanosystems in Staphylococcus aureus biofilms treatment. FEMS Microbiol Rev 43(6):622–641
Quinn J, McFadden R, Chan CW, Carson L (2020) Titanium for orthopedic applications: an overview of surface modification to improve biocompatibility and prevent bacterial biofilm formation. IScience 23(11):101745
Rath S, Bal SCB, Dubey D (2021) Oral biofilm: development mechanism, multidrug resistance, and their effective management with novel techniques. Rambam Maimonides Med J 12(1):e0004
Rather MA, Gupta K, Bardhan P, Borah M, Sarkar A, Eldiehy KS et al (2021a) Microbial biofilm: a matter of grave concern for human health and food industry. J Basic Microbiol 61(5):380–395
Rather MA, Gupta K, Mandal M (2021b) Microbial biofilm: formation, architecture, antibiotic resistance, and control strategies. Braz J Microbiol 52:1–18
Rodríguez-Merchán EC, Davidson DJ, Liddle AD (2021) Recent strategies to combat infections from biofilm-forming bacteria on orthopaedic implants. Int J Mol Sci 22(19):10243
Saccomano SC, Jewell MP, Cash KJ (2021) A review of chemosensors and biosensors for monitoring biofilm dynamics. Sens Actuat Rep 3:100043
Sahli C, Moya SE, Lomas JS, Gravier-Pelletier C, Briandet R, Hémadi M (2022) Recent advances in nanotechnology for eradicating bacterial biofilm. Theranostics 12(5):2383
Sauer K, Stoodley P, Goeres DM, Hall-Stoodley L, Burmølle M, Stewart PS, Bjarnsholt T (2022) The biofilm life cycle: expanding the conceptual model of biofilm formation. Nat Rev Microbiol 20(10):608–620
Schilcher K, Horswill AR (2020) Staphylococcal biofilm development: structure, regulation, and treatment strategies. Microbiol Mol Biol Rev 84(3):10–1128

Schulze A, Mitterer F, Pombo JP, Schild S (2021) Biofilms by bacterial human pathogens: clinical relevance-development, composition and regulation-therapeutical strategies. Microbial Cell 8(2):28

Sen CK, Roy S, Mathew-Steiner SS, Gordillo GM (2021) Biofilm management in wound care. Plast Reconstr Surg 148(2):275e–288e

Sharma S, Mohler J, Mahajan SD, Schwartz SA, Bruggemann L, Aalinkeel R (2023) Microbial biofilm: a review on formation, infection, antibiotic resistance, control measures, and innovative treatment. Microorganisms 11(6):1614

Shineh G, Mobaraki M, Perves Bappy MJ, Mills DK (2023) Biofilm formation, and related impacts on healthcare, food processing and packaging, industrial manufacturing, marine industries, and sanitation–a review. Appl Microbiol 3(3):629–665

Silva NBS, Marques LA, Röder DDB (2021) Diagnosis of biofilm infections: current methods used, challenges and perspectives for the future. J Appl Microbiol 131(5):2148–2160

Singh A, Amod A, Pandey P, Bose P, Pingali MS, Shivalkar S et al (2022) Bacterial biofilm infections, their resistance to antibiotics therapy and current treatment strategies. Biomed Mater 17(2):022003

Srinivasan R, Santhakumari S, Poonguzhali P, Geetha M, Dyavaiah M, Xiangmin L (2021) Bacterial biofilm inhibition: a focused review on recent therapeutic strategies for combating the biofilm mediated infections. Front Microbiol 12:676458

Su Y, Yrastorza JT, Matis M, Cusick J, Zhao S, Wang G, Xie J (2022) Biofilms: formation, research models, potential targets, and methods for prevention and treatment. Adv Sci 9(29):2203291

Thaarup IC, Bjarnsholt T (2021) Current in vitro biofilm-infected chronic wound models for developing new treatment possibilities. Adv Wound Care 10(2):91–102

Thambirajoo M, Maarof M, Lokanathan Y, Katas H, Ghazalli NF, Tabata Y, Fauzi MB (2021) Potential of nanoparticles integrated with antibacterial properties in preventing biofilm and antibiotic resistance. Antibiotics 10(11):1338

Tran HM, Tran H, Booth MA, Fox KE, Nguyen TH, Tran N, Tran PA (2020) Nanomaterials for treating bacterial biofilms on implantable medical devices. Nano 10(11):2253

Vaishnavi KV, Safar L, Devi K (2019) Biofilm in dermatology. J Skin Sex Trans Dis 1(1):3–7

Vickery K (2019) Microbial biofilms in healthcare: formation, prevention and treatment. Materials 12(12):2001

Wang Y (2021) Liposome as a delivery system for the treatment of biofilm-mediated infections. J Appl Microbiol 131(6):2626–2639

Wang H, Christiansen DE, Mehraeen S, Cheng G (2020) Winning the fight against biofilms: the first six-month study showing no biofilm formation on zwitterionic polyurethanes. Chem Sci 11(18):4709–4721

Xiu W, Shan J, Yang K, Xiao H, Yuwen L, Wang L (2021) Recent development of nanomedicine for the treatment of bacterial biofilm infections. Viewpoints 2(1):20200065

Xu Y, Dhaouadi Y, Stoodley P, Ren D (2020) Sensing the unreachable: challenges and opportunities in biofilm detection. Curr Opin Biotechnol 64:79–84

Yin R, Cheng J, Wang J, Li P, Lin J (2022) Treatment of Pseudomonas aeruginosa infectious biofilms: challenges and strategies. Front Microbiol 13:955286

Zea L, McLean RJ, Rook TA, Angle G, Carter DL, Delegard A et al (2020) Potential biofilm control strategies for extended spaceflight missions. Biofilms 2:100026

Zhang K, Li X, Yu C, Wang Y (2020) Promising therapeutic strategies against microbial biofilm challenges. Front Cell Infect Microbiol 10:359

Zhao A, Sun J, Liu Y (2023) Understanding bacterial biofilms: from definition to treatment strategies. Front Cell Infect Microbiol 13:1137947

Prolonging Perfection: Biofilm and Shelf Life

9

M. Menaka, N. B. Brunda, V. Vathsala, Ram Asrey,
P. R. Geetha Lekshmi, S. G. Lekshmi,
and Karan Singh Dhami

Abstract

Biofilms are microbial aggregation membranes that are formed when microorganisms attach to the surfaces of living or nonliving things. Importantly, biofilm provides protection to microorganisms against environmental pressures and enhances their resistance to antimicrobial agents, contributing to food contamination, toxicity, and shorter shelf lives. Furthermore, the formation of biofilms by foodborne pathogens greatly exacerbates the risk of foodborne infections, leading to major public health risks and adverse economic consequences. Thus, it is essential and imperative to understand and control biofilm formation and to reduce the microbial risks associated with it. Food waste due to spoilage within the food industry remains a global challenge to environmental sustainability and the security of food supplies. Consequently, this chapter delves into the existing control strategies, antimicrobial mechanisms, and their applications for extending the shelf life of food. Overall, current strategies for controlling harmful biofilms fall into three main categories: (i) modifying abiotic surface features to prevent biofilm formation; (ii) regulating signaling pathways to inhibit biofilm formation and stimulate biofilm diffusion; and (iii) applying external forces to

M. Menaka (✉) · V. Vathsala · R. Asrey · S. G. Lekshmi
Division of Food Science and Postharvest Technology, ICAR-Indian Agricultural Research Institute, New Delhi, India

N. B. Brunda
Division of Post Harvest Technology and Agricultural Engineering, ICAR-Indian Institute of Horticultural Research, Bengaluru, India

P. R. G. Lekshmi
Division of Post Harvest Management, College of Agriculture, Vellayani, Kerala Agricultural University, Thrissur, India

K. S. Dhami
Horticulture, Mid-West University, Surkhet, Nepal

T. Sarkar et al. (eds.), *Biofilm Applications to Revolutionize Food Technology*,
https://doi.org/10.1007/978-3-031-85205-3_9

eradicate biofilms. Although, in the food industry, the best strategy to eradicate bacterial biofilms is to prevent biofilm formation, this can be achieved by avoiding the formation of bacterial biofilms in key locations via methods such as aseptic processing, regular disinfection cleaning, and the sterilization of equipment before and after use.

Keywords

Contamination · Food hygiene · Food safety · Human health · Sustainability · Antiadhesion

9.1 Introduction to Biofilms and Their Formation

Microorganisms mostly cling to and grow on inanimate and living surfaces in the natural world. These surfaces can be anything from those found in soil and water systems to those found on a range of indwelling medical devices and living tissues such as heart valves, lungs, and middle ear tissues. The development of a biofilm by the cells characterizes this connected growth condition collectively. In order for the various species to support one another, the closeness of the microbes allows for substrate exchange, metabolic product distribution, and toxic end-product removal. Bacteria within biofilm communities are resistant to antimicrobials, mechanical forces, and the immune system due to their structural composition. Biofilm formation refers to the process by which microorganisms adopt a multicellular lifestyle, increasing their survival and adaptability across diverse environments. Biofilms can develop on both living and nonliving surfaces in natural and clinical settings. Their association with chronic infections is significant because biofilms help pathogens evade the body's innate immune defenses, leading to persistent infections (Singh et al. 2019).

Biofilm formation is the process by which microorganisms attach irreversibly to a surface, producing extracellular polymers that facilitate matrix formation and attachment (Donlan 2001). This attachment alters the microorganisms' phenotype, affecting their growth rate and gene expression. Biofilm development is a complex, dynamic, and multistep process involving the transformation of bacteria from a free-swimming planktonic state to a sessile, biofilm-forming state. The process can be divided into five stages: initial reversible attachment (Gupta et al. 2016), irreversible attachment (Jamal et al. 2015; Lohse et al. 2018), maturation (Castiblanco and Sundin 2016), and dispersion (Pang et al. 2018). A variety of external factors, including temperature, pH, hydrodynamic and gravitational forces, Brownian motion, surface properties, quorum sensing, and signaling molecules, influence biofilm formation (Jamal et al. 2015). Overall, biofilm formation can be categorized into four primary stages (Rather et al. 2021).

Attachment: A Phase of Surface Sensing

Biofilm formation begins with the adherence of planktonic microorganisms to surfaces, marking the transition from free-flowing cells to an organized community structure (Haggag 2010). Initially, planktonic bacteria attach loosely and reversibly, orienting themselves polar to the surface. Bacteria eventually become flat, creating a monolayer and secreting an extracellular matrix, or "slime," to shield themselves. This matrix consists of nucleic acids, polysaccharides, structural proteins, and cell detritus. It is referred to as extracellular polymeric substances (EPS). Extracellular DNA dominates early matrix synthesis; subsequently, polysaccharides and proteins take center stage. Microcolonies emerge as biofilm development advances, exhibiting notable proliferation and quorum sensing-based cell-to-cell communication. Three-dimensional growth of the biofilm results in irreversible adhesion. Early in the biofilm formation process, the intracellular signaling molecule Bis-(3′-5′)-cyclic dimeric guanosine monophosphate (c-di-GMP) is essential because it decreases flagella-mediated movement and increases matrix creation (Römling et al. 2013). c-di-GMP levels rise with each attachment/detachment event, mediated by the Pil-Chp surface-sensing system. Early biofilm formation involves the transition from surface-naive to surface-sentient planktonic cells, leading to irreversible attachment and biofilm development (Armbruster and Parsek 2018).

Formation of Microcolony

Microorganisms stick to surfaces and begin to multiply and agglomerate within the self-produced extracellular polymer shell (EPS) shortly after, which results in the creation of microcolonies in the presence of high concentrations of c-di-GMP. For microorganisms to interface with surfaces and for cell-cell aggregations to form microcolonies, respectively, flagella and type IV pili-mediated motilities are critical (Rabin et al. 2015).

Maturation

In addition to helping microbial adhesion, maintaining the three-dimensional structure of the biofilm, organizing cells, and providing defense against oxidative damage, metallic cations, the host immunological response, and antibiotics, EPS is essential for the development of biofilms. Enzymes, metabolic products, and signaling molecules for quorum sensing are also enclosed in EPS (Toyofuku et al. 2016). Mature biofilms may form "mushroom" or "tower" shapes, influenced by aerotolerance and metabolism (Rabin et al. 2015). Surfactants and modulins, through quorum sensing, are vital in Staphylococcal biofilm maturation (Otto 2013). Mature biofilms have three layers: an inner regulating layer, a middle microbial basement layer, and an outer layer with planktonic microorganisms ready to exit (Zhao et al. 2017).

Dispersion

Mature biofilms disperse either through active mechanisms, such as bacterial motility and the breakdown of extracellular polymeric substances (EPS), or passively due to physical forces such as fluid flow. This process is driven by factors such as overgrowth, heightened competition, and nutrient scarcity (Rabin et al. 2015). Enzymatic actions, such as alginate degradation in *Pseudomonas* spp. (Costerton et al. 1999), as well as environmental shifts such as temperature changes, lack of oxygen, and the buildup of metabolites, also play a role. Additionally, gene regulation during dispersion involves the upregulation of genes related to cell motility and EPS degradation, while genes responsible for polysaccharide and fimbriae production are downregulated (McDougald et al. 2012; Rather et al. 2021).

9.2 Types of Biofilms

Biofilms can be classified into two main categories: (i) multispecies biofilms and (ii) single-species biofilms.

I. *Multispecies biofilms*: Multispecies biofilms consist of multiple bacterial species, each inhabiting a distinct ecological niche within the biofilm, determined by its metabolic activities and morphological traits (Stoodley et al. 2002). The composition and function of these mixed-culture biofilms depend on the species involved and their interactions. For instance, one species may engage in anaerobic fermentation within a particular environment of the biofilm, while the same species might switch to aerobic metabolism in a different environment with other neighboring species. Multispecies biofilms are significant both clinically and industrially. Clinically, they are problematic as they are associated with persistent infections, including dental caries, nosocomial infections, and other diseases (Costerton et al. 1999). Industrially, they can be both harmful and advantageous. For example, natural biofilms can contaminate food processing equipment (Carpentier and Cerf 1993), foul ultrafiltration, reverse osmosis membranes (McDonogh et al. 1994), and hamper heat transmission in heat exchangers and cooling towers (Mortensen and Conley 1994). Positively, organic chemicals and heavy metals can be eliminated from wastewater by using multispecies biofilms (Taras et al. 2005; Meyer and Wallis 1997).
II. *Single-species biofilms*: These are utilized in the production of commercially important chemicals (Qureshi and Maddox 1990). While these biofilms are less common in natural environments, they are crucial in industrial settings for their role in controlled chemical production. In these cases, a single species is inoculated into a sterile environment to form a biofilm, which is then used to produce a specific chemical product.

Both multispecies and single-species biofilms can be classified into two types: (a) biofilms that form on different substrates such as bone char, charcoal, resin, concrete, clay bricks, or sand particles and (b) biofilms that arise through the aggregation and flocculation of cells.

9.3 Biofilms in the Food Industry

Biofilms in the food industry present both challenges and opportunities. On the downside, biofilms can cause major problems, such as the contamination of food products and processing equipment. They often harbor harmful bacteria such as *Salmonella* and *Listeria*, which are resistant to standard cleaning methods, leading to foodborne illnesses and spoilage. This can result in economic losses from product recalls, waste, and increased cleaning expenses. Additionally, biofilms can cause equipment fouling, reducing operational efficiency, and increasing maintenance needs. However, biofilms also offer advantages in specific food industry applications. For instance, they are used in biofilm-based wastewater treatment systems to efficiently remove organic pollutants and heavy metals from industrial effluents. Controlled biofilm applications can also be harnessed to create innovative food preservation methods such as antimicrobial packaging, which extends product shelf life and maintains quality. Thus, while biofilms present significant challenges related to contamination and equipment upkeep, they also provide opportunities to improve food safety and sustainability through creative solutions.

9.4 Dark Side of Biofilms: Impact on Shelf Life and Their Mitigation Strategies

The capacity of microbes to stick to and multiply on food and surfaces in contact with food presents a serious problem for food hygiene owing to biofilm formation. Food rotting can result in severe hygienic issues and financial losses due to this attachment (Carpentier and Cerf 1993). Research has also shown that foodborne pathogens can persist on contact surfaces. Notable pathogens include *Escherichia coli* O157 (Doyle 1988), *Campylobacter jejuni* (Stern and Kazmi 1989), *Yersinia enterocolitica* (Kumar and Singh 1994), and *Listeria monocytogenes* (Farber and Peterkin 1991).

Biofilm Development in Food Processing Environments

Biofilms can quickly develop in the environments of the food industry, where various microorganisms grow on food products and equipment, leading to potential biofilm formation. In food processing, biofilms can form on both food-contact and nonfood-contact surfaces. The complexity of modern food processing systems characterized by long production runs, large-scale product output, and vast surface areas creates an ideal environment for biofilm formation (Lindsay and von Holy 2006). As a result, foodborne bacteria can adhere to these surfaces, increasing the risk of bacterial contamination and foodborne illnesses. In fact, biofilms are implicated in 80% of bacterial infections in the USA, many of which are linked to foodborne pathogens (Srey et al. 2013). The formation of mixed-species biofilms is a dynamic process influenced by factors such as the characteristics of the attachment surface (Tang et al. 2011), components of the food matrix (Van Houdt and Michiels 2010), environmental conditions (Govaert et al. 2018), and the type of bacterial cells involved (Makovcova et al. 2017). Surface characteristics such as hydrophobicity, electrostatic charge, roughness, and topography play a crucial role in biofilm formation and overall hygiene (Araújo et al. 2010). However, research findings are inconsistent. Some studies suggest that rougher surfaces promote bacterial attachment (Dhowlaghar et al. 2018), while others find no significant correlation (Jindal et al. 2018). Similarly, while hydrophobic surfaces are generally thought to attract more bacteria, research on hydrophobicity is conflicting, with some studies indicating that hydrophilic surfaces may actually support greater bacterial adherence (Dhowlaghar et al. 2018; Jindal et al. 2018). These variations may result from different methodologies and bacterial strains used in studies, as attachment is likely influenced by multiple factors (Carrascosa et al. 2015). Contaminants in food processing environments typically originate from the surrounding air, machinery, or food surfaces. If not thoroughly cleaned, these areas can become breeding grounds for biofilm formation, leading to food spoilage, health risks, and financial losses (Coughlan et al. 2016; Galie et al. 2018). Furthermore, biofilms pose significant technological challenges for the food industry. They have the ability to increase the frictional resistance of fluids on surfaces, hinder heat transfer through equipment, cause mechanical blockages in fluid handling systems, and speed up the rate of corrosion on metal surfaces, all of which have the potential to reduce productivity (Meesilp and Mesil 2019). Thermophilic and cryophilic bacteria are frequently found in dairy enterprises throughout the pasteurization, preservation, and processing stages. For example, because of its heat-resistant spores, *Geobacillus* spp., which can grow at temperatures as high as 65 °C, presents serious challenges for the manufacturing of milk powder (Palmer et al. 2010). The shelf life of dairy products is shortened by cryophilic bacteria, which proliferate at chilled temperatures and affect dairy storage even more. A common psychrophilic bacterium called *Pseudomonas* frequently causes spoiling by growing biofilms in the walls of pipes and cooling tanks for milk. In the absence of heat treatment, *Pseudomonas* can produce enzymes, build biofilms, and lower lipids' thermal stability by acting as a protease, which can cause milk to sour (Marchand et al. 2009).

Impact on Food Quality and Safety

Biofilms pose a serious threat to food safety and human health owing to diseases associated with them. The food sector has a great deal of challenges as bacteria can quickly attach to surfaces such as wood, polypropylene, glass, plastic, rubber, stainless steel, and even food within minutes. This rapid adhesion allows mature biofilms to form in just days or even hours (Carrascosa et al. 2021). Most of the time, biofilm

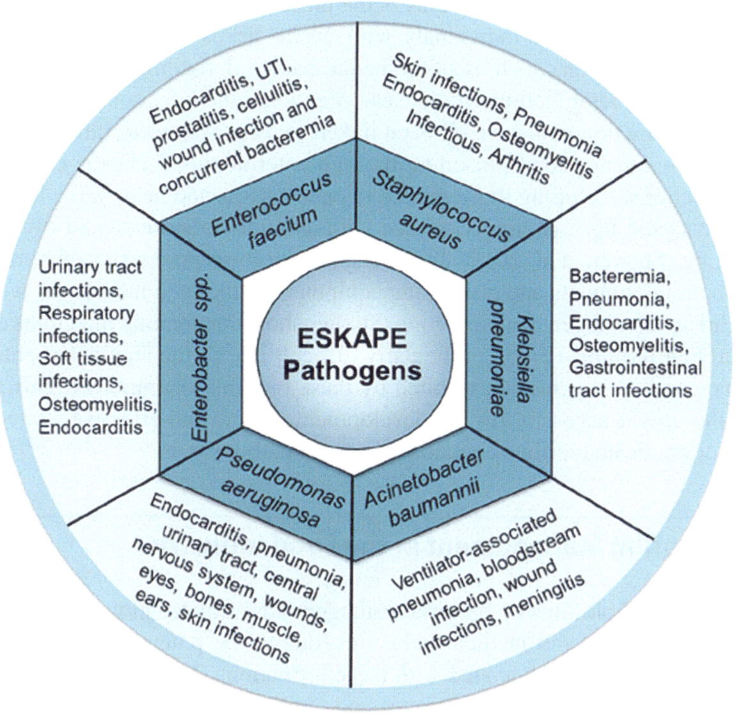

Fig. 9.1 Biofilm-associated infections caused by ESKAPE pathogens. (Picture courtesy: Kumar et al. 2023a)

development is detrimental (Coughlan et al. 2016). Foodborne viruses in the food business have the ability to create biofilms, which can cause food to degrade and endanger consumer health (Galiè et al. 2018). Foodborne bacteria and related biofilms are the main sources of foodborne diseases, and they represent a major risk to public health as well as the expansion of the food industry (Fig. 9.1). Between 40% and 80% of bacteria on Earth have the ability to create biofilms (Flemming and Wuertz 2019). Additionally, it is important to note that some microorganisms that form biofilms in food processing environments can be human pathogens. According to Colagiorgi et al. (2017), these pathogens can form biofilm structures on a range of synthetic substrates, such as rubber, glass, stainless steel, polyethylene, wood, and polypropylene. Numerous studies have emphasized that the development of biofilms on equipment surfaces in the food processing and medical industries can harbor bacteria, posing a persistent risk for infection and cross-contamination. They are significantly implicated in the transmission of the human immunodeficiency virus and foodborne illnesses (Hall-Stoodley et al. 2004). Microbial biofilms are responsible for approximately 60% of foodborne outbreaks globally, highlighting their presence in food processing environments as a major threat to food safety (Galie et al. 2018; Han et al. 2017). A notable example of the role of biofilms in persistent and often deadly infections is the presence of *Pseudomonas aeruginosa* biofilms in the lungs of individuals with cystic fibrosis (Hassett et al. 2010). Furthermore, infections such as otitis media, bacterial vaginitis, gingivitis, dental plaque, urinary tract infections, middle ear infections, and infections related to catheters and prosthetic joints have all been linked to *Staphylococcus aureus*, which is notorious for causing food poisoning through enterotoxins (Lebeaux et al. 2014). Some infections, including those related to endocarditis and heart valve infections, as well as cystic fibrosis, are associated with antibiotic resistance and can be life-threatening (Sharahi et al. 2019). In conclusion, biofilms pose a risk of contamination to both instruments and processing equipment in the food industry as well as a direct threat of pathogen contamination. Studies show that treating infections caused by microbial biofilms can be particularly difficult due to the significant antibiotic resistance and pathogenicity exhibited by these biofilm communities. This underscores the urgent necessity for the development of new and effective strategies to control or eradicate biofilms containing harmful pathogens.

9.5 Biofilm Management in the Food Industry

The formation of biofilms by bacterial pathogens has attracted global attention, particularly in the food and medical industries, due to the potential health risks and public health concerns associated with these formations (Liu et al. 2023). Biofilms not only exhibit mechanical persistence and antibiotic resistance but also produce virulence factors, all of which can lead to severe health complications for humans (Bridier et al. 2015; Giaouris et al. 2015). Processed foods are particularly vulnerable to microbial contamination because biofilms in food processing environments can harbor foodborne pathogens and spoilage organisms. If microorganisms from surfaces that come into contact with food are not thoroughly eliminated, they can facilitate biofilm formation and increase the risk of microbial transfer. To mitigate biofilm formation on food-contact surfaces, a variety of preventive and control strategies can be effectively implemented. These include physical measures combined with hygienic plant layout and equipment design, material selection, and the appropriate use of detergents and disinfectants. Additionally, the role of bacteriocins and enzymes is becoming increasingly important, as they offer unique potential for effective biocontrol and removal of biofilms in the food industry. According to Kumar and Anand (1998), these innovative biocontrol methods are essential for maintaining biofilm-free systems as well as ensuring the safety and quality of food. Current strategies for controlling harmful biofilms can be broadly categorized into three approaches: (i) modifying abiotic surface properties to prevent biofilm formation; (ii) manipulating signaling pathways to inhibit biofilm formation and encourage biofilm dispersal; and (iii) applying external forces to completely remove biofilms (Yin et al. 2021).

Biofilm Elimination in the Food Industry

Biofilms shield microorganisms from disinfectants, boosting their chances of surviving and contaminating food later on. This results in food having a shorter shelf life and poses a danger of spreading disease. Therefore, to lessen the microbiological hazards connected to biofilm formation, it is essential and crucial to comprehend and regulate biofilm production as well as develop ways for biofilm removal. Hospital and food sector settings have been studied for biofilm formation. Due to the fact that surfaces of operating rooms, implants, prostheses, and tools have simpler applications and distinct surface compositions (antibiofilm activity) may contribute to hospitals' greater success in their attempts to remove biofilms. Many attempts have been undertaken to date to lessen the formation of biofilm on surfaces in the food business, but these efforts have mostly relied on novel disinfectants with varying degrees of efficacy. These findings have led to a better understanding of the specific mechanisms involved in initial surface attachment, the formation of group structures and ecosystems, and detachment processes although the results have varied (Palmer et al. 2010).

The most effective strategy for eliminating bacterial biofilms in the food industry is to prevent their formation. This can be accomplished through methods such as aseptic processing, regular cleaning and disinfection, and sterilizing equipment after use to inhibit the growth of biofilms in key areas. Sterilizing every environment in manufacturing areas is not feasible or economical in most circumstances, particularly during food processing. Consequently, further steps need to be taken to reduce the amount of dangerous bacteria and biofilms in manufacturing locations. In order to prevent cells from adhering firmly (reversibly) to contact surfaces, it is imperative to perform routine cleaning and disinfection in order to prevent the formation of biofilms. Disinfectants are now your greatest ally in the fight against biofilms. Quaternary ammonium compounds, hypochlorites, peroxides (peracetic acid and hydrogen peroxide), chloramines, iodine, ozone, aldehydes (formaldehyde, glutaraldehyde, and paraformaldehyde), and phenols are the disinfectants most frequently employed in sanitary disinfection programs in the food business. Programs for disinfection also include mixes of quaternary ammonium, chlorine dioxide, and alkyl amines (Srey et al. 2013). They have detrimental effects on bacterial cells as a result of their reactions with different parts of the cells. Nonetheless, some areas of research have shown promising outcomes, such as the development of phages to attack bacteria that form biofilms and the composition of surfaces for materials to hinder bacterial adhesion. Undoubtedly, the majority of studies have concentrated on the bacteriological biofilm, avoiding the discussion of the theory that filamentous fungi are the cause of biofilm development. Various researchers (Moreno-García et al. 2018; Ogawa et al. 2019) bolster this hypothesis, which attributes *Aspergillus fumigatus's* existence to biofilm formation. In this instance, morphological alterations, the existence of an extracellular polymeric matrix, differential gene expression, and a unique susceptibility to antifungal medications distinguish the marked similarities between bacteriological and filamentous fungal biofilms from diffuse or loosely associated (planktonic) colonies (Carrascosa et al. 2021). More efficient and ecologically friendly control methods for getting rid of or managing the growth of dangerous biofilms have also been found in recent years. For instance, it has been shown that ibuprofen at subinhibitory doses decreases the production of biofilms on abiotic surfaces by *E. coli*, *S. aureus*, *Streptococcus pneumoniae*, and *Candida albicans* (Oliveira et al. 2019). To enhance the control of biofilm formation, phage lysosomes and bacteriophages have been identified as effective antibiofilm agents (Yin et al. 2021). Unlike phages, which can directly kill bacteria, phage lysosomes induce host bacteria to produce enzymes that degrade EPS, facilitating the removal of mature biofilms more rapidly (Melo et al. 2020). Additionally, multiple bacteriophages or phage lysates can be combined with other strategies to achieve a broad spectrum of antibacterial effects. Recently, there has been growing interest in developing or exploring alternative physical methods for decontaminating surfaces to eliminate bacteria from biofilms. These methods include photodynamic inactivation using pulsed UV light, electron beam irradiation, steam heating, irradiation at 405 nm, and surface treatments involving ozone, ultrasound, or gaseous chlorine dioxide (Alvarez-Ordóñez et al. 2019). Together, these innovative approaches hold promise for reducing biofilm formation in the food industry in the future.

Techniques for Monitoring Biofilms

Monitoring and managing biofilms is crucial in various industries, such as healthcare, food processing, and water treatment. Effective biofilm management is vital for maintaining hygiene, preventing contamination, and ensuring operational efficiency. To tackle biofilm-related challenges, several advanced techniques are utilized (Wang et al. 2024).

Optical Biosensors

Graphene-Based Biosensors: These biosensors employ exonuclease III-assisted target recycling and DNA walker cascade amplification to detect agrD expression in methicillin-resistant *Staphylococcus aureus* (MRSA). They are capable of monitoring biofilm formation and clarifying the mechanisms of drug action. Similar sensors have also been developed to measure the expression of agrA, agrB, and agrC, facilitating real-time biofilm monitoring and investigation of drug mechanisms (Ning et al. 2020; Wang et al. 2022).

Fluorescent Biosensors: These utilize gold nanoparticles conjugated with fluorescent proteins to distinguish biofilms based on their physical and chemical characteristics. This approach allows for the rapid identification of bacterial species and strains, making them valuable tools for visualizing and targeting biofilms (Li et al. 2014).

Local Surface Plasmon Resonance (LSPR): LSPR technology detects wavelength shifts to monitor biofilm formation in real-time. Sensors developed by Funari et al. (2018) demonstrate high sensitivity and versatility, making them effective for evaluating biofilm growth and the effects of drugs.

Electrochemical Biosensors

Colloidal Gold-Based Sensors: These sensors are utilized to screen for drugs targeting *Helicobacter pylori* biofilms by integrating biomolecular recognition elements with electrochemical effects. They have been effective in identifying antibiofilm drugs and assessing their efficacy (Xiao et al. 2022).

Point-of-Care (POC) Biosensors: These microbial-sensing platforms are engineered for the rapid and sensitive detection of pathogens such as *E. coli*. They employ techniques such as bacterium-directed click chemistry-based colorimetry, which uses gold nanoparticles modified with azide and acetylene for result visualization on smartphones. This approach provides high sensitivity and is particularly applicable in remote or underdeveloped regions (Mou et al. 2019).

Methods for Detecting Biofilm Formation in the Food Industry

Online monitoring of biofilm adhesion, growth, and removal in industrial environments can significantly lower cleaning costs and minimize production downtime. Traditional techniques, such as agar plating, often fail to detect many biofilm-forming bacteria. For instance, some pathogens such as *L. monocytogenes* can enter a "viable but non-culturable" (VBNC) state, exhibiting low metabolic activity that conventional culture methods cannot identify. PCR amplification can be used to identify VBNC cells (Gião and Keevil 2014), which emphasizes the need for novel detection techniques. Furthermore, the presence of VBNC cells in some biofilms (in the case of agar plating) and the high cost of reactive and equipment (in the case of qRT-PCR and other DNA amplifications) make these techniques ineffective at the industrial level (Li et al. 2017). Advanced methods such as metagenomics and metatranscriptomics offer insights into biofilm community interactions (Mukherjee et al. 2016). For instance, multiple-locus variable number of tandem repeats analysis (MLVA) allows subspecies typing of *S. aureus* by PCR amplification and gel electrophoresis, distinguishing genotypes based on the length of amplified PCR fragments (Rešková et al. 2014; Coughlan et al. 2016). Furthermore, industrial sensors using thermal pulse analysis can detect biofilm deposits as thin as a few micrometers by measuring local thermal conductivity and heat variations (Fratamico et al. 2009). Another commercial approach involves capacitance/impedance calculations and electrochemical tests to measure electric signals. In industrial environments, devices such as BIOX and BIOGEORGE are used to optimize antibiofilm chemical dosages by measuring the coupling current between zinc and stainless steel electrodes (Fratamico et al. 2009).

9.6 Innovations in Biofilm Control

New Antibiotics and Their Carries

Conventional antibiotics, whether taken separately or in combination, aid in the fight against antibiotic resistance. For instance, the volume and motility of *P. aeruginosa* biofilm are decreased by sub-minimum inhibitory doses of ceftazidime (Otani et al. 2018), while colistin reduces *E. coli* biofilms in a concentration-dependent manner (Klinger-Strobel et al. 2017). Gentamicin released from bone grafts can prevent *E. coli* adhesion and eliminate 24-hour-old biofilms (Butini et al. 2018).

Antibiotic Adjuvants Triclosan, a Food and Drug Administration (FDA)-approved antimicrobial, enhances tobramycin's efficacy against *P. aeruginosa* and other bacteria, including *S. aureus* (Maiden et al. 2018). It also shows synergy with gentamicin and streptomycin. However, most antimicrobial compounds lack FDA approval for use in hygiene products and their in vivo efficacy and bioavailability remain under-explored.

Antimicrobial Peptides and Proteins Antimicrobial peptides (AMPs), such as nisin, offer significant antibacterial and antibiofilm activity. Nisin, approved by the FDA, effectively combats several bacterial species and biofilms, including those of MRSA (Shin et al. 2015, 2016). Other AMPs such as garvicin KS and synthetic peptides such as Bac8c and Omiganan have shown promise against biofilms (Ovchinnikov et al. 2016; Zapotoczna et al. 2017). Despite this, in vitro data for many synthetic AMPs is lacking.

Proteolytic Enzymes Proteolytic enzymes can disrupt biofilm matrices, aiding in biofilm control. Proteases such as bromelain and DNAses have demonstrated effectiveness against various dental and pathogenic biofilms (Fleming and Rumbaugh 2017; Mugita et al. 2017). Enzymes such as dextranase and mutanase show potential in preventing dental biofilm formation, especially when used together (Otsuka et al. 2015; Ren et al. 2019).

Bacteriophage Therapy Phage therapy offers a promising alternative or supplement to conventional antibiotics. This method employs phages, phage cocktails, phage-derived enzymes, and genetically engineered phages to effectively target biofilms. For example, phage PELP20 and lysins produced by phages have been demonstrated to significantly decrease biofilm colony-forming units and disrupt bacterial peptidoglycan (Waters et al. 2017; Sharma et al. 2018). Furthermore, combining phage cocktails with antibiotics can enhance the effectiveness of biofilm removal (Maszewska et al. 2018; Tkhilaishvili et al. 2018).

Antiseptics and Disinfectants

Chlorhexidine is a commonly used antiseptic and disinfectant, valued for its strong antimicrobial effects against bacteria, fungi, and viruses (Karpiński and Szkaradkiewicz 2015). As such, it is a favorite ingredient in various dental products such as oral rinses, aerosols, dental floss, gels, and varnishes, where it helps to prevent gingivitis, cavities, and plaque buildup (Karpiński and Szkaradkiewicz 2015). Additionally, earlier literature has shown that a combination of 2% chlorhexidine with trypsin and proteinase K can significantly reduce viable bacterial counts and disrupt nutrient-stressed biofilms formed by bacteria such as *C. acnes, Actinomyces radicidentis, S. epidermidis, Enterococcus faecalis*, and *Streptococcus mitis* in the root canals of single-rooted teeth (Niazi et al. 2015). This combination could improve the treatment of dental infections linked to biofilms. However, for this strategy to be effective, clinical trials must confirm the formulation's efficacy, and the stability and bioavailability of these proteases need to be thoroughly tested.

Small Molecule Antibiofilm Agents

Antibiotics, although effective, can alter microbial flora and contribute to resistance, with biofilm structures often limiting their effectiveness. Recent research suggests using small molecules to target biofilm formation at a molecular level (Qvortrup et al. 2019). These molecules typically work by inhibiting initial biofilm formation, interfering with c-di-GMP signaling, or disrupting quorum sensing (QS).

Inhibition of Biofilm Formation by Pilicides and Curlicides The attachment of bacteria, which is essential for biofilm formation, relies on structures such as pili and curli. Pili are composed of subunits, including adhesins and anchoring units (Busch et al. 2015), while curli are adhesive amyloid fibers that play a vital role in biofilm development (Kikuchi et al. 2005). Targeting these components can help mitigate bacterial adhesion and prevent biofilm formation.

Interfering with c-di-GMP Signaling c-di-GMP is a bacterial second messenger that promotes adhesion by decreasing flagellar motility and increasing the production of bacterial adhesins (Hengge 2009; Purcell and Tamayo 2016). Modulating the levels of c-di-GMP can effectively influence biofilm formation.

Inhibition of Quorum Sensing Quorum sensing (QS) is a mechanism that regulates bacterial behavior and biofilm development through signaling molecules such as acyl-homoserine lactones and autoinducers (Jiang et al. 2019). Anti-QS agents can decrease bacterial virulence without fostering drug resistance. Organic acids such as acetic, citric, and lactic acids have proven effective against biofilms of *E. coli* and *Salmonella* found on fresh produce (Amrutha et al. 2017). Notably, lactic acid inhibits violacein production and reduces both biofilm formation and bacterial motility, and it is FDA-approved and generally recognized as safe.

Surface Treatment and Nanostructure Functionalization

Surface modification techniques such as grinding, polishing, sintering, heat treatment, sandblasting, and etching are frequently used to adjust the roughness and wettability of implant surfaces (Moritz et al. 2019; Zboun et al. 2020). These treatments enhance the physicochemical properties of metals, which in turn affect bacterial adhesion and biofilm formation. For example, a study showed that electrolytic polishing of the Ti-6Al-4V titanium alloy (a common ingredient in medical implants) achieved a surface roughness of about 10.33 nm ± 1.14 (Jhong et al. 2020). Following this polishing process, the water contact angle was measured at 92°, which significantly reduced the adhesion of *S. mutans*. In contrast, when surface roughness increased to 120.05 nm ± 7.89, bacterial adhesion nearly tripled. This illustrates the critical role of precise surface treatment in improving the performance and durability of medical implants.

Surface Modification Strategies

Biofilm formation on medical implants leads to frequent revision surgeries and can be fatal. Recent research focuses on surface modification to prevent bacterial colonization, emphasizing surface roughness and biologically active coatings (Dziadek et al. 2021).

Engineering Surface Roughness and Topography via Laser Treatment Surface roughness affects bacterial attachment. Studies show that features smaller than microorganisms reduce attachment more effectively than smooth surfaces. Laser treatments have successfully introduced nano-textures on medical implants, such as Ti and Ti6Al4V alloys, enhancing surface roughness and reducing bacterial adhesion, such as with *S. aureus* (Chan et al. 2017).

Biomimetic Antiadhesion Coatings Research has advanced biomimetic coatings inspired by natural surfaces (e.g., cicada wings and shark skin) to prevent bacterial adhesion (Li et al. 2020). For example, diamond films mimicking plant leaves reduced bacterial attachment by 90–99% (Wang et al. 2020). Mussel-inspired dendritic polyglycerol coatings, functionalized with amines and catechols, achieved up to 99.7% bacterial attachment reduction and high antibacterial activity with silver nanoparticles (Ag NPs) (Li et al. 2019).

Multifunctional Antibacterial Coatings Recent advances focus on multifunctional coatings that offer antibacterial properties along with other benefits. These include coatings with bacteriostatic or bactericidal properties as well as those supporting eukaryotic cell growth (Cloutier et al. 2015).

Electroactive Coatings Electrochemical methods are explored for biofilm control, potentially damaging bacterial membranes or altering environmental factors such as pH and temperature (Czerwińska-Główka and Krukiewicz 2020). Conducting polymers is of interest for applications such as chemotherapy, tissue scaffolding, and controlled drug delivery (Vallejo-Giraldo et al. 2018).

Switchable Coatings Coatings that can change in response to environmental factors such as pH represent a promising strategy for controlling bacterial growth. For example, silver nanoparticles (Ag NPs) with zwitterionic groups can switch from being cytocompatible to antibacterial in acidic environments, effectively targeting bacteria within biofilms (Qiao et al. 2019). However, a key challenge is to improve their sensitivity to slight pH variations often associated with infections (Simmen and Blaser 1993).

Antibacterial/Antiadhesive Porous Oxide Layers Plasma electrolytic oxidation (PEO) is a process used to generate porous oxide layers on metals such as titanium, enhancing the effectiveness of bone implants. The pore size, which is influenced by various process parameters, promotes bone tissue formation, making PEO-treated surfaces particularly suitable for implant applications (Kazek-Kęsik et al. 2019).

Nanotechnological Approaches for Biofilm Control

Advanced nanotechnology offers significant advantages for inhibiting biofilms, including enhanced bioavailability, selectivity, stability of compounds, and targeted drug delivery. Here is an overview of recent advancements (Kumar et al. 2023b):

Metal and Metal Oxide Nanoparticles Metal nanoparticles (MNPs) such as silver, gold, zinc, copper, and iron are effective in managing biofilms due to their high surface-to-volume ratio and their ability to generate reactive oxygen species (ROS), which can damage bacterial cells and disrupt biofilm formation (Ifijen et al. 2023). These MNPs can penetrate bacterial membranes, interfere with quorum sensing, and disrupt the EPS matrix (Sánchez-López et al. 2020).

Functionalized Nanoparticles Functionalized nanoparticles (FNs) are enhanced with reactive molecules to improve the delivery and effectiveness of antimicrobial agents. They can penetrate EPS barriers more effectively by altering their surface chemistry using surfactants, polymers, or antibodies (Numan et al. 2022). FNs can disrupt bacterial cell walls, membranes, and enzymes (Kumar et al. 2021).

Cyclodextrin-Based Nanomaterials Cyclodextrins, which are cyclic oligosaccharides, are being utilized to enhance drug delivery, stability, and solubility. They can be conjugated with antibiotics, metal nanoparticles, or essential oils to improve their antibiofilm efficacy against a variety of bacteria (Liu et al. 2022).

Polymeric Nanoparticles Polymeric nanoparticles (PNPs) offer advantages such as improved drug stability, bioavailability, and controlled release. They exhibit strong antibiofilm activity by targeting bacterial growth and biofilm formation. PNPs made from materials such as polyethylene glycol (PEG), chitosan, and polylactic-co-glycolic acid (PLGA) have shown effective antibacterial properties (Zhang et al. 2023).

Solid Lipid Nanoparticles (SLNs) Composed of lipids and submicrometer-sized, SLNs are promising for biofilm control due to their biocompatibility, biodegradability, and high drug-loading capacity. They effectively reduce biofilm biomass and deliver antibiotics directly to infection sites (Naik et al. 2023).

9.7 Bright Side of Biofilms: Innovative Solutions for Shelf-Life Extension

Biofilms in Food Processing and Packaging

Biofilms in the food industry pose substantial economic and health threats (Shi and Zhu 2009). These biofilms can form on food matrices, packaging materials, or processing equipment, harboring pathogenic bacteria such as *E. coli*, *L. monocytogenes*, *C. jejuni*, *Salmonella enterica*, *P. aeruginosa*, *Bacillus cereus*, and *S. aureus* (Mafu et al. 2011). Traditional methods of eliminating bacteria through thermal processing and drying can adversely affect food quality (Król and Jarmoluk 2016). Consequently, novel food preservation methods have emerged, including the development of antibacterial food packaging films and coatings (Kumar and Mukherjee 2020). Research indicates that incorporating *Aloe vera* gel and polyvinyl alcohol (PVA) with antifungal and antibacterial agents can significantly extend the shelf life of vegetables (Rengaraju et al. 2023). Furthermore, bioactive packaging films infused with sonorensin, an antimicrobial peptide, have effectively suppressed spoilage bacteria in chicken and tomatoes (Chopra et al. 2015). Recently, there has been a rising demand for biodegradable polymer-based packaging materials (Khan et al. 2017; Munoz-Bonilla et al. 2019). For instance, adding 5% (w/w) titanium dioxide (TiO_2) and silver nanoparticles (Ag NPs) to a polybutylene adipate-co-terephthalate (PBAT) matrix enhances mechanical strength, barrier properties, UV protection, thermal stability, and antibacterial efficacy against *E. coli* and *S. aureus*. These improvements stem from TiO_2's photocatalytic activity and the synergistic effects of TiO_2 and Ag NPs, which accelerate PBAT crystallization, positioning PBAT/TiO_2/Ag composite biofilms as promising active food packaging materials (Cao et al. 2020).

To overcome the limitations of polylactic acid (PLA), various strategies have been investigated, including the addition of plasticizers, polymer blending, the use of coupling agents, co-polymerization, and nanotechnology (Fan et al. 2017, 2018).

Nanotechnology, in particular, offers significant benefits, such as enhanced antibacterial properties and UV shielding. Various metal oxides such as ZnO, MgO, TiO_2, and SiO_2 are being explored for their packaging potential (Marra et al. 2016). Magnesium oxide (MgO) is a naturally occurring, colorless mineral that is economically produced on a large scale and is recognized by the US FDA as generally safe for food contact (Sanuja et al. 2014). Research has shown that PLA/MgO biopolymer films possess antibacterial and UV protection capabilities (Swaroop and Shukla 2018). Due to its lower cost compared to other metal oxides and enhanced properties, MgO nanoparticles within PLA matrices present a promising solution for food packaging (Swaroop and Shukla 2018).

Ensuring food quality during packaging is a significant challenge, especially in light of recent outbreaks that demand innovative packaging solutions to maintain freshness and safety (Atieno et al. 2018). Researchers are exploring active materials such as yeast for use in bioactive food packaging. Notably, *Meyerozyma guilliermondii* shows promising antimicrobial properties along with nutritional benefits (Bertini et al. 2019; Ruas and Guerra-Sá 2020). Incorporating *M. guilliermondii* or its metabolites into packaging materials enhances their antimicrobial effectiveness and prolongs the shelf life of food products (Coda et al. 2013; Leneveu-Jenvrin et al. 2020). Biofilms made from bacterial cellulose (BC), carboxymethyl cellulose (CMC), glycerol, and yeast have shown a high water solubility of 42.86% and strong antimicrobial activity against *E. coli, P. aeruginosa*, and *S. aureus*. Coating fruits such as oranges and tomatoes with these composite materials have been effective in extending their shelf life by shielding them from environmental factors while preserving their texture, aroma, and color (Atta et al. 2021). These composite films can extend the shelf life of oranges by up to 14 days and remain stable for 63 days under refrigeration or 28 days at room temperature.

In another study, coatings developed from quaternized chitosan and CMC showed enhanced tensile strength, thermal stability, and water resistance. These coatings not only prolonged the shelf life of bananas at 30 °C but also exhibited antibacterial activity against *S. aureus* and *E. coli* (Hu et al. 2016). Additionally, integrating 3% (w/w) MgO/Ag NPs into a polybutylene succinate-co-terephthalate (PBST) matrix improved decomposition temperature, elongation at break, tensile strength, water vapor permeability, and oxygen transmission rate. PBST/MgO/Ag films demonstrated superior antibacterial effectiveness against both Gram-positive and Gram-negative bacteria, showing excellent preservation capabilities for cherry tomatoes. After 7 days, tomatoes covered with PBST/MgO/Ag films exhibited nearly five times fewer microorganisms compared to uncovered tomatoes, underscoring the synergistic antibacterial effects of MgO and Ag NPs (Zhu et al. 2016). Furthermore, the PBST/MgO/Ag films outperformed traditional polyethylene and polyethylene terephthalate films in terms of elongation at break, tensile strength, water vapor permeability, and oxygen transmission rate, making them a competitive alternative for extending the shelf life of packaged food products.

Biological Wastewater Treatment Using Biofilms

As the volume of wastewater continues to rise and regulations become more stringent, there is an increasing need for enhanced wastewater treatment solutions, especially given the constraints of limited space. This demand has led to the development of advanced biofilm-based technologies that offer significant treatment capacities (Lazarova and Manem 2000). Gaining a more comprehensive understanding of biofilm formation and interspecies interactions could further improve these biofilm processes. The food processing industry, in particular, produces large amounts of toxic effluents that must be treated prior to discharge. Biofilm technology has proven effective in treating such wastewater streams since its inception with trickling filters in the 1880s (Lazarova and Manem 2000). Biofilm processes in wastewater treatment can be divided into two categories: (i) fixed-medium systems with static biofilm media and (ii) moving-medium systems where the biofilm media are in continuous motion (Rodgers and Zhan 2003). Fixed-medium systems include trickling filters and biological aerated filters, while moving-medium systems encompass rotating biological contactors (RBC), moving bed biofilm reactors (MBBR), vertical flow biofilm reactors (VMBR), and fluidized bed reactors (FBR). Rotating biological contactors use thin aerobic biofilms on rotating cylinders for efficient wastewater treatment (Qureshi et al. 2005). Research on single-strain biofilms has primarily focused on clinically significant bacteria (Mack et al. 2007) or strains associated with food spoilage (Midelet and Carpentier 2004). However, investigations have also extended to high-grade chemical production (Qureshi et al. 2005), microorganisms in water distribution systems (Simões et al. 2007), and bioaugmentation via natural genetic transformation (Hendricks et al. 2000). Dual-species biofilms have proven effective in studying various microbial interactions (Komlos et al. 2005; Simões et al. 2007). Biofilm cells exhibit different physiological characteristics compared to planktonic cells; a distinction that has been the focus of much research. Biofilms are intricate biostructures that develop on surfaces regularly exposed to water, displaying features akin to multicellular organisms and ecosystems. Various biofilm reactors, such as membrane-supported biofilm reactors, moving bed biofilm reactors, granular sludge systems, and integrated fixed-film activated sludge processes, are used for water treatment. Advances in biofilm control and beneficial applications in membrane processes are ongoing. Continued progress in biofilm control and its beneficial applications in membrane processes are underway. Biofilm models are essential for foundational research, reactor design, and architecture although there are notable differences between biofilm modeling and biofilm reactor modeling approaches (Verma et al. 2023). Research has established fundamental principles of biofilms, with biofilm reactors commonly applied in municipal wastewater treatment. Applied research has contributed to understanding biofilm reactor mechanisms, leading to design criteria for these reactors although mathematical models are becoming valuable tools. There remains a gap between the fundamental understanding of biofilms and empirical information from reactor-scale research, highlighting a clear division in the literature between micro- and macro-scales. Microbial activity in biofilms contributes to the biodeterioration of materials such as metals, concrete, and plastics, with microbially influenced corrosion (MIC) alone costing billions annually (Lewandowski and Boltz 2011). Immobilizing microorganisms in biofilms is particularly advantageous for environmental biotechnology processes. Biofilm reactors can achieve high biomass concentrations of up to 50 g/l, but the transfer of substrates from the medium to the biofilm often limits the reaction rate. To enhance efficiency, emerging techniques are utilizing thin biofilms (less than 0.5 mm) with large specific surface areas (over 1000 m^2/m^3 of bioreactor volume) (Van Loosdrecht et al. 1995). The applications of particulate biofilm reactors such as upflow sludge blankets, biofilm fluidized beds, expanded granular sludge blankets, biofilm airlift suspension, and internal circulation reactors for municipal and industrial wastewater treatment, analyzing key design and operation aspects such as biofilm formation, hydrodynamics, mass transfer, and mixing. This technology is considered mature, with established design and scale-up guidelines (Nicolella et al. 2000). Microalgae offer significant advantages for biofuel production over energy crops due to their high growth rates, substantial lipid yields, and ability to grow in wastewater, efficiently removing primary nutrients, heavy metals, and micropollutants. However, the high costs associated with algae harvesting and processing remain a fundamental barrier. Algal biofilms, which simplify harvesting and dewatering, are increasingly explored. Natural microalgal biofilms from various aquatic environments have been characterized. For instance, biofilm is used for bioenergy production through dark fermentation, resulting in notable hydrogen production. Additionally, the fatty acid profiles of biofilm were modified by promoting selected photosynthetic organisms. These biofilms have been utilized for specific wastewater treatments, altering fatty acid compositions, and concentrating commercial microalgal cells (Miranda et al. 2017).

Despite the advantages of algal biofilm-based systems, their widespread application is limited. The performance of these systems, including nutrient removal and biomass productivity, is not thoroughly covered in existing literature, affecting design and scaling. This critical analysis highlights trends in nutrient removal, biomass productivity, growth requirements, and challenges related to algal biofilm technology in wastewater treatment. It focuses on key research areas, such as factors that influence algal growth, mass transport, species selection, algal–bacterial interactions, and the need to validate laboratory results in field conditions to create effective algal biofilm technologies (Kesaano and Sims 2014). Research on Ag-NPs reveals that while wastewater biofilms show high tolerance to Ag-NPs, planktonic bacteria are more susceptible. A study showed minimal reduction in biofilm bacteria viability with 200 mg/L Ag-NPs, while planktonic bacteria were largely eradicated with 1 mg/L Ag-NPs. This indicates that EPS and microbial community interactions in biofilms play crucial roles in Ag-NP resistance. Susceptibility varies among microorganisms, with Thiotrichales being particularly sensitive (Sheng and Liu 2011). A study of 13 bacterial strains in wastewater treatment systems using crystal violet assays demonstrated that medium composition significantly affects biofilm formation. All strains formed biofilms in at least one of four tested media, with some strains forming biofilms in all media. Dual-strain biofilms exhibited synergistic effects, and the strongest biofilm formation occurred with a mixture of all 13 strains. These findings can guide the design of systems for municipal and industrial wastewater degradation (Andersson et al. 2008).

Biofilm Bioreactors and Fermentation

Biochemical reactors are pivotal in the biochemical industry, influencing reaction rates, operational ease, and the longevity of reactor use, all of which impact productivity and process economics. To select the most suitable reactor for an industrial process, it is essential that the reaction rate is high, and the reactor configuration is straightforward. Optimizing parameters such as pH, temperature, substrate concentration, and medium components can enhance reaction rates by increasing cell mass concentration within the reactor. Two primary methods exist for enhancing cell mass concentration: using a permeable membrane to retain cells and employing immobilized cell techniques. Membrane reactors facilitate the flow of liquid, substrate, and product while keeping cells contained, allowing for high cell concentrations. However, their high costs and fouling issues can make them less suitable for certain applications such as wastewater treatment and the production of large quantities of low-cost chemicals such as vinegar or acetic acid (Qureshi et al. 2005). Biofilm reactors present several advantages, including reduced membrane fouling, lower capital expenditures, higher biomass density, and improved operational stability. In these reactors, biofilm formation occurs naturally as microbial cells attach to a support structure without the use of chemical agents. Over recent decades, biofilm reactors have been extensively researched and implemented in wastewater treatment and the production of value-added products. To optimize the performance of biofilm reactors, it is crucial to understand the fundamentals of biofilm formation, the characteristics of the biofilm matrix and the factors influencing biofilm development (Ercan and Demirci 2015). The application of biofilm reactors also extends to the production of valuable compounds, typically utilizing immobilized cells or enzymes rather than suspended microorganisms. For treating wastewater and odors, immobilized cells have proven effective as they are embedded in a matrix of EPS they produce, creating a biofilm that avoids the toxic effects associated with chemical cell entrapment. Although biofilm processes have been thoroughly investigated, their economic impact remains limited (Muffler et al. 2014). Biofilms, composed of layers of microbial cells attached to solid surfaces and embedded in a self-produced exopolysaccharide matrix, provide several advantages over suspended cells, including greater biomass density and stability. Biofilm reactors are particularly well-suited for fermentation and bioconversion processes that require high biomass concentrations.

In recent years, these reactors have effectively produced a variety of value-added products (Cheng et al. 2010). Biofilm formation enhances cell density, achieving concentrations as high as 74 g/L. Reactor configurations can vary from simple batch reactors to more complex systems, including continuously stirred tank reactors, packed bed reactors, fluidized bed reactors (FBR), airlift reactors, and upflow anaerobic sludge blanket (UASB) reactors. UASB reactors utilize granular biofilm particles and have shown superior productivity compared to other types of reactors (Qureshi et al. 2005). In one significant case, an immobilized-cell biofilm reactor was utilized for the continuous production of lactic acid using *Lactobacillus casei* subsp. *rhamnosus* (ATCC 11443). Researchers at Iowa State University developed a unique plastic composite support (PCS) designed to promote biofilm formation. This PCS blend, made from agricultural products and polypropylene, not only enhanced lactic acid production but also regulated biofilm thickness by adjusting agitation speed. The PCS reactor achieved an optimal production rate of 9.0 g/L/h, significantly higher than the 5.8 g/L/h rate observed in the polypropylene control reactor (Cotton et al. 2001). To support industrial growth and sustainability, renewable energy sources are increasingly important. Syngas fermentation offers a sustainable alternative to fossil fuels, requiring less energy and providing flexibility in feedstock and end products. Biological catalysts in biofilm reactors can tolerate impurities in syngas, a significant advantage over metal catalysts. Lanzatech's successful commercial plants utilize CO-rich off-gas from the steel industry. However, low mass transfer rates in the gas-liquid interface pose challenges. Recent research highlights biofilm reactors' potential to improve mass transfer rates and process stability, guiding future research and commercialization of renewable biofuels and biochemicals (Gunes 2021). Continuous and repeat-batch biofilm fermentations using *Actinobacillus succinogenes* demonstrated superior performance in PCS compared to suspended-cell systems. The highest succinic acid concentrations and yields were achieved with PCS bioreactors at a dilution rate of 0.2 per hour. Repeat-batch fermentations with PCS supports showed increased productivity and efficiency over traditional methods (Urbance et al. 2004). Evaluations of plastic supports for lactic acid fermentations revealed that composites made from polypropylene blended with agricultural materials outperformed polypropylene-alone supports. Consistently higher concentrations of lactic acid were achieved using plastic-composite supports in repeated-batch fermentations, indicating that these supports are advantageous for long-term lactic acid production (Demirci and Pometto 1995). For continuous pullulan production, biofilm reactors utilized a pigment-reduced strain of *Aureobasidium pullulans* along with PCS supports. By identifying optimal conditions, researchers were able to attain high concentrations and purity of pullulan. The PCS reactor's ability to retain biomass played a significant role in enhancing productivity (Cheng et al. 2011). A specially designed stirred-tank biofilm reactor incorporating PCS exhibited superior lactic acid productivity and glucose consumption compared to traditional suspended-cell reactors. These PCS reactors not only produced more lactic acid but also required less yeast extract, demonstrating their efficiency in repeated fed-batch fermentations (Vélazquez et al. 2001). Mathematical modeling is instrumental in understanding and optimizing fermentation processes. A study that evaluated ten different models for ethanol fermentation in a biofilm reactor using a carob extract medium found that the W model yielded the best fit for experimental data, providing valuable insights for process optimization (Germec et al. 2020). At Iowa State University, PCS was developed to enhance both biofilm formation and nutrient supply for *Saccharomyces cerevisiae*. Bioreactors utilizing PCS exhibited significantly higher ethanol production compared to those using only polypropylene supports, underscoring the potential of innovative bioreactor designs to boost productivity (Demirci et al. 1997). In another investigation, a biohydrogen production system using packed bed reactors with biofilm consortia on granitic stones demonstrated that granular sludge cultures were more effective at producing hydrogen than those relying solely on stone biofilms. This finding highlights the important role of lactic acid bacteria in hydrogen production and points to potential avenues for improving biohydrogen production processes (Chojnacka et al. 2011).

9.8 Recent Discoveries and Technologies in Biofilms

Research into biofilms has been dynamic and expansive. Here are some recent discoveries and emerging technologies related to biofilms (Dutta et al. 2024):

Transcriptomics in Detecting Microbial Biofilm-Mediated Drug Resistance

Recent studies have shown that advanced transcriptomic techniques can effectively identify drug resistance in microbial biofilms. For instance, RNA sequencing (RNA-seq) analysis has revealed the transcriptomic profiles associated with biofilm production and drug resistance mechanisms in *E. faecalis*. This research has highlighted several key genes, including paiA (involved in sporulation regulation), arsD, and arsR (part of the arsenic resistance operon), adenosine triphosphate-binding cassette drug transporters and penicillin-binding proteins that contribute to biofilm resistance (Seneviratne et al. 2017). In another RNA-seq investigation, it was found that low concentrations of ampicillin promote biofilm formation in *S. aureus*, leading to increased cell viability and the upregulation of genes associated with penicillin-binding proteins and drug resistance pumps (Liu et al. 2018). Additionally, RNA-seq analysis indicated that an aqueous extract of *Gynostemma chinensis* downregulated genes related to biofilm formation and carbohydrate metabolism in MRSA. This extract also influenced multidrug resistance mechanisms by regulating multidrug efflux pumps and penicillin-binding protein expression (Wu et al. 2019). However, further validation of these results is necessary for their translational and post-translational applications.

Transcriptomics in Biofilm Development and Maturation

Biofilm formation consists of several stages, including irreversible adhesion, microcolony development, and maturation. Transcriptomic analysis of *Streptococcus mutans* biofilms conducted through comparative DNA microarray techniques indicated that the surface protein WapA is significantly upregulated in dense biofilms of 200 and 400 mm compared to those that are 100 mm thick (Shemesh et al. 2008).

RNA-seq has also been employed to explore the biochemical characteristics of biofilms and to assess how environmental and genetic factors influence the development of dental plaque biofilms (Peterson et al. 2014). An analysis of 27 genera in dental plaque biofilms revealed key species such as *Streptococcus, Veillonella*, and *Capnocytophaga.* The internalin protein In1J facilitates adhesion between *C. albicans* and *P. gingivalis,* leading to the upregulation of 57 genes via In1J-dependent mechanisms and the type IX secretion system (Peterson et al. 2014). This system amplifies the pathogenicity of *P. gingivalis* in conjunction with *C. albicans.*

Single-Cell Transcriptomics

Current transcriptomic studies often assume biofilm populations as uniform; but, biofilms actually comprise diverse cells with varying metabolic activities and resistance profiles. Single-cell transcriptomics, enabled by next-generation sequencing, offers in-depth insights into antibiotic and drug resistance mechanisms within these microbial communities. For example, research on *S. aureus* persisters revealed that prolonged antibiotic exposure led to decreased metabolic activity, yet transcriptomic reprogramming preserved metabolic functions (Peyrusson et al. 2020). Transcriptomics provides a thorough understanding of biofilm development and drug resistance, moving beyond predefined gene sets. Additionally, advancements in machine learning and sequencing technologies are improving the analysis of genes related to iron acquisition and carbon metabolism in *P. aeruginosa* infections (Cornforth et al. 2018). As transcriptomics represents the initial step in exploring gene expression, integrating this data with proteomics and metabolomics is crucial for a complete understanding of biological complexity.

Proteomics as a Tool for Monitoring Biofilms

Proteomics involves the comprehensive analysis of protein expressions under specific conditions, offering insights into biological changes that go beyond genetic observations. Recent developments have favored gel-free methodologies combined with mass spectrometry (MS) over traditional gel-based techniques (Soufi and Soufi 2016). MS methods can effectively identify and quantify a wide range of proteins, including those that are typically challenging to detect, such as membrane proteins.

Role of Proteomics in Biofilm Research

High-throughput proteomics has produced extensive data or the study of microbial biofilms. Techniques such as tandem mass tag (TMT), isobaric tags for relative and absolute quantitation (iTRAQ), and stable isotope labeling by amino acids in cell culture (SILAC) are commonly used for relative protein quantification. For example, SILAC has been applied to profile proteins in both planktonic and biofilm forms of *Neisseria gonorrhoeae* (Phillips et al. 2012). Meanwhile, TMT and iTRAQ have been used to analyze protein expression differences between antibiotic-sensitive and resistant strains of *Aeromonas hydrophila* and *Candida* (Ellepola et al. 2019). Furthermore, label-free methods have also been employed to study bacterial and fungal biofilms, revealing differences in protein levels among various species (Bao et al. 2015).

Proteomics in Studying Biofilm-Host Interactions

Research on host-biofilm interactions using proteomics is still limited. In one study, label-free proteomics was used to examine proteins released by gingival epithelial cells in a co-culture model involving a subgingival biofilm with ten different species. This interaction led to a decrease in epithelial protein release, heightened inflammation, and activated apoptosis pathways while downregulating pathways related to tissue integrity and turnover. Over a 48-h period, there was further negative regulation of metabolic pathways and complex degradation. Importantly, the "red-complex" bacterial species significantly impacted the host immune response (Bostanci et al. 2015).

Omics Approaches for Detecting Antibiotic Resistance in Complex Biofilms

Metabolomics, studying small molecules involved in cellular metabolism, offers insights into microbial identification and antibiotic resistance. Bacterial metabolites, categorized as primary (essential for healthy development) and secondary (appearing during stationary growth phases), play crucial roles in bacterial survival and environmental adaptation. Exometabolomics, examining metabolites released outside cells, provides a comprehensive view of microbial behavior. Overuse of antibiotics has led to increased resistance, prompting the need for new antibiotics and more precise detection methods. MALDI-TOF-MS combined with multi-omic approaches offers enhanced microbial and resistance detection, with ongoing improvements aimed at reducing diagnostic time and enhancing accuracy. The evolution of MALDI-TOF technology to identify specific protein ions and sequences unique to microorganisms is critical for discovering protein toxins, virulence factors, and resistance pathways (Janiszewska et al. 2022).

9.9 Future Directions in Biofilm and Shelf-Life Studies

The extensive presence of biofilms in various diseases, along with their resistance to many antimicrobial treatments, has led to increased interest in biofilm research. With the rise of antimicrobial resistance, current studies are shifting focus from traditional methods that target bacterial growth or division, leading to cell death or dormancy to innovative strategies. These include promoting the dispersal of biofilms and exploring methods to prevent their initial formation, such as re-engineering surfaces that are susceptible to biofilm development, such as urinary catheters and implants. Advancements in imaging techniques, including super-resolution microscopy and electron tomography, have enhanced our ability to visualize biofilm structures at the nanoscale, which aids in understanding the architecture and microbial diversity within biofilms. Research is also progressing in developing materials with inherent resistance to biofilm formation, such as hydrophilic or superhydrophobic surfaces, which are engineered to reduce microbial adhesion and subsequent biofilm growth. Furthermore, biofilm-based bioreactors are being recognized for their effectiveness in wastewater treatment and bioremediation, with advancements in reactor design optimizing biofilm growth and activity to improve overall process performance. Techniques from synthetic biology are being utilized to create microorganisms with improved biofilm-forming capabilities or to generate agents that disrupt biofilms, allowing for targeted control over biofilm properties and interactions.

References

Amrutha B, Sundar K, Shetty PH (2017) Effect of organic acids on biofilm formation and quorum signaling of pathogens from fresh fruits and vegetables. Microb Pathog 111:156–162. https://doi.org/10.1016/j.micpath.2017.08.016

Alvarez-Ordóñez A, Coughlan LM, Briandet R, Cotter PD (2019) Biofilms in food processing environments: Challenges and opportunities. Annu Rev Food Sci Technol 10(1):173–195

Andersson S, Kuttuva Rajarao G, Land CJ, Dalhammar G (2008) Biofilm formation and interactions of bacterial strains found in wastewater treatment systems. FEMS Microbiol Lett 283(1):83–90. https://doi.org/10.1111/j.1574-6968.2008.01173.x

Araújo EA, de Andrade NJ, da Silva LHM, de Carvalho AF, da Silva CA, Ramos AM (2010) Control of microbial adhesion as a strategy for food and bioprocess technology. Food Bioproc Technol 3(3):321–332. https://doi.org/10.1007/s11947-009-0231-1

Armbruster CR, Parsek MR (2018) New insight into the early stages of biofilm formation. Proc Natl Acad Sci 115(17):4317–4319. https://doi.org/10.1073/pnas.1802654115

Atieno L, Owino W, Ateka EM, Ambuko J (2018) Effect of surface coatings on the shelf life and quality of cassava. J Food Res 7(4):46–60. https://doi.org/10.5539/jfr.v7n4p46

Atta OM, Manan S, Ahmed AAQ, Awad MF, Ul-Islam M, Subhan F, Ullah MW, Yang G (2021) Development and characterization of yeast-incorporated antimicrobial cellulose biofilms for edible food packaging application. Polymers 13(14):2310. https://doi.org/10.3390/polym13142310

Bao K, Bostanci N, Selevsek N, Thurnheer T, Belibasakis GN (2015) Quantitative proteomics reveal distinct protein regulations caused by *Aggregatibacter actinomycetemcomitans* within subgingival biofilms. PLoS One 10(3):e0119222. https://doi.org/10.1371/journal.pone.0119222

Bertini EV, Leguina AC, del Figueroa LIC, Nieto-Penalver CG (2019) Endophytic microorganisms *Agrobacterium tumefaciens* 6N2 and *Meyerozyma guilliermondii* 6N serve as models for the study of microbial interactions in colony biofilms. Rev Argent Microbiol 51(3):286–287. https://doi.org/10.1016/j.ram.2019.03.002

Bostanci N, Bao K, Wahlander A, Grossmann J, Thurnheer T, Belibasakis GN (2015) Secretome of gingival epithelium in response to subgingival biofilms. Mol Oral Microbiol 30(4):323–335. https://doi.org/10.1111/omi.12100

Bridier A, Sanchez-Vizuete P, Guilbaud M, Piard J-C, Naïtali M, Briandet R (2015) Biofilm-associated persistence of food-borne pathogens. Food Microbiol 45:167–178. https://doi.org/10.1016/j.fm.2014.11.003

Busch A, Phan G, Waksman G (2015) Molecular mechanism of bacterial type 1 and P pili assembly. Philos Trans R Soc A Math Phys Eng Sci 373(20130153). https://doi.org/10.1098/rsta.2013.0153

Butini ME, Cabric S, Trampuz A, Di Luca M (2018) In vitro anti-biofilm activity of a biphasic gentamicin-loaded calcium sulfate/hydroxyapatite bone graft substitute. Colloids Surf B: Biointerfaces 161:252–260. https://doi.org/10.1016/j.colsurfb.2017.10.014

Cao C, Wang Y, Zheng S, Zhang J, Li W, Li B, Guo R, Yu J (2020) Poly (butylene adipate-co-terephthalate)/titanium dioxide/silver composite biofilms for food packaging application. LWT 132:109874. https://doi.org/10.1016/j.lwt.2020.109874

Carpentier B, Cerf O (1993) Biofilms and their consequences with particular reference to hygiene in the food industry. J Appl Bacteriol 75(5):499–511. https://doi.org/10.1111/j.1365-2672.1993.tb03316.x

Carrascosa C, Millan R, Jaber JR, Lupiola P, del Rosario-Quintana C, Mauricio C, Sanjuan E (2015) Blue pigment in fresh cheese produced by Pseudomonas fluorescens. Food Control 54:95–102

Carrascosa C, Raheem D, Ramos F, Saraiva A, Raposo A (2021) Microbial biofilms in the food industry—a comprehensive review. Int J Environ Res Public Health 18(4):2014. https://doi.org/10.3390/ijerph18042014

Castiblanco LF, Sundin GW (2016) New insights on molecular regulation of biofilm formation in plant-associated bacteria. J Integr Plant Biol 58(4):362–372. https://doi.org/10.1111/jipb.12428

Chan CW, Carson L, Smith GC, Morelli A, Lee S (2017) Enhancing the antibacterial performance of orthopaedic implant materials by fibre laser surface engineering. Appl Surf Sci 404:67–81. https://doi.org/10.1016/j.apsusc.2017.01.142

Cheng K, Demirci A, Catchmark J (2010) Advances in biofilm reactors for production of value-added products. Appl Microbiol Biotechnol 87(2):445–456. https://doi.org/10.1007/s00253-010-2722-2

Cheng KC, Demirci A, Catchmark JM (2011) Continuous pullulan fermentation in a biofilm reactor. Appl Microbiol Biotechnol 90:921–927. https://doi.org/10.1007/s00253-011-3454-x

Chopra L, Singh G, Kumar Jena K, Sahoo DK (2015) Sonorensin: A new bacteriocin with potential of an antibiofilm agent and a food biopreservative. Sci Rep 5(1):13412

Chojnacka A, Błaszczyk MK, Szczęsny P, Nowak K, Sumińska M, Tomczyk-Żak K, Sikora A (2011) Comparative analysis of hydrogen-producing bacterial biofilms and granular sludge formed in continuous cultures of fermentative bacteria. Bioresour Technol 102(21):10057–10064. https://doi.org/10.1016/j.biortech.2011.08.050

Cloutier M, Mantovani D, Rosei F (2015) Antibacterial coatings: challenges, perspectives, and opportunities. Trends Biotechnol 33(10):637–652. https://doi.org/10.1016/j.tibtech.2015.08.004

Coda R, Rizzello CG, Di Cagno R, Trani A, Cardinali G, Gobbetti M (2013) Antifungal activity of *Meyerozyma guilliermondii*: identification of active compounds synthesized during dough fermentation and their effect on long-term storage of wheat bread. Food Microbiol 33(1):243–251. https://doi.org/10.1016/j.fm.2012.11.012

Colagiorgi A, Bruini I, Di Ciccio PA, Zanardi E, Ghidini S, Ianieri A (2017) *Listeria monocytogenes* biofilms in the wonderland of food industry. Pathogens 6(3):41. https://doi.org/10.3390/pathogens6030041

Cornforth DM, Dees JL, Ibberson CB, Huse HK, Mathiesen IH, Kirketerp-Møller K, Whiteley M (2018) *Pseudomonas aeruginosa* transcriptome during human infection. Proc Natl Acad Sci 115(22):E5125–E5134. https://doi.org/10.1073/pnas.1717903115

Costerton JW, Stewart PS, Greenberg EP (1999) Bacterial biofilms: a common cause of persistent infections. Science 284(5418):1318–1322. https://doi.org/10.1126/science.284.5418.1318

Cotton J, Pometto A, Gvozdenovic-Jeremic J (2001) Continuous lactic acid fermentation using a plastic composite support biofilm reactor. Appl Microbiol Biotechnol 57(5–6):626–630. https://doi.org/10.1007/s002530100714

Coughlan LM, Cotter PD, Hill C, Alvarez-Ordóñez A (2016) New weapons to fight old enemies: novel strategies for the (bio)control of bacterial biofilms in the food industry. Front Microbiol 7:1641. https://doi.org/10.3389/fmicb.2016.01641

Czerwińska-Główka D, Krukiewicz K (2020) A journey in the complex interactions between electrochemistry and bacteriology: from electroactivity to electromodulation of bacterial biofilms. Bioelectrochemistry 131:107401. https://doi.org/10.1016/j.bioelechem.2020.107401

Demirci A, Pometto AL (1995) Repeated-batch fermentation in biofilm reactors with plastic-composite supports for lactic acid production. Appl Microbiol Biotechnol 43:585–589

Demirci A, Pometto AL III, Ho KG (1997) Ethanol production by *Saccharomyces cerevisiae* in biofilm reactors. J Ind Microbiol Biotechnol 19(4):299–304

Dhowlaghar N, Bansal M, Schilling MW, Nannapaneni R (2018) Scanning electron microscopy of *Salmonella* biofilms on various food-contact surfaces in catfish mucus. Food Microbiol 74:143–150. https://doi.org/10.1016/j.fm.2018.01.003

Donlan RM (2001) Biofilm formation: a clinically relevant microbiological process. Clin Infect Dis 33(8):1387–1392

Doyle MP (ed) (1988) Foodborne bacterial pathogens. Marcel Dekker

Dutta B, Chatterjee D, Sarkar N, Lahiri D, Nag M, Ray RR (2024) Multi-omics technology in detection of multispecies biofilm. Microbe 100128. https://doi.org/10.1016/j.microbe.2024.100128

Dziadek M, Dziadek K, Checinska K, Zagrajczuk B, Golda-Cepa M, Brzychczy-Wloch M, Menaszek E, Kopec A, Cholewa-Kowalska K (2021) PCL and PCL/bioactive glass biomaterials as carriers for biologically active polyphenolic compounds: comprehensive physicochemical and biological evaluation. Bioact Mater 6:1811–1826. https://doi.org/10.1016/j.bioactmat.2020.11.014

Ellepola K, Truong T, Liu Y, Lin Q, Lim TK, Lee YM, Seneviratne CJ (2019) Multi-omics analyses reveal synergistic carbohydrate metabolism in *Streptococcus mutans-Candida albicans* mixed-species biofilms. Infect Immun 87(10):e01128–e01118. https://doi.org/10.1128/IAI.01128-18

Ercan D, Demirci A (2015) Current and future trends for biofilm reactors for fermentation processes. Crit Rev Biotechnol 35(1):1–14. https://doi.org/10.3109/07388551.2014.952511

Fan Z, Fu M, Xu Z, Zhang B, Li H, Li X, Zhou X, Liu Y, Duan PH, Lin PH, Duann P, Xie J, Ma Z, Liu J, Guan J (2017) Sustained release of a peptide-based matrix metalloproteinase-2 inhibitor to attenuate adverse cardiac remodeling and improve cardiac function following myocardial infarction. Biomacromolecules 18(9):2820–2829. https://doi.org/10.1021/acs.biomac.7b00760

Fan Z, Xu Z, Niu H, Gao N, Guan Y, Li C, Dang Y, Cui XL, Liu Y, Duan H, Li X, Zhou PH, Lin J, Ma J, Guan J (2018) An injectable oxygen release system to augment cell survival and promote cardiac repair following myocardial infarction. Sci Rep 8:1371. https://doi.org/10.1038/s41598-018-19906-w

Farber JM, Peterkin P (1991) *Listeria monocytogenes*, a food-borne pathogen. Microbiol Rev 55(3):476–511

Fleming D, Rumbaugh K (2017) Approaches to dispersing medical biofilms. Microorganisms 5(1):15. https://doi.org/10.3390/microorganisms5010015

Flemming HC, Wuertz S (2019) Bacteria and archaea on earth and their abundance in biofilms. Nat Rev Microbiol 17(5):247–260. https://doi.org/10.1038/s41579-019-0158-9

Fratamico PM, Annous BA, Guenther NW (2009) Biofilms in the food and beverage industries. Woodhead Publishing

Funari R, Bhalla N, Chu KY, Söderström B, Shen AQ (2018) Nanoplasmonics for real-time and label-free monitoring of microbial biofilm formation. ACS Sensors 3(8):1499–1509. https://doi.org/10.1021/acssensors.8b00238

Galiè S, García-Gutiérrez C, Miguélez EM, Villar CJ, Lombó F (2018) Biofilms in the food industry: health aspects and control methods. Front Microbiol 9:898. https://doi.org/10.3389/fmicb.2018.00898

Galie S, Garcia-Gutierrez C, Miguelez EM, Villar CJ, Lombo F (2018) Biofilms in the food industry: health aspects and control methods. Front. Microbiolo 9:898
Germec M, Cheng KC, Karhan M, Demirci A, Turhan I (2020) Application of mathematical models to ethanol fermentation in biofilm reactor with carob extract. Biomass Convers Biorefinery 10:237–252. https://doi.org/10.1007/s13399-019-00490-3
Gião MS, Keevil CW (2014) *Listeria monocytogenes* can form biofilms in tap water and enter into the viable but non-cultivable state. Microb Ecol 67:603–611. https://doi.org/10.1007/s00248-014-0468-8
Giaouris E, Heir E, Desvaux M, Hébraud M, Møretrø T, Langsrud S, Doulgeraki A, Nychas GJ, Kačániová M, Czaczyk K et al (2015) Intra- and inter-species interactions within biofilms of important foodborne bacterial pathogens. Front Microbiol 6:1–26. https://doi.org/10.3389/fmicb.2015.00016
Govaert M, Smet C, Baka M, Janssens T, Van Impe J (2018) Influence of incubation conditions on the formation of model biofilms by *Listeria monocytogenes* and *Salmonella Typhimurium* on abiotic surfaces. J Appl Microbiol 125:1890–1900. https://doi.org/10.1111/jam.13939
Gunes B (2021) A critical review on biofilm-based reactor systems for enhanced syngas fermentation processes. Renew Sust Energ Rev 143:110950. https://doi.org/10.1016/j.rser.2021.110950
Gupta P, Sarkar S, Das B, Bhattacharjee S, Tribedi P (2016) Biofilm, pathogenesis and prevention—a journey to break the wall: a review. Arch Microbiol 198:1–15. https://doi.org/10.1007/s00203-015-1157-2
Haggag W (2010) The role of biofilm exopolysaccharides on biocontrol of plant diseases. In: Biopolymers. IntechOpen, pp 271–284. https://doi.org/10.5772/10266
Hall-Stoodley L, Costerton JW, Stoodley P (2004) Bacterial biofilms: from the natural environment to infectious diseases. Nat Rev Microbiol 2(2):95–108. https://doi.org/10.1038/nrmicro821
Han Q, Song X, Zhang Z, Fu J, Wang X, Malakar PK, Liu H, Pan Y, Zhao Y (2017) Removal of foodborne pathogen biofilms by acidic electrolyzed water. Front Microbiol 8:988. https://doi.org/10.3389/fmicb.2017.00988
Hassett DJ, Korfhagen TR, Irvin RT, Schurr MJ, Sauer K, Lau GW, Sutton MD, Yu H, Hoiby N (2010) *Pseudomonas aeruginosa* biofilm infections in cystic fibrosis: insights into pathogenic processes and treatment strategies. Expert Opin Ther Targets 14(1):117–130. https://doi.org/10.1517/14728220903454988
Hendricks SK, Kwok C, Shen M, Horbett TA, Ratner BD, Bryers JD (2000) Plasma-deposited membranes for controlled release of antibiotic to prevent bacterial adhesion and biofilm formation. J Biomed Mater Res 50(2):160–170. https://doi.org/10.1002/(SICI)1097-4636(200002)50:2<160::AID-JBM1>3.0.CO;2-X
Hengge R (2009) Principles of c-di-GMP signalling in bacteria. Nat Rev Microbiol 7(4):263–273. https://doi.org/10.1038/nrmicro2102
Hu D, Wang H, Wang L (2016) Physical properties and antibacterial activity of quaternized chitosan/carboxymethyl cellulose blend films. LWT Food Sci Technol 65:398–405. https://doi.org/10.1016/j.lwt.2015.08.042
Ifijen IH, Maliki M, Udokpoh NU, Odiachi IJ, Atoe B (2023) TMS 2023 152nd annual meeting and exhibition supplemental proceedings. Springer
Jamal M, Tasneem U, Hussain T, Andleeb S (2015) Bacterial biofilm: its composition, formation, and role in human infections. Res Rev J Microbiol Biotechnol 4:1–14
Janiszewska D, Szultka-Młyńska M, Pomastowski P, Buszewski B (2022) "Omic" approaches to bacteria and antibiotic resistance identification. Int J Mol Sci 23(17):9601. https://doi.org/10.3390/ijms23179601
Jhong YT, Chao CY, Hung WC, Du JK (2020) Effects of various polishing techniques on the surface characteristics of the Ti-6Al-4V alloy and on bacterial adhesion. Coatings 10:1057. https://doi.org/10.3390/coatings10121057
Jiang Q, Chen J, Yang C, Yin Y, Yao K, Song D (2019) Quorum sensing: a prospective therapeutic target for bacterial diseases. Biomed Res Int 2019:2015978. https://doi.org/10.1155/2019/2015978

Jindal S, Anand S, Metzger L, Amamcharla J (2018) A comparison of biofilm development on stainless steel and modified-surface plate heat exchangers during a 17-h milk pasteurization run. J Dairy Sci 101:2921–2926. https://doi.org/10.3168/jds.2017-13828
Karpiński T, Szkaradkiewicz AK (2015) Chlorhexidine—pharmaco-biological activity and application. Eur Rev Med Pharmacol Sci 19:1321–1326
Kazek-Kęsik A, Nosol A, Płonka J, Śmiga-Matuszowicz M, Gołda-Cępa M, Krok-Borkowicz M, Brzychczy-Włoch M, Pamuła E, Simka W (2019) PLGA-amoxicillin-loaded layer formed on anodized Ti alloy as a hybrid material for dental implant applications. Mater Sci Eng C 94:998–1008. https://doi.org/10.1016/j.msec.2018.09.014
Kesaano M, Sims RC (2014) Algal biofilm based technology for wastewater treatment. Algal Res 5:231–240. https://doi.org/10.1016/j.algal.2014.01.003
Khan B, Niazi MBK, Samin G, Jahan Z (2017) Thermoplastic starch: a possible biodegradable food packaging material—a review. J Food Process Eng 40(3):e12447. https://doi.org/10.1111/jfpe.12447
Kikuchi T, Mizunoe Y, Takade A, Naito S, Yoshida SI (2005) Curli fibers are required for development of biofilm architecture in *Escherichia coli* K-12 and enhance bacterial adherence to human uroepithelial cells. Microbiol Immunol 49:875–884. https://doi.org/10.1111/j.1348-0421.2005.tb03789.x
Klinger-Strobel M, Stein C, Forstner C, Makarewicz O, Pletz MW (2017) Effects of colistin on biofilm matrices of *Escherichia coli* and *Staphylococcus aureus*. Int J Antimicrob Agents 49:472–479. https://doi.org/10.1016/j.ijantimicag.2016.12.014
Komlos J, Cunningham AB, Camper AK, Sharp RR (2005) Interaction of Klebsiella oxytoca and Burkholderia cepacia in dual-species batch cultures and biofilms as a function of growth rate and substrate concentration. Microb Ecol 49:114–125
Król Z, Jarmoluk A (2016) The effects of using a direct electric current on the chemical properties of gelatine gels and bacterial growth. J Food Eng 170:1–7. https://doi.org/10.1016/j.jfoodeng.2015.05.018
Kumar CG, Anand SK (1998) Significance of microbial biofilms in food industry: a review. Int J Food Microbiol 42(1–2):9–27. https://doi.org/10.1016/S0168-1605(98)00061-2
Kumar S, Mukherjee A (2020) Chitosan based nanocomposite films and coatings: emerging antimicrobial food packaging alternatives. Trends Food Sci Technol 97:196–209. https://doi.org/10.1016/j.tifs.2020.01.027
Kumar CG, Singh RS (1994) *Yersinia enterocolitica* as an emerging foodborne pathogen—a review. Indian J Dairy Sci 47:537–544
Kumar H, Bhardwaj K, Cruz-Martins N, Nepovimova E, Oleksak P, Dhanjal DS et al (2021) Applications of fruit polyphenols and their functionalized nanoparticles against foodborne bacteria: a mini review. Molecules 26(11):3447. https://doi.org/10.3390/molecules26113447
Kumar L, Bisen M, Harjai K, Chhibber S, Azizov S, Lalhlenmawia H, Kumar D (2023a) Advances in nanotechnology for biofilm inhibition. ACS Omega 8(24):21391–21409. https://doi.org/10.1021/acsomega.3c04832
Kumar S, Shukla MK, Sharma AK, Jayaprakash GK, Tonk RK, Chellappan DK et al (2023b) Metal-based nanomaterials and nanocomposites as promising frontier in cancer chemotherapy. MedComm 4(2):e253. https://doi.org/10.1002/mdc3.253
Lazarova V, Manem J (2000) Innovative biofilm treatment technologies for water and wastewater treatment. ChemInform 31(32):no-no
Lebeaux D, Ghigo J-M, Beloin C (2014) Biofilm-related infections: bridging the gap between clinical management and fundamental aspects of recalcitrance toward antibiotics. Microbiol Mol Biol Rev 78:510–543. https://doi.org/10.1128/MMBR.00013-14
Leneveu-Jenvrin C, Charles F, Barba FJ, Remize F (2020) Role of biological control agents and physical treatments in maintaining the quality of fresh and minimally-processed fruit and vegetables. Crit Rev Food Sci Nutr 60:1–19. https://doi.org/10.1080/10408398.2018.1522731
Lewandowski Z, Boltz JP (2011) Biofilms in water and wastewater treatment. In: Treatise on water science. Elsevier

Li X, Kong H, Mout R, Saha K, Moyano DF, Robinson SM et al (2014) Rapid identification of bacterial biofilms and biofilm wound models using a multichannel nanosensor. ACS Nano 8(12):12014–12019. https://doi.org/10.1021/nn5066093

Li Y, Yang L, Fu J, Yan M, Chen D, Zhang L (2017) The novel loop-mediated isothermal amplification-based confirmation methodology on the bacteria in viable but non-culturable (VBNC) state. Microb Pathog 111:280–284. https://doi.org/10.1016/j.micpath.2017.08.013

Li M, Schlaich C, Willem Kulka M, Donskyi IS, Schwerdtle T, Unger WES, Haag R (2019) Mussel-inspired coatings with tunable wettability for enhanced antibacterial efficiency and reduced bacterial adhesion. J Mater Chem B 7:3438–3445. https://doi.org/10.1039/C9TB00536A

Li S, Liu Y, Tian Z, Liu X, Han Z, Ren L (2020) Biomimetic superhydrophobic and antibacterial stainless-steel mesh via double-potentiostatic electrodeposition and modification. Surf Coat Technol 403:126355. https://doi.org/10.1016/j.surfcoat.2020.126355

Lindsay D, von Holy A (2006) What food safety professionals should know about bacterial biofilms. Br Food J 108:27–37. https://doi.org/10.1108/00070700610646200

Liu J, Yang L, Hou Y, Soteyome T, Zeng B, Su J et al (2018) Transcriptomics study on *Staphylococcus aureus* biofilm under low concentration of ampicillin. Front Microbiol 9:2413. https://doi.org/10.3389/fmicb.2018.02413

Liu Y, Chen Y, Gao X, Fu J, Hu L (2022) Application of cyclodextrin in the food industry. Crit Rev Food Sci Nutr 62(10):2627–2640. https://doi.org/10.1080/10408398.2021.1955065

Liu X, Yao H, Zhao X, Ge C (2023) Biofilm formation and control of foodborne pathogenic bacteria. Molecules 28(6):2432. https://doi.org/10.3390/molecules28062432

Lohse MB, Gulati M, Johnson AD, Nobile CJ (2018) Development and regulation of single- and multi-species *Candida albicans* biofilms. Nat Rev Microbiol 16:19. https://doi.org/10.1038/nrmicro.2017.107

Mack D, Davies A, Harris L, Rohde H, Horstkotte M, Knobloch J (2007) Microbial interactions in *Staphylococcus epidermidis* biofilms. Anal Bioanal Chem 387(1):399–408. https://doi.org/10.1007/s00216-006-0801-4

Mafu AA, Plumety C, Deschênes L, Goulet J (2011) Adhesion of pathogenic bacteria to food contact surfaces: influence of pH of culture. Int J Microbiol 2011:972494. https://doi.org/10.1155/2011/972494

Maiden MM, Agostinho Hunt AM, Zachos MP, Gibson JA, Hurwitz ME, Mulks MH, Waters CM (2018) Triclosan is an aminoglycoside adjuvant for eradication of *Pseudomonas aeruginosa* biofilms. Antimicrob Agents Chemother 62(7):e00146–e00118. https://doi.org/10.1128/AAC.00146-18

Makovcova J, Babak V, Kulich P, Masek J, Slany M, Cincarova L (2017) Dynamics of mono- and dual-species biofilm formation and interactions between *Staphylococcus aureus* and gram-negative bacteria. Microb Biotechnol 10(4):819–832. https://doi.org/10.1111/1751-7915.12586

Marchand S, Vandriesche G, Coorevits A, Coudijzer K, De Jonghe V, Dewettinck K, De Vos P, Devreese B, Heyndrickx M, De Block J (2009) Heterogeneity of heat-resistant proteases from milk *Pseudomonas* species. Int J Food Microbiol 133(1–2):68–77. https://doi.org/10.1016/j.ijfoodmicro.2009.04.027

Marra A, Silvestre C, Duraccio D, Cimmino S (2016) Polylactic acid/zinc oxide biocomposite films for food packaging application. Int J Biol Macromol 88:254–262

Maszewska A, Zygmunt M, Grzejdziak I, Różalski A (2018) Use of polyvalent bacteriophages to combat biofilm of *Proteus mirabilis* causing catheter-associated urinary tract infections. J Appl Microbiol 125(5):1253–1265. https://doi.org/10.1111/jam.13986

McDonogh R, Schaule G, Flemming HC (1994) The permeability of biofouling layers on membranes. J Membr Sci 87(2):199–217. https://doi.org/10.1016/0376-7388(94)00127-I

McDougald D, Rice SA, Barraud N, Steinberg PD, Kjelleberg S (2012) Should we stay or should we go: mechanisms and ecological consequences for biofilm dispersal. Nat Rev Microbiol 10(1):39–50. https://doi.org/10.1038/nrmicro2695

Meesilp N, Mesil N (2019) Effect of microbial sanitizers for reducing biofilm formation of *Staphylococcus aureus* and *Pseudomonas aeruginosa* on stainless steel by cultivation with UHT milk. Food Sci Biotechnol 28(1):289–296. https://doi.org/10.1007/s10068-018-0448-4

Melo LDR, Oliveira H, Pires DP, Dabrowska K, Azeredo J (2020) Phage therapy efficacy: a review of the last 10 years of preclinical studies. Crit Rev Microbiol 46(1):78–99. https://doi.org/10.1080/1040841X.2019.1692222

Meyer A, Wallis FM (1997) Development of microbial biofilms on various surfaces for the treatment of heavy metal containing effluents. Biotechnol Tech 11(12):859–863. https://doi.org/10.1023/A:1018379602230

Midelet G, Carpentier B (2004) Impact of cleaning and disinfection agents on biofilm structure and on microbial transfer to a solid model food. J Appl Microbiol 97(2):262–270. https://doi.org/10.1111/j.1365-2672.2004.02216.x

Miranda AF, Ramkumar N, Andriotis C, Höltkemeier T, Yasmin A, Rochfort S et al (2017) Applications of microalgal biofilms for wastewater treatment and bioenergy production. Biotechnol Biofuels 10(1):1–23. https://doi.org/10.1186/s13068-017-0860-4

Moreno-García J, García-Martinez T, Moreno J, Mauricio JC, Ogawa M, Luong P, Bisson LF (2018) Impact of yeast flocculation and biofilm formation on yeast-fungus coadhesion in a novel immobilization system. Am J Enol Vitic 69(3):278–288. https://doi.org/10.5344/ajev.2018.17019

Moritz J, Abram A, Čekada M, Gabor U, Garvas M, Zdovc I, Dakskobler A, Cotič J, Ivičak-Kocjan K, Kocjan A (2019) Nanoroughening of sandblasted 3Y-TZP surface by alumina coating deposition for improved osseointegration and bacteria reduction. J Eur Ceram Soc 39(13):4347–4357. https://doi.org/10.1016/j.jeurceramsoc.2019.04.034

Mortensen KP, Conley SN (1994) Film fouling in counterflow cooling towers: mechanisms and design. Cool Technol Inst J 15(1):10–25

Mou XZ, Chen XY, Wang J, Zhang Z, Yang Y, Shou ZX et al (2019) Bacteria-instructed click chemistry between functionalized gold nanoparticles for point-of-care microbial detection. ACS Appl Mater Interfaces 11(26):23093–23101. https://doi.org/10.1021/acsami.9b07724

Muffler K, Lakatos M, Schlegel C, Strieth D, Kuhne S, Ulber R (2014) Application of biofilm bioreactors in white biotechnology. In: Productive biofilms. Springer, pp 123–161

Mugita N, Nambu T, Takahashi K, Wang PL, Komasa Y (2017) Proteases, actinidin, papain and trypsin reduce oral biofilm on the tongue in elderly subjects and in vitro. Arch Oral Biol 82:233–240. https://doi.org/10.1016/j.archoralbio.2017.06.005

Mukherjee N, Bartelli D, Patra C, Chauhan BV, Dowd SE, Banerjee P (2016) Microbial diversity of source and point-of-use water in rural Haiti–a pyrosequencing-based metagenomic survey. PLoS One 11(12):e0167353. https://doi.org/10.1371/journal.pone.0167353

Munoz-Bonilla A, Echeverria C, Sonseca A, Arrieta MP, Fernandez-Garcia M (2019) Bio-based polymers with antimicrobial properties towards sustainable development. Materials 12(4):641. https://doi.org/10.3390/ma12040641

Naik SS, Anusha G, Leela KV, Ravi S (2023) Promising approaches in drug delivery against resistant bacteria. In: Advances in novel formulations for drug delivery, pp 219–229

Niazi SA, Al-Ali WM, Patel S, Foschi F, Mannocci F (2015) Synergistic effect of 2% chlorhexidine combined with proteolytic enzymes on biofilm disruption and killing. Int Endod J 48(12):1157–1167. https://doi.org/10.1111/iej.12408

Nicolella C, Van Loosdrecht MCM, Heijnen JJ (2000) Wastewater treatment with particulate biofilm reactors. J Biotechnol 80(1):1–33. https://doi.org/10.1016/S0168-1656(00)00343-5

Ning Y, Chen S, Hu J, Li L, Cheng L, Lu F (2020) Fluorometric determination of agrA gene transcription in methicillin-resistant *Staphylococcus aureus* with a graphene oxide–based assay using strand-displacement polymerization recycling and hybridization chain reaction. Microchim Acta 187(1):1–10. https://doi.org/10.1007/s00604-020-04452-w

Numan A, Singh S, Zhan Y, Li L, Khalid M, Rilla K et al (2022) Advanced nanoengineered—customized point-of-care tools for prostate-specific antigen. Microchim Acta 189(1):27. https://doi.org/10.1007/s00604-021-05081-w

Ogawa M, Bisson LF, García-Martínez T, Mauricio JC, Moreno-García J (2019) New insights on yeast and filamentous fungus adhesion in a natural co-immobilization system: proposed advances and applications in the wine industry. Appl Microbiol Biotechnol 103(10):4723–4731. https://doi.org/10.1007/s00253-019-09853-7

Oliveira IM, Borges A, Borges F, Simões M (2019) Repurposing ibuprofen to control *Staphylococcus aureus* biofilms. Eur J Med Chem 166:197–205. https://doi.org/10.1016/j.ejmech.2018.10.067

Otani S, Hiramatsu K, Hashinaga K, Komiya K, Umeki K, Kishi K, Kadota JI (2018) Sub-minimum inhibitory concentrations of ceftazidime inhibit *Pseudomonas aeruginosa* biofilm formation. J Infect Chemother 24(6):428–433. https://doi.org/10.1016/j.jiac.2017.10.010

Otsuka R, Imai S, Murata T, Nomura Y, Okamoto M, Tsumori H, Kakuta E, Hanada N, Momoi Y (2015) Application of chimeric glucanase comprising mutanase and dextranase for prevention of dental biofilm formation. Microbiol Immunol 59(1):28–36. https://doi.org/10.1111/1348-0421.12200

Otto M (2013) Staphylococcal infections: mechanisms of biofilm maturation and detachment as critical determinants of pathogenicity. Annu Rev Med 64:175–188. https://doi.org/10.1146/annurev-med-042711-140023

Ovchinnikov KV, Chi H, Mehmeti I, Holo H, Nes IF, Diep DB (2016) Novel group of leaderless multipeptide bacteriocins from Gram-positive bacteria. Appl Environ Microbiol 82(18):5216–5224. https://doi.org/10.1128/AEM.01331-16

Palmer JS, Flint S, Schmid J, Brooks JD (2010) The role of surface charge and hydrophobicity in the attachment of *Anoxybacillus flavithermus* isolated from milk powder. J Ind Microbiol Biotechnol 37(11):1111–1119. https://doi.org/10.1007/s10295-010-0758-x

Pang Z, Raudonis R, Glick BR, Lin TJ, Cheng Z (2018) Antibiotic resistance in *Pseudomonas aeruginosa*: mechanisms and alternative therapeutic strategies. Biotechnol Adv 37(1):177–192. https://doi.org/10.1016/j.biotechadv.2018.11.013

Peterson SN, Meissner T, Su AI, Snesrud E, Ong AC, Schork NJ, Bretz WA (2014) Functional expression of dental plaque microbiota. Front Cell Infect Microbiol 4:108. https://doi.org/10.3389/fcimb.2014.00108

Peyrusson F, Varet H, Nguyen TK, Legendre R, Sismeiro O, Coppée JY et al (2020) Intracellular *Staphylococcus aureus* persisters upon antibiotic exposure. Nat Commun 11(1):2200. https://doi.org/10.1038/s41467-020-16073-1

Phillips NJ, Steichen CT, Schilling B, Post DM, Niles RK, Bair TB et al (2012) Proteomic analysis of *Neisseria gonorrhoeae* biofilms shows shift to anaerobic respiration and changes in nutrient transport and outer-membrane proteins. PLoS One 7(6):e38303. https://doi.org/10.1371/journal.pone.0038303

Purcell EB, Tamayo R (2016) Cyclic diguanylate signaling in Gram-positive bacteria. FEMS Microbiol Rev 40(5):753–773. https://doi.org/10.1093/femsre/fuw016

Qiao Z, Yao Y, Song S, Yin M, Luo J (2019) Silver nanoparticles with pH induced surface charge switchable properties for antibacterial and antibiofilm applications. J Mater Chem B 7:830–840. https://doi.org/10.1039/C9TB00564D

Qureshi N, Maddox IS (1990) Novel bioreactors for the ABE fermentation using cells of *Clostridium acetobutylicum* immobilized by adsorption onto bonechar. In: Yu PL (ed) Fermentation technologies: industrial applications. Elsevier Applied Science Publishers, pp 343–348

Qureshi N, Annous BA, Ezeji TC, Karcher P, Maddox IS (2005) Biofilm reactors for industrial bioconversion processes: employing potential of enhanced reaction rates. Microb Cell Factories 4:1–21. https://doi.org/10.1186/1475-2859-4-1

Qvortrup K, Hultqvist LD, Nilsson M, Jakobsen TH, Jansen CU, Uhd J, Andersen JB, Nielsen TE, Givskov M, Tolker-Nielsen T (2019) Small molecule anti-biofilm agents developed on the basis of mechanistic understanding of biofilm formation. Front Chem 7:742. https://doi.org/10.3389/fchem.2019.00742

Rabin N, Zheng Y, Opoku-Temeng C, Du Y, Bonsu E, Sintim H (2015) Biofilm formation mechanisms and targets for developing antibiofilm agents. Future Med Chem 7(4):493–512. https://doi.org/10.4155/fmc.15.6

Rather MA, Gupta K, Mandal M (2021) Microbial biofilm: formation, architecture, antibiotic resistance, and control strategies. Braz J Microbiol 52:1701–1718. https://doi.org/10.1007/s42770-021-00624-x

Ren Z, Kim D, Paula AJ, Hwang G, Liu Y, Li J, Daniell H, Koo H (2019) Dual-targeting approach degrades biofilm matrix and enhances bacterial killing. J Dent Res 98(3):322–330. https://doi.org/10.1177/0022034518822925

Rengaraju B, Jeevanantham M, Shanmugaraja S, Selvakumar A, Tamilselvi S (2023) Evaluation of mechanical and antimicrobial properties for indigenously developed biofilm for food packaging industries. In: IOP conference series: materials science and engineering, vol 012001. IOP Publishing. https://doi.org/10.1088/1757-899X/1181/1/012001

Rešková Z, Koreňová J, Kuchta T (2014) Effective application of multiple locus variable number of tandem repeats analysis to tracing *Staphylococcus aureus* in food-processing environment. Lett Appl Microbiol 58(4):376–383. https://doi.org/10.1111/lam.12198

Rodgers M, Zhan XM (2003) Moving-medium biofilm reactors. Rev Environ Sci Biotechnol 2(3):213–224. https://doi.org/10.1023/A:1022789907827

Römling U, Galperin MY, Gomelsky M (2013) Cyclic di-GMP: the first 25 years of a universal bacterial second messenger. Microbiol Mol Biol Rev 77(1):1–52. https://doi.org/10.1128/MMBR.00043-12

Ruas FAD, Guerra-Sá R (2020) In silico prediction of protein–protein interaction network induced by manganese II in *Meyerozyma guilliermondii*. Front Microbiol 11:236. https://doi.org/10.3389/fmicb.2020.00236

Sánchez-López E, Gomes D, Esteruelas G, Bonilla L, Lopez-Machado AL, Galindo R, Souto EB (2020) Metal-based nanoparticles as antimicrobial agents: an overview. Nano 10(2):292. https://doi.org/10.3390/nano10020292

Sanuja S, Agalya A, Umapathy MJ (2014) Studies on magnesium oxide reinforced chitosan bionanocomposite incorporated with clove oil for active food packaging application. Int J Polym Mater Polym Biomater 63(11):733–740. https://doi.org/10.1080/00914037.2013.844748

Seneviratne CJ, Suriyanarayanan T, Swarup S, Chia KHB, Nagarajan N, Zhang C (2017) Transcriptomics analysis reveals putative genes involved in biofilm formation and biofilm-associated drug resistance of *Enterococcus faecalis*. J Endod 43(6):949–955. https://doi.org/10.1016/j.joen.2016.12.023

Sharahi JY, Azimi T, Shariati A, Safari H, Tehrani MK, Hashemi A (2019) Advanced strategies for combating bacterial biofilms. J Cell Physiol 234(10):14689–14708. https://doi.org/10.1002/jcp.28225

Sharma U, Vipra A, Channabasappa S (2018) Phage-derived lysins as potential agents for eradicating biofilms and persisters. Drug Discov Today 23(4):848–856. https://doi.org/10.1016/j.drudis.2017.12.007

Shemesh M, Tam A, Kott-Gutkowski M, Feldman M, Steinberg D (2008) DNA-microarrays identification of *Streptococcus mutans* genes associated with biofilm thickness. BMC Microbiol 8:1–12. https://doi.org/10.1186/1471-2180-8-91

Sheng Z, Liu Y (2011) Effects of silver nanoparticles on wastewater biofilms. Water Res 45(18):6039–6050. https://doi.org/10.1016/j.watres.2011.08.012

Shi X, Zhu X (2009) Biofilm formation and food safety in food industries. Trends Food Sci Technol 20(9):407–413. https://doi.org/10.1016/j.tifs.2009.05.005

Shin JM, Ateia I, Paulus JR, Liu H, Fenno JC, Rickard AH, Kapila YL (2015) Antimicrobial nisin acts against saliva-derived multi-species biofilms without cytotoxicity to human oral cells. Front Microbiol 6:617. https://doi.org/10.3389/fmicb.2015.00617

Shin JM, Gwak JW, Kamarajan P, Fenno JC, Rickard AH, Kapila YL (2016) Biomedical applications of nisin. J Appl Microbiol 120(6):1449–1465. https://doi.org/10.1111/jam.13259

Simmen HP, Blaser J (1993) Analysis of pH and pO2 in abscesses, peritoneal fluid, and drainage fluid in the presence or absence of bacterial infection during and after abdominal surgery. Am J Surg 166(1):24–27. https://doi.org/10.1016/S0002-9610(05)80276-2

Simões LC, Simões M, Vieira MJ (2007) Biofilm interactions between distinct bacterial genera isolated from drinking water. Appl Environ Microbiol 73(19):6192–6200. https://doi.org/10.1128/AEM.00758-07

Singh AK, Gaur V, Maurya AK (2019) The role of biofilm in originating, mediating, and proliferating infectious diseases. In: Kumar S, Chandra N, Singh L, Hashmi M, Varma A (eds) Biofilms in human diseases: treatment and control. Springer, pp 23–43. https://doi.org/10.1007/978-3-030-23764-2_2

Soufi Y, Soufi B (2016) Mass spectrometry-based bacterial proteomics: focus on dermatologic microbial pathogens. Front Microbiol 7:181. https://doi.org/10.3389/fmicb.2016.00181

Stoodley P, Sauer K, Davies DG, Costerton JW (2002) Biofilms as complex differentiated communities. Annu Rev Microbiol 56(1):187–209

Srey S, Jahid IK, Ha SD (2013) Biofilm formation in food industries: a food safety concern. Food Control 31(2):572–585. https://doi.org/10.1016/j.foodcont.2012.10.001

Swaroop C, Shukla M (2018) Nano-magnesium oxide reinforced polylactic acid biofilms for food packaging applications. Int J Biol Macromol 113:729–736

Stern NJ, Kazmi SU (1989) Campylobacter jejuni. In: Doyle MP (ed) Foodborne bacterial pathogens. Marcel Dekker, pp 71–110

Taras M, Håkansson K, Guieysse B (2005) Continuous acetonitrile degradation in packed-bed bioreactor. Appl Microbiol Biotechnol 66(5):567–574. https://doi.org/10.1007/s00253-004-1743-1

Tang L, Pillai S, Revsbech NP, Schramm A, Bischoff C, Meyer RL (2011) Biofilm retention on surfaces with variable roughness and hydrophobicity. Biofouling 27(1):111–121

Tkhilaishvili T, Lombardi L, Klatt AB, Trampuz A, Di Luca M (2018) Bacteriophage Sb-1 enhances antibiotic activity against biofilm, degrades exopolysaccharide matrix and targets persisters of *Staphylococcus aureus*. Int J Antimicrob Agents 52(6):842–853. https://doi.org/10.1016/j.ijantimicag.2018.08.018

Toyofuku M, Inaba T, Kiyokawa T, Obana N, Yawata Y, Nomura N (2016) Environmental factors that shape biofilm formation. Biosci Biotechnol Biochem 80(1):7–12. https://doi.org/10.1080/09168451.2015.1058701

Urbance SE, Pometto AL, DiSpirito AA, Denli Y (2004) Evaluation of succinic acid continuous and repeat-batch biofilm fermentation by *Actinobacillus succinogenes* using plastic composite support bioreactors. Appl Microbiol Biotechnol 65(6):664–670. https://doi.org/10.1007/s00253-004-1683-8

Vallejo-Giraldo C, Krukiewicz K, Calaresu I, Zhu J, Palma M, Fernandez-Yague M, McDowell BW, Peixoto N, Farid N, O'Connor G et al (2018) Attenuated glial reactivity on topographically functionalized poly(3,4-ethylenedioxythiophene) sulfonate (PEDOT) neuroelectrodes fabricated by microimprint lithography. Small 14(18):1800863. https://doi.org/10.1002/smll.201800863

Van Houdt R, Michiels CW (2010) Biofilm formation and the food industry: a focus on the bacterial outer surface. J Appl Microbiol 109(3):1117–1131. https://doi.org/10.1111/j.1365-2672.2010.04658.x

Van Loosdrecht MCM, Eikelboom D, Gjaltema A, Mulder A, Tijhuis L, Heijnen JJ (1995) Biofilm structures. Water Sci Technol 32(8):35–43. https://doi.org/10.2166/wst.1995.0604

Vélazquez AC, Pometto AL, Ho KLG, Demirci A (2001) Evaluation of plastic composite-supports in repeated-fed-batch biofilm lactic acid fermentation by *Lactobacillus casei*. J Appl Biotechnol Microbiol 55(6):434–441. https://doi.org/10.1134/1.1410442

Verma S, Kuila A, Jacob S (2023) Role of biofilms in wastewater treatment. Appl Biochem Biotechnol 195(9):5618–5642. https://doi.org/10.1007/s12010-023-04884-4

Wang T, Huang L, Liu Y, Li X, Liu C, Handschuh-Wang S, Xu Y, Zhao Y, Tang Y (2020) Robust biomimetic hierarchical diamond architecture with a self-cleaning, antibacterial, and antibiofouling surface. ACS Appl Mater Interfaces 12(21):24432–24441. https://doi.org/10.1021/acsami.0c05628

Wang X, Liu S, Xiao R, Hu J, Li L, Ning Y, Lu F (2022) Graphene-oxide-based bioassay for the fluorometric determination of agrC gene transcription in methicillin-resistant *Staphylococcus aureus* that uses nicking-enzyme-assisted target recycling and a hybridization chain reaction. Talanta 250:123714. https://doi.org/10.1016/j.talanta.2022.123714

Wang X, Chen C, Hu J, Liu C, Ning Y, Lu F (2024) Current strategies for monitoring and controlling bacterial biofilm formation on medical surfaces. Ecotoxicol Environ Saf 282:116709. https://doi.org/10.1016/j.ecoenv.2023.116709

Waters EM, Neill DR, Kaman B, Sahota JS, Clokie MRJ, Winstanley C, Kadioglu A (2017) Phage therapy is highly effective against chronic lung infections with *Pseudomonas aeruginosa*. Thorax 72(8):666–667. https://doi.org/10.1136/thoraxjnl-2017-210323

Wu C, Zhou F, Ren J, Li X, Jiang Y, Ma S (2019) A selective review of multi-level omics data integration using variable selection. High-Throughput 8(1):4. https://doi.org/10.3390/highthroughput8010004

Xiao S, Shang K, Zhang L, Li W, Wang X (2022) A rapid anti-*Helicobacter pylori* biofilm drug screening biosensor based on AlpB outer membrane protein and colloidal gold/nanoporous gold framework. Biosens Bioelectron 215:114599. https://doi.org/10.1016/j.bios.2022.114599

Yin W, Xu S, Wang Y, Zhang Y, Chou S-H, Galperin MY, He J (2021) Ways to control harmful biofilms: prevention, inhibition, and eradication. Crit Rev Microbiol 47(1):57–78. https://doi.org/10.1080/1040841X.2020.1854967

Zapotoczna M, Forde É, Hogan S, Humphreys H, O'gara JP, Fitzgerald-Hughes D, Devocelle M, O'Neill E (2017) Eradication of *Staphylococcus aureus* biofilm infections using synthetic antimicrobial peptides. J Infect Dis 215(7):975–983. https://doi.org/10.1093/infdis/jix043

Zboun M, Arısan V, Topcuoglu N, Kuruoglu F, Sener LT, Sarcan F (2020) In vitro comparison of titanium surface conditioning via boron-compounds and sand-blasting acid-etching. Surf Interf 21:100703. https://doi.org/10.1016/j.surfin.2020.100703

Zhang Y, Chen J, Shi L, Ma F (2023) Polymeric nanoparticle-based nanovaccines for cancer immunotherapy. Mater Horiz 10(2):361–392. https://doi.org/10.1039/D3MH00101E

Zhao X, Zhao F, Wang J, Zhong N (2017) Biofilm formation and control strategies of foodborne pathogens: food safety perspectives. RSC Adv 7(47):36670–36683. https://doi.org/10.1039/C7RA02497E

Zhu X, Wu D, Wang W, Tan F, Wong PK, Wang X et al (2016) Highly effective antibacterial activity and synergistic effect of Ag-MgO nanocomposite against *Escherichia coli*. J Alloys Compd 684:282–290. https://doi.org/10.1016/j.jallcom.2016.05.179

Elevating Flavours: Biofilm Innovation 10

Rajdeep Mohanta, Souvik Chakraborty, and Payal Maiti

Abstract

Biofilms are intricate ecosystems consisting of one or more types of microorganisms encased in an extracellular matrix that they build on their own and stick to surfaces or interfaces. They can form on both biotic and abiotic surfaces and may constitute a microbiome or be a portion of it. Biofilms may have both positive and negative impacts on food products. In a controlled environment, biofilms play an important and positive role in the food industry, particularly in the flavour development of mostly fermented and non-fermented foods. However, in an uncontrolled environment, they also possess challenges in terms of the safety and sanitation of food products. Biofilms in the food industry are used for various purposes such as food fermentation, bioremediation and wastewater treatment, preparation of probiotics, biofertilizers, and flavour development. This chapter explores the impact of biofilms on the flavour enhancement of food products, the functions of biofilms throughout the process of fermentation, and their effects on the enrichment of flavour. It highlights the use of genetically engineered and biotechnologically modified microbial biofilms to improve the sensory attributes of fermented and non-fermented food products. This chapter also covers the impact of contemporary worldwide trends in flavour innovation on consumer preferences. The potential obstacles and opportunities of biofilm innovation in flavour enhancement as well as the future in this field are also discussed here. This chapter serves as a valuable resource for those interested in the intersection of biofilm science and flavour innovations.

R. Mohanta (✉)
Department of Agriculture, Brainware University, Barasat, Kolkata, West Bengal, India

S. Chakraborty · P. Maiti
Department of Post-Harvest Management, Bidhan Chandra Krishi Viswavidyalaya, Mohanpur, Nadia, West Bengal, India

T. Sarkar et al. (eds.), *Biofilm Applications to Revolutionize Food Technology*,
https://doi.org/10.1007/978-3-031-85205-3_10

Keywords

Biofilms · Flavour enhancement · Food industry · Beneficial microbes · Fermentation

10.1 Introduction

Microorganisms cling to surfaces in three-dimensional aggregations called biofilms. Within the biofilm, the bacteria encircle one another in a protective matrix. It is estimated that biofilms, as opposed to individual organisms (planktonic cells) floating around in the environment, account for up to 90% of microbial populations. For a long time, the food sector has been researching biofilm. The initial paperwork was finished in 1943, which was roughly 75 years ago.

Because of their extraordinary adaptability, bacteria can thrive in a vast range of conditions. The capacity of microorganisms to create biofilms, which facilitates adaptation to adverse environmental conditions, is one of the most crucial microbial traits that enable these adaptations.

Because bacteria may stick to almost any surface, including plastic, metal, glass, soil particles, wood, and food products, biofilms are a typical occurrence in the food business and should be taken seriously (Anderson et al. 2012; Gandhi and Chikindas 2007). Economic losses are brought on by bacterial adhesion to food goods or product contact surfaces, which also raises the risk of bacterial foodborne illnesses. For instance, 59% of foodborne illness outbreaks that were looked into in France were caused by biofilm contamination of equipment (Midelet and Carpentier 2004). Because they are shielded from environmental stressors including UV light, dehydration, and treatment with antimicrobial and sanitizing chemicals, bacteria in biofilms offer survival advantages to their members, making their removal a significant problem (Anderson et al. 2012). On the other side, microbes in biofilms can help the food industry through their usage in biotechnological applications, the creation of fermented food, flavour enhancement, and other processes (Stier 2005).

Foods that undergo fermentation procedures sometimes experience nutritional and biochemical alterations in comparison to their original constituents. Fermented foods are extraordinarily complex ecosystems formed by the combination of microorganisms' metabolic activities and raw material enzymes. Microorganisms that ferment food offer a unique method of enhancing the physical and biochemical properties of fermented meals, hence enhancing their durability. Foods that have undergone fermentation may offer advantages to consumers in terms of antioxidants, peptide synthesis, probiotic and antibacterial qualities, and organoleptic benefits over simpler foods. Moreover, it lowers toxic and antinutrient levels. Depending on the production process, storage circumstances, and durability of the fermented food, different microbial communities have different qualities and quantities (Sharma et al. 2020).

Biofilms have the power to dramatically enhance food products' flavours. Through the production of particular secondary metabolites, microorganisms in biofilms can enhance the quality of food. Food items can have better biochemical quality, flavour, and texture thanks to these microbial interactions in biofilms. It has been demonstrated that various strains, including *Lactobacillus* sp., *Enterococcus* sp., *Pseudomonas*, *Acetobacter aceti*, some fungi, and *Bacillus* sp., can create advantageous biofilms. Applications for these biofilms in the food sector are numerous and include the manufacturing of fermented foods including kimchi, vinegar, kombucha, kefir, wine, miso, and cheese. Many traditional fermented foods have unique qualities that come from the biofilm that yeast and LA bacteria produce.

10.2 Science Behind Biofilm

Biofilms are sessile microbial communities where microbes live together in association with each other on biotic or abiotic substrates which are bounded by extracellular polysaccharides, proteins, lipids, and DNA. In short, most natural surfaces are colonized by biofilms, which are an essential form of bacterial life (Singh et al. 2019).

The preference of microorganisms residing in biofilms versus freely floating planktonic stages can be attributed to their numerous advantages. Among the possible advantages are as follows:

1. Microorganisms found in biofilms exhibit resilience against an extensive array of environmental stressors, such as extremes in pH, oxygen, osmotic shock, temperature, UV radiation, and predator activity. They also show a strong resistance to antibiotics.
2. The extracellular polymeric matrix formed by the released exo-polysaccharides (EPS) improves water binding and lowers the risk of dehydration in bacterial cells (Bhosale et al. 2020).
3. The adhering microbial cells in biofilms facilitate the quick interchange of genetic material, metabolites, and nutrients.

Polysaccharides make up the bulk of extracellular polymeric compounds. EPSs serve as the structural framework of the biofilm. They cling fiercely to the surface below and are extremely hydrated (98% water). 'Water channels' in the biofilm carry vital nutrients and oxygen to the cells that are developing there. As a filter, biofilm collects a range of particles, including minerals and host elements such as platelets, red blood cells, and fibrin. Divalent cations, metal ions, and other macromolecules (including proteins, DNA, lipids, and even humic compounds) can all interact with EPSs.

Comprising intricate bacterial colonies, primarily of mixed species, a biofilm is a persistent bacterial adhesion to a variety of surfaces. The matrix that surrounds this structure is made up mostly of polysaccharide material and a combination of proteins, lipids, and nucleic acids that can create three-dimensional structures or a single layer. Biofilms vary in their electrical and chemical composition. Furthermore, planktonic and biofilm bacteria express different genes. Through intercellular communication using molecular signalling molecules, cells can cooperate with each other.

In comparison to bacteria found in planktonic settings, antimicrobial drugs are likewise noticeably less efficient against biofilm-associated microbes. Biofilms get harder to remove from surfaces that come into contact with food as a result (Singh et al. 2019). According to Cowan et al. (2000), biofilms are ordered communities of bacteria that are attached to surfaces, interfaces, and each other. They are also embedded in a matrix of self-produced exo-polysaccharides. The first fixation of planktonic cells to the surface, the growth of extracellular polymeric matrix (EPS), the establishment of microcolonies and the release of chemical signals, the development of biofilm architecture, and the dispersion of cells are all steps in their formation.

Biofilm development is influenced by EPS, surface properties, and environmental variables. In the food processing sector, the physicochemical characteristics of solid and bacterial surfaces have a significant impact on early cell attachment, which in turn influences the formation of biofilms. Some examples of surface characteristics are hydrophobicity, electrical charge, cation bridging, roughness, and topography. The range of surface materials (metal, plastic, glass, granite, and marble) is frequently used in residential kitchens or in the food processing industry for the retention of foodborne pathogens. There is disagreement over whether bacteria can attach to hydrophilic or hydrophobic surfaces.

Flowchart for biofilm formation. (Hooshdar et al. 2020)

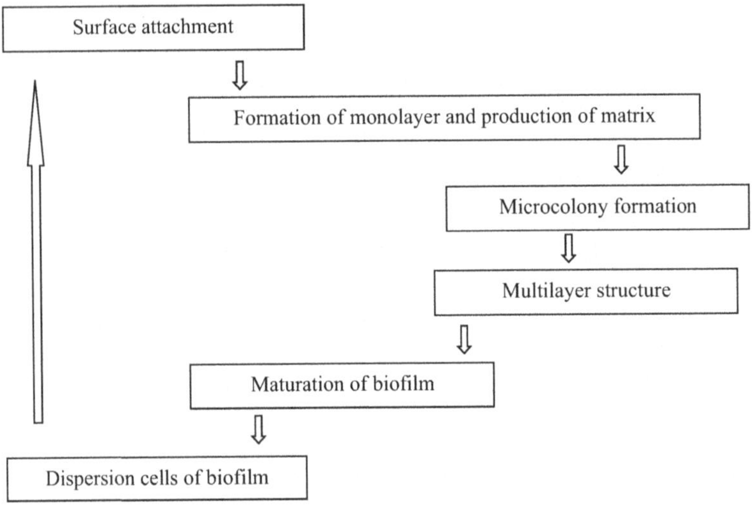

Different components of the cellular membrane, such as protein and lipopolysaccharide, appendages, such as flagella, pili, fimbriae, and curli fibres, and sporulation are used by bacteria to sense and attach to surfaces (Barken et al. 2008; Biswas et al. 2010). Moreover, bacterial population characteristics and environmental variables, including pH, temperature, and nutritional composition of the medium, affect cell wall transformation and the physicochemical characteristics of solid surfaces

(Hancock et al. 2011). Between species and strains within the same species, there can be substantial differences in the volume, content, chemical, and physical characteristics of EPS (Flemming and Wingender 2010; Decker et al. 2011; Renner and Weibel 2011). Among gram-positive bacteria, Staphylococci are good producers of extracellular polymeric substances (EPS). They form a slime consisting of modest amounts of proteins, teichoic acid, polysaccharide intercellular adhesion (PIA), and a polycationic exopolysaccharide called poly-N-acetylglucosamine (PNAG) (Otto 2008). Comparably, PIA-like polymers that are produced by gram-negative bacteria, including *E. coli*, act as adhesion molecules (Wang et al. 2004).

10.3 Biofilm and Fermentation

Fermentation is a long-standing food preservation technique that relies on the recurring formation of diverse microbial communities. Microorganisms break down large organic molecules into smaller ones through a process called fermentation. For example, yeast enzymes hydrolyze proteins into peptides and amino acids, whereas sugars and starches undergo fermentation to produce alcohol. The fermentation process, driven by microbial or enzymatic activity, leads to biochemical transformations that fundamentally alter the composition and characteristics of the food. Enhancing vitamins, proteins, critical amino acids, anti-nutrients, food appearance, flavour, and scent naturally is possible through fermentation. In addition to creating a safer product, fermentation lowers the energy needed for cooking (Nkhata et al. 2018; Xiang et al. 2019). As a result, through causing modifications to the chemical and physical characteristics of food, microbial activity contributes significantly to food fermentation.

Over millennia, humans have cultivated ideal environmental conditions to foster the growth of specific microbial communities essential for fermentation. The safety, flavour, texture, and aroma of fermented foods are intricately influenced by the metabolic activities of these microbial populations (Hutkins 2006). A diverse array of fermented products, including cheese, beer, wine, chocolate, sourdough, sauerkraut, kimchi, and miso, can be produced by modulating factors such as temperature, salinity, and moisture. Consuming these fermented foods offers numerous benefits (Melini et al. 2019; Sanlier et al. 2019):

1. The shelf life of fermented foods is longer than that of their original products.
2. The refinement of organoleptic properties; for example, the flavour of the cheese is elevated to a greater degree than that of its original milk-based substance. Eliminating toxic or undesired components from raw materials; for instance, cyanide concentration in cassava is decreased during the manufacturing of Garri, and soybean flatulence causes are eliminated by fermentation.
3. Nutritional qualities are enhanced by fermentation. For instance, the nutritional value of Garri is derived from lactic acid bacteria and yeast in bread.
4. Food requires less time to cook when it is fermented. For instance, fermented products of maize and soybeans are used to make the West African dish ogi.

5. In vitro, antioxidant activity is higher in the fermented products. For example, yoghurt and fermented milk produce biopeptides once milk proteins (α-casein, α-lactalbumin, and β-lactoglobulin) are broken down, which is why they have higher antioxidant qualities than milk.

These foods' unique fermentation-related microbial communities have several qualities that make them ideal candidates for the creation of model ecosystems that can be studied in an experimental setting. Experiments can be conducted in a variety of environments because fermented food microbial communities offer a broad range of community formation models. Multispecies biofilms on surfaces (such as cheese rinds), suspended biofilms in liquid (such as kombucha, kefir, and vinegar), dispersed growth in liquid (such as lambic beers, natural wines, and yogurts), and semi-solid substrates (such as kimchi and miso) are just a few examples of the various forms these communities can take.

The primary elements influencing fermentation are the chemical compositions of the various substrates employed and the microorganisms that are fermenting. Moreover, food fermentation is influenced by food processing duration and food treatment (Okafor 2009). According to Mokoena (2017), the majority of fermenting microorganisms are L.A.*B. (Enterococcus, Streptococcus, Leuconostoc, Lactobacillus, and Pediococcus* spp.) as well as yeasts and moulds (*Debaryomyces, Kluyveromyces, Saccharomyces, Geotrichium, Mucor, Penicillium,* and *Rhizopus* spp.) (Blandino et al. 2003; Doyle and Meng 2006; Anukam and Reid 2009; Vagelas et al. 2011). In addition to adding useful components during fermentation, food microbes also aid in preventing the development of certain harmful substances and germs. These microorganisms are also responsible for the synthesis of new digestive enzymes (Table 10.1).

Foods that have undergone fermentation, such as wine, vinegar, yoghurt, sausage, and vegetables, are widely consumed across the globe. A biofilm develops on the surface of fermented meals and solutions during the fermentation process. The features of the substrate and the bacterial cells as well as environmental parameters, including pH, temperature, and nutrient components, all have an impact on biofilm formation, in addition to genetic bases and their regulation (Shi and Zhu 2009).

10.4 Flavour Enhancement by Microbial Biofilms

Microorganisms and chemical substances combine to form biofilm, and the microbes within can release viscous metabolic products that lead to the accumulation of microbial cells. Taste is generally a key indicator when assessing fermented meals. By shielding acetic acid bacteria from acetic acid damage, biofilm aids in the conversion of substrates to acids and the creation of mellow products during the vinegar-brewing process (Stier 2005). Similarly, it has been demonstrated that biofilm increases fermentation microbes' tolerance and metabolic capacity (Bastard et al. 2016).

Table 10.1 List of some fermented foods and beverages with microbial biofilms causing the fermentation processes

Fermented foods/beverages	Substrates used	Microbial biofilms involved
Dairy products: curd, yogurt, cheese, yakult, kefir	Milk and milk casein	*Lactobacillus bulgaricus*, *Lactococcus lactis*, *L. acidophilus*, *L. cremoris*, *L. casei*, *L. paracasei*, *L. thermophilus*, *L. kefiri*, *L. caucasicus*, *Penicillium camemberti*, *P. roqueforti*, *Acetobacter lovaniensis*, *Kluyveromyces lactis*, *Saccharomyces cerevisiae*
Vegetable products: kimchi, tempeh, natto, miso, sauerkraut	Soybean, cabbage, ginger, cucumber, broccoli, radish	*Leuconostoc mesenteroides*, *Aspergillus* sp., *Rhizopus oligosporus*, *R. oryzae*, *Rhodotorula rubra*, *L. plantarum*, *Thermotoga* sp., *L. hokkaidonensis*, *Leuconostoc carnosum*, *L. rhamnosus*, *L. sakei*, *Bifidobacterium dentium*, *Enterococcus faecalis*, *Weissella confusa*, *Candida sake*
Cereals: bahtura, ambali, chilra, dosa, kunu-zaki, marchu	Wheat, maize, sorghum, millet, rice	*L. pantheris*, *L. plantarum*, *Penicillium* sp., *S. cerevisiae*, *L. mesenteroides*, *E. faecalis*, *Trichosporon pullulans*, *Pediococcus acidilactici*, *P. cerevisiae*, *Delbrueckii hansenii*, *D. Tamari*
Beverages: wine, beer, kombucha, sake	Grapes, rice, cereals	*Aspergillus oryzae*, *Zygosaccharomyces bailii*, *S. cerevisiae*, *Acetobacter pasteurianus*, *Gluconacetobacter* sp., *Acetobacter xylinus*, *Komagataeibacter xylinus*
Meat products: sucuk, salami, arjia, jama, nham	Meat	*L. curvatus*, *L. plantarum*, *Leuconostoc carnosum*, *L. sakei*, *Leuconostoc gelidium*, *B. licheniformis*, *E. faecalis*, *E. hirae*, *E. durans*, *Bacillus subtilis*, *L. divergens*, *L. carnis*, *E. cecorum*, *B. lentus*

Source: Sharma et al. (2020)

Although global knowledge and understanding of microbial function remain limited, fermented foods have been integral to the human diet for millennia. This enduring presence is attributed to the transformation of these foods through fermentation, which enhances their flavour and significantly augments their nutritional profile. Because it can enhance the biochemical quality, texture, flavour, and taste of food products, biofilm production can be a positive process (Malik et al. 2019). Microbial biofilms improve the flavour of many meals, such as dairy, meat, fish, and fruits and vegetables. After fermentation, catabolic and anabolic factors—the breakdown of complex chemicals and the synthesis of complex nutrients found in the food—are the key contributors to food flavour. The chemical components are altered by microbial biofilms, improving nutrition and flavours.

Fermentation of Fruits and Vegetables

During the fermentation of fruits and vegetables, macronutrients may be first broken down by enzymes derived from fermented microbes (Xiang et al. 2019). Food's nutritional quality can be improved by fermentation in a number of ways, such as by raising nutrient density and bioavailability and enhancing flavour and nutrient availability. The latter can be achieved through predigesting specific dietary components, affecting nutrient uptake by the mucosa, destroying antinutritional factors, and synthesising promoters for absorption (Nkhata et al. 2018). The solubility of proteins is enhanced by lactic acid fermentation, which also increases the availability of certain minerals and limiting amino acids.

Food's texture, flavour, fragrance, and general organoleptic qualities are all enhanced by fermentation in a regulated setting with beneficial microbe biofilms. Similar to this, food fermentation increases the availability of vitamins including riboflavin, thiamine, folic acid, and niacin, which has a direct therapeutic impact and enhances consumer health (Melini et al. 2019). Dissolving complicated materials into inorganic iron and vitamin C also enhances the utilization of iron (Nkhata et al. 2018). By eliminating indigestible plant material such as cellulose, hemicelluloses, and polygalacturonic and glucuronic acids, food fermentation increases the bioavailability of minerals and trace elements (Gupta et al. 2015).

Wine colour and flavour were significantly affected by the fermentation of wild yeast strains of grape epidermis, which was facilitated by the creation of yeast biofilms, which also strengthened the grape's resistance to external invasion (Sidari et al. 2014). However, it was discovered that film-forming yeasts were responsible for the unwanted yeasty flavour observed in some fermented vegetables, such as kimchi (Hung and Kyu 2006). Moreover, during the early stages of spoiling, film-forming microorganisms change the chemical composition of cucumbers, giving them an off-taste and off-flavour (Franco and Pérez-Díaz 2013). The alcoholic fermentation of the must, which is usually followed by malolactic fermentation, is the first step in the winemaking process.

This is primarily achieved by the lactic acid bacterium *Oenococcus oeni*, which is employed to enhance wine quality, particularly when a reduction in acidity is needed. Additionally, it refines the wine's organoleptic profile and safeguards against microbiological spoilage (Bastard et al. 2016). These results imply that biofilm contributes differently to the flavour profiles of different fermented foods; yet, it is still unclear how biofilm influences these flavours. According to Mokoena (2017), biofilms of lactic acid bacteria predominate in all reported fermented foods and beverages. This is assumed to be the primary element determining the health benefits associated with fermented meals and beverages prepared from fruits and vegetables. *Rhizopus oligosporus* or related moulds are used to soak, dehull, and partially cook soya beans for the Indonesian delicacy 'tempe'. By weaving cotyledons into slices and frying them, this mould creates a sturdy cake. Proteins undergo partial hydrolysis, while lipids are broken down into their component stachyose (a tetrasaccharide indigestible by humans) and fatty acids. Additionally, riboflavin levels double, niacin increases sevenfold, and a fermenting bacterium, in conjunction with the requisite mould, synthesizes vitamin B-12—a nutrient often deficient in plant-based diets. Many cereal/legume mixes have their texture considerably improved, digestibility increased, and cooking times shortened because of the tempe production method (Nout and Kiers 2005). Yeast biofilm contributes significantly to the fermentation of table olives, resulting in the emergence of diverse sensory characteristics. When the enzyme β-glucosidase is present, the yeast biofilm is crucial to the debittering process of olives. By releasing different forms of inorganic phosphorus, this enzyme aids in the breakdown of phytic substances (Zara et al. 2020). *Candida boidinii* forms polymicrobial biofilms on the olive surface by co-aggregating with yeast and lactic acid bacteria (LAB). This helps olive oil ferment more effectively.

Mexican Pulque is made by fermenting the fluids of the agave cactus plant, which is the first known alcohol-containing beverage on the American continent. Due to its rich flavour, pulque is a common food among low-income Mexican communities. Alcoholic drinks, such as kafir beer, have a nice sour flavour and a thin, gruel consistency (Tamang et al. 2020).

A traditional beverage of the Bantu people of South Africa, kafir has an alcohol concentration ranging from 1% to 8%. According to Amoa-Awua et al. (2007), palm sap wine is a transparent, sweet, plump, and colourless liquid with a milky white suspension of bacteria and yeasts that contains between 10% and 12% fermentable sugar. Because of its rich taste and sweet taste that comes from the fermenting process, it is well liked in tropical regions. Numerous alcoholic beverages also include yeast biofilm, which supports the oxidative metabolic process. The process by which yeast promotes the oxidation of ethanol to produce acetaldehyde is a crucial precursor molecule in the formation of some aged wine sensory attributes (Table 10.2).

Cereals Fermentation

Worldwide, cereals constitute the most significant source of vitamins, minerals, fibre, carbs, and dietary proteins (Laskowski et al. 2019). However, when compared

Table 10.2 List of essential enzymes used in commercially fermented foods and beverages production

Substrates	Enzymes	Microbial biofilms	Enzymatic processes
Beverages	Protease	*A. niger*, *A. orzyae*, *B. subtilis*	Cheese production
	Catalase	*S. boydii*, *Bacillus* sp.	Removing H_2O_2
	Lactase	*B. subtilis*	Lactose-free milk
Cereals	Amylase	*B. licheniformis*, *B. subtilis*	Malting, mashing, liquefaction, and production of flavour esters
	Protease	*A. niger*	
	Pentosanase	*Trichoderma* sp.	
	Glucose oxidase	*P. notatum*	
	Phytase	*A. niger*	
	Pullulanase	*B. acidopullulyticus*	
	Xylanase	*A. oryzae*, *B. subtilis*	
	Lipases	*Aspergillus niger*	
	Β-glucanase	*B. subtilis*, *A. niger*, *P. funiculosum*	
	Α-acetolactate-decarboxylase	*B. subtilis*	
	Amyloglucosidase	*A. niger*, *A. flavus*	
	Cellulase	*T. longibrachiatum*	
	Pectinase	*A. niger*	
Dairy	Glucose oxidase	*P. notatum*	Clarification of juices
	Tannase	*A. niger*	Removing O_2 hydrolysis of easters
Meat	Papain	*S. aureus*	Tenderization of meat
	Protease	*T. longibrachiatum*, *A. niger*, *A. oryzae*, *B. subtilis*	

Source: Sharma et al. (2020)

with milk and milk products, the biggest problem with cereals is customer acceptance regarding their sensory qualities and nutritional value. According to Verni et al. (2019), the low protein content, absence of certain critical amino acids (such as lysine), low carbohydrate availability, presence of anti-nutrients (such as phytic acid, tannins, and polyphenols), and coarse texture of the grains are the causes of this. The main component of the majority of traditional fermented dishes and drinks made worldwide is cereal. While some of these are staples of the human diet, others are used as drinks, colourants, spices, and breakfast items. According to Erkmen and Bozoglu (2016), natural cereal fermentation typically lowers the amount of carbohydrates with indigestible oligo and polysaccharides, improving the availability of vitamin B groups and amino acid synthesis. Similarly, fermenting grains such as sorghum, millet, maize, and other cereals raises the levels of lysine and increases the quality of the protein (Singh et al. 2015). Foods that have undergone fermentation have better or enhanced flavour, texture, taste, aroma, and—most importantly—longer shelf lives. Numerous volatile chemicals are created during the fermentation

Fig. 10.1 (**a**) Camembert-style cheese displaying rind development or the surface LAB biofilm that forms as the cheese ages and adds to the flavour and aroma. (**b**) Rind biofilm plated on regular lab medium displaying a subset of prokaryotic (Proteobacteria) and mixed eukaryotic (Filamentous fungus) biofilm community members. (Source: Wolfe and Dutton (2015))

of cereals, giving the products a diverse flavour combination (Erkmen and Bozoglu 2016). Apart from their taste, fermented cereal foods are enhanced by aroma-producing substances including diacetyl acetic acid and butyric acid (Blandino et al. 2003). Most fermented goods made from cereals are made using natural or hybrid cultures of yeasts, fungi, and bacteria. During fermentation, these microbes either concurrently or sequentially alter the dominant microbiota (Nkhata et al. 2018) (Fig. 10.1).

Milk Products Fermentation

Due to their fermentation, fermented dairy products—such as sour milk, yoghurt, dahi, kumiss, acidophilus milk, and other comparable milk products—have much higher nutritional and organoleptic characteristics. Fermented milk has different quality and amounts of proteins, vitamins, carbs, and some fat, but its mineral composition doesn't change (Mehta 2015). The sour milk's quality is determined by the material that is created during the biochemical reactions between the fermenting microorganisms and the milk's sourness. These materials include vitamins, carbon dioxide, alcohol, lactic acid, and antibiotics. The process of lactose hydrolysis, which gives milk products their distinct flavour and aroma, is carried out by certain bacteria found in milk. Compared to the average of 5% lactose found in milk, lactose hydrolysis produces 0.6–0.8% glucose, 16–20% galactose, and 45–50% lactose. Lactose is hydrolyzed by β-galactosidase, which is produced by LAB lactic acid production, which enhances organoleptic qualities and requires lactose hydrolysis. In addition, lactic acid is required for the absorption of calcium, reducing the pH of the colon and preventing the proliferation of putrefaction microorganisms (Rosolen et al. 2015) (Fig. 10.2).

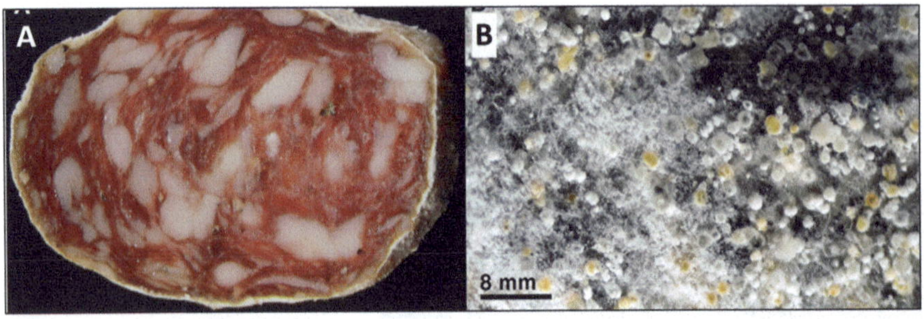

Fig. 10.2 (**a**) Salami meat fermentation caused by lactic acid bacteria (LAB). (**b**) A variety of bacteria, yeast, and filamentous fungi, including *Penicillium* sp., were apparent as white and yellow colonies on the salami surface. (Source: Wolfe and Dutton (2015))

Meat-Based Products Fermentation

The gram-positive, acidogenic (lactic acid) bacteria *Lactobacillus*, *Streptococcus*, *Pediococcus*, *Leuconostoc*, *Lactococcus*, and *Enterococcus* spp. are facultative/obligate anaerobes. With differing degrees of effectiveness, these bacteria are used to convert saccharides into lipids, amino acids, alcohols, and aliphatic chemicals. These microorganisms perform three simultaneous functions in fermented sausages: they produce nitric oxide by reducing nitrate and nitrite, which gives cured sausages their colour and they also lower pH by producing DL-lactic acid from glucose through anaerobic glycolysis (Bintsis 2018). The primary process driving lactic acid metabolism is the fermentation of carbohydrates in combination with the required phosphorylation levels. Usually, unsaturated fats are associated with autocatalytic hydrolytic and oxidative modifications that generate offensive flavours and odours.

So far, fermented meat products have been shown to have 200 volatile chemicals that release odours. For example, the aroma of vinegar is produced by acetic acid, while the aroma of butter is produced by acetoin and diacetyl at quantities of 1300 and 1400 mg/g. Aldehydes are produced when phenylalanine breaks down; methyl aldehyde smells delicious and malty, whereas methyl acids contribute to a sweaty and cheesy aroma. Lipid degradation results in the production of alkanes, alkenes, and straight-chain aldehydes, which are all green in colour, metallic, fruity, and rotten in flavour. *Staphylococcus carnosus* is linked to methyl ketones, which give off a musty, fruity, and cheese-like aroma; ethanol (55 ppm) and 3-methylbutanal, on the other hand, give off a malty flavour and a fruity, mushroom-like aroma. Hexanal and methyl ketones, which are also produced when linoleic acid oxidizes, have a green colour and a metallic, fruity, rotten, musty, and cheese-like fragrance. 2-Pentanone, green in colour and with a fruity and mushroom-like scent, converts ketones into 2-nonanone. The growth of mould encourages the formation of ketone compounds by bacteria and chemical fermentations, mainly ethyl esters that are metabolized from ethanol. According to Gianelli et al. (2012), esters add to the fruity flavours present in fermented sausages.

10.5 Innovation in Flavour Enhancement Using Biofilm

Biofilms are communities of bacteria that adhere to surfaces and are immersed in a self-produced matrix. They can interact with food items and the metabolites of fermenting microbes to facilitate growth. This interaction may alter the biochemical and nutritional characteristics of the fermented food items, enhancing their flavours. Indeed, developments in biofilm technology have opened up new avenues for enhancing food flavour. Indeed, new directions for food flavour enhancement have been made possible by advancements in biofilm technology.

Engineered biofilms are one such invention. Engineered biofilms are made from microorganisms that have been genetically and biotechnologically modified to increase their ability to produce biofilms. This could lead to the manufacture of certain amino acids, such as L-proline and L-threonine, by modified bacteria. The altered genetic expression of the biofilm may lead to novel biochemical pathways that affect the chemical and organoleptic properties linked to consumer acceptance. Moreover, a few nutritional and environmental elements as well as the cooperative relationships between multispecies biofilms can encourage the growth of biofilm. This may lead to improvements in the food products' biochemical quality, texture, flavour, and taste.

Most often, bacteria adhere to surfaces to form biofilms, which are made up of an ordered, organized network of the bacteria and a self-generated matrix of exopolysaccharides (EPS) (Mariani et al. 2011; Bastard et al. 2016; Lebleux et al. 2020; Lahiri et al. 2021). This sessile bacterial life form has long been used by microbes as a well-organized survival strategy when battling external stress conditions that are commonly encountered by bacteria in both natural and man-made settings. This is because the biofilm matrix causes physiological changes as well as the formation of a protective shield. These structures, or biofilm-forming elements, can be biologically engineered and used as a component to enhance fermented food quality. The food and agriculture sectors can benefit from the many applications of biofilm (Shah 2018). The high biomass level of the biofilm is essential for several types of metabolic activities (Marapatla 2014). The industries are currently using various engineered biofilms to create a range of product kinds (Winkelströter et al. 2014). Biofilm is helpful in a range of situations due to its many beneficial characteristics. Biofilms are believed to be involved in the development of fermented foods, including but not limited to cheese, vinegar, kombucha, kefir, wine, lambic beer, miso, and kimchi (Wolfe and Dutton 2015). However, due to microbial activity, fermentation breaks down organic macromolecules into smaller molecules. For example, yeast enzymes convert proteins to amino acids and peptides, but they also convert starch and sugar to alcohol. Food components experience fermentation due to microbial enzymatic activity, which produces the required biochemical alterations and a noticeable change in the food. Fermentation is a natural way to boost essential amino acids, vitamins, proteins, antioxidants, flavours, food appearance, and enhanced fragrances. Because of natural modifications that enhance their nutritional content and flavour by merely knowing the functions that various bacteria play in the environment, fermented foods have been a mainstay of human diets for thousands of years.

An immobilized fermentation method based on biofilm for large-scale industrial synthesis of amino acids has several disadvantages. The biologically engineered strains have the ability to significantly accelerate the formation of biofilms under industrial cultivation conditions. This method can also be utilized in repeated batch (continuous) immobilized fermentation. Analysis of transcriptome profiles revealed that regulating the genes involved in the growth and generation of biofilms can improve biofilm formation (Ghosh et al. 2022).

Next-generation sequencing, or NGS, has become a widely used technique in several sectors, such as forensics, diagnostics, antimicrobial resistance, outbreak investigations, and food authenticity, in the last 10 years, from its beginnings as a strictly scientific tool. The speed at which this instrument is developing is improving the cost-effectiveness and quality of food. Furthermore, food microbiology is significantly impacted by it (Ghosh et al. 2022). In food microbiology, NGS is applied in two separate ways:

1. Whole genome sequencing (WGS): This method determines the complete genome sequence of a single isolated culture, such as a bacterial colony, virus, or other organism.
2. Meta-genomics, in which all or the majority of the microorganisms found in a biological sample are produced by applying NGS to the material and running several sequences through it.

The application of NGS to enhance food quality has the potential to completely transform the world's food supply chain. It is imperative to bridge the technological gap between the theoretical and actual uses of these NGS techniques as soon as possible. The growth of NGS impacts will also result in decreased prices for the food industries. This is because it will get much cheaper to produce genetically modified bacterial genome sequences; in a few years, the cost of using NGS technology will be more than the cost of microbial cultures and the physiological analyses needed to support them. The decrease in costs is another benefit that this new technology is meant to offer (Ghosh et al. 2022). Traditional microbiological typing and characterization tools are gradually being replaced by the WGS approach, which produces faster and more precise findings. Although it is still in its early stages, applying meta-genomics to improve the quality and safety of food presents new opportunities for altering microbial biofilms, predicting changes in microbial communities, and supporting the identification of unknown microbiota.

10.6 Conclusion

Most natural surfaces are colonized by biofilms, an essential form of bacterial life. It is estimated that 90% of microbial populations do not reside as individual planktonic cells floating around in the environment but rather as biofilms. Members of bacteria in biofilms have an advantage in survival because they are shielded from external stresses. Because they are far more obstinate than individual planktonic bacteria, biofilms—which are prevalent in the food industry—cause concern.

However, biofilms can help the food business through food fermentation, biotechnological applications, and flavour enhancement. Biofilms have the ability to produce particular secondary metabolites that can greatly enhance the flavour of fermented food products. Since ancient times, these microbial biofilms have been used to produce a wide range of food products with improved texture, taste, aroma, and flavours, including fruits, vegetables, cereals, milk, and meat. Developments in biofilm technology have opened up new avenues for food flavour enhancement. New directions for food flavour enhancement have been made possible by advancements in biofilm technology. Engineered biofilms are made from microorganisms that have been genetically and biotechnologically modified to increase their ability to produce biofilms. The altered genetic expression of the biofilm may lead to novel biochemical pathways that affect the chemical and organoleptic properties linked to consumer acceptance. This may lead to improvements in the food products' biochemical quality, texture, flavour, and taste. Next-generation sequencing (NGS) has the potential to transform the global food sector through two methods: whole genome sequencing (WGS) and meta-genomics. Traditional microbiological typing and characterization tools are gradually being replaced by these methods, which produce faster and more precise findings when it comes to creating different flavour enhancers for food products worldwide.

References

Amoa-Awua WK, Sampson E, Tano-Debrah K (2007) Growth of yeasts, lactic, and acetic acid bacteria in palm wine during tapping and fermentation from felled oil palm (*Elaeis guineensis*) in Ghana. J Appl Microbiol 102(5):599–606

Anderson JM, Lin Y, Gillman AN, Parks PJ, Schlievert PM, Peterson ML (2012) Alpha-toxin promotes Staphylococcus aureus mucosal biofilm formation. Front Cell Infect Microbiol 2:64–69

Anukam KC, Reid G (2009) African traditional fermented foods and probiotics. J Med Food 12:1177–1184

Barken KB, Pamp SJ, Yang L, Gjermansen M, Bertrand JJ, Klausen M, Tolker-Nielsen T (2008) Roles of type IV pili, flagellum-mediated motility and extracellular DNA in the formation of mature multicellular structures in *Pseudomonas aeruginosa* biofilms. Environ Microbiol 10:2331–2343

Bastard A, Coelho C, Briandet R, Canette A, Gougeon R, Alexandre H et al (2016) Effect of biofilm formation by *Oenococcus oeni* on malolactic fermentation and the release of aromatic compounds in wine. Front Microbiol 7:613

Bhosale S, Desale RJ, Fulpagare YG (2020) Biofilm: an overview with respect to dairy industry. Int J Curr Microbiol Appl Sci 9(10):150–160

Bintsis T (2018) Lactic acid bacteria as starter cultures: an update in their metabolism and genetics. AIMS Microbiol 4:665

Biswas R, Agarwa RK, Bhilegaonkar KN, Kumar A, Nambiar P, Rawat S, Singh M (2010) Cloning and sequencing of biofilm-associated protein (bapA) gene and its occurrence in different serotypes of Salmonella. Lett Appl Microbiol 52:138–143

Blandino A, Al-Aseeri ME, Pandiella SS, Cantero D, Webb C (2003) Cereal-based fermented foods and beverages. Food Res Int 36:527–543

Cowan SE, Gilbert E, Khlebnikov A, Keasling JD (2000) Dual labeling with green fluorescent proteins for confocal microscopy. Appl Environ Microbiol 66:413–418

Decker EM, Dietrich I, Klein C, Von OC (2011) Dynamic production of soluble extracellular polysaccharides by *Streptococcus* mutants. Int J Dent 46:1–6
Doyle MP, Meng J (2006) Bacteria in food and beverage production. In: Dworkin M (ed) Prokaryotes, vol 1. Springer, New York, pp 797–811
Erkmen O, Bozoglu TF (2016) Fermented cereal and grain products. In: Food microbiology: principles into practice. John Wiley & Sons, Ltd, Chichester
Flemming HC, Wingender J (2010) The biofilm matrix. Nat Rev Microbiol 8:623–633
Franco W, Pérez-Díaz IM (2013) Microbial interactions associated with secondary cucumber fermentation. J Appl Microbiol 114:161–172
Gandhi M, Chikindas ML (2007) Listeria: a foodborne pathogen that knows how to survive. Int J Food Microbiol 113:1–15
Ghosh S, Nag M, Lahiri D, Sarkar T, Pati S, Kari ZA, Nirmal NP, Edinur HA, Ray RR (2022) Engineered biofilm: innovative nextgen strategy for quality enhancement of fermented foods. Front Nutr 9:808630
Gianelli MP, Salazar V, Mojica L, Friz M (2012) Volatile compounds present in traditional meat products (charqui and longaiza sausage) in Chile. Braz Arch Biol Technol 55:603–612
Gupta RK, Gangoliya SS, Singh NK (2015) Reduction of phytic acid and enhancement of bioavailable micronutrients in food grains. J Food Sci Technol 52:676–684
Hancock V, Witsø IL, Klemm P (2011) Biofilm formation as a function of adhesin, growth medium, substratum and strain type. Int J Med Microbiol 301:570–576
Hooshdar P, Kermanshahi RK, Ghadam P, Khosravi-Darani K (2020) A review on production of exopolysaccharide and biofilm in probiotics like lactobacilli and methods of analysis. Biointerface Res Appl Chem 10(5):6058–6075
Hung LD, Kyu HK (2006) Inhibition of yeast film formation in fermented vegetables by materials derived from garlic using cucumber pickle fermentation as a model system. Food Sci Biotechnol 15:469–473
Hutkins RW (2006) Microbiology and biotechnology of fermented foods. Blackwell Publishing, Hoboken
Lahiri D, Nag M, Sarkar T, Dutta B, Ray RR (2021) Anti-biofilm activity of α-amylase from *Bacillus subtilis* and prediction of the optimized conditions for biofilm removal by response surface methodology (RSM) and artificial neural network (ANN). Appl Biochem Biotechnol 193:1853–1872
Laskowski W, Gorska-Warsewicz H, Rejman K, Czeczotko M, Zwolinska J (2019) How important are cereals and cereal products in the average Polish diet? Nutrients 11:679. https://doi.org/10.3390/nu11030679
Lebleux M, Abdo H, Coelho C, Basmaciyan L, Albertin W, Maupeu J et al (2020) New advances on the *Brettanomyces bruxellensis* biofilm mode of life. Int J Food Microbiol 318:108464
Malik A, Erginkaya Z, Erten H (eds) (2019) Health and safety aspects of food processing technologies. Springer, Cham
Marapatla NP (2014) Study of biofilmed biofertilizers to improve crop production and disease control in chickpea
Mariani C, Oulahal N, Chamba JF, Dubois-Brissonnet F, Notz E, Briandet R (2011) Inhibition of *Listeria monocytogenes* by resident biofilms present on wooden shelves used for cheese ripening. Food Control 22:1357–1362
Mehta BM (2015) Chemical composition of milk and milk products. In: Cheung P, Mehta B (eds) Handbook of food chemistry. Springer, Berlin, Heidelberg. https://doi.org/10.1007/978-3-642-36605-5_31
Melini F, Melini V, Luziatelli F, Ficca AG, Ruzzi M (2019) Health-promoting components in fermented foods: an up-to-date systematic review. Nutrients 11:1189
Midelet G, Carpentier B (2004) Impact of cleaning and disinfection agents on biofilm structure and on microbial transfer to a solid model food. J Appl Microbiol 97:262–270
Mokoena MP (2017) Lactic acid bacteria and their bacteriocins: classification, biosynthesis, and applications against uropathogens: a mini-review. Molecules 22:1255

Nkhata SG, Ayua E, Kamau EH, Shingiro JB (2018) Fermentation and germination improve nutritional value of cereals and legumes through activation of endogenous enzymes. Food Sci Nutr 6:2446–2458

Nout MJR, Kiers JL (2005) Tempe fermentation, innovation, and functionality: update into the third millennium. J Appl Microbiol 98:789–805

Okafor N (2009) Fermented foods and their processing. In: Horst W, Doelle J, Rokem S, Berovic M (eds) Biotechnology, vol 8. EOLSS Publishers

Otto M (2008) Staphylococcal biofilms. Curr Top Microbiol Immunol 322:207–228

Renner LD, Weibel DB (2011) Physico-chemical regulation of biofilm formation. MRS Bull 36:347–355

Rosolen MD, Gennari A, Volpato G, de Souza CFV (2015) Lactose hydrolysis in milk and dairy whey using microbial β-galactosidases. Enzyme Res 2015:806240

Sanlier N, Gokcen BB, Sezgin AC (2019) Health benefits of fermented foods. Crit Rev Food Sci Nutr 59:506–527

Shah MP (2018) Bioremediation-waste water treatment. J Bioremed Biodegr 9:1–10

Sharma R, Garg P, Kumar P, Bhatia SK, Kulshrestha S (2020) Microbial fermentation and its role in quality improvement of fermented foods. Fermentation 6:106

Shi X, Zhu X (2009) Biofilm formation and food safety in food industries. Trends Food Sci Technol 20:407–413

Sidari R, Caridi A, Howell KS (2014) Wild *Saccharomyces cerevisiae* strains display biofilm-like morphology in contact with polyphenols from grapes and wine. Int J Food Microbiol 189:146–152

Singh AK, Rehal J, Kaur A, Jyot G (2015) Enhancement of attributes of cereals by germination and fermentation: a review. Crit Rev Food Sci Nutr 55:1575–1589

Singh P, Basak G, Sharma B, Jain U, Mishra R, Vaishali (2019) Biofilm: an alarming niche in dairy industry. Int J Livest Res 9(4):10–24

Stier RF (2005) Beating back biofilms in food processing. Food Safety Magazine 11(1):31–34

Tamang JP, Cotter PD, Endo A, Han NS, Kort R, Liu SQ, Mayo B, Westerik N, Hutkins R (2020) Fermented foods in a global age: east meets west. Compr Rev Food Sci Food Saf 19:184–217

Vagelas I, Gougoulias N, Nedesca ED, Liviu G (2011) Bread contamination with fungus. Carpathian J Food Sci Technol 3:1–6

Verni M, Rizzello CG, Coda R (2019) Fermentation biotechnology applied to cereal industry by-products: nutritional and functional insights. Front Nutr 6:42

Wang X, Preston JF III, Romeo T (2004) The pgaABCD locus of Escherichia coli promotes the synthesis of a polysaccharide adhesion required for biofilm formation. J Bacteriol 186:2724–2734

Winkelströter LK, Teixeira FB, Silva EP, Alves VF, De Martinis EC (2014) Unraveling microbial biofilms of importance for food microbiology. Microb Ecol 68:35–46

Wolfe BE, Dutton RJ (2015) Fermented foods as experimentally tractable microbial ecosystems. Cell 161:49–55

Xiang H, Sun-Waterhouse D, Waterhouse GI, Cui C, Ruan Z (2019) Fermentation-enabled wellness foods: a fresh perspective. Food Sci Human Wellness 8:203–243

Zara G, Budroni M, Mannazzu I, Fancello F, Zara S (2020) Yeast biofilm in food realms: occurrence and control. World J Microbiol Biotechnol 36:134

Approaches to Cultivating Beneficial Biofilms in Cheese Production for Enhanced Flavor and Safety

11

Abdullah Ahmed Butt and Hafiza Saima

Abstract

This chapter discusses the two most important microbial activities in cheesemaking: acid production and flavor generation, as well as other potential benefits of growing beneficial biofilm. It then considers how biofilms influence the particular taste characteristics of some cheeses by describing the microbial ecology responsible for flavor development and environmental conditions that foster biofilm growth. We cover a few ways in which you can select for better biofilm formers, or to continue with the agricultural analogy—good starter cultures comprise this into this chapter. This chapter also covers biofilms cultivated for safety and gives a view of how barriers can be formed by the structure found in pathogenic biofilms. Proposed strategies for securing safety and quality during biofilm-assisted cheese manufacturing are elaborated in the sections. An analysis of challenges and opportunities in biofilm cultivation, pinpointing difficulties relative to consistency and scale have been highlighted in the chapter. In addition, this chapter investigates emerging biofilm-based cheese manufacturing innovations. Moreover, the last sections detail success stories of cheese brands employing biofilm processes, shining a light on their efforts and takeaways. This chapter ends with a perspective in which the key results are resumed and future research is suggested, highlighting how biofilms can be used for other dairy products.

Keywords

Biofilm · Dairy products · Cheese · Milk · Food safety

A. A. Butt (✉) · H. Saima
Department of Food Science and Technology, Government College University, Faisalabad, Pakistan

T. Sarkar et al. (eds.), *Biofilm Applications to Revolutionize Food Technology*,
https://doi.org/10.1007/978-3-031-85205-3_11

11.1 Introduction

Biofilms, in general, are colonies of microorganisms that stick to a surface and form a protective co-operative network which accounts well for the cheese production. These communities contain not only bacteria, but also yeasts and molds which interact with each other in order to help produce the variety of flavors found within cheese (Fröhlich-Wyder et al. 2019). Biofilm growth on cheese surfaces or in the cheese production environment can greatly affect both the flavor and safety of final product (Ghosh et al. 2021). This means biofilms are absolutely vital in forming cheese flavor. Together, the microorganisms in these films create enzymes that release proteins and fats within the cheese, resulting in flavor compounds essential to different flavors of cheeses (Zheng et al. 2021). Cheese could not have the complexity and flavor depth that appeal to consumers without biofilms being active. In addition to being biofilms, cheese rind consortia differ in composition depending on the type of milk and other specific practices used for creating certain types of cheeses as well as the ripening conditions provided in the area where starters are added (Amin et al. 2016). Beyond their role in flavor, biofilms could also be a key to cheese safety. Beneficial microorganisms or probiotics can form the microbial community on biofilms, which masks and prevents other pathogenic bacteria from growing. In addition, this natural protection may be particularly advantageous for producers of artisanal and traditional cheeses, as the use of chemical preservatives is often restricted or unwarranted (Ünal Turhan et al. 2019).

Not all biofilms are good. The production of cheese is challenging because it requires that we form protective biofilms while inhibiting the formation of harm-inducing bacteria. Pathogenic biofilms can be the source of spoilage and contamination by rendering significant risks to both cheese quality and safety (Seale et al. 2015). Thus, any cheesemakers desiring to produce very high quality and safe cheese with well-defined flavor profiles would require an understanding of how biofilms are formed and controlled. In this chapter, we will examine different ways to foster advantageous biofilms in cheesemaking. Here, we will explore how to foster the growth of beneficial microbes as a biofilm separated from undesirable bacteria or microalgae, monitor that culture makeup and structure over time so you can manage it effectively, and limit colonization by pathogenic species in your process units (e.g., cooling towers) where they produce disease- and spoilage-causing artwork. Cheesemakers are able to improve the flavors and the safety of these products all via the use of biofilms assisting in meeting consumer demands as well meeting food standards (Fig. 11.1).

Overview of Cheese Production Processes Involving Biofilms

Cheese is the result of a complex multiple-stage process during which biofilms might take up important roles. Biofilms even alter the flavor and safety of cheese from first milk processing down to final ripening stage. In discovering the ways that those biofilms tend to behave differently in different portions of cheesemaking,

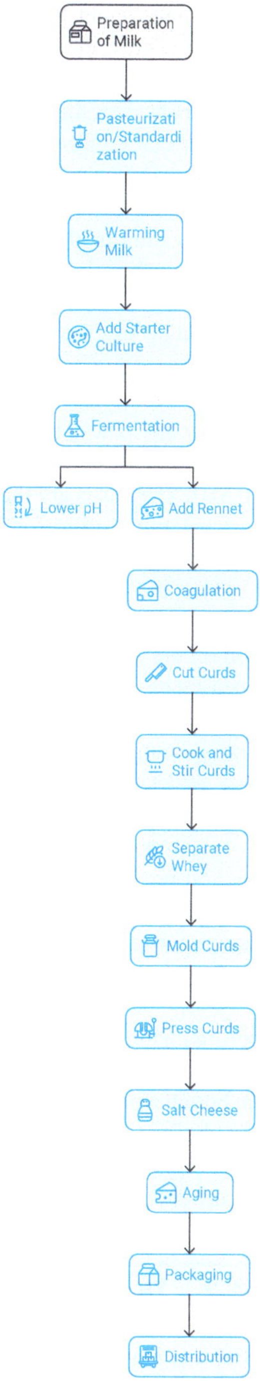

Fig. 11.1 General cheesemaking process

cheesemakers can figure out how best to take advantage of their behaviors and, if necessary, prevent microbes from settling on equipment. It starts from the treatment of milk for food processing such as pasteurization to kill harmful bacteria (Ramesh 2020). However, this step is insufficient for sterilizing the milk, allowing the growth of beneficial cultures, which then contribute to the cultivation of biofilms. Biofilm formation on the surface of equipment had another implication: biofilms formed by these helpful microorganisms could lead to flavor development during cheesemaking. Raw milk is often used in traditional cheesemaking, which brings more diverse microorganism that are able to form biofilms (Pappa et al. 2019). Once the milk is prepared, it is inoculated with starter cultures, which are specific strains of bacteria that initiate the fermentation process (Parente et al. 2017). Prepared milk is then inoculated using starter cultures, which are particular strains of bacteria that start the fermentation. These bacteria can grow in biofilms on the milk surface and within cheese curd. These biofilms generate lactic acid, which in turn lowers the pH of the milk and causes it to coagulate into curds and whey during fermentation (Li et al. 2020a). The curds are then subsequently cut and cooked, expelling more whey that enriches the biofilm-forming species (Bulut-Solak and Akin 2019), which further concentrates the biofilm-forming microorganisms.

Once the curds are done and they have drained out whey, the salting process begins to be turned into molds (Fox et al. 2017b). Within this stage, biofilm formation on the cheese surface still proceeds. These biofilms are an essential part of the ripening process since they secrete enzymes that will break down proteins and fats, giving aged cheeses their complex flavors and textures (Fröhlich-Wyder et al. 2019). The ripening environment, including the presence of moisture and temperature, can also affect both biofilm composition and activity thereby potentially affecting cheese quality in a final preparation (Akoğlu 2020).

It is important to monitor and control biofilm formation in the entire cheese production progress. While good biofilms can contribute to desirable flavors and safe food, bad biofilm development results in spoilage or contamination. Cleaning the equipment and maintaining strict sanitation practices along with using various microbial cultures can help control biofilm development (Kukhtyn et al. 2017). This plays into how much cheesemakers will allow beneficial biofilm to form on the rind before coating it in a less permeable, more microbe-resistant layer or aging condition. In the end, biofilms are two-faced in cheese production: they impact positively on both flavor and safety of the dairy. Biofilms are crucial to cheese quality throughout the processes of milk treatment, ripening, and biofilm formation/comment. Knowing how to control and eradicate biofilms will greatly improve the operations of any cheese factory, resulting in better products for consumers (Fig. 11.2).

11.2 The Role of Biofilms in Flavor Development

Biofilms play a complex and intriguing role in the flavor development of cheese. Biofilms, in which microorganisms attach to a surface and cooperate, can facilitate the necessary complex biochemical reactions that transform milk into cheese with

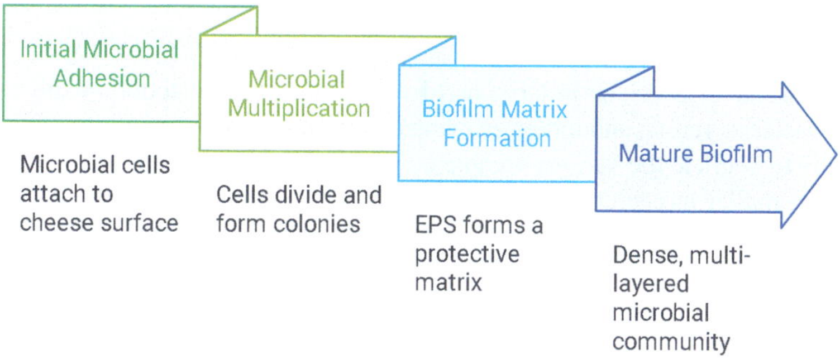

Fig. 11.2 Overview of key biofilm formation stages

its characteristic flavors. These include not only the bacteria, yeasts, and molds that among them produce a wide variety of flavors by breaking down proteins, fats, and carbohydrates in the cheese (Kumar et al. 2021).

The evolution of cheese flavor is very complex and depends on various factors, such as the type of milk used in the cheesemaking process, strains of microorganisms present to create that particular one, and time of ripening or storage under which these cheeses are kept. In the process, biofilms appear on the surface and even in a matrix of cheese which starts to produce enzymes that help catalytically decompose components of milk. This is said to be an important factor in giving high-quality cheese something to power its rich and complex flavors. These volatile compounds contribute to the characteristic aroma and taste of cheese, being one of the main features in flavor development mediated by biofilms (Bertuzzi et al. 2018). Microbial metabolism is the source of their secondary metabolites.

Generally, the nature of these volatile compounds is not constant and changes with the types of bacteria or microorganisms that grow on the cheese surface during the ripening period and the environmental conditions (Salum et al. 2018). Bizarre as it may sound, not every biofilm contributes a positive impact on the flavor. Biofilms arising from undesirable microorganisms that adhere to cheese and packaging materials can emanate off-flavors or spoilage substances that may ruin the cheese. Hence, cheesemakers have to take charge of the microbial populations that contribute to the formation of biofilms, such that it facilitates desirable flavors but avoids contamination. To more thoroughly evaluate the role of biofilm in essential flavor generation, it advanced itself in trying to understand the interaction that exists between different microbial organisms and other possible pathways of metabolism. Recent developments in microbiology and biotechnology have opened up new ways of studying, understanding, and modifying these microbial communities for targeted flavor enhancement. Seeders even more importantly, cheesemakers by seeding beneficial biofilms and reducing the harmful ones can enhance the flavor profiles of their products in accordance with the preferences of contemporary consumers who prefer distinct cheeses produced through artisanal processes.

Biofilms and Flavor Compounds in Cheese

The cheese flavor is driven in large part by biofilms. In a collection of tiny organisms—bacteria, yeasts, and molds growing as communities—the cultures act synergistically to degrade the various organic components within the milk (e.g., proteins, fats) into smaller molecules that give cheeses their balanced flavor. These biofilms have several beneficial effects that are necessary for the production of cheese with high organoleptic quality.

Biofilms add the flavor, assisted by beneficial enzymes, which breakdown milk proteins into peptides and amino acids (Nahar et al. 2018). New flavor compounds, a number of which are sulfur-based (such as ones that contribute to cheddar's sharpness) and others including glutamate (Zhao et al. 2016). Because a biofilm contains so many different kinds of microbes, each one offers enzymes that play varied roles in forming flavor. In addition to proteins in milk, the biofilms also produce enzymes that break down lipids (Thierry et al. 2017). This breaks down fats by lipolysis to produce free fatty acids which are important flavor precursors. This helps to remove the fats through lipolysis which yields free fatty acids, and these acids act as important flavor precursors. These free fatty acids are then processed into volatile components such as esters, ketones, and aldehydes (Bertuzzi et al. 2018). As lactose gets fermented during the making of the cheese, a compound called diacetyl is formed, and this is responsible for the nice buttery scent in these soft rind cheeses (Gebreyowhans et al. 2020).

Such differentiation is also due to the specific strains of microorganisms within the biofilm, which also cause the flavor change in the cheese. For every type of cheese, there are common communities that are *Lactobacillus* species-specific, such as the type of milk used, goat, or cow, the environmental factors, and the traditional ways of making cheese. It is a fundamental part of the biofilms for aroma development and involves fermentation where metabolic activities of different microorganisms in biofilm casement. Lactic acid bacteria, found in cheese biofilms and used for food preservation since antiquity as a result of efforts to keep fresh the products of high-value production processes (such as those required by cheesemaking), so recently might produce an acidic environment helping curd formation while spicing up flavors such feta or yogurt-type cheeses (Bintsis 2018a; Hu et al. 2022). In addition, these bacteria produce a range of other metabolites including acetoin and 2,3-butanediol that further contribute to the spectrum of flavor compounds (Ruggirello et al. 2018).

The field of cheese microbiology has advanced significantly in this area, and cheesemakers are now learning to use these promising biofilms for better flavor development. In order to functionally guide biofilms being favorable in flavor development on the cheese, specific volatile metabolites of perfumed final cheeses are identified by selecting individual microbial cultures and ripening conditions. As a result, this leads to not only better cheese sensory properties but also permits the development of cheeses that are consumer-preference driven and different from those already produced. In general, biofilms are an essential part of cheese manufacturing and help to enhance the final flavor and aroma. What makes biofilms even better is that the subtle interplay of these microbial populations and their metabolic processes are what result in the perfect taste profile for each type of cheese—therefore making biofilms a true ally to those who craft artisanal cheeses.

Factors Influencing Biofilm-Induced Flavor Profiles

Some advantageous factors that contribute to the creation of flavorsome biofilms in cheese production provide rich and complex taste profiles. By understanding these factors, cheesemakers are enabled to control conditions that maximize the desirable properties of biofilms in order to produce higher quality and more unique cheeses.

Type of milk is one of the most important factors. The microbial community of the biofilm is considerably influenced by different milk types (cow, goat, sheep, and buffalo) (Ngo et al. 2023). These different proteins, fats, and sugars in the milk give the microorganisms an environment that ferment them into unique substrates to metabolize resulting in diverse flavors. For example, biofilms that metabolize the unique fatty acids found in goat milk can endow cheeses with a goaty tang as they break down those lipids (Talavera and Chambers 2016), due to the specific fatty acids present that are metabolized by the biofilms.

The temperature and humidity of the ripening process are also important. Different controlled ripening environments can have huge effects on biofilm composition and activity (Walsh et al. 2020). This results in the growth of characteristic bacteria and molds that break down fats and proteins into their aromatic compounds, leading to strongly flavored mature cheeses: a time-consuming process that takes many months or even years (Leclercq-Perlat et al. 2019).

Another important aspect is the microbial cultures that we add to the cheese. The starter and adjunct cultures are very carefully selected for establishment of biofilm, which supports the flavor profiles. First, there are several lactic acid bacteria (which include *Lactococcus* and *Lactobacillus*) that produce the compound lactic acid as part of their normal metabolism; this not only helps with curd formation but also imparts a tangy flavor to our final cheese (Allam et al. 2017). In this case, lactic acid percentage is calculated as follows.

$$C_{\mathrm{LA}} = C_{\mathrm{S}} - \frac{C\mathrm{so}}{Y_{\mathrm{LA/s}}}$$

C_{LA} is concentration of lactic acid
C_{s} is final substrate concentration
C_{so} is initial substrate concentration
$Y_{\mathrm{LA/s}}$ is lactic acid per unit substrate

The addition of secondary cultures like *P. roqueforti* (blue cheeses) or *B. linens* (washed-rind cheese) will create additional aromatic and flavor notes, anything from earthy mushroom to nutty pungent (Caron et al. 2021). Salt, in general, is critical in flavor profiles from biofilms. The salt effect is not merely a preservative; it modulates the growth and activity of bacteria in the biofilm (Youn and Seo 2022). Salt in the right proportions will help start those flavor-producing enzymes and hinder microorganisms that we do not want growing, helping guide full-bodied, not off-flavor-profiled cheese to develop (Hashempour-Baltork and Farshi 2022). Salt is involved in the development of a rind (also called crust) and can develop especially on the surface to growing specific biofilms contributing unique tastes often related to cheeses in their place locale (Bisig 2017).

The amount of time you age your cheese is probably one of the most important factors. Maturation allows biofilms all that much more time to break bigger proteins and fats into saucy bits via enzymatic activity (Khattab et al. 2019). Flavors in aged cheeses like Parmigiano-Reggiano or Aged Gouda are complex and high in umami flavor which develop as a consequence of the large protein degradation to free amino acids and peptides (D'Incecco et al. 2020).

The choice of cheesemaking techniques or traditions used by the cheesemakers plays one of the most important roles in shaping biofilm-induced flavors (Papademas and Bintsis 2017). Methods like aging the product on wooden boards or following unique turning and washing protocols bring in different microbe communities as well as differences in the environment that help ensure your cheese tastes a certain way (Licitra et al. 2017). Cultivation of useful biofilms which bear particular importance in bringing out the many attributes as well as flavors have been studied appreciated (Paxson 2016; Ramírez-Rivera et al. 2017). Given these factors, cheesemakers are harvesting these biofilms from different hosts, and in turn, enhancing the sensory quality of their products through an elaborate range.

11.3 Techniques for Cultivating Beneficial Biofilms

Effective cheese biofilms are not made in a snap; rather, they are produced by people who have vast knowledge of the microbial world and cheesemaking processes. Biofilms are composed of a variety of microbes and play a crucial role in developing the delicate flavor of this product as well as its safety. This is a simple matter, and so is poor quality targeting of helpful microorganisms at a price that is high enough, and low enough to allow bad stuff. With the development of new application techniques and the creation of new layers of the cheese industry, many contemporary cheesemakers have an arsenal of ways to build up protective biofilms on purpose (Chourasia et al. 2021). They leverage the information on microbial interactions, environmental conditions, and cheese-producing processes to promote the growth of favorable microorganisms, enhancing both the quality of flavors and, in particular, increased safety for the consumers. Taking this approach, it is presumably possible to utilize biofilm by cheesemakers to produce a new kind of fresh cheese with tweaked quality considerations controlled by practices and state-of-the-art supervision. Encouraging the growth of desirable biofilms in this production of cheese brings with it the hassle of picking the right microbial strains and manipulating the ripening conditions.

Selection of Microbial Cultures for Biofilm Formation

It is crucial to choose the right type of microbial culture as it enables the formation of desirable biofilms during the process of cheese production. The users' preferences regarding flavoring cultures of this type are not dependent on any utilitarian aspect, but rather on the emotional satisfaction brought by the tastes and smells. Biofilms that improve the texture and safety of cheese can be intentionally selected by choosing specific cultures (Mayo et al. 2021).

An important aspect of choosing a microbial culture is its ability to release enzymes capable of cleaving milk ingredients into potent flavor compounds. Lactic acid bacteria of the genera *Lactococcus* and *Lactobacillus* are commonly employed. The bacteria naturally produce lactic acid, during fermentation which assists in curd formation and leads to the tangy flavor of cheese (Cui et al. 2018). Furthermore, many lactic acid bacteria strains produce proteolytic enzymes that degrade proteins into their constituent amino acids or shorter peptides adding to the flavor of cheese (García-Cano et al. 2019; Kieliszek et al. 2021). Consequently, certain yeast strains are also beneficial in biofilm formation along with lactic acid bacteria (Fröhlich-Wyder et al. 2019). Metabolization by yeast, which decomposes milk sugars and converts them into alcohol, that is, other volatile compounds that give the cheese its unique flavors. For instance, different strains of *D. hansenii* are capable of synthesizing manhole and fruity type alcohols which contribute special character traits in the cheese aroma profile (Alzobaay and Alzobaay 2018). However, a general equation is reported in the literature for calculating microbial growth.

$$N_t = N_{0e}{}^{\mu t}$$

Where N_t is microbial population at time t
N_0 is initial microbial population
μ is specific growth rate
t is time

In addition, molds are also a critical part of the formation of biofilms; this is important with respect to surface-ripened cheeses. Some molds or yeasts, such as *P. candidum* and *G. candidum*, form the rind of traditional soft cheeses like Brie and Camembert (Li et al. 2020b). The molds in these varieties of cheese add to the appearance and additionally create compounds that break down fats and proteins which contribute to a variety of flavors (Galli et al. 2016). Besides its capacity to singly or in mixed culture grow on a range of substrates and secrete an array of enzymes, the final choice about which microbe to select for ripening also depends on compatibility with other cultures form suitability at different environmental conditions (e.g., temperature). Using them selectively with other cultures that help develop desired attributes, cheesemakers can encourage the presence of these biofilms in their cheeses and improve taste/flavor/texture/safety. The choice of bacteria that represent the microbial cultures is crucial to obtaining a cheese with certain

Table 11.1 Microbial cultures used for beneficial biofilm formation in cheese production

Microbial cultures	Types of culture	Applicable for cheeses	Role for biofilm formation
Lactobacillus plantarum	Bacterial-based	Gouda, Cheddar	Promotes stable biofilm matrix
Lactococcus lactis	Bacterial-based	Brie, Camembert	Encourages early biofilm development
Geotrichum candidum	Yeast-based	Soft cheeses (e.g., Brie)	Forms surface biofilms
Penicillium roqueforti	Fungal-based	Blue cheeses	Establishes complex biofilm networks

qualitative and macroscopic characteristics. Using the powers of necessary cultures, cheesemakers are able to make unique cheeses that taste good and which keep consumers coming back for more; thus, ensuring a solid future in small cheese creation (Table 11.1).

Environmental Conditions Favoring Biofilm Growth

In industrial cheesemaking, encouraging beneficial biofilms to form is achieved by ensuring the right environmental conditions are in place during cheese production. Such factors are crucial to determine the population dynamics of microorganisms in a cheese system and by that help to characterize the odor, structure, and hygienic status of a given product. Cheesemakers can use this knowledge to help guide the successful growth of biofilms which aid in production quality.

An important environmental factor affecting biofilm development is temperature. The objective is to help beneficial microorganisms usually have a moderate range temperature (proper temperatures are normally 10–18 °C) (Morin-Sardin et al. 2016). It is vital to maintain the appropriate conditions for their reproduction. This is a range where lactic acid bacteria and other desirable cultures truly do well, which means that the fermentation process will happen faster (therefore chances for bad contamination) and more fast creation of tasty compounds. Cold temperature also slows the growth of harmful bacteria, helping to keep food safe and fresh (Soodam et al. 2017).

Biofilm formation is proportional to the relative humidity, as well. The moisture percentage, typically between 80% and as high as 95% in some cases, helps keep the cheese moist while encouraging beneficial molds and yeasts to form (Bettera et al. 2020; Yangılar and Oğuzhan Yıldız 2016). These bacteria play a role in creating unique flavors and textures, especially seen in varieties of surface-ripened cheese. In addition, the elevated humidity further prohibits the activity of detrimental biofilms such as those causing spoilage or off-flavors (Galli et al. 2016).

Another significant environmental parameter is proper aeration for biofilm establishment. To make sure there is decent airflow so that aerobic microorganisms such as molds, and some bacteria can have access to the oxygen they need for forming beneficial biofilms (Carpino et al. 2017). Proper ventilation, temperature control,

Table 11.2 Key environmental factors favoring beneficial biofilm growth

Environmental factors	Optimal range	Effect on biofilm formation
Temperature	12–15 °C (ripening)	Promotes balanced microbial activity
pH Level	4.6–6.0	Encourages growth of beneficial microbes
Humidity	85–95%	Maintains hydrated biofilm structure
Airflow	Moderate	Reduces unwanted mold growth

and humidity are necessary for an optimal environment that allows microbial multiplication. But too much airflow can disturb the forming biofilm and consequent ripen erratically, so it is very crucial to hit a sweet spot (Corrieu et al. 2018). In addition, acidity (as measured by its pH) plays an important role in dictating which microbes are present and what they will be up to. A slightly acidic pH, usually at 4.5–5.5, is conducive to the growth of lactic acid bacteria and biofilm formation. These bacteria form lactic acid as a waste product (an example of fermentation) which lowers the pH and helps to both produce specific cheese flavors and render other mild microbiologic life unfavorable while favoring environmental conditions that encourage the growth of still more desirable microorganisms (Table 11.2).

Use of Starter Cultures to Initiate Beneficial Biofilms

Crucial to getting the beneficial biofilms started are selection strategies for starter cultures during cheese production, an essential step that helps promote not only flavor but also texture development in biofilm-rich cheeses (Fox et al. 2017c; Parente et al. 2017).

Starter cultures are a mixture of different microorganisms, each selected for their distinct impacts on fermentation and the formation of biofilm. One of the important ingredients in starter cultures is lactic acid bacteria (LAB) like *L. lactis* and *Streptococcus thermophilus*. Due to the ability of these bacteria, they are able to ferment lactose, which is a milk sugar into lactic acid. This leads to the acidification of milk which helps the milk to clot and due to this coagulation, simultaneously a conducive acidic environment is formed that promotes yeasts further (Thomas Bintsis 2018a, b). In addition, some lactic acid bacteria strains have proteolytic enzymes that cleave proteins to peptides and amino acids during the aging of the cheese with a more complex flavor (van Mastrigt et al. 2018).

Adjunct culture: This refers to specific yeast and mold strains used in starter cultures. Yeasts, such as *D. hansenii*, ferment sugars in the milk producing ethanol and volatile compounds (and probably some organic acids also) thereby influencing the aroma and taste development of the cheese (Sevinc-Demircan and Ozturkoglu-Budak 2023). Surface-ripened cheeses require common molds, such as *P. candidum* and *G. candidum*, to interact with the outside of the cheese for a rind that will form over it—these surface molds help produce flavors or textures otherwise not possible as they secrete enzymes that break down lipids and proteins in their environment (Yalçin and Ozbas 2017).

In addition, some bacterial strains including *B. linens* are used to encourage healthy biofilms in washed-rind cheeses. *B. linens* love the wet, salty world created by washing down a cheese while it ripens. This microbe carries out proteolysis and lipolytic activities, contributes to biofilm formation (it can produce exoenzymes releasing proteins and lipids), and brings characteristic organoleptic properties as fatty flavors aroma compounds in the cheese rind imprinted by its orange color (Gonçalves de Melo 2020). This is followed by a discussion of the strategic choice and use of these starter cultures as building blocks for designing more complex biofilms with disruptive properties, but also flavor components in cheese manufacture (Bunyard 2021) in cheese production. With the help of their strains and teams of cheesemakers, can create highly unique flavor profiles in cheeses as well as precise levels of texture, allowing for consistently satisfying the product that will keep consumers coming back over and over again. In all these cases, biofilm formation rate can be calculated by the following equation:

$$B_t = B_0 + \left(r^* t\right)$$

Where B_t is biofilm mass/ thickness
B_0 is initial biofilm mass
r is biofilm growth rate
t is time

11.4 Safety and Quality Assurance with Biofilms

The establishment of biofilms in cheese production is a monumental contributor to safety and quality. Comprising a variety of microbial populations, biofilms can aid in flavor development and at the same time be hazardous to food quality (Speranza and Corbo 2017). Therefore, there needs to be a strict no-no, and effective safety and quality assurance measures must be in place. On the one hand, useful biofilms aid in flavor formation and texture enhancement or natural preservation. Conversely, detrimental biofilms can cause spoilage through contamination or the production of toxins—compromising the safety and quality of cheese (Alvarez-Ordóñez et al. 2019).

Proper management of biofilms has to be holistic and incorporates strict hygiene, careful selection and use of cultures, as well as control over environmental conditions. Cheese manufacturers can reduce the likelihood of deleterious biofilm formation greatly while also enabling access to beneficial films merely by aggressive quality control. In this regard, biofilms pose a risk to safety and quality assurance; however, it should be mentioned that cheese production processes can also benefit from the retention of positive properties related to some types of biofilm (Biofilms working for us). Well-balanced flavors to enhance the natural flavor of cheese while ensuring optimum safety.

Biofilms as a Barrier to Pathogens in Cheese Production

In cheese production, biofilms act as a critical defense mechanism against pathogens present on the surface and act as the first line and a natural barrier to protect final products. These diverse bacterial communities create a defensive barrier against the colonization and growth of pathogenic bacteria.

Biofilms also prevent other bacteria from reaching these pathogens through competition (competitive exclusion) (Son et al. 2016). Biofilm-associated beneficial microorganisms outcompete pathogenic bacteria for Gal space as well, thus reducing their ability to reside and thrive. This provides them with a competitive advantage and leads to better stability of the microbial ecosystem in that environment needed for cheesemaking, thus lowering contamination risk. Biofilms also produce antimicrobial compounds that actively inhibit the growth of pathogens (Yoplac et al. 2022). These microbes are capable of manufacturing a range of antimicrobial agents, including organic acids, hydrogen peroxide, and bacteriocins. On top of this, these components also eliminate the moisture and food sources for pathogens but even produce it in a condition unfavorable for pathogenic activity thereby drastically eliminating chances of contamination or spoilage (Aljohani et al. 2023).

Biofilms also act as an extra layer against pathogenic microorganisms. The EPS that many microbes synthesize is a foraging term that encompasses various types of parasitically drifting shields that prevent onslaughts of certain surfaces, especially those underneath from microorganisms (Tiwari et al. 2020). It is an obstructive structure that interferes with the attachment of agents which causes infections on surfaces and avoids biofilm formation thereby reducing their interference with the safety of cheese (Patlan-Velázquez et al. 2024).

Although these protective actions are advantageous, biofilms can also cause problems in cheesemaking if they become contaminated with pathogenic bacteria. Improper hygiene, environmental conditions, and handling of the cheesemaking equipment will allow harmful biofilms to develop which poses a threat to food safety (Johnson 2017). The identification of biofilms as a pathogen barrier is key for the advancement and quality improvement in cheese production. However, cheese manufacturers can help their consumers feel at ease when buying the final product by encouraging beneficial biofilms and maintaining good hygienic practices in order to lessen risks associated with pathogenic contamination.

Strategies to Maintain Cheese Safety During Biofilm Cultivation

Balancing cheese safety with biofilm development is crucial, and this will need to be specifically addressed in terms of hygienic factors, at each step of the overall series for making cheeses. While a good thing for flavor, biofilms can also contain dangerous bacteria if not handled correctly. Since then all hygiene practices must be followed to ensure that the environment is not contaminated and is safe.

A common tactic is to implement stringent sanitation procedures for cheese production equipment and surfaces. The routine cleaning and sanitization of cheesemaking equipment will decrease the risk that biofilms can form which are capable of housing pathogenic organisms (Wirtanen 2018; Bhosale et al. 2020). Since cleanliness is crucial, it will greatly reduce the chance of contamination by using sanitizing agents that are safe for food contact surfaces and established cleaning schedules (Giaouris and Simões 2018). A second important strategy is the control of environmental conditions in the cheesemaking premises. These ideal temperature, humidity, and airflow levels can provide an environment that will not promote the growth of harmful bacteria. Conditions of pH and moisture content can also be managed so that the growth of pathogens is suppressed while beneficial microorganisms appear.

A well-constructed food safety plan is necessary for pinpointing and avoiding any kind of risks at the level best as all in the entire cheesemaking process (Vågsholm et al. 2020). This involves routine inspection, analysis of microbial contaminants, and staff training on the maintenance of impeccable hygiene conditions (Shuvo 2018). Moreover, by having proper protocols for dealing with waste materials, we can decrease cross-contamination and the spread of pathogens (Marriott et al. 2018). When implemented within a cheesemaking system, these approaches can tip the balance in favor of safe biofilm development. Cheesemakers can maintain these high-quality standards by focusing on hygiene and the environment as well as utilizing the best food safety practices to provide consumers with a quality, safe product that they can trust.

Monitoring and Quality Control of Biofilm-Based Cheese Production

Appropriate standards to regulate all these are necessary so that biofilm-based cheese can be produced and monitored following quality requirements as well as food safety constraints. While biofilms are in fact very important for the development and enhancement of the flavor, they could potentially be harmful if not properly managed. That is why it is easy for things to go wrong and therefore strict monitoring and control should be ensured at all stages of cheese production. One of the most vital elements in biofilm-based cheese production monitoring is scheduled sampling and testing. Cheesemakers sample batches of cheese during different stages of the process to test for populations of microbes present and any deviation from their expected levels. Microbiological tests have also had a major role to play in cheese manufacture, with it being used as an indicator of contamination (by means such as total bacterial counts and presence of certain pathogens) by microbiologists for the all-round safety testing that cheese can be subject to (Kızanlık 2018).

There is a requirement for monitoring within the cheesemaking facility of environmental conditions essential for biofilm control. By monitoring temperature, humidity, and airflow levels, you can discover factors that may create an environment more conducive to the development of dangerous biofilms (de Castro et al. 2017). Detect visible signs of biofilm formation by regular equipment and surface inspections. Slimy, discolored surfaces can be a clue to biofilms, but the last stage will always result in harsh penalties.

The Effect of Biofilm Formation and Hygiene on Cheese Safety
Strict hygiene and sanitation measures must be applied in the facility to control biofilms from forming (Gray et al. 2018). Cleaning and disinfection of equipment, surfaces, and production areas should be conducted regularly to avoid biofilms as a medium for pathogens. One of the most common ways to reduce contamination is through using sanitizing agents, making sure they are acceptable for use on food contact surfaces and restaurants that follow a proper cleaning schedule (Skåra and Rosnes 2016).

Quality control is employed at every step of the cheesemaking process to ensure that final products conform to established standards. Monitoring cheese properties including texture, flavor, and visual/styling attributes; sensory analysis for global quality assessment is therefore important (Quintieri et al. 2021). Cheesemakers keep a detailed record of all the production processes and quality control tests conducted, followed by periodic audits to see how they are performing over time so that weak areas can be detected. Consistent testing and monitoring, attention to environmental conditions, and strict sanitation practices along with quality control from start to finish are important for cheesemakers wishing to meet consumer expectations while remaining within regulatory guidelines.

11.5 Challenges and Opportunities in Biofilm Cultivation

The biofilm map is nefarious for cheesemakers, and it creates a challenge and opportunity landscape. The role of biofilms, which consist of diverse microbial communities, is pivotal for texture enhancement and safety assurance as well aid the complex fermentative processes during cheese production. Nonetheless, with the great power of biofilms comes the responsibility to manage the risks associated as well.

Controlling microbial diversity is one of the most difficult problems to solve in biofilm cultivation. Bacteria, yeasts, and molds are initially used to actively spoil the milk, but they also play a role during different steps of cheesemaking; not only do beneficial microorganisms help in flavor development and safety, but harmful pathogens can also make high-quality cheese if they are controlled properly (Jahromi 2012). The challenge is obtaining the right balance between promoting beneficial biofilms and preventing harmful ones, which means there are no black-and-white answers when it comes to managing microbes.

For cheese producers, one hurdle in cultivating a stable biofilm growth is to maintain the most ideal environmental conditions. Important parameters such as temperature, relative humidity, and pH have to be continuously controlled in order to establish the right conditions for a healthy biofilm settlement (Fox et al. 2017a). Environmental conditions play a role in microbial activity and biofilm development;

hence, there is some room for improvements till the point of this technology is optimized.

For cheese manufacturers, improving the reproducibility and consistency of biofilm cultivation is a major challenge. Therefore, variability in raw materials, production practices, and environmental conditions which can have an impact on the identity of biofilms may result in flavor and texture variations as well as quality alterations from cheese batch to cheese batch. All of these challenges also give rise to many other potential areas for innovation and advancement in cheesemaking using biofilm culture. New possibilities to capitalize on biofilm development, and subsequently cheese quality, are fueled by modern pathways in microbiology, biotechnology as well as cheese manufacture techniques (Khanal et al. 2019). Using these opportunities, cheesemakers can harness biofilms more fully to achieve new cheeses that exhibit improved flavor and texture attributes while also presenting enhanced safety options.

Challenges in Consistency and Reproducibility

While cheese producers can struggle to obtain consistent and reproducible biofilm growth, which is important for subsequent processing, this is another challenge that leaves the quality of the resulting product up in the air. This complexity is representative of how dynamic biofilms are formed among microbial communities interacting with environmental factors and the cheesemaking process—very real challenges to applications in air conveyance wherein trustful accurate bacteria identification is required. This chapter will describe seven problems of consistency and reproducibility.

1. Microbial Variability: The presence of various types of microorganisms in biofilms can make flavors, textures, and safety among batches produced on fresh cheese lots differ by inherent diversity (Fagerlund et al. 2021). However, obtaining uniform responses has been challenging in the control of microbial species, and their activity is easier said than done (as seen in artisanal cheese manufacture with natural fermentation).
2. Environmental Fluctuations: Changes in the environment such as temperature, moisture, and pH can interfere with microbial growth and biofilm creation (Wirtanen and Salo 2016). Consistent conditions at various steps during cheesemaking are important to get the same outcomes over and over again, but this is quite difficult in a variety of production environments.
3. Raw Material Variability: The heterogeneity of raw materials, for example, milk composition and quality can affect microbial activity and biofilm formation (Nahusenay et al. 2023). This can lead to uncertainty in the cheesemaking process and make it impossible for you to achieve uniform end products due to variations that milk from different sources or seasons may introduce (Yap et al. 2024).

4. Production Process Variability: The modulations in cheesemaking techniques such as fermentation duration time, maturation surroundings, and handling can influence biofilm formation as well as the quality of cheeses (Rossi et al. 2016). The above steps are required to be followed to have uniform production practices across batches, the process of extraction needs standardization.
5. Hygiene and Sanitation Challenges: Keeping hygienic standards and sanitation very high is particularly important to prevent contamination, as well as maintain uniformity of our biofilms (Lee et al. 2021). No limits were imposed on the order of loading of the panels in the forming of integrated builders and curd promoters and handlers, neither mixing adding growth enhancement with equipment cleaning and sanitation nor with personal cleanliness (Pal et al. 2016).
6. Interactions with Other Microorganisms: The quality of cheese is subjective and differs because of the dynamic interaction between biofilm-forming microorganisms and other nonmilk-inhabiting microorganisms that might be present during the process of cheesemaking, which consequently affects the biofilm development. Such interactions have to be understood and controlled in order to produce consistent outcomes over time. Incorporating more complex biofilm control procedures and those for monitoring cheese quality and safety parameters brings an additional layer of difficulty into the process of manufacture.
7. Quality control complexity: Very effective and robust quality control strategies to evaluate biofilm structure, cheese quality, and safety measures as well as parameters have added another layer in the production chain management. Consistency and Reproducibility: This is where detailed testing protocols come into play, along with powerful data analysis techniques (Nasr et al. 2017; Thomareis and Chatziantoniou 2022).

That means it demands an integrated view of science and technology, along with the knowledge that spans all aspects of best-practice cheesemaking. In doing so, cheese producers can further strengthen and add more consistency to the quality of their products.

Addressing Biofilm-Related Issues in Cheese Production

In cheese production, biofilms can cause a variety of problems and result in substantial cost damages if not properly controlled. To prevent these negative impacts on the quality or safety aspects of cheesemaking processes, there are several solutions that all benefit to maintain high-quality standards for this popular food product line. These solutions can effectively tackle the issues of biofilm in cheese production, and at last, produce high-quality cheeses that meet consumer expectations and regulatory requirements. Here are strategies:

1. Enhanced Sanitation Practices: Following strict cleaning and sanitation schedules for equipment, surfaces, and production areas consistently prevents biofilms from forming. With adequate rinse and a regular scrub of sanitizing agents

approved for bioburden, most biofilm residues can be removed which in turn decreases the contaminants (Faille et al. 2018).

2. Optimized Environmental Conditions: Controlling the growth of biofilm is dependent on keeping environmental conditions stable ranging particularly from temperatures, humidity, and airflow (Leggieri et al. 2018). Natural cheese production is a balancing act that requires monitoring and adjusting these factors during the transformation of milk into cheese to make an environment less favorable for biofilm formation and other microbiological hazards.
3. Selective Microbial Cultures: The importance of several specific microbial cultures with known beneficial properties argues for careful selection and use to promote desirable biofilm formation while minimizing the presence or growth of these pathogens. The selection of cultivars with demonstrated efficacy in flavor standardization, improving texture, and safety can contribute to this aim's consistency (Sun and D'Amico 2021).
4. Regular Monitoring and Testing: Regular tracking and testing routines of the microbial populations, environmental conditions, and cheese characteristics allow cheesemakers to rapidly identify problems associated with biofilms. Cheeses are tested microbiologically and tested regularly throughout the aging process to make sure that they meet quality and safety standards (Short et al. 2021).
5. Hygienic Design of Equipment and Facilities: The use of hygienic design principles in the manufacture and construction of cheesemaking equipment and facilities can significantly reduce biofilm formation potential, thereby minimizing contamination risks. Biofilm growth is largely a result of having smooth, non-porous surfaces that can be cleaned and disinfected easily; proper drainage to remove water after the cleaning process further reduces any moisture that may remain (İpek and Zorba 2018).
6. Education and Training: Offering in-depth education and training on hygiene, sanitation, and biofilm management to staff would be a much-needed way of ensuring the standards are met for quality and safety (Machado and Cutter 2018). Training staff on proper food hygiene practices and ensuring they follow through is vital.
7. Continuous Improvement Practices: A culture of continuous improvement emphasizes promoting innovation and standardization in biofilm management within every cheese production facility (van Assen 2021). Reviewing the performance metric and customer feedback periodically for the production process will help in identifying a lot of scope to improve cheese safety with quality (Suchánek and Králová 2019).

Emerging Opportunities for Innovation in Biofilm-Based Cheesemaking

Advanced Microbial Cultures: Advancements in microbiology and biotechnology offer the possibility of developing new, custom-made culture strains specifically for cheesemaking relying on biofilms (Hu et al. 2019). Such types of cultures are likely to show enhanced flavors, improved texture, and greater ability to prevent pathogens than the already established colonial genera (Afshari et al. 2020).

Precision Environmental Control: Controlling these parameters means that cheesemakers will regulate the conditions that are conducive to the biofilm formation, and this will quiet the quality of the cheese since the same kind of characteristics will always be available (Centeno and Carballo 2023). It could be possible to have biofilm-mediated cheese production by designing consortia of microbes with the necessary composition and interaction.

Microbial Consortia Engineering: By developing biofilms of consortia with distinguished metabolic activities, cheesemakers can practice the production of many types of flavoring compounds without compromising the safety and stability of the product (Ben-Harb et al. 2020).

Bio-preservation Techniques: The new novel concept of bio-preservation using either bacteriocins or bacteriophages specifically targeted to the microbial inhabitants of the biofilms (Pujato et al. 2019). These natural antibiotics work by targeting pathogens leaving places for healthy biofilms to establish themselves uninhibited by infections (Gutiérrez et al. 2016).

Sensor and Data Analytics Integration: This future paradigm within the science of biofilm development vertically arranges both cheese development and real-time monitoring and analysis of the biofilm by incorporation of sensors particularly in the area of sensor technologies and data analytics platforms (Kamilaris et al. 2017). Using these tools, cheesemakers can see how biofilm dynamics occur and find out possible improvement areas so that they can optimize product quality and effectiveness (Danezis et al. 2016).

11.6 Exemplary Applications

Technological Innovations in Biofilm Applications

Modern technological developments have expanded the usage of biofilms in many ways including cheese production. Such improvements allowed to achieve new heights in flavor, texture and safety of cheese products (Sachan and Karnwal 2022). A prime example is the construction of elaborate systems that help the cheesemakers monitor biofilm formation and microbial activities at any point in time (González-Rivas et al. 2018). These interventional systems are knowledge-based and employ a combination of sensors and data intelligence to analyze the biofilm processes for better cheese manufacturing output. Bioreactors have emerged as another interesting technology in biofilm cultivation (Germec et al. 2020). This specialized apparatus makes it easier for cheesemakers to know the right conditions to alter for a particular microbial activity (Mandenius 2016). Conditions, such as temperature, pH, and nutrient usage, make it easy to produce even better cheeses with uniform quality. Furthermore, the utilization of modern biotechnology positioned the production of designer biofilms as well. It has now become possible to alter the genes

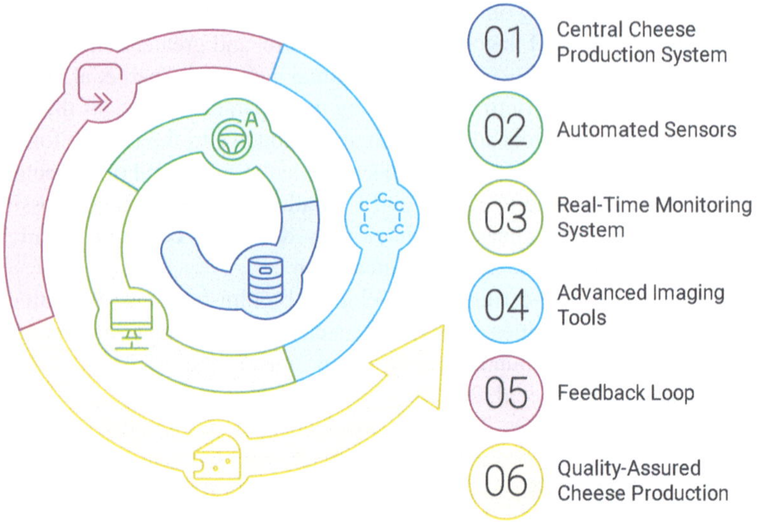

Fig. 11.3 Technological innovations in biofilm formation

of bacteria that produce flavors, or bacteria that have been engineered to outcompete pathogenic bacteria (Wang et al. 2023). Biofilms like these could lead to new cheeses being developed—which adds even more potential for innovation within the cheese industry (Ghosh et al. 2022). The advancements in technology will certainly help increase the range of biofilm applications, and this will influence cheese production in years to come. Application of such effective tools as sensors, bioreactors, or genetic engineering toward the process of cheesemaking can bring forth new infusions, textures, or reassuring food safety practices in the products (Fig. 11.3).

Benefits and Outcomes of Biofilm Application

There has been an evolution in how biofilms are exploited in the field of cheese production, thanks to technological advancements. These improvements have created new opportunities for the enhancement of taste, texture, and safety in the manufacture of cheese (Carpino et al. 2017). For instance, newer monitoring systems computing cheesemaking, targeting the biofilm formation and microbial performance during manufacture, can be traced.

This offers more precision in cheese production processes as these systems incorporate sensors and analytics for the analysis of biofilm behavior (Mohamed and Altowaijri 2023). The use of biofilms could also improve the safety as well as the shelf life of the food products (Bai et al. 2021). The beneficial organisms present in the biofilm outcompete the pathogens which are harmful to the product thus preventing the contamination of food and spoilage. Biofilms are capable of secretion of substances that are toxic to pathogens and spoilage, thus prolonging the time that food products are safe for eating (Cui et al. 2018). Another application for the concept of biofilm involves process optimization or process efficiency and Eco compatibility. Another area of biofilm application considers the optimization or improvement of production processes or their efficiency in an ecologically friendly manner (Vishwakarma 2020). Making use of the biologically active substances in biofilms makes it possible for industries to optimize their industrial processes and churn out lesser wastes and do less damage to the ecosystem. As a result, there come benefits in terms of savings and resources being used rendering the use of biofilm desirable in many sectors of the economy (Ünal Turhan et al. 2019). Biofilms provide shorter cycle times and improved textures and tastes among the products. Thanks to the biomass biofilms, the economy can produce very high-class products while promoting creative ideas in support of sustainability.

Consumer Preferences and Market Trends

Furthermore, customer tastes and choices as well as market dynamics are critical aspects which guide the rate at which biofilm-aided processes are embraced in cheese manufacturing. This market is facilitated by high customer demand for new taste and high-quality artisan and specialty cheeses (Ouyang et al. 2021). As such, these markets will continue increasing often the demand of bio-resistant packaging materials. Furthermore, it fulfills such customer expectations who prefer to order unique involved customized premium "restaurant quality" cheese that contributes additional pleasantness and texture, as such cheese does (Yao et al. 2022). Awareness and understanding of biofilm usage across various product applications, including cheese, and the impact of biofilm on the end product has improved (Bhosale et al. 2020). Cheesemakers have also begun to capitalize on this perception of increased awareness and added value on their products as the industry continues to expand. There is an emerging consumer demand for artisanal, craft cheeses; producers can use biofilm-based culture techniques to market their cheeses as specialty artisan products.

There are still hurdles in helping consumers accept biofilm-aided cheeses as well as compliance with policies and regulations. Cultural differences and false perceptions about biofilms as well as food safety issues can all act as hurdles that stand in the way to full market penetration (Gilmore 2023). There is a proper understanding that education of the end users on the biofilm-assisted cheese production and the quality assurance and control practiced by the manufactures has to come for the market segregation and acceptability to take place. In many ways, consumer tastes and market circumstances provide a promising prospect for biofilm-aided cheeses due to the increased appetite for novelty, premium, and handmade products. It enhances their capacity to serve biofilm-assisted cheeses to the target market, which is increasingly expanding by catering for these preferences and educating the consumer on their concerns.

11.7 Conclusion

This restates the content of the section and emphasizes the contributions of the biofilms on the flavors and textures of the cheese products, as well as the safety during the cheese processing. Reading through things like what biofilms are, the role of biofilms (biofilm producers) concerning cheesemaking, and some observations found on their challenges in production, this assertion is appreciated as it is clear that a holistic view reveals that biofilm-assisted cheesemaking has gone from fashion trends to enablers with deep-seated consequences for industry, respectively. We explored some applications and roles of biofilms in cheesemaking that demonstrate their importance in the synthesis of flavor compounds as well as their protective function against pathogens. We have also explored the technological advancements underpinning biofilm production and control progressions, illustrating how state-of-the-art monitoring technologies, bioreactors, and genetic manipulation can be harnessed to improve cheese manufacture. In this way, our dialogue points to consumer taste and market trends related to biofilm-assisted cheeses being in increasing demand when it concerns the artisanal premium quality output. Cheese producers that align with these preferences and educate the consumer by being transparent about biofilm-assisted methods can adapt to maturing market expectations, and create stronger points of differentiation in an increasingly competitive landscape. Biofilm promoted cheesemaking is an emerging technology, but despite the obvious advantages of this methodology, challenges exist such as standardization, regulatory compliance, and consumer acceptance. Solving these challenges will necessitate collective efforts from the industry, researchers, and regulators to realize safe and sustainable biofilm-assisted approaches. The journey into creating cheese with biofilm is a concocted story of ingenuity, exploration, and prospect. Through the manipulation of biofilms, cheese producers can transform their products in new ways flavor or texture at one end of a molecular spectrum and on the other, food safety for the pleasure they will engender among consumers, while this brings about an ongoing advance where cheesemaking is concerned.

References

Afshari R, Pillidge CJ, Dias DA, Osborn AM, Gill H (2020) Cheesomics: the future pathway to understanding cheese flavour and quality. Crit Rev Food Sci Nutr 60(1):33–47

Akoğlu A (2020) The effect of some environmental conditions on planktonic growth and biofilm formation by some lactic acid bacteria isolated from a local cheese in Turkey. Biotechnol Lett 42:481–492

Aljohani AB, Al-Hejin AM, Shori AB (2023) Bacteriocins as promising antimicrobial peptides, definition, classification, and their potential applications in cheeses. Food Sci Technol 43:e118021

Allam MG, Darwish AM, Ayad EH, Shokery ES, Darwish SM (2017) Lactococcus species for conventional Karish cheese conservation. LWT-Food Sci Technol 79:625–631

Alvarez-Ordóñez A, Coughlan LM, Briandet R, Cotter PD (2019) Biofilms in food processing environments: challenges and opportunities. Annu Rev Food Sci Technol 10:173–195

Alzobaay AHH, Alzobaay WTS (2018) Role of Debaryomyces hansenii yeast in improving the microbial and sensory properties of Monterey cheese. AJVS 11(1):8–14

Amin A, Nazem A, Hassan E (2016) Problems of biofilms in processed cheese factory. Suez Canal Veterin Med J 21(2):97–107

Bai X, Nakatsu CH, Bhunia AK (2021) Bacterial biofilms and their implications in pathogenesis and food safety. Food Secur 10(9):2117

Ben-Harb S, Irlinger F, Saint-Eve A, Panouillé M, Souchon I, Bonnarme P (2020) Versatility of microbial consortia and sensory properties induced by the composition of different milk and pea protein-based gels. LWT 118:108720

Bertuzzi AS, McSweeney PL, Rea MC, Kilcawley KN (2018) Detection of volatile compounds of cheese and their contribution to the flavor profile of surface-ripened cheese. Compr Rev Food Sci Food Saf 17(2):371–390

Bettera L, Alinovi M, Mondinelli R, Mucchetti G (2020) Ripening of nostrano valtrompia PDO cheese in different storage conditions: influence on chemical, physical and sensory properties. Food Secur 9(8):1101

Bhosale S, Desale R, Fulpagare Y (2020) Biofilm: an overview with respect to dairy industry. Int J Curr Microbiol App Sci 9(10):150–160

Bintsis T (2018a) Lactic acid bacteria as starter cultures: an update in their metabolism and genetics. AIMS Microbiol 4(4):665

Bintsis T (2018b) Lactic acid bacteria: their applications in foods. J Bacteriol Mycol 6(2):89–94

Bisig W (2017) The importance of salt in the manufacturing and ripening of cheese. IDF Factsheet 1:2017–2003

Bulut-Solak B, Akin N (2019) Impact of cooking pH values on the textural and chemical properties for processed cheeses with/without the use of traditional village cheese during storage. Food Sci Anim Resour 39(4):541

Bunyard E (2021) The art of cheese. In: Delicious: the evolution of flavor and how it made us human. Princeton University Press, p 182

Caron T, Le Piver M, Péron A-C, Lieben P, Lavigne R, Brunel S et al (2021) Strong effect of Penicillium roqueforti populations on volatile and metabolic compounds responsible for aromas, flavor and texture in blue cheeses. Int J Food Microbiol 354:109174

Carpino S, Randazzo CL, Pino A, Russo N, Rapisarda T, Belvedere G, Caggia C (2017) Influence of PDO Ragusano cheese biofilm microbiota on flavour compounds formation. Food Microbiol 61:126–135

Centeno JA, Carballo J (2023) Current advances in cheese microbiology. Food Secur 12(13):2577

Chourasia R, Abedin MM, Chiring Phukon L, Sahoo D, Singh SP, Rai AK (2021) Biotechnological approaches for the production of designer cheese with improved functionality. Compr Rev Food Sci Food Saf 20(1):960–979

Corrieu G, Perret B, Kakouri A, Pappas D, Samelis J (2018) Positive effects of sequential air ventilation on cooked hard Graviera cheese ripening in an industrial ripening room. J Food Eng 222:162–168

Cui X, Shi Y, Gu S, Yan X, Chen H, Ge J (2018) Antibacterial and antibiofilm activity of lactic acid bacteria isolated from traditional artisanal milk cheese from Northeast China against enteropathogenic bacteria. Probiot Antimicrob Prot 10:601–610

D'Incecco P, Limbo S, Hogenboom J, Rosi V, Gobbi S, Pellegrino L (2020) Impact of extending hard-cheese ripening: a multiparameter characterization of Parmigiano reggiano cheese ripened up to 50 months. Food Secur 9(3):268

Danezis GP, Tsagkaris AS, Camin F, Brusic V, Georgiou CA (2016) Food authentication: techniques, trends & emerging approaches. TrAC Trends Anal Chem 85:123–132

de Castro MR, da Silva Fernandes M, Kabuki DY, Kuaye AY (2017) Biofilm formation on stainless steel as a function of time and temperature and control through sanitizers. Int Dairy J 68:9–16

Fagerlund A, Langsrud S, Møretrø T (2021) Microbial diversity and ecology of biofilms in food industry environments associated with Listeria monocytogenes persistence. Curr Opin Food Sci 37:171–178

Faille C, Cunault C, Dubois T, Benezech T (2018) Hygienic design of food processing lines to mitigate the risk of bacterial food contamination with respect to environmental concerns. Innovative Food Sci Emerg Technol 46:65–73

Fox PF, Guinee TP, Cogan TM, McSweeney PL, Fox PF, Guinee TP et al (2017a) Overview of cheese manufacture. In: Fundamentals of cheese science. Springer, pp 11–25

Fox PF, Guinee TP, Cogan TM, McSweeney PL, Fox PF, Guinee TP et al (2017b) Salting of cheese curd. In: Fundamentals of cheese science. Springer, pp 251–277

Fox PF, Guinee TP, Cogan TM, McSweeney PL, Fox PF, Guinee TP et al (2017c) Starter cultures. In: Fundamentals of cheese science. Springer, pp 121–183

Fröhlich-Wyder MT, Arias-Roth E, Jakob E (2019) Cheese yeasts. Yeast 36(3):129–141

Galli BD, Martin JGP, da Silva PPM, Porto E, Spoto MHF (2016) Sensory quality of Camembert-type cheese: relationship between starter cultures and ripening molds. Int J Food Microbiol 234:71–75

García-Cano I, Rocha-Mendoza D, Ortega-Anaya J, Wang K, Kosmerl E, Jiménez-Flores R (2019) Lactic acid bacteria isolated from dairy products as potential producers of lipolytic, proteolytic and antibacterial proteins. Appl Microbiol Biotechnol 103:5243–5257

Gebreyowhans S, Zhang S, Pang X, Yang B, Wang T, Wu Z et al (2020) Changes in texture, composition and sensory characteristics of Camembert cheese made from a mixture of goat milk and cow milk during ripening. Int J Dairy Technol 73(3):604–615

Germec M, Demirci A, Turhan I (2020) Biofilm reactors for value-added products production: an in-depth review. Biocatal Agric Biotechnol 27:101662

Ghosh S, Sarkar T, Chakraborty R (2021) Formation and development of biofilm-an alarming concern in food safety perspectives. Biocatal Agric Biotechnol 38:102210

Ghosh S, Nag M, Lahiri D, Sarkar T, Pati S, Kari ZA et al (2022) Engineered biofilm: innovative nextgen strategy for quality enhancement of fermented foods. Front Nutr 9:808630

Giaouris EE, Simões MV (2018) Pathogenic biofilm formation in the food industry and alternative control strategies. In: Foodborne diseases. Elsevier, pp 309–377

Gilmore BF (2023) Microbial biofilms: consequences for health. In: Hugo and Russell's pharmaceutical microbiology, 8, 8.2

Gonçalves de Melo A (2020) Bacteriophages of Brevibacterium aurantiacum: diversity, host interactions, and impact in washed rind cheeses

González-Rivas F, Ripolles-Avila C, Fontecha-Umaña F, Ríos-Castillo AG, Rodríguez-Jerez JJ (2018) Biofilms in the spotlight: detection, quantification, and removal methods. Compr Rev Food Sci Food Saf 17(5):1261–1276

Gray JA, Chandry PS, Kaur M, Kocharunchitt C, Bowman JP, Fox EM (2018) Novel biocontrol methods for Listeria monocytogenes biofilms in food production facilities. Front Microbiol 9:344705

Gutiérrez D, Rodríguez-Rubio L, Martínez B, Rodríguez A, García P (2016) Bacteriophages as weapons against bacterial biofilms in the food industry. Front Microbiol 7:189912

Hashempour-Baltork F, Farshi P (2022) Enzymes in flavor development and food preservation. In: Novel food grade enzymes: applications in food processing and preservation industries. Springer, pp 317–356

Hu M-X, Li J-N, Guo Q, Zhu Y-Q, Niu H-M (2019) Probiotics biofilm-integrated electrospun nanofiber membranes: a new starter culture for fermented milk production. J Agric Food Chem 67(11):3198–3208

Hu Y, Zhang L, Wen R, Chen Q, Kong B (2022) Role of lactic acid bacteria in flavor development in traditional Chinese fermented foods: a review. Crit Rev Food Sci Nutr 62(10):2741–2755

İpek D, Zorba NND (2018) Microbial load of white cheese process lines after CIP and COP: a case study in Turkey. LWT 90:505–512

Jahromi FA (2012) Microbial properties of traditional cheese (jar cheese) during ripening. Adv Environ Biol 6:2456–2460

Johnson M (2017) A 100-year review: cheese production and quality. J Dairy Sci 100(12):9952–9965

Kamilaris A, Kartakoullis A, Prenafeta-Boldú FX (2017) A review on the practice of big data analysis in agriculture. Comput Electron Agric 143:23–37

Khanal BKS, Pradhan M, Bansal N (2019) Cheese: importance and introduction to basic technologies. J Food Sci Technol Nepal 11:14–24
Khattab AR, Guirguis HA, Tawfik SM, Farag MA (2019) Cheese ripening: a review on modern technologies towards flavor enhancement, process acceleration and improved quality assessment. Trends Food Sci Technol 88:343–360
Kieliszek M, Pobiega K, Piwowarek K, Kot AM (2021) Characteristics of the proteolytic enzymes produced by lactic acid bacteria. Molecules 26(7):1858
Kızanlık PK (2018) Microbiological quality evaluation of various types of cheese. Erciyes Üniversitesi Veteriner Fakültesi Dergisi 15(2):86–93
Kukhtyn M, Berhilevych O, Kravcheniuk K, Shynkaruk O, Horiuk Y, Semaniuk N (2017) Formation of biofilms on dairy equipment and the influence of disinfectants on them. Восточно-Европейский журнал передовых технологий 5(11):26–33
Kumar M, Tierney J, Wilkinson M (2021) Enzymatic disruption of biofilms during cheese manufacturing: a mini review. Front Microbiol 12:791061
Leclercq-Perlat M-N, Saint-Eve A, Le Jan E, Raynaud S, Morge S, Lefrileux Y, Picque D (2019) Physicochemical and sensory evolutions of the lactic goat cheese Picodon in relation to temperature and relative humidity used throughout ripening. J Dairy Sci 102(6):5713–5725
Lee JC, Daraba A, Voidarou C, Rozos G, Enshasy HAE, Varzakas T (2021) Implementation of food safety management systems along with other management tools (HAZOP, FMEA, Ishikawa, Pareto). The case study of Listeria monocytogenes and correlation with microbiological criteria. Food Secur 10(9):2169
Leggieri MC, Decontardi S, Battilani P (2018) Modelling the sporulation of some fungi associated with cheese, at different temperature and water activity regimes. Int J Food Microbiol 278:52–60
Li J, Huang Q, Zheng X, Ge Z, Lin K, Zhang D et al (2020a) Investigation of the lactic acid bacteria in kazak cheese and their contributions to cheese fermentation. Front Microbiol 11:228
Li Y, Zhang X, Yang JJ, Ma XY, Jia XD, Du P, Li AL (2020b) Influence of the addition of Geotrichum candidum on the microbial, chemical, textural, and sensory features of soft soy cheese. J Food Proc Preserv 44(11):e14823
Licitra G, Caccamo M, Valence F, Lortal S (2017) Traditional wooden equipment used for cheesemaking and their effect on quality. In: Global cheesemaking technology: cheese quality and characteristics. Wiley, pp 157–172
Machado RA, Cutter CN (2018) Training hard-to-reach Pennsylvanian cheesemakers about food safety, using a low-tech training tool. Food Prot Trends 38(4):266
Mandenius C-F (2016) Bioreactors: design, operation and novel applications. Wiley
Marriott NG, Schilling MW, Gravani RB, Marriott NG, Schilling MW, Gravani RB (2018) Food contamination sources. In: Principles of food sanitation, pp 83–91
Mayo B, Rodríguez J, Vázquez L, Flórez AB (2021) Microbial interactions within the cheese ecosystem and their application to improve quality and safety. Food Secur 10(3):602
Mohamed SHS, Altowaijri SA (2023) Prolong shelf life of Kareish cheese: a review. Egypt J Food Sci 51:151
Morin-Sardin S, Rigalma K, Coroller L, Jany J-L, Coton E (2016) Effect of temperature, pH, and water activity on Mucor spp. growth on synthetic medium, cheese analog and cheese. Food Microbiol 56:69–79
Nahar S, Mizan MFR, Ha AJW, Ha SD (2018) Advances and future prospects of enzyme-based biofilm prevention approaches in the food industry. Compr Rev Food Sci Food Saf 17(6):1484–1502
Nahusenay H, Tola A, Sisay Tessema T, Vipham J, Woldegiorgis AZ (2023) Seasonal comparison of microbial hygiene indicators in raw and pasteurized Milk and cottage cheese collected across dairy value chain in three regions of Ethiopia. Food Secur 12(24):4377
Nasr N, El-Gizawy S, Zahra M (2017) Application of quality assurance programs in small dairy plants. J Nutr Health Food Eng 7(1):214–220
Ngo TP, Bui HL, Pham TBN (2023) Identifying biofilm forming bacteria in cow milk in Mekong Delta, Viet Nam. CTU J Innov Sustain Dev 15(1):84–90

Ouyang H, Li B, McCarthy M, Miao S, Kilcawley K, Fenelon M et al (2021) Understanding preferences for and consumer behavior toward cheese among a cohort of young, educated, internationally mobile Chinese consumers. J Dairy Sci 104(12):12415–12426
Pal M, Mulu S, Tekle M, Pintoo SV, Prajapati J (2016) Bacterial contamination of dairy products. Bever Food World 43(9):40–43
Papademas P, Bintsis T (2017) Global cheesemaking technology: cheese quality and characteristics. Wiley
Pappa EC, Kondyli E, Samelis J (2019) Microbiological and biochemical characteristics of Kashkaval cheese produced using pasteurised or raw milk. Int Dairy J 89:60–67
Parente E, Cogan TM, Powell IB (2017) Starter cultures: general aspects. In: Cheese. Elsevier, pp 201–226
Patlan-Velázquez L-F, González-Olivares L-G, García-Garibay M, Alatorre-Santamaría S, Gómez-Ruiz L, Rodríguez-Serrano G, Cruz-Guerrero A (2024) Effect of biogenic exopolysaccharides in characteristics and stability of a novel Requeson-type cheese. Food Biosci 59:103896
Paxson H (2016) Craftsmanship and quality in artisanal cheesemaking. In: Food and architecture: at the table. Bloomsbury Academic, p 191
Pujato S, Quiberoni A, Mercanti DJ (2019) Bacteriophages on dairy foods. J Appl Microbiol 126(1):14–30
Quintieri L, Caputo L, Brasca M, Fanelli F (2021) Recent advances in the mechanisms and regulation of QS in dairy spoilage by Pseudomonas spp. Food Secur 10(12):3088
Ramesh M (2020) Pasteurization and food preservation. In: Handbook of food preservation. CRC Press, pp 599–608
Ramírez-Rivera EDJ, Ramón-Canul LG, Díaz-Rivera P, Juárez-Barrientos JM, Herman-Lara E, Prinyawiwatkul W, Herrera-Corredor JA (2017) Sensory profiles of artisan goat cheeses as influenced by the cultural context and the type of panel. Int J Food Sci Technol 52(8):1789–1800
Rossi C, Chaves-López C, Serio A, Goffredo E, Goga BTC, Paparella A (2016) Influence of incubation conditions on biofilm formation by Pseudomonas fluorescens isolated from dairy products and dairy manufacturing plants. Italian J Food Safe 5(3):5793
Ruggirello M, Giordano M, Bertolino M, Ferrocino I, Cocolin L, Dolci P (2018) Study of Lactococcus lactis during advanced ripening stages of model cheeses characterized by GC-MS. Food Microbiol 74:132–142
Sachan RSK, Karnwal A (2022) Advancement in cheese production technology. In: Advances in dairy microbial products. Elsevier, pp 191–208
Salum P, Govce G, Kendirci P, Bas D, Erbay Z (2018) Composition, proteolysis, lipolysis, volatile compound profile and sensory characteristics of ripened white cheeses manufactured in different geographical regions of Turkey. Int Dairy J 87:26–36
Seale B, Bremer P, Flint S, Brooks J, Palmer J (2015) Overview of the problems resulting from biofilm contamination in the dairy industry. In: Biofilms in the dairy industry. Wiley, pp 49–64
Sevinc-Demircan B, Ozturkoglu-Budak S (2023) Use of yeast isolates of cheese origin as adjunct culture in Beyaz cheese: influence on sensorial, textural and quality characteristics. J Food Sci Technol 60(10):2670–2680
Short EC, Kinchla AJ, Nolden AA (2021) Plant-based cheeses: a systematic review of sensory evaluation studies and strategies to increase consumer acceptance. Food Secur 10(4):725
Shuvo SD (2018) Assessing food safety and associated food hygiene and sanitary practices in food industries: a cross-sectional study on biscuit industry of Jessore, Bangladesh. Nutr Food Sci 48(1):111–124
Skåra T, Rosnes JT (2016) Emerging methods and principles in food contact surface decontamination/prevention. In: Innovation and future trends in food manufacturing and supply chain technologies. Woodhead Publishing, pp 151–172
Son H, Park S, Beuchat LR, Kim H, Ryu J-H (2016) Inhibition of Staphylococcus aureus by antimicrobial biofilms formed by competitive exclusion microorganisms on stainless steel. Int J Food Microbiol 238:165–171
Soodam K, Ong L, Powell IB, Kentish SE, Gras SL (2017) Effect of elevated temperature on the microstructure of full fat Cheddar cheese during ripening. Food Struct 14:8–16

Speranza B, Corbo MR (2017) The impact of biofilms on food spoilage. In: The microbiological quality of food. Elsevier, pp 259–282

Suchánek P, Králová M (2019) Customer satisfaction, loyalty, knowledge and competitiveness in the food industry. Econ Res-Ekonomska istraživanja 32(1):1237–1255

Sun L, D'Amico DJ (2021) Composition, succession, and source tracking of microbial communities throughout the traditional production of a farmstead cheese. Msystems 6(5):00830-00821. https://doi.org/10.1128/msystems

Talavera M, Chambers DH (2016) Flavor lexicon and characteristics of artisan goat cheese from the United States. J Sens Stud 31(6):492–506

Thierry A, Collins YF, Mukdsi MA, McSweeney PL, Wilkinson MG, Spinnler HE (2017) Lipolysis and metabolism of fatty acids in cheese. In: Cheese. Elsevier, pp 423–444

Thomareis AS, Chatziantoniou SE (2022) Quality and testing of processed cheese: defects, QA, and QC. In: Processed cheese science and technology. Elsevier, pp 349–426

Tiwari ON, Sasmal S, Kataria AK, Devi I (2020) Application of microbial extracellular carbohydrate polymeric substances in food and allied industries. 3 Biotech 10:1–17

Ünal Turhan E, Erginkaya Z, Korukluoğlu M, Konuray G (2019) Beneficial biofilm applications in food and agricultural industry. In: Health and safety aspects of food processing technologies, pp 445–469

Vågsholm I, Arzoomand NS, Boqvist S (2020) Food security, safety, and sustainability—getting the trade-offs right. Front Sustain Food Syst 4:16

van Assen MF (2021) Training, employee involvement and continuous improvement–the moderating effect of a common improvement method. Prod Plan Control 32(2):132–144

van Mastrigt O, Gallegos Tejeda D, Kristensen MN, Abee T, Smid EJ (2018) Aroma formation during cheese ripening is best resembled by Lactococcus lactis retentostat cultures. Microb Cell Factories 17:1–8

Vishwakarma V (2020) Impact of environmental biofilms: industrial components and its remediation. J Basic Microbiol 60(3):198–206

Walsh EA, Diako C, Smith DM, Ross CF (2020) Influence of storage time and elevated ripening temperature on the chemical and sensory properties of white Cheddar cheese. J Food Sci 85(2):268–278

Wang D, Zhao X, Wang H (2023) Recent advances of bacteriophage-derived strategies for biofilm control in the food industry. Food Biosci 54:102819

Wirtanen G (2018) Biofilm risks and prevention in food and biotech processing-a review. Eur J Parent Pharm Sci 23(3):99

Wirtanen G, Salo S (2016) Biofilm risks. In: Handbook of hygiene control in the food industry. Elsevier, pp 55–79

Yalçin SK, Ozbas ZY (2017) Yeasts from traditional cheeses for potential applications. In: Global food security and wellness. Springer Science+Business Media, New York, pp 277–293

Yangılar F, Oğuzhan Yıldız P (2016) Casein/natamycin edible films efficiency for controlling mould growth and on microbiological, chemical and sensory properties during the ripening of Kashar cheese. J Sci Food Agric 96(7):2328–2336

Yao S, Hao L, Zhou R, Jin Y, Huang J, Wu C (2022) Multispecies biofilms in fermentation: biofilm formation, microbial interactions, and communication. Compr Rev Food Sci Food Saf 21(4):3346–3375

Yap M, O'Sullivan O, O'Toole PW, Sheehan JJ, Fenelon MA, Cotter PD (2024) Seasonal and geographical impact on the Irish raw milk microbiota correlates with chemical composition and climatic variables. Msystems 9(4):e01290–e01223

Yoplac I, Hidalgo A, Vargas L (2022) Antimicrobial biofilms with microencapsulated citral and sodium caseinate to extend the shelf life of fresh cheese. Food Packag Shelf Life 34:100932

Youn H-Y, Seo K-H (2022) Isolation and characterization of halophilic Kocuria salsicia strains from cheese brine. Food Sci Anim Resour 42(2):252

Zhao CJ, Schieber A, Gänzle MG (2016) Formation of taste-active amino acids, amino acid derivatives and peptides in food fermentations–a review. Food Res Int 89:39–47

Zheng X, Shi X, Wang B (2021) A review on the general cheese processing technology, flavor biochemical pathways and the influence of yeasts in cheese. Front Microbiol 12:703284

Brewing Wonders: Biofilm in Beverages 12

Saikat Mazumder, Debasmita Bhattacharya,
Dibyajit Lahiri, and Moupriya Nag

Abstract

Brewery biofilms have the potential to harbour spoiling bacteria that might harm the finished product. The majority of bacteria that cause beer deterioration are believed to rely on several interactions with the surrounding microbiota. Microbial deterioration is a persistent risk in the beer brewing sector that needs to be watched over and managed to guarantee the final product's palatability. While many of the common bacteria that cause beer spoiling have been found and studied, little is known about how they become contaminated and survive. In retail environments, where there is much potential for microbial deterioration, many beers are supplied post-production as kegs connected to draft delivery systems. As a result, when beer draft lines are severely polluted, restaurants and bars may have to pay a large amount of money and lose time cleaning. Although spoilage monitoring in the retail beer business is sometimes disregarded, if invading bacteria significantly alter a beer's flavour profile, this might pose one of the biggest dangers to the beverage's profitability. The most prevalent bacterial cause of beer spoiling in the world is *Lactobacillus brevis*. It becomes apparent that there are contaminants throughout any aspect of the brewing process,

S. Mazumder
Department of Biotechnology, University of Engineering and Management,
Kolkata, West Bengal, India

Department of Food Technology, Guru Nanak Institute of Technology,
Kolkata, West Bengal, India

D. Bhattacharya
Department of Basic Science and Humanities, Institute of Engineering and Management,
Kolkata, West Bengal, India

D. Lahiri (✉) · M. Nag (✉)
Department of Biotechnology, University of Engineering and Management,
Kolkata, West Bengal, India
e-mail: dibyajit.lahiri@uem.edu.in; moupriya.nag@uem.edu.in

T. Sarkar et al. (eds.), *Biofilm Applications to Revolutionize Food Technology*,
https://doi.org/10.1007/978-3-031-85205-3_12

from storage to filtering to primary and secondary fermentation to packing. Nonetheless, the ability of *L. brevis* strains to deteriorate varies and they may grow in a wide variety of beers. Weak wort filtering performance is linked to many bacterial groupings. The formation of biofilm on the kernel surface layers during malting that contained extracellular polymeric substances was a significant component in wort separation. The study focuses on yeast-induced biofilm development, biofilm-forming microorganisms in breweries, and potential biofilm prevention in brewing operations.

Keywords

Biofilms · Cheese ripening · Microbial consortia · Food safety · Flavor development

12.1 Introduction

Due to its exceptional microbiological stability, beer is often considered harmless when it comes to foodborne diseases. In actuality, beer's high carbon dioxide concentration, low pH, low oxygen content, bitter hop chemicals (ethanol), traces of glucose, maltose, and maltotriose, and high carbon dioxide content all prevent infections and a broad spectrum of other bacteria from growing. Nevertheless, because of the wort's abundant nutrient environment and the extra growth stimuli created by brewing yeast, bacteria may become contaminated throughout the fermentation process by spoiling microorganisms (Maifreni et al. 2015; Sakamoto and Konings 2003). In beer, the effects of microbiological contamination are quite severe. The majority of beer is produced in closed systems using cleaning-in-place (CIP) techniques, which eliminate the need to disassemble equipment. However, due to their beneficial economics, quick cleaning schedules and lengthy cleaning intervals are common. As a result, these systems are vulnerable to the time-dependent process of surface attachment of bacteria and the subsequent generation of microbial biofilms (Maifreni et al. 2015).

Breweries use biofilms, which contain the majority of contaminating microorganisms. These biofilms serve as a breeding ground and reservoir for bacteria that are known to cause beer deterioration as well as germs that are safe to be present in bottled beer (Mohammadi and Saris 2022). Biofilms can create serious public health problems and/or financial losses when they are found on surfaces used in the production of food. Pathogenic bacteria that attack healthy people or, occasionally, just immunocompromised patients may be present in biofilms (Galié et al. 2018). There is mixed information about *Saccharomyces cerevisiae var. boulardii's* potential opportunistic pathogenicity, particularly in individuals with central venous catheters, despite the fact that the bacterium has been widely employed in medical treatments and gastrointestinal problems (Nikolaos et al. 2008). The multicellular formations known as yeast biofilms are intricate and diverse, providing excellent protection for the cells within against harsh environmental factors. Yeast forms biofilms by a series of sequential processes that involve simultaneous cell-to-surface and cell-to-cell contacts that lead to colonisation and adherence. Yeast cells attach themselves to the surface, begin to divide, and form microcolonies in the first stage. The yeast cells then continue to proliferate, producing an extracellular matrix (ECM) and differentiating to form hyphae and elongated pseudohyphae that can survive in low-nutrient settings. Eventually, throughout the maturation stage, there is an upsurge in ECM (Richard et al. 2005).

Biofilm development seldom involves the study of adhesion genes, which are involved in cell-to-cell or cell-biotic/abiotic interactions (Stringer and Keely 2001; Yang et al. 2018). FLO1, FLO9, FLO5, FLO11, and FLO10 are a group of FLO genes found in *Saccharomyces cerevisiae* that encode different sticky surface glycoproteins (Teunissen and Steensma 1995). FLO11 confers the development of biofilms, pseudohyphae, and cell-surface adhesion (flo11p) (Verstrepen and Klis 2006; Zara et al. 2005). In reaction to environmental stimuli, the expression of FLO11 is regulated by three signalling cascades: glucose repression, the Ras/cAMP route, and the mitogen-activated protein kinase (MAPK) pathway. Stress-related variables and/or environmental stimuli can activate the expression of the aforementioned genes and cell-surface adherence (Verstrepen and Klis 2006). Although numerous genes have been identified as being important in the development of biofilms, yeast species exhibit a variety of adhesins (Douglas et al. 2007).

Biofilm Formation: The Causes

Theprimary drivers for biofilm growth (Fig. 12.1) and the ensuing tainting of beer are frequently due to inadequate equipment cleaning and disinfection, especially in the harder-to-clean locations such as joints, curves, edges, and dead ends in pipes. Notably, microbial biofilms exhibit greater resistance to antimicrobial agents than do planktonic cells. Consequently, eliminating biofilms from food processing operations is a significant problem (Gilbert et al. 2002). Microbial biofilms have the potential to contaminate beer both during the production process and the bottling process, which is thought to be the primary cause of microbial issues in beer production. Along with the air and raw ingredients, equipment during manufacturing frequently harbours germs that cause beer to degrade, including lactic acid-producing bacteria, wild yeast, and anaerobic bacteria. These microbes multiply when wastes are present, polluting the entire production process. They can thrive in hard-to-reach places such as seals, joints, and valves. Other crucial areas include heat exchangers (because of their difficult-to-clean structure), fermentation, and maturation tanks (due to the presence of unfermented sugar and the initial lack of ethanol) (Timke et al. 2005b). When bacteria that produce biofilm are allowed to proliferate in beer, it can lead to unfavourableness or opacity in the finished product due to the formation of metabolites and sediment. This is why biofilm creation in the brewing environment is problematic. Very little is known about the makeup of brewery biofilms and the capacity of biofilm isolates of bacterial species to establish themselves in beer despite the fact that a substantial quantity of investigation has

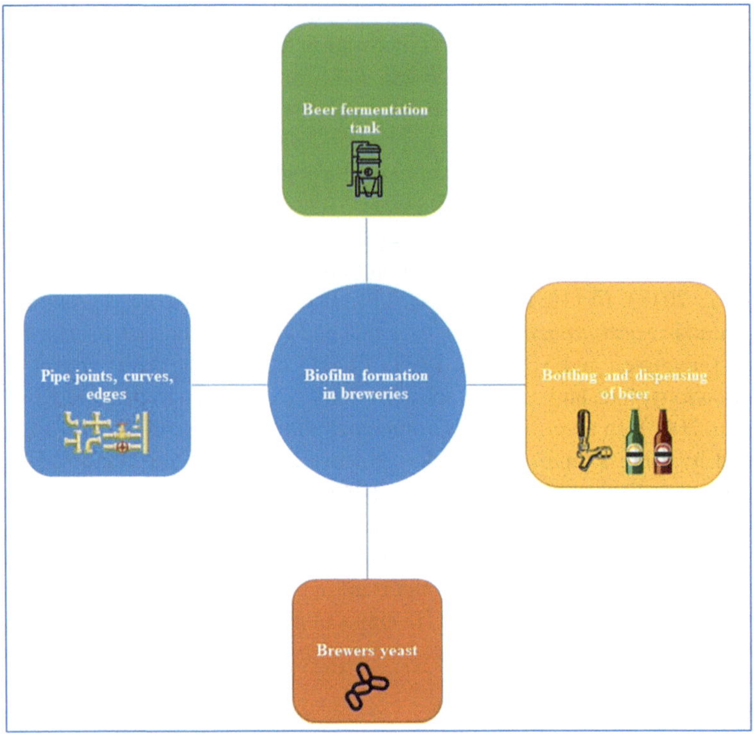

Fig. 12.1 Development of biofilm in breweries

been devoted to the detection of bacteria that spoil beer. The two main categories of contaminations found in a brewery are secondary contaminations that come from canning, bottling, or kegging, and primary contaminations that come from the yeast, wort, the process of fermentation, maturation, or pressure tanks. In breweries without tunnel pasteurisation, secondary contaminations account for at least half of the cases of microbiological deterioration. All areas that come into touch, whether directly or indirectly, with cleaned or filled, unopened bottles might therefore be contaminated. Microbiological issues seldom occur overnight; instead, they are the consequence of microbial populations gradually accumulating at crucial locations (Storgårds and Priha 2009).

12.2 Beer Firm and Its Biofilm: A Conversation

Under fermentation, which occurs in storage tanks, valves, pumps, and fillers, biofilm formation in the brewing industry is started (Heller 1995). *Leuconostoc oenos* and other *Lactobacillus* and *Pediococcus* species thrive in the wine, which causes a bad odour to develop. By attaching sterile stainless steel coupons to various filler key points, biofilm development in beer bottling operations was investigated. It was shown that surfaces that were horizontal were more likely to accumulate microorganisms (Storgårds et al. 2006b). Researchers investigated the elimination of microbial biofilms from the dispensing apparatus using an enzyme mixture comprising enzymes such as pectinase, β-amylase, β-glucuronidase, glucose oxidase, dextranase, and protease. It was determined by temperature and duration, and the complexity of treating each biofilm was evaluated (Vishwakarma 2020). There exist numerous methods to impede the growth of bacteria, including the application of sulphur dioxide, filtration (such as physical removal), dimethyl carbonate, lysozyme and bacteriocins (cell lysis), ultrahigh-pressure (cytoplasmic membrane damage), ultrasound techniques (localised heating and thinning of cell membrane), UV radiation (damages DNA), and pulsed electric field (the decomposition of cell membranes) (Bartowsky 2009).

The majority of microorganisms find beer to be a hostile habitat; therefore, only a small number of species may genuinely cause spoiling by developing in beer. These mostly consist of yeasts from the *Saccharomyces* or *Dekkera* genera and bacteria from the *Lactobacillus*, *Pediococcus*, *Pectinatus*, or *Megasphaera* genera. Acidic environments are preferred or tolerated by the microaerophiles, also known as anaerobes, that cause beer deterioration. Thought to benefit from the many interactions within complex communities, it is considered that the beer-spoilage microbes themselves are not the major colonisers in biofilms. By lowering the oxygen content and creating an acidic environment, slime-producing acetic acid-producing bacteria are thought to be the main colonisers in brewery biofilms, where they contribute to creating ideal circumstances for organisms that cause beer deterioration (Back 2003). Additionally, a variety of other gram-negative bacteria, including *Pseudomonas* species and members of the *Enterobacteriaceae* family, are commonly found in breweries. The most severely polluted areas in the brewery filling area are the track systems next to the fillers and the can and bottle warmers. Moreover, biofilms have been discovered on side rails, worn strips, drop plates, struts connecting the chains, sides of conveyor carriages, and the spaces between and beneath chain links. *Pseudomonas*, *Enterobacter*, *Klebsiella*, *Alcaligenes*, *Flavobacterium*, *Lactobacillus*, *Bacillus*, and *Arthrobacter* were the most common genera of bacteria found in biofilms connected to conveyor tracks, bottle and can warmers, and other related surfaces. It has also been found that moulds and yeasts belonging to the genera *Rhodotorula*, *Trichosporon*, *Cladosporium*, *Penicillium*, *Geotrichum*, *Trichoderma*, *Mucor*, *Aureobasidium*, and *Paecilomyces* were present (Banner 1994). Moreover, biofilms in beer distribution system lines have been seen time and again (Casson 1985; Harper 1981). It has been discovered that brewing and wild yeasts, acetic acid bacteria, and *Pediococcus* species stick to dispensing lines (Storgårds et al. 2006b) A research involving the cultivation/isolation strategy and culture-independent approaches to analyse the microbial makeup of brewery biofilms was published. According to the findings, within hours of the commencement of manufacturing, main colonisers gathered on fresh stainless steel surfaces. Daily routine cleaning only temporarily decreased the amount of bacteria on the surfaces. Comparatively speaking to bottling machines, canning machines showed a much lower propensity for microbial buildup. Initially, the surfaces were

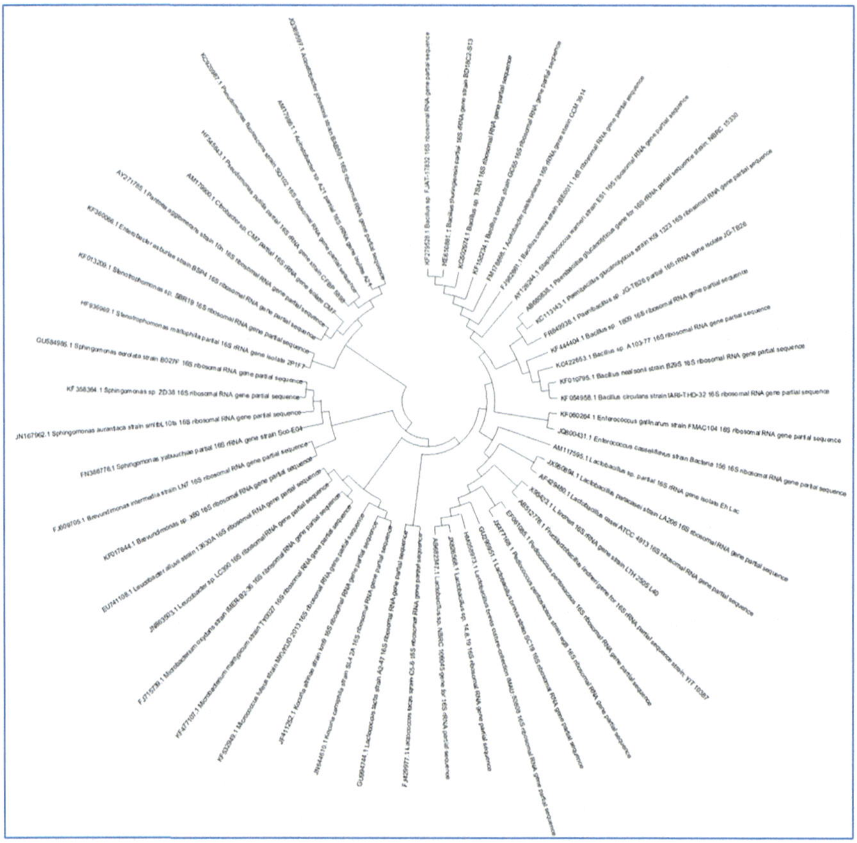

Fig. 12.2 Maximum likelihood tree of the microorganisms linked to the microbrewery producing biofilms

colonised by yeasts, moulds, and gram-negative bacteria. Sweeteners and sugars boosted the attachment of main colonisers to stainless steel. It is best to avoid using horizontal surfaces in construction since they might lead to the buildup of microorganisms (Storgårds et al. 2006b; Timke et al. 2004).

Different microorganisms involved in biofilm development are shown by another investigation on different kinds of bacteria isolated from brewing surfaces. These genera of microorganisms have the highest likelihood of forming biofilms, as shown in Fig. 12.2 (Maifreni et al. 2015).

Microbes Involved in the Brewing Process

The majority of microorganisms find beer to be a hostile habitat; therefore, only a small number of species may genuinely cause spoiling by developing in beer

(Storgårds et al. 2006a). These mostly consist of yeasts from the *Saccharomyces* or *Dekkera* genera and bacteria from the *Lactobacillus*, *Pediococcus*, *Pectinatus*, or *Megas phaera* genera. The microaerophiles, or anaerobes, that cause beer spoiling either prefer or can withstand acidic environments. Beer is impervious to microbial contamination due to the inhibitors it contains, including alcohol, carbon dioxide, sulphur dioxide, hop substances, low pH, and a lack of nutrients and oxygen. Additionally, methods such as pasteurisation, low-temperature storage, and filtering lessen infection. Conversely, species acclimated to the brewery ecosystem have frequently not been isolated elsewhere even though the unique environment of the brewing process limits the variety of bacteria likely to cause spoiling to relatively few species (Haikara and Helander 2006). These organisms may persist indefinitely in process niches, most likely away from the initial product stream, without posing a risk of contamination. Subsequently, they could abruptly contaminate the entire procedure due to technical malfunctions or inadequate cleaning.

As new techniques are introduced into the brewing industry, there is a steady description of new species that have become accustomed to the brewery environment. *Lactobacillus paracollinoides* (Suzuki et al. 2004), *Pediococcus claussenii* (Dobson et al. 2002), *Pectinatus haikarae*, *Megasphaera paucivorans*, and *Megasphaera sueciensis* (Juvonen and Suihko 2006) are the 'newcomers' in the beer spoiling bacterial community. Despite this, *Lactobacillus brevis* remains the most prevalent bacteria that causes beer to deteriorate (Ashtavinayak and Elizabeth 2003). Even though they cannot thrive in beer, a variety of gram-negative bacteria, mostly those belonging to the *Enterobacteriaceae* family and *Pseudomonas* species, are also commonly found at breweries. If enough of them are present, some of them may proliferate early in the process and produce off-flavours in the finished product. Standard beer products have not been documented to harbour pathogens despite the possibility that the pollutants present might result in quality flaws (Righelato 2004).

Biofilms in the Bottling of Beer

Complex machinery with lots of space for biofilm growth makes up beer bottling operations. The bottling department is responsible for the majority of microbiological spoiling issues that occur in breweries. Utilising jets that cause the bottle to overfoam in order to lower the oxygen level in the container headspace, nutrients are continuously supplied to the site. Because conveyors come into frequent touch with beer, microbes may be more likely to adapt to it. It is also believed that within the bottling factory, conveyors disperse bacteria. Biofilms have been discovered in the filling division of the brewery on a variety of surfaces, including conveyor systems, pasteurisers, points near fillers and can or bottle warmer environments, side rails, wear strips, both the interior and the exterior of conveyor carriages, drip plates, and struts connecting the chains, as well as the bottom of and in between chain links. It was discovered that after 2–12 h of the production beginning, primary biofilm colonisers accumulated on sterile surfaces made of stainless steel on the filler, and that regular, thorough cleaning only temporarily decreased the amount of bacteria on the surfaces. *Acinetobacter* species were responsible for up to 71% of all cells found in a research that examined first-attached biofilm organisms on disinfected brewery surfaces (Storgårds and Priha 2009). Additionally, tests revealed that *Acinetobacter* species stuck to stainless steel more securely and quickly than the majority of other microbes. This suggests that *Acinetobacter* species may serve as a foundation for other microorganisms to settle on and multiply in order to build a fully fledged biofilm. Furthermore, it has been suggested that four novel species of *Chryseobacterium* are responsible for the quick recolonisation of cleansed surfaces and the production of biofilms in beer bottling factories (Herzog et al. 2008). There were filling machine locations that had 100% biofilm cover in just 1 week, according to an 8-week follow-up investigation. Sites subjected to routine cleaning stress were found to have biofilms 1–2 mm high 7 months following the most recent thorough cleaning (Timke et al. 2005a).

It has been demonstrated that in brewery biofilms, bacteria predominate, with fewer than 3% of the total cell count belonging to the Eukarya and Archaea orders (Timke et al. 2005a). It is considered that the bacteria responsible for beer deterioration do not serve as the main colonisers in biofilms but rather get an advantage from the many interactions that occur within intricate communities. Since they lower oxygen levels and provide an acidic environment, slime-producing bacteria that produce acetic acid were thought to be the main colonisers in breweries' biofilms. However, further evidence refuted this theory (Storgårds et al. 2006b). While certain brewery biofilms do include acetic acid bacteria, they do not appear to be a particularly large population. Because they appear to prefer the identification of gammaproteobacteria, cultivation-based approaches provide only a partial view of the population (Timke et al. 2005a). As a result, *Pseudomonas* species and *Enterobacteriaceae* are frequently isolated from brewery surfaces (Timke et al. 2004).

Primary colonisers facilitate the incorporation of secondary species, such as microbes that damage products and create biofilm, as is frequently observed in practice. Beer-spoiling bacteria might therefore be present in brewery biofilms. It was discovered that bacteria capable of multiplying in beer with 4.8% ethanol were present in around 25% of brewery biofilms. Snacking bacteria were rarely discovered in microbial diversity investigations of brewery biofilms, which runs counter to this finding. Anaerobic beer spoilage bacteria such as *Pectinatus* spp. often make up a very small percentage of the brewery biofilm community despite the fact that the population's oxygen-requiring metabolism is creating anaerobic microenvironments. In severely contaminated areas, though, they may proliferate and endanger the beer's microbiological stability (Storgårds et al. 2006b; Timke et al. 2005a).

Despite the fact that bacteria outweigh yeasts in brewery biofilms, yeasts are often isolated from brewery locations. The most prevalent yeasts found in brewery specimens have been identified as belonging to *Debaryomyces* spp., *Saccharomyces* spp., and *Candida* spp. Beer was proven to support the growth of every strain of *Saccharomyces cerevisiae* that was isolated from brewery biofilms (Timke et al. 2008). The presence of yeast in the brewery biofilms, as reported by Timke et al.

(2005b), likely contributed to their propensity for spoiling. Among the yeast species, most frequently seen in brewery biofilms, *S. cerevisiae* colonises preexisting biofilm, increasing the likelihood that beer may deteriorate, whereas *Candida pelliculosa* (teleomorph *Pichia anomala*) is likely involved in adhesion and initial biofilm production in beer bottling workplaces. Despite rigorous cleaning procedures, brewery biofilms have been shown to contain a surprisingly wide variety of bacteria (Timke et al. 2008). Furthermore, these biofilms showed no signs of spatial concentration, and even in different locations inside the bottling facility, no conventional brewery biofilm community might have been seen. As an alternative, these communities appeared to be experiencing population shifts (Timke et al. 2005b). The biofilm population can therefore adapt to shifting cleaning and disinfection procedures because of the presence of a sizable gene pool. Since *Methylobacterium* species are frequently found in brewery biofilms, it is possible that their presence contributes to the biofilms' susceptibility to disinfection since they may break down formaldehyde-based sanitisers (Timke et al. 2005a). In light of current understanding, attempts to eliminate existing biofilms by thorough cleaning should be avoided in favour of preventing microbial attachment and biofilm growth.

Beer Dispenser Lines and Biofilms

Compared to the brewing operation or bottled beer, a larger variety of microorganisms might cause issues with beer-distributing equipment. This is mostly because the system's temperature regulation is less stringent and the dispensing tubes have larger oxygen concentrations. Additionally, the pipes may be rather oxygen permeable. Aerobic conditions are present at the keg tapping head and the distributing tap. There could be a non-cooled section of the dispensing lines near the tap if they are not always fully chilled.

The aforementioned circumstances facilitate the growth of aerobic wild yeast, modest quantities of coliforms, and acetic acid bacteria along with the oxygen-tolerant beer-spoiling organisms commonly seen in brewery environments. In investigations of beer dispensing platforms, a number of different genera of bacteria and yeasts have been isolated: *Kloeckera*, *Pediococcus*, *Rhodotorula*, *Zymomonas*, *Obesumbacterium*, *Debaryomyces*, *Brettanomyces/Dekkera*, *Pichia*, *Saccharomyces* (brewing and wild yeast strains), *Torulopsis*, *Lactobacillus* (including *L. brevis*), and *Gluconobacter*. During the installation of sterile, stainless steel coupons on key filler locations in three brewing facilities, microbial adhesion and biofilm growth in brewery bottling operations were investigated. In order to identify the main colonisers in the biofilm development process, microbiological samples were extracted from the coupons. Partial 16S rDNA sequencing, carbohydrate fermentation assays, and ribotyping were used to identify the bacterium (Storgårds and Priha 2009).

Epifluorescence microscopy was used to examine how sweets and sweeteners affected the main colonisers' ability to adhere to stainless steel. Using epifluorescence microscopy, the test coupons were gradually removed from each sample location over the course of 8 weeks in order to examine the biofilm production rate. The findings demonstrated that within hours of the commencement of manufacturing, primary colonisers collected on freshly plated stainless steel surfaces. The amount of bacteria on the surfaces was only slightly decreased by regular daily cleaning. Compared to bottling equipment, canning machines were noticeably less prone to the buildup of microbes. The first organisms to colonise the surfaces were moulds, yeasts, and gram-negative bacteria. Sugars and sweeteners boosted the primary colonisers' affinity for stainless steel. It is best to avoid using horizontal surfaces in buildings since they are prone to microbiological collection. Moreover, several surfaces still developed biofilms after regular washing and disinfection (Storgårds et al. 2006b).

During the process of brewing probiotic beer with *Saccharomyces cerevisiae var. boulardii*, researchers observed that the yeast may form biofilm in glass bottles when the bottles get cloudy. Probiotic beer was made in this research using *S. cerevisiae var. boulardii* CNCM I-745 as a starting culture. In order to record the creation and structure of the biofilm, we employed light and electron microscopy techniques to examine the biofilm characteristics in conjunction with the expression of FLO11 mRNA. The beer matured from a sugar-rich condition to a sugar-limited state, and the brewing process's stress factors (pH decrease and produced metabolites) increased the mass of the biofilm, according to the study's results. The viable count, however, stayed mostly stable at roughly 7.1 log10 cells/mL. Comparing biofilm *S. boulardii* cells to planktonic cells, the former demonstrated noticeably more FLO11 mRNA expression during the process of exponential growth and stationary stages. Thus, the results of this investigation demonstrate that throughout the ageing process of beer bottles, *S. cerevisiae var. boulardii* forms biofilm on glass surfaces. The relevance of this study is highlighted by the effects of issues resulting from biofilm formation on returnable bottles (Mohammadi and Saris 2022).

12.3 In Brewing Environments, Yeast Biofilm: Prevalence

It is astonishing how many substrates and hosts fungal species can stick to and grow on. These substrates and hosts include matrices of food, fuel lines, bare rocks, and human and plant tissues (Fanning and Mitchell 2012; Rola et al. 2016) The implications of fungi such as *Cryptococcus*, *Pneumocystis*, *Aspergillus*, and *Coccidioides* in human disease have sparked the majority of the information gathered on fungal biofilms (Fanning and Mitchell 2012). Similarly, the research on yeast biofilms has been prompted by health concerns associated with the growth of *Candida albicans* (Lohse et al. 2018). It has proven difficult to distinguish between yeasts as well as dimorphic filamentous fungi that frequently generate large amounts of growth that resemble yeast. Despite this potential ambiguity, yeasts such as *ascomycetes* or *basidiomycetes* have been described as single-cell microorganisms that are mostly known for their asexual reproduction by budding or fission as well as their sexual states not being contained within fruiting bodies (Kurtzman et al. 2011). Similar to other microbiological biofilms, the yeast-generated biofilm is a highly organised microbial community that is connected to a surface and surrounds itself with an extracellular matrix that is formed by the microbial cells. There have also been descriptions of beneficial yeast biofilms in the food domain, mostly related to *Saccharomyces cerevisiae*, including those needed for the ageing and maturity of speciality wines, fermented olives, and dairy products (Wu et al. 2017).

Genetic factors that influence the production of yeast biofilms: During the initial stage of biofilm formation, cells of yeast attach themselves to both surfaces that are biotic and abiotic (Bojsen et al. 2012). Adhesins are specialised adhesion molecules that facilitate cell attachment through hydrophobic or amyloid-like acquaintances (Lipke 2018). Several proteins have been identified as being involved in the production of biofilms in *S. cerevisiae*, including Hsp12p (Zara et al. 2002), Flo11p (Zara et al. 2005), and Ccw14p (Moreno-García et al. 2018). Flo11p has a direct role in the production of biofilms on semisolid or solid agar, plastic surfaces, and the contact between air and liquid (Zara et al. 2005). The number of sequences that repeat and the transcriptional levels of FLO11 have been linked to variations in the capacity of *S. cerevisiae* strains to generate biofilms (Zara et al. 2009). The MAPK cascade, the cAMP-PKA pathway, the TOR system, and the Cyc8p/Tup1p complex are just a few of the pathways that the FLO11 transcription activates in response to dietary and environmental cues (Van Nguyen et al. 2018; Vinod et al. 2008). Using whole-genome resequencing, 155 loci differ significantly among 110 *S. cerevisiae* isolates that can generate biofilm and those that cannot have been found (Coi et al. 2017). The MAP-kinases STE7, KDX1, and RGA2, as well as the key regulators cAMP and IRA1, are located at these loci. The MAPK and cAMP pathways, in addition to transcriptional regulators such as Bcr1p and Tec1p, are the main regulatory mechanisms in Candida that control the formation of biofilms (Fox et al. 2015; Gulati and Nobile 2016).

Yeast cells multiply, develop into filamentous forms (hyphae and pseudohyphae), and begin to assemble an extracellular matrix (ECM) as soon as they begin to stick to one another and a surface (Vopálenská et al. 2010). The ability of cells to build a network of the hyphae or pseudohyphae is demonstrated by their increased adherence to inert surfaces, such as plastic and stainless steel. Particularly, because pseudohyphal cells have a greater chitin content, there is a direct correlation between the pseudohyphal substance and biofilm strength and resilience (Paramonova et al. 2009). According to Faria-Oliveira et al. (2015) and Flemming and Wingender (2010), the components of ECM include typically carbohydrates, proteins, lipids, and nucleic acids. In mature *S. cerevisiae* biofilms, Flo11p is thought to be a component of the extracellular matrix (ECM) where it supports cellular movement through a process known as cellular lubrication (Karunanithi et al. 2010). Many benefits, including cohesiveness, adhesion, and mechanical qualities, as well as food supplies, enzymatic activity, and protection, are bestowed upon yeast cells by the extracellular matrix (ECM) (Flemming and Wingender 2010).

12.4 Regulating the Emergence of Biofilm in Breweries: A Methodical Approach

Bacterial Biofilms Linked to Barley

The significant metabolic potential of barley is exploited during the natural physiological process of malting. A varied microbiota that naturally lives in the outermost parts of grains is another active ingredient in the malting process, which also involves the germination of barley (Raulio et al. 2009; Tsitko et al. 2018). The environment surrounding malting is dynamic and constantly changing. For microbial activity, the most critical stages of the process are, in particular, the first couple of hours of soaking and kilning (Noots et al. 1999; Tsitko et al. 2018). The microbiota associated with barley engages in active interactions with grains, impacting the end product's nutritional value, safety, and organoleptic qualities (Laitila et al. 2007; van Nierop et al. 2004). It is commonly recognised that microorganisms can lead to variations in both process and product quality. Retarded wort filtering efficiency has been linked to the overgrowth of bacteria that produce exopolysaccharides and the production of biofilms during malting. Previous research found that even when regular, premium malting barley was utilised as the raw material, the performance of wort separation increased when gram-negative microbes were inhibited throughout the malting process.

The porosity of the bed of filters that forms during the lautering process may be impacted by microorganisms' secretion of flocculants onto the malt's surface, according to a 1993 Anderson report (Tsitko et al. 2018). According to Raulio et al. (2009), steeping induced the microorganisms that were latent to develop into biofilms. The ineffectiveness of wort filtering is linked to many bacterial groupings. The changes in the bacterial population during malting were demonstrated by next-generation sequencing in research by Tsitko et al. (2018). The objective was to identify the bacterial groups causing the barley damaged by a worse mash filterability. The process of biofilm development in the outermost layers of barley was seen using field-emitting scanning electron microscopy. Beginning cultures of *Lactobacillus plantarum* VTT E-78076 (E76) were added to the steeping process in order to modify and inhibit the development of the bacteria. More specifically, wort filtering efficiency was shown to be lowered in barley-associated *Leuconostoc*. This study demonstrated that wort separation was significantly influenced by the formation of biofilm comprising polymeric substances that are extracellular (EPS) on the kernel layer surfaces during malting. During malting, the bacterial community was effectively managed by using *L. plantarum* E76, which also enhanced wort filtering. Consequently, control of barley-associated biofilms amid malting might lead to better wort separation and increased extract recovery.

Disinfection, Cleaning, and Hygienic Design

Tanks, pipes, couplings, and other food processing equipment must be designed with hygiene in mind in order to minimise the possibility of biotransfer and/or regulate the growth of biofilm. The specifications for sanitary design are well documented and specify exactly how machinery should be designed so that it is simple to wipe any surface that comes into contact with food or drink. The European Hygienic Engineering and Design Group (EHEDG) or similar US-based groups have provided guidelines on appropriate equipment, materials, and accessories, as well as proper construction, process architecture, and process automation. It is generally recommended that all surfaces that come into touch with the product be smooth (ideally with a Ra of less than 0.8 μm), free of sharp edges, pits, and dead ends and that all pipes and equipment drain themselves. The same consideration for sanitary design applies to beer-distributing systems as it does to beer-producing apparatus development. Nevertheless, a number of weak areas in these systems have been found, such as the joints, fittings, and tap armature, as well as the dispensing tap. Generally speaking, the dispensing lines are more vulnerable to biofilm development and microbial colonisation than the stainless steel system components (Walker et al. 2007). The construction of the dispensing systems should ensure that all pipes, pumps, and refrigeration components self-drain and leave no room for gas pockets or dead ends (Davies et al. 2015).

Cleaning-in-Place (CIP): A Useful Approach

In enclosed processing lines used in the brewing process, cleaning-in-place (CIP) methods are utilised. Biofilm removal has been demonstrated to depend critically on mechanical cleaning input (Carpentier and Cerf 1993; Salo et al. 2006). Although turbulent flow in the pipes and spray nozzles in the cylindrical tanks can provide mechanical force, there are really certain process steps where there is minimal mechanical activity, such as the enormous cylindroconical fermentation tanks (Salo et al. 2006). It has been discovered that biofilms cannot be removed by low cleaning temperatures (Holah and Gibson 1999); however, cleaning brewing containers with extreme temperatures can only be partially accomplished. The best cleansers for eliminating biofilms of species connected to breweries in CIP were discovered to be chlorinated alkaline detergents (Czechowski and Banner 1992).

Thin film cleaning or low-pressure foam systems are typically used in the brewery to clean exposed surfaces such as filling machines, bottle inspectors, and bottling hall conveyor belts. For safety reasons, it is best to avoid using hot solutions or powerful chemicals; nonetheless, disinfectants work well in cold climates and can be used to maintain cleanliness. Following each manufacturing shift, foam cleaning and then dousing with an antiseptic are typically advised, sometimes even during breaks (Paludan-Muller 2003; Wershofen 2003). However, due to the aerosols created during pressure washing, caution must be exercised to prevent the spread of rotting organisms. Apart from routine sanitation and decontamination, periodic extensive equipment overhauls, which involve disassembling parts that pose challenges for visual inspection while whole, are crucial.

Disinfection is used to lessen the amount of viable bacteria on the surface following washing and to stop microbial growth while surfaces are in the interproduction phase. Since germs are largely eradicated during the cleaning stage, the line separating cleaning from disinfection is fairly blurry. Certain detergents have bactericidal properties, whereas certain disinfectants such as hydrogen peroxide and chlorine depolymerise EPS and cause biofilms to separate from surfaces (Carpentier and Cerf 1993). The most often used cleaning solution in CIP, sodium hydroxide, has been demonstrated to exhibit microbicidal action against microorganisms found in the brewery. *Pediococcus* and *Lactobacillus* brevis reduced by five logarithmic units in 3 min and 2 min, respectively, in a solution containing 0.5% sodium hydroxide at 20 °C and *Pediococcus damnosus* in 5 min (Donhauser et al. 1991). In general, chemical cleansers work better than disinfectants to remove adherent microorganisms from surfaces.

Moreover, in the dearth of organic material, disinfectants are often most effective (Czechowski and Banner 1992). Consequently, the surface must be cleaned with detergent and then treated with a disinfectant in order to manage and inactivate adhering bacteria or biofilms.

12.5 Conclusion

The creation of biofilms in breweries is significant at every stage of production. The biofilm developed in manufacturing and distributing pipe junctions, curves, edges, and dead ends. Biofilm growth was also seen in the packing and bottling of beer. Beer fermented with *S. boulardii* developed a coating on the bottle wall during storage, giving the bottle a foggy look. The outcomes shown here demonstrated that *S. boulardii* may stick to glass surfaces in beer and create a biofilm. Microbiofilms found in brewing settings may contain bacteria, such as *L. brevis*, *L. lindneri*, and *P. pentosaceus*, that can proliferate and contaminate craft beer. Because of the diversity of bacteria found in these biofilms, it appears that there may be a sizable gene consortium present, which would enable the microflora to withstand standard sanitation practices. Most of the information about fungal biofilms has been acquired via the use of fungi such as *Pneumocystis*, *Aspergillus*, *Cryptococcus*, and *Coccidioides* in human illness. Hygiene must be considered throughout the design of tanks, pipelines, couplings, and other food processing equipment in order to reduce the risk of biotransfer and/or control the development of biofilm. Alkaline detergents with chlorination were shown to be the most effective cleaners for removing biofilms of species associated with breweries in CIP. Nevertheless, additional investigation is needed to find other methods for biofilm removal in breweries.

References

Ashtavinayak P, Elizabeth HA (2003) In: Priest FG, Campbell I (eds) Brewing microbiology. Springer US. https://doi.org/10.1007/978-1-4419-9250-5

Back W (2003) Biofilm in der Brauerei und Getränkeindustrie. Brauwelt 24:766–777

Banner MJ (1994) Perspectives on conveyor track treatment. Tech Q Master Brew Assoc Am 31:142–148

Bartowsky EJ (2009) Bacterial spoilage of wine and approaches to minimize it. Lett Appl Microbiol 48(2):149–156. https://doi.org/10.1111/j.1472-765X.2008.02505.x

Bojsen RK, Andersen KS, Regenberg B (2012) *Saccharomyces cerevisiae* — a model to uncover molecular mechanisms for yeast biofilm biology. FEMS Immunol Med Microbiol 65(2):169–182. https://doi.org/10.1111/j.1574-695X.2012.00943.x

Carpentier B, Cerf O (1993) Biofilms and their consequences, with particular reference to hygiene in the food industry. J Appl Bacteriol 75(6):499–511. https://doi.org/10.1111/j.1365-2672.1993.tb01587.x

Casson D (1985) Microbiological problems of beer dispense. Brewer 71:417–425

Coi AL, Bigey F, Mallet S, Marsit S, Zara G, Gladieux P, Galeote V, Budroni M, Dequin S, Legras JL (2017) Genomic signatures of adaptation to wine biological ageing conditions in biofilm-forming flor yeasts. Mol Ecol 26(7):2150–2166. https://doi.org/10.1111/mec.14053

Czechowski MH, Banner M (1992) Control of biofilms in breweries through cleaning and sanitizing. Master Brew Assoc Am Tech Q 29:86–88

Davies S, Sykes T, Philips M, Hancock J (2015) Hygienic design and Cleaning-In-Place (CIP) systems in breweries. In: Brewing microbiology. Elsevier, pp 221–239. https://doi.org/10.1016/B978-1-78242-331-7.00010-1

Dobson CM, Deneer H, Lee S, Hemmingsen S, Glaze S, Ziola B (2002) Phylogenetic analysis of the genus Pediococcus, including Pediococcus claussenii sp. nov., a novel lactic acid bacterium isolated from beer. Int J Syst Evol Microbiol 52(6):2003–2010. https://doi.org/10.1099/00207713-52-6-2003

Donhauser S, Wagner D, Geiger E (1991) Zur Wirkung von Desinfektionsmitteln in der Brauerei. Brauwelt 131:604–616

Douglas LM, Li L, Yang Y, Dranginis AM (2007) Expression and characterization of the Flocculin Flo11/Muc1, a *Saccharomyces cerevisiae* Mannoprotein with homotypic properties of adhesion. Eukaryot Cell 6(12):2214–2221. https://doi.org/10.1128/EC.00284-06

Fanning S, Mitchell AP (2012) Fungal biofilms. PLoS Pathog 8(4):e1002585. https://doi.org/10.1371/journal.ppat.1002585

Faria-Oliveira F, Carvalho J, Belmiro CL, Ramalho G, Pavão M, Lucas C, Ferreira C (2015) Elemental biochemical analysis of the polysaccharides in the extracellular matrix of the yeast *Saccharomyces cerevisiae*. J Basic Microbiol 55(6):685–694. https://doi.org/10.1002/jobm.201400781

Flemming H-C, Wingender J (2010) The biofilm matrix. Nat Rev Microbiol 8(9):623–633. https://doi.org/10.1038/nrmicro2415

Fox EP, Bui CK, Nett JE, Hartooni N, Mui MC, Andes DR, Nobile CJ, Johnson AD (2015) An expanded regulatory network temporally controls C andida albicans biofilm formation. Mol Microbiol 96(6):1226–1239. https://doi.org/10.1111/mmi.13002

Galié S, García-Gutiérrez C, Miguélez EM, Villar CJ, Lombó F (2018) Biofilms in the food industry: health aspects and control methods. Front Microbiol 9. https://doi.org/10.3389/fmicb.2018.00898

Gilbert P, Allison DG, McBain AJ (2002) Biofilms *in vitro* and *in vivo*: do singular mechanisms imply cross-resistance? J Appl Microbiol 92:98S–110S. https://doi.org/10.1046/j.1365-2672.92.5s1.5.x

Gulati M, Nobile CJ (2016) Candida albicans biofilms: development, regulation, and molecular mechanisms. Microbes Infect 18(5):310–321. https://doi.org/10.1016/j.micinf.2016.01.002

Haikara A, Helander I (2006) Pectinatus, megasphaera and zymophilus. In: The prokaryotes. Springer US, pp 965–981. https://doi.org/10.1007/0-387-30744-3_32
Harper DR (1981) Microbial contamination in draught beer in public houses. Process Biochem 16:2–7
Heller A (1995) Chemistry and applications of photocatalytic oxidation of thin organic films. Acc Chem Res 28(12):503–508. https://doi.org/10.1021/ar00060a006
Herzog P, Winkler I, Wolking D, Kampfer P, Lipski A (2008) Chryseobacterium ureilyticum sp. nov., Chryseobacterium gambrini sp. nov., Chryseobacterium pallidum sp. nov. and Chryseobacterium molle sp. nov., isolated from beer-bottling plants. Int J Syst Evol Microbiol 58(1):26–33. https://doi.org/10.1099/ijs.0.65362-0
Holah J, Gibson H (1999) Biofouling: cleanability problems in food processing? Proceedings of the 30th R3 -Nordic contamination control symposium. Espoo, VTT Symposium, pp 177–186
Juvonen R, Suihko M-L (2006) Megasphaera paucivorans sp. nov., Megasphaera sueciensis sp. nov. and Pectinatus haikarae sp. nov., isolated from brewery samples, and emended description of the genus Pectinatus. Int J Syst Evol Microbiol 56(4):695–702. https://doi.org/10.1099/ijs.0.63699-0
Karunanithi S, Vadaie N, Chavel CA, Birkaya B, Joshi J, Grell L, Cullen PJ (2010) Shedding of the mucin-like Flocculin Flo11p reveals a new aspect of fungal adhesion regulation. Curr Biol 20(15):1389–1395. https://doi.org/10.1016/j.cub.2010.06.033
Kurtzman CP, Fell JW, Boekhout T (2011) The yeasts, a taxonomic study, 5th edn. Elsevier Science Ltd, Amsterdam
Laitila A, Kotaviita E, Peltola P, Home S, Wilhelmson A (2007) Indigenous microbial community of barley greatly influences grain germination and malt quality. J Inst Brew 113(1):9–20. https://doi.org/10.1002/j.2050-0416.2007.tb00250.x
Lipke P (2018) What we do not know about fungal cell adhesion molecules. J Fungus 4(2):59. https://doi.org/10.3390/jof4020059
Lohse MB, Gulati M, Johnson AD, Nobile CJ (2018) Development and regulation of single- and multi-species Candida albicans biofilms. Nat Rev Microbiol 16(1):19–31. https://doi.org/10.1038/nrmicro.2017.107
Maifreni M, Frigo F, Bartolomeoli I, Buiatti S, Picon S, Marino M (2015) Bacterial biofilm as a possible source of contamination in the microbrewery environment. Food Control 50:809–814. https://doi.org/10.1016/j.foodcont.2014.10.032
Mohammadi K, Saris PEJ (2022) Biofilm formation of probiotic Saccharomyces cerevisiae var. boulardii on glass surface during beer bottle ageing. Beverages 8(4):77. https://doi.org/10.3390/beverages8040077
Moreno-García J, Coi AL, Zara G, García-Martínez T, Mauricio JC, Budroni M (2018) Study of the role of the covalently linked cell wall protein (Ccw14p) and yeast glycoprotein (Ygp1p) within biofilm formation in a flor yeast strain. FEMS Yeast Res 18(2). https://doi.org/10.1093/femsyr/foy005
Nikolaos L, Dimitrios V, Hellen M, Sofia K, Aristea V, Charis T, Chronis T, Angellos P, Ioannis P (2008) Saccharomyces boulardii fungaemia in an intensive care unit patient treated with caspofungin. Crit Care 12(2):414. https://doi.org/10.1186/cc6843
Noots I, Delcour JA, Michiels CW (1999) From field barley to malt: detection and specification of microbial activity for quality aspects. Crit Rev Microbiol 25(2):121–153. https://doi.org/10.1080/10408419991299257
Paludan-Muller T (2003) Selection of the best system for filling lines for soft drinks of varying pH, viscosity and hygienic requirements is not easy. Scand Brew Rev 60(6):180–200
Paramonova E, Krom BP, van der Mei HC, Busscher HJ, Sharma PK (2009) Hyphal content determines the compression strength of Candida albicans biofilms. Microbiology 155(6):1997–2003. https://doi.org/10.1099/mic.0.021568-0
Raulio M, Wilhelmson A, Salkinoja-Salonen M, Laitila A (2009) Ultrastructure of biofilms formed on barley kernels during malting with and without starter culture. Food Microbiol 26(4):437–443. https://doi.org/10.1016/j.fm.2009.02.004

Richard ML, Nobile CJ, Bruno VM, Mitchell AP (2005) *Candida albicans* biofilm-defective mutants. Eukaryot Cell 4(8):1493–1502. https://doi.org/10.1128/EC.4.8.1493-1502.2005

Righelato R (2004) In: Hughes PS, Baxter ED (eds) Beer: quality, safety and nutritional aspects. Royal Society of Chemistry, Cambridge, 2001. 138 pp ISBN 0-85404-588-0. J Sci Food Agric, 84(6), 608–608. https://doi.org/10.1002/jsfa.1576

Rola K, Osyczka P, Kafel A (2016) Different heavy metal accumulation strategies of epilithic lichens colonising artificial post-smelting wastes. Arch Environ Contam Toxicol 70(2):418–428. https://doi.org/10.1007/s00244-015-0180-5

Sakamoto K, Konings WN (2003) Beer spoilage bacteria and hop resistance. Int J Food Microbiol 89(2–3):105–124. https://doi.org/10.1016/S0168-1605(03)00153-3

Salo S, Johnsen S, Jensen BBB, Storgårds E, Friis A (2006) The mechanical effect of rinsing in cleaning of fermentation tank bottoms. In: Fouling, cleaning & disinfection in food processing. University of Cambridge, pp 162–171

Storgårds E, Priha O (2009) Biofilms and brewing. In: Biofilms in the food and beverage industries. Elsevier, pp 432–454. https://doi.org/10.1533/9781845697167.4.432

Storgårds E, Haikara A, Juvonen R (2006a) Brewing control systems: microbiological analysis. In: *Brewing*. Elsevier, pp 391–426. https://doi.org/10.1533/9781845691738.391

Storgårds E, Tapani K, Hartwall P, Saleva R, Suihko M-L (2006b) Microbial attachment and biofilm formation in brewery bottling plants. J Am Soc Brew Chem 64(1):8–15. https://doi.org/10.1094/ASBCJ-64-0008

Stringer JR, Keely SP (2001) Genetics of surface antigen expression in *Pneumocystis carinii*. Infect Immun 69(2):627–639. https://doi.org/10.1128/IAI.69.2.627-639.2001

Suzuki K, Funahashi W, Koyanagi M, Yamashita H (2004) Lactobacillus paracollinoides sp. nov., isolated from brewery environments. Int J Syst Evol Microbiol 54(1):115–117. https://doi.org/10.1099/ijs.0.02722-0

Teunissen AWRH, Steensma HY (1995) The dominant flocculation genes of *Saccharomyces cerevisiae* constitute a new subtelomeric gene family. Yeast 11(11):1001–1013. https://doi.org/10.1002/yea.320111102

Timke M, Wolking D, Wang-Lieu NQ, Altendorf K, Lipski A (2004) Microbial composition of biofilms in a brewery investigated by fatty acid analysis, fluorescence in situ hybridisation and isolation techniques. Appl Microbiol Biotechnol 66(1):100–107. https://doi.org/10.1007/s00253-004-1601-y

Timke M, Wang-Lieu NQ, Altendorf K, Lipski A (2005a) Community structure and diversity of biofilms from a beer bottling plant as revealed using 16S rRNA gene clone libraries. Appl Environ Microbiol 71(10):6446–6452. https://doi.org/10.1128/AEM.71.10.6446-6452.2005

Timke M, Wang-Lieu NQ, Altendorf K, Lipski A (2005b) Fatty acid analysis and spoilage potential of biofilms from two breweries. J Appl Microbiol 99(5):1108–1122. https://doi.org/10.1111/j.1365-2672.2005.02714.x

Timke M, Wang-Lieu NQ, Altendorf K, Lipski A (2008) Identity, beer spoiling and biofilm forming potential of yeasts from beer bottling plant associated biofilms. Antonie Van Leeuwenhoek 93(1–2):151–161. https://doi.org/10.1007/s10482-007-9189-8

Tsitko I, Manninen J, Smart K, James S, Laitila A (2018) Management of barley-associated bacterial biofilms: a key to improving wort separation. J Inst Brew 124(4):325–335. https://doi.org/10.1002/jib.506

Van Nguyen P, Hlaváček O, Maršíková J, Váchová L, Palková Z (2018) Cyc8p and Tup1p transcription regulators antagonistically regulate Flo11p expression and complexity of yeast colony biofilms. PLoS Genet 14(7):e1007495. https://doi.org/10.1371/journal.pgen.1007495

van Nierop SNE, Cameron-Clarke A, Axcell BC (2004) Enzymatic generation of factors from malt responsible for premature yeast flocculation. J Am Soc Brew Chem 62(3):108–116. https://doi.org/10.1094/ASBCJ-62-0108

Verstrepen KJ, Klis FM (2006) Flocculation, adhesion and biofilm formation in yeasts. Mol Microbiol 60(1):5–15. https://doi.org/10.1111/j.1365-2958.2006.05072.x

Vinod PK, Sengupta N, Bhat PJ, Venkatesh KV (2008) Integration of global signaling pathways, cAMP-PKA, MAPK and TOR in the regulation of FLO11. PLoS One 3(2):e1663. https://doi.org/10.1371/journal.pone.0001663

Vishwakarma V (2020) Impact of environmental biofilms: industrial components and its remediation. J Basic Microbiol 60(3):198–206. https://doi.org/10.1002/jobm.201900569

Vopálenská I, Št'ovíček V, Janderová B, Váchová L, Palková Z (2010) Role of distinct dimorphic transitions in territory colonizing and formation of yeast colony architecture. Environ Microbiol 12(1):264–277. https://doi.org/10.1111/j.1462-2920.2009.02067.x

Walker SL, Fourgialakis M, Cerezo B, Livens S (2007) Removal of microbial biofilms from dispense equipment: the effect of enzymatic pre-digestion and detergent treatment. J Inst Brew 113:61–66

Wershofen T (2003) Practical experiences in the handling of cold aseptic filling lines from the viewpoint of detergent and disinfectant manufacturers. Brauwelt 143(44):1434–1440

Wu S, Wang Y, Liu N, Dong G, Sheng C (2017) Tackling fungal resistance by biofilm inhibitors. J Med Chem 60(6):2193–2211. https://doi.org/10.1021/acs.jmedchem.6b01203

Yang L, Zheng C, Chen Y, Ying H (2018) FLO genes family and transcription factor MIG1 regulate Saccharomyces cerevisiae biofilm formation during immobilized fermentation. Front Microbiol 9. https://doi.org/10.3389/fmicb.2018.01860

Zara S, Antonio Farris G, Budroni M, Bakalinsky AT (2002) *HSP12* is essential for biofilm formation by a Sardinian wine strain of *S. cerevisiae*. Yeast 19(3):269–276. https://doi.org/10.1002/yea.831

Zara S, Bakalinsky AT, Zara G, Pirino G, Demontis MA, Budroni M (2005) *FLO11* -based model for air-liquid interfacial biofilm formation by *Saccharomyces cerevisiae*. Appl Environ Microbiol 71(6):2934–2939. https://doi.org/10.1128/AEM.71.6.2934-2939.2005

Zara G, Zara S, Pinna C, Marceddu S, Budroni M (2009) FLO11 gene length and transcriptional level affect biofilm-forming ability of wild flor strains of Saccharomyces cerevisiae. Microbiology 155(12):3838–3846. https://doi.org/10.1099/mic.0.028738-0

Sustainable Valorization of Food Waste Through Bacterial Biofilm: A Pathway Towards Environmental Safety and Economic Viability

13

Arindam Ganguly, Ritwick Konar, Ponmurugan Karuppiah, Uma Eswaranpillai, Mousumi Ganguly, Manasi Chattaraj, Debasis Mitra, and Sourav Chattaraj

Abstract

The management of food waste presents a significant challenge in modern society, where a substantial portion of edible resources is discarded, contributing to environmental pollution and resource depletion. In recent years, the valorization of food waste through bacterial biofilm has emerged as a promising strategy to address this issue. Bacterial biofilms, consisting of complex microbial

A. Ganguly · R. Konar
Department of Microbiology, Bankura Sammilani College, Bankura, West Bengal, India

P. Karuppiah
Department of Botany and Microbiology, College of Science, King Saud University, Riyadh, Saudi Arabia
e-mail: pkaruppiah@ksu.edu.sa

U. Eswaranpillai
Department of Botany, PSGR Krishnammal College for Women, Coimbatore, Tamil Nadu, India
e-mail: euma@psgrkcw.ac.in

M. Ganguly
Department of Anthropology, Vidyasagar University, Midnapore, West Bengal, India

M. Chattaraj
Department of Geography, Bankura University, Bankura, West Bengal, India

D. Mitra
Department of Microbiology, Graphic Era (Deemed to be University), Dehradun, Uttarakhand, India

S. Chattaraj (✉)
Centre for Industrial Biotechnology Research, School of Pharmaceutical Sciences, Siksha 'O' Anusandhan Deemed to be University, Bhubaneswar, Odisha, India

T. Sarkar et al. (eds.), *Biofilm Applications to Revolutionize Food Technology*,
https://doi.org/10.1007/978-3-031-85205-3_13

communities adhering to surfaces, possess inherent capabilities to degrade organic matter efficiently. The study explores the potential of bacterial biofilms in the valorization of food waste, focusing on their ability to convert diverse food substrates into valuable products such as biofuels, enzymes, and biopolymers. Key factors influencing biofilm formation, substrate specificity, and metabolic pathways involved in food waste degradation are discussed. Moreover, the utilization of engineered biofilms and consortia to enhance the degradation efficiency and product yields is highlighted. The environmental implications, economic viability, and scalability of biofilm-based food waste valorization processes are considered. Additionally, the integration of Omics approaches enhances the understanding of the metabolic networks and genetic pathways involved in the valorization of food waste by bacterial biofilms, offering insights for optimizing bioprocesses and improving efficiency. Overall, this study highlights the importance of bacterial biofilms as a viable strategy for food waste valuation, providing guidance for future studies and useful applications in resource recovery and waste management.

Keywords

Bacterial biofilm · Food waste · Omics · Resource recovery · Sustainability · Valorization

13.1 Introduction

Food waste management has become a major global concern, with significant environmental impacts, including greenhouse gas emissions, water pollution, and land degradation (Bellemare et al. 2017). Food waste refers to any edible food that is rendered unusable due to inadequate post-harvest management techniques or through the intentional disposal of consumable food items. In order to quantify these losses and inefficiencies, the Food and Agriculture Organization (FAO) has established two distinct sub-indicators: the Food Loss Index and the Food Waste Index. Among the overall food waste, upstream activities (including harvesting, drying, storage, and transportation) account for 54%, while downstream activities (encompassing processing, distribution, and consumption) represent 46% of total food wastage (Pandey 2021). The horticultural sector, particularly the fruit and vegetable industry, produces a significantly higher volume of waste in comparison with other food processing domains, representing approximately 25–30% of the overall waste generated. This waste can encompass a variety of materials, including but not limited to peels, rinds, seeds, cores, residual plant parts, stones, pods, vines, skins, and pomaces (Ajila et al. 2010). The processing of horticultural products results in a substantial volume of waste, and the appropriate management of this refuse elevates total production expenses. Consequently, in numerous circumstances, residuals from fruits and vegetables are released into the ecosystem to mitigate production costs (Thiviya et al. 2022). Due to its biodegradable nature, food waste undergoes rapid decomposition, leading to the release of unpleasant and noxious odors. If such waste is not addressed in a timely manner, it can result in considerable environmental and public health challenges. Improper disposal of food remnants may foster the proliferation of pathogens and attract vectors of disease, including flies and mosquitoes. Furthermore, this waste has the potential to contaminate both surface and groundwater, leading to phytotoxicity and various serious health concerns (Nayak and Bhushan 2019; Sadhu et al. 2018). Food products confer unique health benefits due to their rich content of bioactive compounds, which encompass phenolic compounds, terpenes, terpenoids, alkaloids, flavonoids, carotenes, and various other bioactives (Azmir et al. 2013). Conversely, primary bioactive metabolites such as proteins, carbohydrates, and lipids are present in relatively low concentrations, which enable plants to endure environmental stressors, pathogens, and solar radiation (Jha and Sit 2022). Agro-industrial by-products are characterized by the presence of phenolic substances and various hazardous compounds, which have the potential to adversely impact environmental integrity upon their release into ecological systems (Mussatto et al. 2012). Certain animal-derived foodstuffs exhibit a notably high concentration of distinct bioactive compounds. For example, meat is abundant in taurine, L-carnitine, and conjugated linoleic acid, whereas eggs serve as significant reservoirs of choline and carotenoids, including lutein and zeaxanthin (Prasow et al. 2019). Additionally, dairy products are characterized by the presence of bioactive peptides and essential fatty acids (Kathuria et al. 2019). Through the implementation of highly effective recycling methodologies and the innovative valorization processes, which are designed to maximize the utility of various materials, food waste, which has historically been viewed as a mere by-product of consumption, can be transformed into a range of valuable and usable resources, thereby playing a crucial role in the achievement of broader sustainability objectives and goals that are increasingly being prioritized in contemporary environmental discourse (Sarker et al. 2024).

The traditional disposal methods, such as landfilling and incineration, are not only unsustainable but also economically burdensome, warranting an urgent need for innovative solutions. In this context, bacterial biofilms, comprising complex microbial communities adhering to surfaces, emerge as a promising solution for food waste valorization. By leveraging the potential of these biofilms, it is possible to break down organic matter efficiently, produce valuable by-products, and mitigate the environmental impacts associated with food waste disposal, paving the way for a more sustainable and circular food system. Biofilms represent complex and naturally occurring assemblages of diverse microorganisms that have established a firm attachment to various surfaces, thereby demonstrating exceptionally robust adhesive characteristics that enable them to resist detachment and persist in a myriad of environmental conditions (Olanbiwoninu and Popoola 2023). The hydrolysis of organic waste is an essential preliminary phase in the anaerobic digestion procedure, during which intricate organic compounds are deconstructed into fundamental sugars and fatty acids. The utilization of bacterial biofilms significantly enhances the efficacy of the hydrolysis and immobilization processes (Ali et al. 2023a). A variety of ecological variables significantly influence the development of bacterial biofilms in organic waste. Certain temperature parameters facilitate microbial proliferation, whereas ideal pH conditions augment the attachment and functionality of bacteria on surfaces. Sufficient moisture is critical for facilitating biofilm formation by providing a hydrated milieu that promotes bacterial proliferation (García-Gonzalo and Pagán 2015). Moreover, organic matter functions as crucial substrates that facilitate bacterial colonization and longevity, thereby markedly increasing food contamination risks (Muhammad et al. 2020). Microbial interaction and biochemical mechanisms also play crucial roles in stability and resilience of biofilm within food degradation environment (García-Gonzalo and Pagán 2015; Muhammad et al. 2020). This comprehensive investigation underscores the critical significance of bacterial biofilms, which are complex aggregations of microorganisms, as a highly effective and promising strategy for the valuation of food waste, thereby not only offering valuable insights and guidance for subsequent research endeavors but also presenting practical applications that can be instrumental in the realms of resource recovery and sustainable waste management practices.

13.2 Sources and Availability of Food Waste

Food waste refers to nutritious food designated for human consumption that is ultimately discarded, lost, degraded, or consumed by pests at any stage of the food supply chain. It constitutes a critical global challenge impacting multiple domains such as domestic settings, food service establishments, agricultural practices, and production industries. Each year, an estimated 1.3 billion metric tons of food is discarded globally. This figure represents roughly one-third of the total food generated worldwide (Verma 2015). What is especially alarming is that the volume of food wasted has the potential to nourish two billion individuals. This figure exceeds twice the number of people suffering from malnutrition across the globe (Verma 2015). Globally, approximately 1.13 million tons of edible food are wasted daily, averaging 178 g per capita, with significant variation across nations. High-income countries exhibit a mean waste of 307 g per capita daily, which starkly contrasts with lower-income regions, where waste figures are considerably lower, such as 43 g in low-income countries (Chen et al. 2020). According to the 2024 Food Waste Index Report by the UN Environment Programme (UNEP), India produces approximately 78.2 million tons of food waste annually, averaging 55 kg per capita, while the global average stands at 79 kg. This underscores a significant contradiction: Despite a considerable population facing hunger, India ranks 111th out of 125 countries on the Global Hunger Index, 2023, indicating severe hunger levels, with food wastage data categorized under "Medium Confidence" due to sampling limitations.

Food waste is prevalent across various categories, with vegetables (25%), cereals (24%), and fruits (12%) being the primary contributors; other groups include roots and tubers (9%), dairy/eggs (7%), meat/fish (6%), and miscellaneous foods (18%) (Chen et al. 2020). North American consumers discard 23% of food entering their households, while Europe, East Asia and Pacific, and China waste 17%, 15%, and

15% respectively; in contrast, South Asia and Africa exhibit post-consumption food waste rates below 5%, and the Middle East and Latin America report waste levels of approximately 8–10% (Gustavsson et al. 2011).

The origins of food waste are multifaceted, prominently featuring manufacturing facilities, transportation and distribution networks, retail businesses, restaurants and institutions, and households. Households are the predominant source of food waste in developed nations, primarily due to inadequate storage practices, excessive purchasing, misinterpretation of expiration dates, and the preparation of excessive quantities leading to discarded leftovers whereas agricultural food waste is predominantly influenced by biological and environmental variables, which encompass pest and disease damage, adverse weather conditions, inadequate harvesting techniques, stringent supermarket cosmetic standards resulting in the rejection of edible produce, and the overproduction linked to contractual obligations and market demand fluctuations leading to unharvested crops. Restaurants significantly contribute to global food waste due to oversized portions often left uneaten and ineffective inventory management resulting in surplus food preparation. The phenomenon of food waste that is generated throughout the various stages of the supply chain reaches a staggering figure of 46.37% when assessed specifically at the consumer level, and it is particularly noteworthy that within this considerable amount of waste, fruits and vegetables alone account for an impressive 46%, highlighting the significant impact these categories have on overall food waste statistics (Caldeira et al. 2019). Food wastage transpires during the cultivation and transformation phases, which can be considerable, particularly in nations with developing economies. Principal concerns are post-harvest depletion, insufficient storage, and conveyance infrastructure result in considerable food wastage subsequent to harvesting. Processing by-products such as excess food produced during the processing phase frequently remains unutilized (Council 1978). Retail establishments, including grocery stores and supermarkets, play a pivotal role in the production of food waste, which is approximated to account for nearly 10% of overall waste. Retailers frequently maintain inventories that exceed the volume of food that can be effectively sold, resulting in considerable spoilage. Items that fail to conform to aesthetic standards are frequently disposed of, notwithstanding their suitability for consumption.

Mitigating food waste necessitates an in-depth comprehension of its origins and the processes involved in recovery. By focusing on the primary sources—domestic settings, retail establishments, food service sectors, and production facilities—stakeholders are positioned to formulate efficient strategies aimed at reducing waste, bolstering food security, and alleviating ecological repercussions.

13.3 Nutritional Content in Food Waste

Food losses and waste refer to the reduction or loss of essential nutrients and micronutrients throughout the food supply chain, resulting in a decrease in the nutritional value of food that is available for human consumption (García-Herrero et al. 2019).

This encompasses losses occurring during production, processing, distribution, and consumption stages, ultimately leading to a significant decrease in the ingestion of vital nutrients. Worldwide, an individual wastes 65 kg of food annually, with vegetables accounting for 25% of the waste, grains for 24%, and fruits for 12% (Chen et al. 2020). Significant quantities of essential micronutrients, including vitamin C, K, zinc, copper, manganese, and selenium, are lost through food waste daily, accounting for 25–50% of their respective daily recommended intake values. The primary food groups responsible for this nutrient waste are cereals, fruits, and vegetables, which contribute substantially to the waste of various vitamins and minerals (Chen et al. 2020). Additionally, meat, dairy, and egg products are significant contributors to the waste of calcium, choline, riboflavin, zinc, and vitamin B12, highlighting the need for targeted strategies to minimize nutrient losses throughout the food supply chain. The global average annual per capita food waste is characterized by significant losses of essential nutrients, with zinc exhibiting the highest waste nutrient density value followed by manganese, selenium, iron, copper, and vitamin C. Spiker et al. (2017) conducted a detailed analysis of nutritional waste in the United States, examining 213 commodities and 27 nutrients at the retail and consumer levels. García-Herrero et al. (2019) quantified the nutritional losses embedded in Spanish food losses and waste, revealing a daily per capita loss of 4251 kJ (1016 kcal) of energy, along with significant amounts of protein, dietary fiber, and essential vitamins and minerals, including vitamin A, vitamin C, and calcium. Chua et al. (2019) conducted a study that revealed significant variations in nutrient content among food waste samples from diverse sources. Their findings highlighted the substantial differences in nutrient profiles, which can have implications for food waste management, nutrient recovery, and potential applications in agriculture or energy production. The European Union is confronted with a significant food waste challenge, with an estimated 88 million tons of waste generated annually (Stenmarck et al. 2016). This considerable amount includes both edible and inedible portions of food, underscoring the importance of implementing comprehensive food waste reduction and recovery initiatives to mitigate environmental pressures and foster a more circular food economy.

13.4 Bacterial Biofilms: Characteristics and Functionality

Biofilms are complex communities of microorganisms that adhere to surfaces and are embedded in a self-produced extracellular matrix. This matrix, composed of polysaccharides, proteins, and nucleic acids, provides structural stability and protection to the microbial community, enabling it to survive in harsh environmental conditions and resist antimicrobial agents. Biofilms can form on a wide variety of surfaces, including natural environments, medical devices, industrial equipment, and food processing surfaces. Among the diverse microorganisms (e.g., *Pseudomonas aeruginosa*, *Staphylococcus aureus*, *Escherichia coli*, *Klebsiella pneumonia*, *Streptococcus mutans*, *Bacillus subtilis*) that can form biofilms, several bacterial species are particularly well known and extensively studied due to their significance in health, industry, and the environment (Karygianni et al. 2020; Sharma et al. 2023).

However, not all biofilms are harmful. In natural environments, biofilms play essential roles in nutrient cycling, bioremediation, waste management, and the maintenance of ecological balance. In agriculture, biofilm-forming plant growth-promoting rhizobacteria (PGPR) can enhance soil fertility and plant health. Understanding the common bacterial species involved in biofilm formation and the mechanisms by which they develop and persist is crucial for developing strategies to control harmful biofilms and harness beneficial ones (Ali et al. 2023b; Bhattacharyya et al. 2023). This knowledge can inform the design of new methodologies for waste management practices across various food-related industries.

13.5 Mechanisms of Food Waste Valorization Through Bacterial Biofilms

Bacterial biofilms offer an innovative and sustainable approach to the valorization of food waste, transforming it into valuable products such as bioenergy, bioplastics, and biofertilizers (Fig. 13.1). This process leverages the natural abilities of bacteria to degrade organic matter and produce bioproducts, while the biofilm structure enhances the efficiency and stability of these microbial activities. The mechanisms by which bacterial biofilms contribute to food waste valorization involve a combination of enzymatic degradation, metabolic processes, and the physical properties of the biofilm itself (Muhammad et al. 2020; Faria et al. 2023).

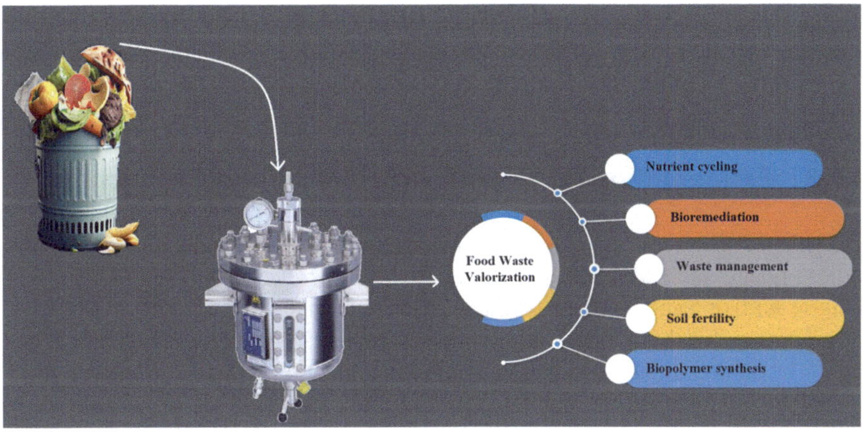

Fig. 13.1 Food waste valorization through bacterial biofilms

13.6 Enzymatic Degradation of Organic Matter

The first step in food waste valorization through bacterial biofilms is the enzymatic degradation of organic matter. Bacterial biofilms consist of microbial communities that secrete a wide range of extracellular enzymes capable of breaking down complex organic compounds found in food waste. These enzymes include proteases, lipases, cellulases, hemi-cellulases, and amylases. These enzymes act synergistically to hydrolyze the complex macromolecules in food waste, converting them into simpler compounds that can be taken up and metabolized by the bacteria within the biofilm. The biofilm structure facilitates the concentration and retention of these enzymes at the site of degradation, enhancing the overall efficiency of the process (Zhao et al. 2023; Mandal et al. 2024).

13.7 Metabolic Pathways for Bioproduct Formation

Once the organic matter is broken down into simpler molecules, the bacteria within the biofilm utilize various metabolic pathways to convert these substrates into valuable products. The specific bioproducts generated depend on the bacterial species present in the biofilm and the environmental conditions.

13.8 Anaerobic Digestion for Biogas Production

In anaerobic conditions, bacterial biofilms can facilitate the conversion of organic matter into biogas, a mixture of methane and carbon dioxide. This process involves a series of metabolic steps carried out by different bacterial groups, including hydrolytic bacteria (which break down organic polymers), acidogenic bacteria (which produce volatile fatty acids), acetogenic bacteria (which convert acids into acetic acid), and methanogenic archaea (which produce methane). The biofilm structure enhances the stability and efficiency of these metabolic interactions, leading to higher biogas yields (Harirchi et al. 2022; Wang et al. 2023).

13.9 Fermentation for Organic Acid and Alcohol Production

Bacterial biofilms can also carry out fermentation processes, converting sugars and other organic substrates into organic acids (such as lactic acid, acetic acid, and butyric acid) or alcohols (such as ethanol). These fermentation products can be used as precursors for the synthesis of bioplastics, biofuels, and other industrial chemicals. The biofilm environment supports the accumulation of high concentrations of metabolites, which can be harvested and purified for industrial use (Arhin et al. 2023; Zhang et al. 2023).

13.10 Phosphate Solubilization for Biofertilizer Production

Certain bacteria within biofilms possess the ability to solubilize insoluble phosphate compounds, making phosphorus available for plant uptake. This process involves the secretion of organic acids and other chelating agents that dissolve mineral phosphates, releasing phosphate ions into the surrounding environment. The biofilm structure helps concentrate these solubilizing agents, enhancing the efficiency of phosphate release. The resulting biofertilizers can improve soil fertility and crop yields (Sharma et al. 2013; Thepbandit and Athinuwat 2024).

13.11 Biopolymer Synthesis for Bioplastics

Some bacterial species within biofilms can produce biopolymers, such as polyhydroxyalkanoates (PHAs), from simple carbon substrates derived from food waste. PHAs are biodegradable plastics that can serve as sustainable alternatives to petroleum-based plastics. The biofilm matrix provides a stable environment for accumulating these biopolymers, which can be extracted and processed into various bioplastic products (Acharjee et al. 2023).

13.12 Synergistic Interactions Within Biofilms

One of the unique advantages of using bacterial biofilms for food waste valorization is the synergistic interactions within the microbial community. In a biofilm, different bacterial species can work together to achieve more efficient degradation and conversion of organic matter than they could individually (Sharma et al. 2023).

13.13 Techno-Economic Analysis

Kwan et al. (2015) developed a pilot-scale facility to handle 1 metric ton of food waste per day, with an operational lifespan of 20 years. Two different product scenarios were considered: Scenario I, which produced plasticizer and lactic acid, and Scenario II, which produced plasticizer and animal feed. Only Scenario I proved to be economically viable. It yielded annual net profits of $422,699, a net present value of $3,028,000, a payback period of 7.56 years, and an internal rate of return of 18.98%. A techno-economic analysis was conducted on an integrated process for producing polyhydroxyalkanoates (PHAs) and biofuels (biohydrogen, bioethanol, and 2,3-butanediol) from food waste. The process, based on previous research, generates 2.01 MT of PHAs, 0.29 MT of biohydrogen, 4.79 MT of bioethanol, and 6.79 MT of 2,3-butanediol daily from 50 MT of food waste. It has a net present value of $4.47 million, a 13.68% return on investment, a payback period of 7.31 years, and an internal rate of return of 11.95%. Sensitivity analysis revealed that the minimum selling price (MSP) of PHAs is currently $4.83/kg, with a potential lowest MSP of

\$2.41/kg at 30% solid loading. Solid loading during hydrolysis and by-product prices significantly affects the economic viability and MSP of PHAs (Rajendran and Han 2022). Again, Demichelis et al. (2018) evaluated the economic feasibility of an integrated biorefinery process for producing lactic acid and biogas sequentially from food waste. This integrated process was compared to separate processes for each product. The assessment, using real biorefinery data and SuperPro Designer® 8.0, covered catchment areas ranging from 2000 to one million people. The study found that the integrated process was more efficient in energy and material use than individual processes and was profitable for areas with populations over 20,000 to 50,000. Similarly, Kwan et al. (2016) designed a biorefinery with a capacity to process 10 metric tons (MT) per hour of food waste powder, operating for 20 years. They explored three different product scenarios: Scenario I for lactic acid, Scenario II for lactide, and Scenario III for poly (lactic acid) (PLA). The mass balance results indicated that from 10 MT of food waste powder, the conversion yields were 3.1 MT of 80% lactic acid, 1.7 MT of lactide, and 1.3 MT of PLA. All scenarios proved economically viable, but Scenario I was the most profitable, generating annual net profits of \$22,184,169, an internal rate of return of 31.1%, a net present value of \$234,803,060, and the shortest payback period of 5.1 years at a 5% discount rate. The minimum selling prices were \$943 per MT for lactic acid, \$2073 per MT for lactide, and \$3330 per MT for PLA. Sensitivity analysis revealed that the prices of these products were the key factors influencing profitability, with by-product sales, such as animal feed, also playing a crucial role. This study provides valuable insights into the techno-economic feasibility of sustainable food waste treatment using a decentralized approach in urban settings.

13.14 Challenges

Logistics play a crucial role in the profitability of food waste valorization, affecting substrate conversion efficiency. Banerjee et al. (2018) highlighted logistical challenges in utilizing pineapple waste from both farms and processing stages. Similarly, Massaya et al. (2019) noted that transportation costs for spent coffee grounds could reach about \$10.25 per ton, presenting a significant expense for a biorefinery with a 10,000-ton annual capacity and a capital cost of around \$four million (Caldeira et al. 2020). Administrative incentives are deemed essential for making food waste recovery economically viable, particularly in commercial and industrial sectors (Mak et al. 2018). Muriana (2017) emphasized that forecasting food waste and losses while considering product shelf life could optimize the food waste supply chain, enhancing collection and valorization. Bottani et al. (2019) further explored logistics, finding that extended storage in distribution centers could optimize transportation to treatment facilities, reducing costs and environmental impacts. However, they noted constraints due to microbiological and sensory degradation, even under refrigeration. Bacci et al. (2019) highlighted microbiological concerns, particularly with meat waste, which could impact future valorization efforts. Despite an increase in studies on food waste valorization profitability and techno-economics, literature remains sparse compared to the extensive research on valorization techniques. Key aspects such as waste stream availability, logistical challenges, and biorefinery location often remain underexplored. Previous Life Cycle Assessment (LCA) studies focused on plant-based and other industrial food waste valorization, aiming to compare technologies, assess environmental performance, and evaluate waste management alternatives (Caldeira et al. 2020). The large-scale development of biorefineries encounters significant obstacles including challenges in mobilizing biomass resources, ensuring economic viability, and understanding environmental impacts. Modern biorefineries must achieve environmental sustainability, aligning with economic growth and social and environmental goals. Effective biorefinery design and scaling require a clear understanding of the trade-offs between economic and environmental performance. The integrated Technical, Economic, and Environmental Assessment (ETEA) framework is useful here, as it merges technical, economic, and environmental indicators to assess the broader impacts of biorefineries. Technical and Economic Assessment (TEA) is crucial for evaluating the feasibility of biorefinery systems before they are scaled up (Pérez-Almada et al. 2023). TEA involves process modeling and simulation using tools such as Aspen Plus® and Aspen Hysys®. These simulations estimate the capital and operational costs (CAPEX and OPEX) of the processes, helping to evaluate economic indicators like Net Present Value (NPV) and payback periods. Given the variability of biomass resources, TEA provides insight into the long-term economic impacts and helps address challenges related to economic viability. Life Cycle Assessment (LCA) offers a standardized method to evaluate the environmental impacts of biorefineries from the entire lifecycle perspective (Pérez-Almada et al. 2023). LCA examines impacts from biomass harvesting to the end-of-life stage or up to the biorefinery plant gate. It is used to compare different process alternatives or assess the best uses of biomass feedstock. While LCA focuses on various environmental impacts, such as climate change, eutrophication, and biodiversity, it also faces challenges due to the multi-product nature of biorefinery systems. Integrating TEA and LCA into the ETEA framework provides a comprehensive tool for stakeholders to make informed decisions that balance sustainability with economic and environmental considerations. This approach helps navigate the complexities of biorefinery development, ensuring that both economic and environmental goals are met (Pérez-Almada et al. 2023).

13.15 Future Prospective

Bioconverting food waste into value-added products offers both economic and environmental advantages. This process tackles two significant societal challenges: creating valuable products and managing waste. By aiming to produce multiple products, overall process costs can be reduced. Various bacterial species are potential candidates for efficient bioconversion and multi-product production. However, the heterogeneous nature and high moisture content of food waste pose significant challenges. Using mixed cultures or natural microbial communities can help address these issues more effectively. Additionally, employing different fermentation techniques for food waste valorization can improve resource utilization and contribute to solving environmental problems. Although awareness of food loss and waste is increasing, it often overlooks specific types of food. One method to reduce waste at home is fermenting carbohydrate-rich leftovers through solid-state fermentation (SSF) to create new food products. This process involves using microbes to ferment items like bread, which can then be prepared in various ways. To promote this practice, it is crucial to share fermentation methods and highlight the benefits of products made from surplus resources. Motivating individuals to engage with microbe-based products or ferment leftovers depends on perceived value, such as contributing to societal movements, personal satisfaction, or nutritional benefits (Awasthi et al. 2022). Fermentation offers individuals a way to become less dependent on supply chains and grocery stores by allowing them to produce their own food. This is particularly valuable in areas where access to nutrient-dense, protein-rich foods is limited. Fermentation can transform inexpensive, carbohydrate-rich foods into nutritious, protein-rich fungal biomass. This method, which uses minimal water, not only helps reduce food waste but also addresses hunger and malnutrition, especially in famine-stricken regions. Additionally, ethical considerations arise in research involving microbial food products from rescued substrates, including tasting studies, which must adhere to varying ethical review requirements based on local legislation.

13.16 Conclusion

The sustainable valorization of food waste through bacterial biofilms represents a promising pathway to enhance environmental safety and economic viability. By utilizing bacterial biofilms, food waste can be transformed into valuable products, addressing pressing issues of waste management and resource efficiency. However, this approach faces challenges such as optimizing biofilm efficiency, managing the heterogeneous nature of food waste, and scaling up the technology. Future research should focus on improving the stability and productivity of bacterial biofilms, developing cost-effective methods for large-scale implementation, and understanding the long-term environmental impacts. Socio-economic considerations include ensuring accessibility and affordability of this technology for diverse communities, promoting public awareness, and supporting policies that encourage waste valorization. By overcoming these challenges, bacterial biofilm technology can contribute significantly to a circular economy, reduce environmental impact, and create economic opportunities, aligning with broader sustainability goals.

References

Acharjee SA, Bharali P, Gogoi B, Sorhie V, Walling B, Alemtoshi (2023) PHA-based bioplastic: a potential alternative to address microplastic pollution. Water Air Soil Pollut 234(1):21. https://doi.org/10.1007/s11270-022-06029-2

Ajila C, Aalami M, Leelavathi K, Rao UP (2010) Mango peel powder: a potential source of antioxidant and dietary fiber in macaroni preparations. Innovative Food Sci Emerg Technol 11(1):219–224. https://doi.org/10.1016/j.ifset.2009.10.004

Ali AM, Zahangir MA, Abdoul-Latif FM, Jami MS, Mohamed J, Ainane T (2023a) Hydrolysis of food waste with immobilized biofilm as a pretreatment method for the enhancement of biogas production. Sustain For 15(4):3316. https://doi.org/10.3390/su15043316

Ali A, Zahra A, Kamthan M, Husain FM, Albalawi T, Zubair M, Alatawy R, Abid M, Noorani MS (2023b) Microbial biofilms: applications, clinical consequences, and alternative therapies. Microorganisms 11(8):1934. https://doi.org/10.3390/microorganisms11081934

Arhin SG, Cesaro A, Di Capua F, Esposito G (2023) Acidogenic fermentation of food waste to generate electron acceptors and donors towards medium-chain carboxylic acids production. J Environ Manag 348:119379

Awasthi MK, Harirchi S, Sar T, Vigneswaran VS, Rajendran K, Gómez-García R, Hellwig C, Binod P, Sindhu R, Madhavan A, Kumar AA (2022) Myco-biorefinery approaches for food waste valorization: present status and future prospects. Bioresour Technol 360:127592

Azmir J, Zaidul I, Rahman M, Sharif K, Mohamed A, Sahena F, Jahurul M, Ghafoor K, Norulaini N, Omar A (2013) Techniques for extraction of bioactive compounds from plant materials: a review. J Food Eng 117(4):426–436. https://doi.org/10.1016/j.jfoodeng.2013.01.014

Bacci C, Vismarra A, Dander S, Barilli E, Superchi P (2019) Occurrence and antimicrobial profile of bacterial pathogens in former foodstuff meat products used for pet diets. J Food Prot 82(2):316–324

Banerjee S, Ranganathan V, Patti A, Arora A (2018) Valorisation of pineapple wastes for food and therapeutic applications. Trends Food Sci Technol 82:60–70

Bellemare MF, Çakir M, Peterson HH, Novak L, Rudi J (2017) On the measurement of food waste. Am J Agric Econ 99(5):1148–1158. https://doi.org/10.1093/ajae/aax034

Bhattacharyya A, Mavrodi O, Bhowmik N, Weller D, Thomashow L, Mavrodi D (2023) Bacterial biofilms as an essential component of rhizosphere plant-microbe interactions. Methods Microbiol 53:3–48. https://doi.org/10.1016/bs.mim.2023.05.006

Bottani E, Vignali G, Mosna D, Montanari R (2019) Economic and environmental assessment of different reverse logistics scenarios for food waste recovery. Sustain Prod Consum 20:289–303

Caldeira C, De Laurentiis V, Corrado S, Van Holsteijn F, Sala S (2019) Quantification of food waste per product group along the food supply chain in the European Union: a mass flow analysis. Resour Conserv Recycl 149:479–488. https://doi.org/10.1016/j.resconrec.2019.06.011

Caldeira C, Vlysidis A, Fiore G, De Laurentiis V, Vignali G, Sala S (2020) Sustainability of food waste biorefinery: a review on valorisation pathways, techno-economic constraints, and environmental assessment. Bioresour Technol 312:123575

Chen C, Chaudhary A, Mathys A (2020) Nutritional and environmental losses embedded in global food waste. Resour Conserv Recycl 160:104912. https://doi.org/10.1016/j.resconrec.2020.104912

Chua GK, Tan FHY, Chew FN, Mohd-Hairul AR (2019) Nutrients content of food wastes from different sources and its pre-treatment. AIP Conf Proc 2124(1):020031. https://doi.org/10.1063/1.5117091

Council NR (1978) Postharvest food losses in developing countries. In: National Academies Press eBooks. https://doi.org/10.17226/20028

Demichelis F, Fiore S, Pleissner D, Venus J (2018) Technical and economic assessment of food waste valorization through a biorefinery chain. Renew Sust Energ Rev 94:38–48

Faria DJ, Carvalho APAD, Conte-Junior CA (2023) Valorization of fermented food wastes and byproducts: bioactive and valuable compounds, bioproduct synthesis, and applications. Fermentation 9(10):920

García-Gonzalo D, Pagán R (2015) Influence of environmental factors on bacterial biofilm formation in the food industry: a review. Postdoc J 3(6). https://doi.org/10.14304/surya.jpr.v3n6.2

García-Herrero I, Margallo M, Laso J, Batlle-Bayer L, Bala A, Fullana-i-Palmer P, Vázquez-Rowe I, González MJ, Amo-Setién F, Durá MJ, Sarabia C, Abajas R, Quiñones A, Irabien A, Aldaco R (2019) Nutritional data management of food losses and waste under a life cycle approach: case study of the Spanish agri-food system. J Food Compos Anal 82:103223. https://doi.org/10.1016/j.jfca.2019.05.006

Gustavsson J, Cederberg C, Sonesson U, Van Otterdijk R, Meybeck A (2011) Global food losses and food waste: extent, causes and prevention. In: *FAO eBooks*. https://agris.fao.org/agris-search/search.do?recordID=XF2013001423

Harirchi S, Wainaina S, Sar T, Nojoumi SA, Parchami M, Parchami M, Varjani S, Khanal SK, Wong J, Awasthi MK, Taherzadeh MJ (2022) Microbiological insights into anaerobic digestion for biogas, hydrogen or volatile fatty acids (VFAs): a review. Bioengineered 13(3):6521–6557. https://doi.org/10.1080/21655979.2022.2035986

Jha AK, Sit N (2022) Extraction of bioactive compounds from plant materials using combination of various novel methods: a review. Trends Food Sci Technol 119:579–591. https://doi.org/10.1016/j.tifs.2021.11.019

Karygianni L, Ren Z, Koo H, Thurnheer T (2020) Biofilm matrixome: extracellular components in structured microbial communities. Trends Microbiol 28(8):668–681

Kathuria D, Gautam S, Sharma S, Sharma K (2019) Animal based bioactives for health and wellness. Food Nutr J 4:203. https://gavinpublishers.com/articles/review-article/Food-Nutrition-Journal/animal-based-bioactives-for-health-and-wellness

Kwan TH, Pleissner D, Lau KY, Venus J, Pommeret A, Lin CSK (2015) Techno-economic analysis of a food waste valorization process via microalgae cultivation and co-production of plasticizer, lactic acid and animal feed from algal biomass and food waste. Bioresour Technol 198:292–299

Kwan TH, Yunzi HU, Lin CSK (2016, May) Techno-economic analysis of a food waste valorisation process for poly (lactic acid) production from food waste. In: BioEnergy & Biorefinery conference–Southeast Asia

Mak TM, Iris KM, Tsang DC, Hsu SC, Poon CS (2018) Promoting food waste recycling in the commercial and industrial sector by extending the theory of planned behaviour: a Hong Kong case study. J Clean Prod 204:1034–1043

Mandal M, Roy A, Popek R, Sarkar A (2024) Micro-and nano-plastic degradation by bacterial enzymes: a solution to 'White Pollution'. Microbe 100072:100072

Massaya J, Pereira AP, Mills-Lamptey B, Benjamin J, Chuck CJ (2019) Conceptualization of a spent coffee grounds biorefinery: a review of existing valorisation approaches. Food Bioprod Process 118:149–166

Muhammad MH, Idris AL, Fan X, Guo Y, Yu Y, Jin X, Qiu J, Guan X, Huang T (2020) Beyond risk: bacterial biofilms and their regulating approaches. Front Microbiol 11:928. https://doi.org/10.3389/fmicb.2020.00928

Muriana C (2017) A focus on the state of the art of food waste/losses issue and suggestions for future researches. Waste Manag 68:557–570

Mussatto SI, Ballesteros LF, Martins SLF, Teixeira JA (2012) Use of agro-industrial wastes in solid-state fermentation processes. In: InTech eBooks. https://doi.org/10.5772/36310

Nayak A, Bhushan B (2019) An overview of the recent trends on the waste valorization techniques for food wastes. J Environ Manag 233:352–370. https://doi.org/10.1016/j.jenvman.2018.12.041

Olanbiwoninu A, Popoola B (2023) Biofilms and their impact on the food industry. Saudi J Biolo Sci 30(2):103523. https://doi.org/10.1016/j.sjbs.2022.103523

Pandey A (2021) Food wastage: causes, impacts and solutions. Sci Herit J 5(1):17–20. https://doi.org/10.26480/gws.01.2021.17.20

Pérez-Almada D, Galán-Martín Á, del Mar Contreras M, Castro E (2023) Integrated techno-economic and environmental assessment of biorefineries: review and future research directions. Sustain Energ Fuels 7(17):4031–4050

Prasow M, Domaradzki P, Litwińczuk A, Kowalczyk M (2019) Bioactive compounds in meat and their importance in human nutrition. MedycynaOgólna I Nauki O Zdrowiu 25(3):170–180. https://doi.org/10.26444/monz/112261

Rajendran N, Han J (2022) Techno-economic analysis of food waste valorization for integrated production of polyhydroxyalkanoates and biofuels. Bioresour Technol 348:126796

Sadhu SD, Garg M, Kumar A (2018) Major environmental issues and new materials. In: Elsevier eBooks, pp 77–97. https://doi.org/10.1016/b978-0-12-811033-1.00004-4

Sarker A, Ahmmed R, Ahsan SM, Rana J, Ghosh MK, Nandi R (2024) A comprehensive review of food waste valorization for the sustainable management of global food waste. Sustain Food Technol 2(1):48–69. https://doi.org/10.1039/d3fb00156c

Sharma SB, Sayyed RZ, Trivedi MH, Gobi TA (2013) Phosphate solubilizing microbes: sustainable approach for managing phosphorus deficiency in agricultural soils. Springer Plus 2:587. https://doi.org/10.1186/2193-1801-2-587

Sharma S, Mohler J, Mahajan SD, Schwartz SA, Bruggemann L, Aalinkeel R (2023) Microbial biofilm: a review on formation, infection, antibiotic resistance, control measures, and innovative treatment. Microorganisms 11(6):1614. https://doi.org/10.3390/microorganisms11061614

Spiker ML, Hiza HAB, Siddiqi SM, Neff RA (2017) Wasted food, wasted nutrients: nutrient loss from wasted food in the United States and comparison to gaps in dietary intake. J Acad Nutr Diet 117(7):1031–1040.e22. https://doi.org/10.1016/j.jand.2017.03.015

Stenmarck A, Jensen C, Quested T, Moates G (2016) Estimates of European food waste levels. Swedish Environmental Research Institute, Stockholm

Thepbandit W, Athinuwat D (2024) Rhizosphere microorganisms supply availability of soil nutrients and induce plant defense. Microorganisms 12(3):558. https://doi.org/10.3390/microorganisms12030558

Thiviya P, Gamage A, Kapilan R, Merah O, Madhujith T (2022) Single cell protein production using different fruit waste: a review. Separations 9(7):178. https://doi.org/10.3390/separations9070178

Verma M (2015) Food wastage—energy wasted. In: *Springer briefs in food, health and nutrition*, pp 23–26. https://doi.org/10.1007/978-3-319-16781-7_4

Wang Z, Hu Y, Wang S, Wu G, Zhan X (2023) A critical review on dry anaerobic digestion of organic waste: characteristics, operational conditions, and improvement strategies. Renew Sust Energ Rev 176:113208

Zhang K, Zhang TT, Guo RR, Ye Q, Zhao HL, Huang XH (2023) The regulation of key flavor of traditional fermented food by microbial metabolism: a review. Food Chem 19:100871. https://doi.org/10.1016/j.fochx.2023.100871

Zhao A, Sun J, Liu Y (2023) Understanding bacterial biofilms: from definition to treatment strategies. Front Cell Infect Microbiol 13:1137947. https://doi.org/10.3389/fcimb.2023.1137947

Future Vision: Biofilm Global Impact

14

Joyeeta Bhattacharya, Subhasis Chakrabarty, Asim Halder, Partha Roy, and Soma Das

Abstract

In the present scenario, the microbial biofilm has bagged much limelight owing to its huge application in translational research. Biofilms are poised to have a significant global impact in diverse sectors, including healthcare, agriculture, and environmental sustainability. With the increasing threat of antibiotic-resistant bacteria, biofilms offer a potential solution as they possess inherent resistance against antibiotics. This has sparked a growing interest in developing biofilm-based therapies for various infections. Moreover, the application of biofilms in agriculture holds promise for enhancing crop growth and protecting plants against diseases, thus reducing the need for harmful pesticides. Additionally, the ability of biofilms to degrade pollutants and remediate contaminated sites makes them a valuable tool for promoting environmental sustainability. With ongoing research and development, it is clear that biofilms have the potential to revolutionize many aspects of our lives and contribute to a healthier, more sustainable future. Nevertheless, advancements in optimizing this technology are essential in order to minimize the suspected and unintended risks of the biofilms toward the society.

Keywords

Biofilm · Present scenario · Future aspects · Global impact

J. Bhattacharya · S. Chakrabarty
Department of Pharmaceutical Technology, DmbH Institute of Medical Science, Rasulpur, West Bengal, India

A. Halder · S. Das (✉)
Department of Pharmaceutical Technology, ADAMAS University, Kolkata, West Bengal, India

P. Roy
GITAM School of Pharmacy, GITAM University, Kolkata, West Bengal, India

T. Sarkar et al. (eds.), *Biofilm Applications to Revolutionize Food Technology*,
https://doi.org/10.1007/978-3-031-85205-3_14

14.1 Introduction

Biofilms are made up of multi-layered community of bacteria encapsulated in a polymeric matrix that adheres to one another on biological or non-biological surfaces. A single species or several species from one or more kingdoms, such as fungi, bacteria, and algae, can make up the microbial community (Sentenac et al. 2022; Edwards and Kjellerup 2013). Microbial biofilms are the most common form of life for microorganisms in all systems and can be found everywhere. They thus offer a plethora of options for translation across several sectors. The last two decades have witnessed a gradual development in the genetical studies of biofilm formation. The intracellular cell signaling and the mechanism of gene transcription are different from the planktonic cell. Thus, it also opens up a different perspective of biofilm study and helps the scientists to correlate the developmental process with other microbial organisms. Researches have used the polysaccharide stains such as Ruthenium red and Osmium tetroxide, to comprehend the basic chemistry of extracellular polymeric substance (EPS). The genetical studies have explained the mechanism of biofilm formation which involve multiple steps such as streamlined layer formation, adhesion of the microbes, evolution, and proliferation. The adhesion can take place on different types of surfaces namely plastic, glass, metals, living cells or tissues, water, and food. In case of bacterial biofilms, the attachment is facilitated by the EPS, flagella, pili, and fimbriae of the biofilm forming bacteria with the contact surface. More than one kind of bacterial species or one kind of microbial organisms (bacteria, fungi, algae) can coexist within a biofilm, ultimately giving a three-dimensional complex structure. These biofilms have a significant negative influence on the health of people and animals, present problems for food safety, impede the output of oil and gas wells, and contaminate drinking water supplies. However, they can also be helpful in other contexts, such as wastewater treatment procedures, and boosting the bioavailability of nutrients in the soil. Translational research on the prevention, detection, management, and engineering of biofilms has a lot of potential because of the widespread impact that biofilms have on several industries (Hemdan et al. 2021; Abdelfattah et al. 2020; Vigneron et al. 2018; Hall-Stoodley et al. 2004; Pandit et al. 2020). In this chapter, we have illustrated both the efficacious and the negative impact of biofilms on biotic and abiotic environment.

14.2 Impact of Biofilms on Human Health

Antibiotic Resistance

Antibiotic resistance is currently one of the most challenging setback faced by the medical fraternity of the world. Antibiotic resistance is also associated with high morbidity and mortality rate (Von Wintersdorff et al. 2016). Biofilms are responsible for several infectious diseases in human caused by both gram-positive and gram-negative bacteria. The multicellular compartments of microorganisms embedded in the matric of polymers decrease the efficacy of the antibiotics which target these microorganisms. This can lead to delay in the treatment of microbial infections. For the complete abolition of the cluster of microorganisms in the biofilm, the concentration of the antibiotic would be 1000 times higher than the conventional concentration of any antibiotics. The literature has highlighted various mechanisms by which the biofilms reduce the susceptibility toward antibiotics. The primary mechanism is based on the electrostatic interactions of the biofilm layer with the extracellular matrix that inhibits the easy the passage of antibiotics molecules through the biofilm layer. Due to the three-dimensional structure of the biofilm, there arises an unequal distribution of nutrients, pH, waste products, oxygen concentration, and other survival factors. As a consequence, the microbial population experience a depletion of nutrition with accumulation of waste products and other physiological stresses which in turn decrease the metabolic activities and growth. The fact that antibiotics target the growth factors and the cellular components of the microbes is thus repressed. Additionally, the mutation of the microbes within the biofilms also leads to antimicrobial resistance. Therefore, it is highly crucial to innovate effective strategies for biofilms associated infections (Connell et al. 2010; Stewart and Franklin 2008; Tack and Sabath 1985; Stewart 2002; Saint-Ruf et al. 2014; Perry and Tan 2023).

A significant impact of biofilms is seen in patients with cystic fibrosis. The lungs and the digestive system of the patients with cystic fibrosis are mostly damaged. The treatment becomes more challenging owing to the antibiotic resistance impacting the global economic with cost reaching up to $7509 m per annum (Domingue et al. 2020; Cystic Fibrosis Foundation 2021).

Device-Related Infections

In the hospital setup, the medical devices are rigorously used for several purposes such as infusion of chemotherapeutics, dialysis, treatment of urinary incontinence, and cardiovascular disorders. With respect to biofilms, they also grow on non-biological surfaces such as medical devices; 65% of hospital-borne infectious diseases are attributed to biofilms associated with medical devices such as urinary catheters, contact lenses, pacemakers, mechanical heart valves, and prosthetic joints (Srivastava and Bhargava 2016).

Biofilms are a common cause of catheter-related infections. The primary method of infection for central venous catheters (CVCs) is migration along the catheter's surface, which allows the intravascular section of the catheter to get colonized without passing through the skin or subcutaneous tissue. Due to the existence of biofilms, decontaminating CVCs in situ is challenging. In certain countries, the average duration of stay of patients in intensive care units has increased, which significantly worsens the prevalence of infection. An estimated $11.5 billion is spent annually worldwide on infections brought on by CVCs (Gominet et al. 2017; Ielapi et al. 2020). Biofilm generation has a significant cost influence on catheter-associated urinary tract infections (UTIs), which affect approximately 150 million individuals annually. Urinary catheterization is thought to be used in 15–25% of hospitalized patients, and it is responsible for around 75% of UTIs in hospital. An estimated $1 billion is spent globally each year on urinary tract infections linked to catheter use (European Centre for Disease Prevention and Control 2018).

Slow Wound Healing

Almost 90% chronic wounds provide a favorable environment for the growth of bacterial and fungi biofilms. The debris of necrotic tissues in chronic wounds is highly susceptible to microbial growth that not only weakens the immune system of the host but also slows down the process of wound healing. In a biopsy study of acute and chronic wound, it was observed that 60% of chronic wound consist of biofilm in comparison with acute wounds that contain only 6% of biofilms. Thus, providing therapy for chronic wounds becomes a challenging task for the health care providers. Several studies have confirmed the presence of biofilms in chronic wounds. Zhao et al. have conducted a study in a diabetic mouse model. A preformed biofilm of *Pseudomonas aeruginosa* presented in the wounds of diabetic mouse which showed delayed healing when compared uninfected control group (Zhao et al. 2012). In another study, Schierle et al. have inoculated Staphylococcus biofilms in wounds of non-diabetic mouse that also showcased slower tissue growth (Schierle et al. 2009). Pastar et al. have studied the impact of multispecies biofilms on chronic wound healing. Complex of *Pseudomonas aeruginosa* and *Staphylococcus aureus* biofilms were inoculated in porcine wounds, which suppressed the expression of keratinization growth factor and re-epithelialization of wound healing process when compared to a single strain biofilms (Pastar et al. 2013).

Carcinogenesis

Extensive studies have been carried out to evaluate the role of biofilms in the carcinogenesis of cancer related to gastrointestinal tract. Research done by Dejea et al. and Drewes et al. has reported the correlation of colorectal cancer with the bacterial biofilms present in the intestinal epithelium. The possible mechanism behind this is attributed to the increased proliferation of intestinal epithelial cells in biofilm-positive intestinal tissues, along with damaging the localization of E-cadherin (Drewes et al. 2017). *Escherichia coli* and *Bacteroides fragilis* biofilms are responsible for secreting oncogenic toxins in the pre-oncogenic lesions in patients with familial adenomatous polyposis (Pleguezuelos-Manzano et al. 2020). A human case-control research found a link between the prevalence of mouth cancer and the presence and severity of periodontal disease, indicating that pathogenic oral biofilms may similarly cause pro-oncogenic alterations in surrounding tissues (Komlós et al. 2021). Incubation of oral squamous cell carcinoma cell lines with periodontal pathogens resulted in enhanced cell motility and tumor development, but commensal mouth bacteria had no effect (Kamarajan et al. 2020). One of the most well-known risk factors for gastric cancer is *Helicobacter pylori*, which produces biofilms in the stomach mucosa. While a definitive correlation between *Helicobacter pylori* biofilm formation and the onset or advancement of gastric cancer remains to be established, strains of the bacteria that carry a particular cytotoxin-associated gene pathogenicity island exhibit enhanced biofilm formation capacities in vitro and are linked to an increased probability of gastric cancer (Wong et al. 2016; Huang et al. 2003).

Dental Hygiene

Biofilms that grow between the teeth and the gums are responsible for the formation of oral plaque. Oral plaques increase bacterial resistance to both the body's defense and oral cleaning treatments. Cavities and gingivitis are the result of the demineralization of teeth brought on by oral plaque. Studies have revealed that the primary microorganism responsible for the formation of dental caries is *Streptococcus mutants*. The toxin produced by the bacteria as a result of fermentation of sucrose erodes the tooth enamel causing dental caries. The maintenance of oral hygiene requires a regular cleaning of teeth and removal of plaques which impose a significance cost on the global economic (Larsen and Fiehn 2017; Gurenlian 2007; Grand View Research 2021; Lasserre et al. 2018).

14.3 Impact of Biofilms on Food Industry

The extracellular matrix of the biofilms adheres to the surface of the food and equipment associated with food (such as surfaces of meat, fruits, equipment in food industry). This strong adhesion is the outcome of the various process involved in food processing such as oxygen diffusion and enzymes interacting with the extracellular matrix of the biofilm. Importantly, food-associated bacteria can coexist with pathogenic bacteria in both dual- and multiple-species biofilms (Galié et al. 2018). In vitro studies have demonstrated that the foodborne pathogens are incapable of formation of biofilms and therefore they establish a symbiotic relationship with the microbes of the biofilms which help them to cling to the surface of food and equipment of food industry.

The major challenge is to arrest the development of these pathogens on biofilms, which have a major impact on human health in terms food-food contamination and spoilage, thereby leading to food poisoning. For instance, one such foodborne pathogen is *Listeria monocytogenes*, the causative factor for Listeriosis. The *Listeria monocytogenes* is capable of forming biofilm on plastics, stainless steels, and other various kind of surface materials in the food processing industries. In the year 2019, 142 cases of listeriosis were reported in the United Kingdom that has resulted in 23 deaths and 8 miscarriages (Public Health England 2021; Yuan et al. 2022; Chitlapilly Dass and Wang 2022). As per the record, the economic impact in the United States have spiked up to $78 billion a year with a number of 128,000 hospitalizations each year, due to the foodborne pathogens (Bai et al. 2021).

14.4 Impact of Biofilms on Wastewater Treatment

The amount of pollution produced by industry has increased along with global urbanization. While pollution of all kinds poses a serious threat to the environment, wastewater treatment has long piqued the curiosity of scientists around the world. Numerous contaminants can be removed from wastewater using physical, biological, or chemical techniques. Technologies for biological wastewater treatment are thought to be cutting-edge substitutes for traditional methods, and they are vital to enhancing both the quality of the water and human health. Organic pollution is a major cause of various phenomena, including eutrophication, dissolved oxygen reduction, and toxin buildup, all of which have an impact on water quality. As a result, it is necessary to remove organic contaminants from wastewater. The employment of bioremediation as an environmentally friendly technology for the decontamination of dangerous chemicals for environmental sustainability and safety has gained favor due to the various shortcomings (cost, operation, efficiency, and start-up) of standard wastewater treatment plants. The process utilizes the biological communities of bacteria, fungi, and algae for elimination and neutralization of toxic contaminants from wastewater. The application of biofilms in the bioremediation process has gained much popularity owing to the affinity of microbes of biofilms toward nutrients in comparison with the natural planktonic cells. These microbes are capable to withstand extreme environmental conditions and robust toward harsh chemicals. This survival strategy of the microbes has caught the eyes of the global scientists, who are now focused in manipulating the gene expressions of microbes, so as to prepare biofilm communities that are efficiently medicated toward elimination of pollutants from water (Saini et al. 2023).

Human activities, including natural gas production, crude oil extraction, and mining, have contributed significantly toward environmental contamination. Dyes, oils, pesticides, heavy metals, herbicides, phenolic chemicals, and polyaromatic hydrocarbons (PAHs) are some of the most hazardous and refractory xenobiotic wastes produced by these operations. These substances are classed as hazardous and carcinogenic due to their harmful impact on the environment and human health. Petrochemical chemicals polluted over 67% of the 5 million sites on Earth (Stasik et al. 2015; Ibrahim and Al-Hossainy 2020).

One of the toxic chemicals present in the industrial wastewater is acetonitrile. This chemical compound after metabolization is converted to hydrogen cyanide and ethylene oxide which is highly hazardous to aquatic and terrestrial species. Li et al., in their study, prepared a biofilm using three bacterial strains, *B. subtilis* N4, E2, and E3, along with *Rhodococcus rhodochrous*, to diminish the acetonitrile toxicity in an effective manner (Li et al. 2013). Polycyclic aromatic hydrocarbons (PAHs) are one of the most harmful contaminants in the environment due to their aromatic ring structures containing carbon, which have mutagenic and carcinogenic implications in humans. Several bacterial species were employed in the bioremediation of PAHs. These species were used in the pollutant trapping inside extracellular polymeric substance (EPS) by biosorption, as a source of nutrition for the growth of the biofilm and the protection against PAH toxicity (Edwards and Kjellerup 2013). Biofilm-mediated reactors have successfully bioremediated a variety of insecticides and herbicides, including 2-propionic acid, diclofop, 2,4-dichlorophenoxyacetic acid, carben-dazim, chlorpropham, methomyl, diazinon, carbamate compounds, carbaryl, and carbofuran (Chen et al. 2015; Horemans et al. 2016; Tien et al. 2013). Synthetic dyes and toxic heavy metals comprise the inorganic pollutants of the wastewater that have shown detrimental effects toward marine and human health. The bacterial biofilms play a significant role in the bioremediation of these inorganic pollutants from wastewater by chelating with heavy metal ions (Dey and Paul 2018; Parellada et al. 2017). For instance, bacterial biofilms produced using *Pseudochrobactrum mendonica* sp., *Pseudochrobactrum sacchaolyicum* sp., and *Arthrobacter* sp. have been used to remove heavy metals and xenobiotic chemicals (Long et al. 2015). A pure bacterial biofilm or a combination of *Cellulosimicrobium* sp. and *Halomonasaquamarina* TA-04208 cultivated on various matrices (such as rubber, polyvinyl chloride, sand, or stone) may reduce 500 g mL^{-1} of chromate by more than 90% in all circumstances. Several heavy metal ions, such as Cu(II), Zn(II), Fe(II), Fe(III), and Al(III), have been found to accumulate in the B. subtilis biofilm. The metal adsorption properties of biofilm have been used in biotechnology to extract uranium from groundwater (Naeem et al. 2013; Focardi et al. 2012; Grumbein et al. 2014; Cao et al. 2011). The electronics, pharmaceutical, and cosmetics sectors release harmful micropollutants into the environment, causing water contamination. The use of microbes in biofilms is advantageous for micropollutant bioremediation since they may break down or absorb a broad spectrum of highly toxic pollutants through many catabolic pathways (Edwards and Kjellerup 2013). Torresi et al. investigated the sorption of several micropollutants in moving bed biofilm reactors (MBBR) carriers with varying biofilm thicknesses, and only eight micropollutants, including roxithromycin, clarithromycin, erythromycin, venlafaxine, citalopram, propranolol, metoprolol, and atenolol, demonstrated significant binding and sorption via carrier-electrostatic interactions (Torresi et al. 2017). Titanium dioxide nanoparticles can be found in a wide range of home products, including cosmetics and paints. Consequently, a significant portion of these nanoparticles find their way into wastewater that is released from residential areas worldwide. Titanium dioxide nanoparticles may pose a significant risk to the microenvironment because titania is known to be photocatalytically active. To reduce the toxicity of nanoparticles, biofilm-mediated absorption and elimination were investigated. The presence of humic and surfactant molecules in the wastewater made nanoparticle remediation easier. The uptake of pollutants by biofilm bacteria is influenced by nutritional, temperature, and osmotic pressure conditions, as well as diffusion transport. The dissolved and organic matter components in wastewater serve as a foundation for absorption of micropollutant (Mathur et al. 2017).

14.5 Impact of Biofilms on Space

A very pertinent concern of the present era is the hazardous effect of biofilms on the health of astronauts. The space agencies are increasingly concerned about the formation of biofilms in space-related settings. In addition to growing on the surfaces of space hardware, biofilms have also been observed to grow on other vital parts, including air ducts, water ducts, and other life support systems, which can be harmful to human health. As the duration of space flights increases, options for growing food become more prominent. Biofilms can cause a wide range of issues with space-based plant growth systems, such as contaminating food, water, and plant growth medium. Additionally, these systems contain a range of particular hardware parts, electrical circuits, and tubing that are prone to contamination and corrosion. It is difficult to remove biofilms from the materials used in these spacecrafts, and biofilm growth must be avoided by constant observation. One significant obstacle in the study of biofilms connected to space travel is the need to recreate the space environment in order to create preventative measures against biofilm formation. (Gu et al. 1998; Castro et al. 2004; Koenig and Pierson 1997). A study by Kim et al. showcased that the biofilm formed by Pseudomonas aeruginosa is denser under microgravity conditions than it is on earth. The rate of diffusion of nutrients and oxygen within the biofilms under microgravity has majorly impacted the cell density of the biofilm. Combating the formation of biofilms on the spacecrafts has opened several avenues of research. Studies incorporating molecular techniques, enzymatic techniques and other chemical methods, are being applied by numerous scientists with an approach to control the biofilm formation and to improve the health of astronauts in longer space missions (Kim et al. 2013a, b).

14.6 Impact of Biofilms on Soil Health

Numerous microorganisms exist in the soil environment in either sessile or biofilm forms. In natural environments, more than 99% of bacteria create biofilms. Microbial biofilms formed on soil are made up of single or several species. They are arranged on many interfaces and substrates, including the surfaces of minerals, pore spaces, and plant roots. Extracellular compounds produced by bacteria serve as a biofilm matrix, ensuring biofilm cohesion. Soil biofilms have various benefits that enhance soil characteristics. The physicochemical qualities of the local soil are improved by the formation of soil biofilms. The biofilm-induced bioclogging of soil pores increases water permeability. Exopolysaccharides produced by soil biofilms strengthen the internal cohesiveness of soil microaggregates. They also help plants by preserving the moisture in nutrients. A key factor in forming soil microniches is the capacity of microorganisms to survive unfavorable environments. Certain bacterial biofilms that grow on rhizosphere plant roots prevent plant diseases from colonizing by reducing the amount of nutrients that are available. The way biofilm communities interact affects climate change, bioremediation, plant protection, plant nutrition, and soil quality. Therefore, biofilms play a crucial role in the advancement of environmentally friendly and sustainable agriculture (Anju et al. 2021).

14.7 Conclusion

The aforementioned implications on the economy and society demonstrate that biofilms play a key role in some of the most significant global issues and that there is a significant potential for growth and benefit from using and managing the complex biofilms' activities. Nonetheless, there are still a lot of unsolved scientific problems, and progress in these fields will depend on advancements in fundamental and applied research. Even if there are still a lot of unanswered concerns, the development of intricate in vitro experimental models will give researchers new tools to sort through in order to understand how the host environment influences biofilm phenotypes and how certain biofilm components cause host responses. This could have a significant impact on human health by enhancing our capacity to identify and modify tissue-associated biofilms in situ.

References

Abdelfattah A, Hossain MI, Cheng L (2020) High-strength wastewater treatment using microbial biofilm reactor: a critical review. World J Microbiol Biotechnol 36(5):75

Anju VT, Siddhardha B, Dyavaiah M (2021) Effect of soil biofilms on ecological function and impact on soil properties. In: Microbes in land use change management. Elsevier, pp 91–107

Bai X, Nakatsu CH, Bhunia AK (2021) Bacterial biofilms and their implications in pathogenesis and food safety. Food Secur 10(9):2117

Cao B, Ahmed B, Kennedy DW, Wang Z, Shi L, Marshall MJ et al (2011) Contribution of extracellular polymeric substances from Shewanella sp. HRCR-1 biofilms to U (VI) immobilization. Environ Sci Technol 45(13):5483–5490

Castro VA, Thrasher AN, Healy M, Ott CM, Pierson DL (2004) Microbial characterization during the early habitation of the International Space Station. Microb Ecol 47:119–126

Chen CS, Wu TW, Wang HL, Wu SH, Tien CJ (2015) The ability of immobilized bacterial consortia and strains from river biofilms to degrade the carbamate pesticide methomyl. Int J Environ Sci Technol 12:2857–2866

Chitlapilly Dass S, Wang R (2022) Biofilm through the looking glass: a microbial food safety perspective. Pathogens 11(3):346

Connell JL, Wessel AK, Parsek MR, Ellington AD, Whiteley M, Shear JB (2010) Probing prokaryotic social behaviors with bacterial "lobster traps". MBio 1(4):10–1128

Cystic Fibrosis Foundation (2021) About cystic Fibrosi. https://www.cff.org/What-is-CF/About-Cystic-Fibrosis

Dey S, Paul AK (2018) Influence of metal ions on biofilm formation by Arthrobacter sp. SUK 1205 and evaluation of their Cr (VI) removal efficacy. Int Biodeterior Biodegradation 132:122–131

Domingue JC, Drewes JL, Merlo CA, Housseau F, Sears CL (2020) Host responses to mucosal biofilms in the lung and gut. Mucosal Immunol 13(3):413–422

Drewes JL, White JR, Dejea CM, Fathi P, Iyadorai T, Vadivelu J et al (2017) High-resolution bacterial 16S rRNA gene profile meta-analysis and biofilm status reveal common colorectal cancer consortia. NPJ Biofilms Microbiomes 3(1):34

Edwards SJ, Kjellerup BV (2013) Applications of biofilms in bioremediation and biotransformation of persistent organic pollutants, pharmaceuticals/personal care products, and heavy metals. Appl Microbiol Biotechnol 97:9909–9921

European Centre for Disease Prevention and Control (2018) Incidence and attributable mortality of healthcare-associated infections in intensive care units in Europe, 2008–2012. European Centre for Disease Prevention and Control

Focardi S, Pepi M, Landi G, Gasperini S, Ruta M, Di Biasio P, Focardi SE (2012) Hexavalent chromium reduction by whole cells and cell free extract of the moderate halophilic bacterial strain Halomonas sp. TA-04. Int Biodeterior Biodegradation 66(1):63–70

Galié S, García-Gutiérrez C, Miguélez EM, Villar CJ, Lombó F (2018) Biofilms in the food industry: health aspects and control methods. Front Microbiol 9:898

Gominet M, Compain F, Beloin C, Lebeaux D (2017) Central venous catheters and biofilms: where do we stand in 2017? APMIS 125(4):365–375

Grand View Research (2021) Oral Care Market Size, Share & Trends Analysis Report By Product (Toothbrush, Toothpaste), By Product Type (Countertop, Cordless), By Application (Home, Dentistry), By Region, And Segment Forecasts. 128, https://www.grandviewresearch.com/industry-analysis/oral-care-market

Grumbein S, Opitz M, Lieleg O (2014) Selected metal ions protect Bacillus subtilis biofilms from erosion. Metallomics 6(8):1441–1450

Gu JD, Roman M, Esselman T, Mitchell R (1998) The role of microbial biofilms in deterioration of space station candidate materials. Int Biodeterior Biodegradation 41(1):25–33

Gurenlian JR (2007) The role of dental plaque biofilm in oral health. American Dental Hygienists' Association 81(suppl 1):116–116

Hall-Stoodley L, Costerton JW, Stoodley P (2004) Bacterial biofilms: from the natural environment to infectious diseases. Nat Rev Microbiol 2(2):95–108

Hemdan BA, El-Taweel GE, Goswami P, Pant D, Sevda S (2021) The role of biofilm in the development and dissemination of ubiquitous pathogens in drinking water distribution systems: an overview of surveillance, outbreaks, and prevention. World J Microbiol Biotechnol 37:1–18

Horemans B, Albers P, Springael D (2016) The biofilm concept from a bioremediation perspective

Huang JQ, Zheng GF, Sumanac K, Irvine EJ, Hunt RH (2003) Meta-analysis of the relationship between cagA seropositivity and gastric cancer. Gastroenterology 125(6):1636–1644

Ibrahim SM, Al-Hossainy AF (2020) Kinetics and mechanism of oxidation of bromothymol blue by permanganate ion in acidic medium: application to textile industrial wastewater treatment. J Mol Liq 318:114041

Ielapi N, Nicoletti E, Lorè C, Guasticchi G, Avenoso T, Barbetta A et al (2020) The role of biofilm in central venous catheter related bloodstream infections: evidence-based nursing and review of the literature. Rev Recent Clin Trials 15(1):22–27

Kamarajan P, Ateia I, Shin JM, Fenno JC, Le C, Zhan L et al (2020) Periodontal pathogens promote cancer aggressivity via TLR/MyD88 triggered activation of integrin/FAK signaling that is therapeutically reversible by a probiotic bacteriocin. PLoS Pathog 16(10):e1008881

Kim W, Tengra FK, Young Z, Shong J, Marchand N, Chan HK, Collins CH (2013a) Spaceflight promotes biofilm formation by Pseudomonas aeruginosa. PLoS One 8(4):e62437

Kim W, Tengra FK, Shong J, Marchand N, Chan HK, Young Z, Collins CH (2013b) Effect of spaceflight on Pseudomonas aeruginosa final cell density is modulated by nutrient and oxygen availability. BMC Microbiol 13:1–10

Koenig DW, Pierson DL (1997) Microbiology of the space shuttle water system. Water Sci Technol 35(11–12):59–64

Komlós G, Csurgay K, Horváth F, Pelyhe L, Németh Z (2021) Periodontitis as a risk for oral cancer: a case–control study. BMC Oral Health 21(1):640

Larsen T, Fiehn NE (2017) Dental biofilm infections – an update. APMIS 125(4):376–384

Lasserre JF, Brecx MC, Toma S (2018) Oral microbes, biofilms and their role in periodontal and peri-implant diseases. Materials 11(10):1802

Li C, Li Y, Cheng X, Feng L, Xi C, Zhang Y (2013) Immobilization of Rhodococcus rhodochrous BX2 (an acetonitrile-degrading bacterium) with biofilm-forming bacteria for wastewater treatment. Bioresour Technol 131:390–396

Long D, Zou L, Hashmi MZ, Cai K, Tang X, Chen G, Shi J (2015) Determination of the accumulation, spatial distribution and reduction of Cr in unsaturated Pseudochrobactrum saccharolyticum LY10 biofilms by X-ray fluorescence and absorption methods. Chem Eng J 280:763–770

Mathur A, Parashar A, Chandrasekaran N, Mukherjee A (2017) Nano-TiO2 enhances biofilm formation in a bacterial isolate from activated sludge of a waste water treatment plant. Int Biodeterior Biodegradation 116:17–25

Naeem A, Batool R, Jamil N (2013) Cr (VI) reduction by Cellulosimicrobium sp. isolated from tannery effluent. Turk J Biol 37(3):315–322
Pandit A, Adholeya A, Cahill D, Brau L, Kochar M (2020) Microbial biofilms in nature: unlocking their potential for agricultural applications. J Appl Microbiol 129(2):199–211
Parellada EA, Igarza M, Isacc P, Bardón A, Ferrero M, Ameta KL, Neske A (2017) Squamocin, an annonaceous acetogenin, enhances naphthalene degradation mediated by Bacillus atrophaeus CN4. Rev Argent Microbiol 49(3):282–288
Pastar I, Nusbaum AG, Gil J, Patel SB, Chen J, Valdes J et al (2013) Interactions of methicillin resistant Staphylococcus aureus USA300 and Pseudomonas aeruginosa in polymicrobial wound infection. PLoS One 8(2):e56846
Perry EK, Tan MW (2023) Bacterial biofilms in the human body: prevalence and impacts on health and disease. Front Cell Infect Microbiol 13:1237164
Pleguezuelos-Manzano C, Puschhof J, Huber AR, Hoeck A, Wood HM, Nomburg J et al (2020) Mutational signature in colorectal cancer caused by genotoxic pks+ E. coli. Nature 580:269–273. https://doi.org/10.1038/s41586-020-2080-8
Public Health England (2021) Listeriosis in England and Wales: summary for 2019. Public Health England
Saini S, Tewari S, Dwivedi J, Sharma V (2023) Biofilm-mediated wastewater treatment: a comprehensive review. Mater Adv 4(6):1415–1443
Saint-Ruf C, Garfa-Traoré M, Collin V, Cordier C, Franceschi C, Matic I (2014) Massive diversification in aging colonies of Escherichia coli. J Bacteriol 196(17):3059–3073
Schierle CF, De la Garza M, Mustoe TA, Galiano RD (2009) Staphylococcal biofilms impair wound healing by delaying re-epithelialization in a murine cutaneous wound model. Wound Repair Regen 17(3):354–359
Sentenac H, Loyau A, Leflaive J, Schmeller DS (2022) The significance of biofilms to human, animal, plant and ecosystem health. Funct Ecol 36(2):294–313
Srivastava S, Bhargava A (2016) Biofilms and human health. Biotechnol Lett 38:1–22
Stasik S, Wick LY, Wendt-Potthoff K (2015) Anaerobic BTEX degradation in oil sands tailings ponds: impact of labile organic carbon and sulfate-reducing bacteria. Chemosphere 138:133–139
Stewart PS (2002) Mechanisms of antibiotic resistance in bacterial biofilms. Int J Med Microbiol 292(2):107–113
Stewart PS, Franklin MJ (2008) Physiological heterogeneity in biofilms. Nat Rev Microbiol 6(3):199–210
Tack KJ, Sabath LD (1985) Increased minimum inhibitory concentrations with anaerobiosis for tobramycin, gentamicin, and amikacin, compared to latamoxef, piperacillin, chloramphenicol, and clindamycin. Chemotherapy 31(3):204–210
Tien CJ, Lin MC, Chiu WH, Chen CS (2013) Biodegradation of carbamate pesticides by natural river biofilms in different seasons and their effects on biofilm community structure. Environ Pollut 179:95–104
Torresi E, Polesel F, Bester K, Christensson M, Smets BF, Trapp S et al (2017) Diffusion and sorption of organic micropollutants in biofilms with varying thicknesses. Water Res 123:388–400
Vigneron A, Head IM, Tsesmetzis N (2018) Damage to offshore production facilities by corrosive microbial biofilms. Appl Microbiol Biotechnol 102:2525–2533
Von Wintersdorff CJ, Penders J, Van Niekerk JM, Mills ND, Majumder S, Van Alphen LB et al (2016) Dissemination of antimicrobial resistance in microbial ecosystems through horizontal gene transfer. Front Microbiol 7:173
Wong EHJ, Ng CG, Chua EG, Tay ACY, Peters F, Marshall BJ et al (2016) Comparative genomics revealed multiple Helicobacter pylori genes associated with biofilm formation in vitro. PLoS One 11(11):e0166835
Yuan L, Wang H, Liu W, Sadiq FA, Zhao Y (2022) Multi-species biofilms in the food industry. Front Microbiol 13:1023428
Zhao G, Usui ML, Underwood RA, Singh PK, James GA, Stewart PS et al (2012) Time course study of delayed wound healing in a biofilm-challenged diabetic mouse model. Wound Repair Regen 20(3):342–352

Food Forensics

15

Somnath Roy and Abir Chakravorty

Abstract

Nowadays, the food business generates enormous profits, making it a prime candidate for fraud. This brings to mind the necessity of strong and dependable diagnostic methods in order to create analytical procedures for fraud detection. A new area of forensic chemistry called "food forensics" steps in to help when food is tainted or mislabeled. It is beneficial to verify food quality and safety, among other things. In India, food fraud, namely adulteration, is becoming a bigger issue. All food products, whether they are processed or raw, are frequently contaminated. Although a lot of money is spent on analyses as well as control measures, food safety is becoming more and more vital in the current world due to the various foodborne illnesses associated to infections, toxins, pesticides, adulterants, colorants, and other pollutants. It has long been believed that certain biogenic amines are the primary indicator of food quality in processed foods, serving as a watchdog over the deterioration of fresh food and food high in protein. Typical methods for detecting hazardous analytes that have been employed; however, array-based sensing strategies are becoming more and more common in order to create an exceptionally precise and accurate analytical procedure. Hence, we briefly describe and evaluate the recently published array-based sensor systems supported by machine learning and multivariate analytics here in order to maintain food quality in the field of food forensics. The fact that protein toxins, prions, and other tiny compounds that can contaminate our food can all be securely detected and quantified using the potential of multiple reaction monitoring (MRM).

Both are the first authors of the manuscript

S. Roy · A. Chakravorty (✉)
Agricultural & Food Engineering Department, Indian Institute of Technology Kharagpur, Kharagpur, India
e-mail: abir.28kh@agfe.iitkgp.ac.in

T. Sarkar et al. (eds.), *Biofilm Applications to Revolutionize Food Technology*,
https://doi.org/10.1007/978-3-031-85205-3_15

Keywords

Biofilm technology · Global applications · Sustainability · Industrial innovation · Health and safety

15.1 Analytical Systems Driven by Machine Learning

Food forensics is the study of food items using scientific techniques to establish their legitimacy and traceability. Testing for different food contaminants, sugar levels, and nucleic acid is usually used to confirm the authenticity of food (Su et al. 2022). Food contamination can be categorized into three main parts: physical, chemical, and biological (Fig. 15.1). There is a chance of both direct and indirect chemical contamination, and these can entail a range of hazardous materials such as pesticides, heavy metals, poisons, fertilizers, and food additives.

An Overview of Machine Learning for Analytical Use in Food Science

Advances in machine learning-based methodologies and multivariate analytical techniques have made array-based sensing possible. The prevalent problem of comprehending the complex response patterns generated by array-based solutions can be solved with the help of these techniques. As a result, machine learning-based neural networks enable sensor arrays to provide fast, dependable, and accurate analyte detection. Consequently, the main emphasis these days is on creating methodologies based on electrical and chemical sensor arrays for commercial applications. Due to the numerous unique analytes in a complicated matrix that might provide

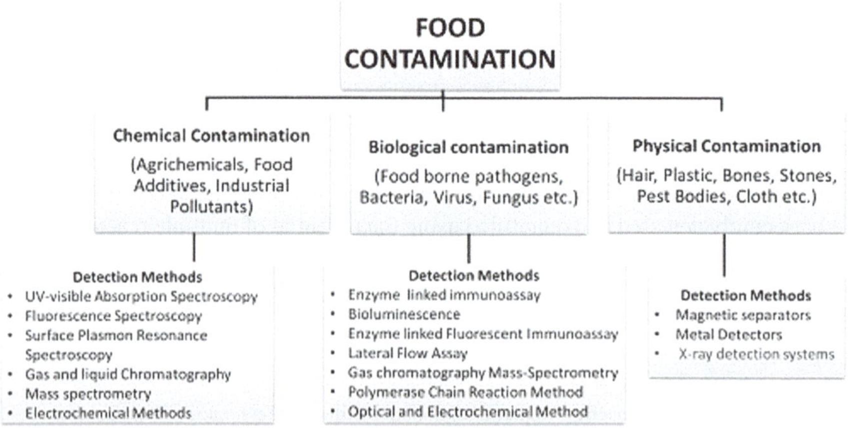

Fig. 15.1 Diagrammatic illustration of food contamination categories and associated methods of detection. (Reprinted with permission (Ranbir et al. 2022))

false-positive results, food analysis is challenging. To reduce false-positive results, multiplexed sensors targeting many analytes are needed. Machine learning (ML) has been extensively employed in environmental monitoring and food forensics because of its consistency in identifying and differentiating between analytes. Lots of optical and fluorescence-based machine-learning techniques are often employed, including principal component analysis, linear discriminant analysis, partial least-squares discriminant analysis, multiple linear regression, and hierarchical clustering analysis (Ranbir et al. 2022).

The intricacy of the investigation in the realm of food forensics is usually the main disadvantage of standard analytical techniques. It is not always possible to identify a specific analyte since different analytes share structural similarities. In particular, for applications in complex matrices like food forensics, bioanalytical systems, etc., the fabrication of these incredibly selective sensors seems to be costly and time-consuming. In order to deal with these issues, array sensors are the most efficient computing methods with conventional analytical instruments. Because array sensors can detect multiple analytes in a complex matrix system at the same time, they are very helpful in food forensics. These systems work to distinguish multiple identical analytes by using different cross-reactive components. Nonetheless, the results produced by the interaction of sensors and analytes exhibit a particular pattern called a fingerprint. Array sensors are unable to selectively identify the target analyte (Keith et al. 2021; Frades and Matthiesen 2010).

Several machine-learning approaches have been used to further refine the fingerprint responses that are obtained. Within the field of artificial intelligence, machine learning employs an approach to problem-solving that departs from conventional methods. A machine-learning methodology uses optimization and fresh data to train a model to become used to its parameters, or it learns from existing models. We call this method "model-based learning." After that, the model produces predictions. In daily life, it is useful to need to be able to swiftly, precisely, and automatically detect dietary qualities. To ascertain food qualities, modern methodologies are including as spectroscopy, spectral imaging, and electronic noses are frequently employed. These technologies allow for the collection of a large amount of digital data on food properties. Since there is a lot of unnecessary and duplicated information in the massive volume of data, data evaluation is crucial to these strategies. It is difficult to implement these approaches in real situations because managing such an enormous amount of data and extracting valuable features what emerges from the data is a critical problem. As a result, many methods for data inspection and analysis—both supervised and unsupervised, as seen in Fig. 15.2—have emerged to handle the enormous amount of data. Machine-learning tools can be trained in a variety of ways, each with unique benefits and drawbacks. Unsupervised learning is a technique that focuses on issues without labels or output and trains models on training data sets with no supervision. It is divided into challenges with clustering and association. The most popular clustering technique for forming clusters from unlabeled datasets is Kmeans. Principal component analysis (PCA) is another unsupervised descriptive tactics that maintains its discriminating power even when the data's

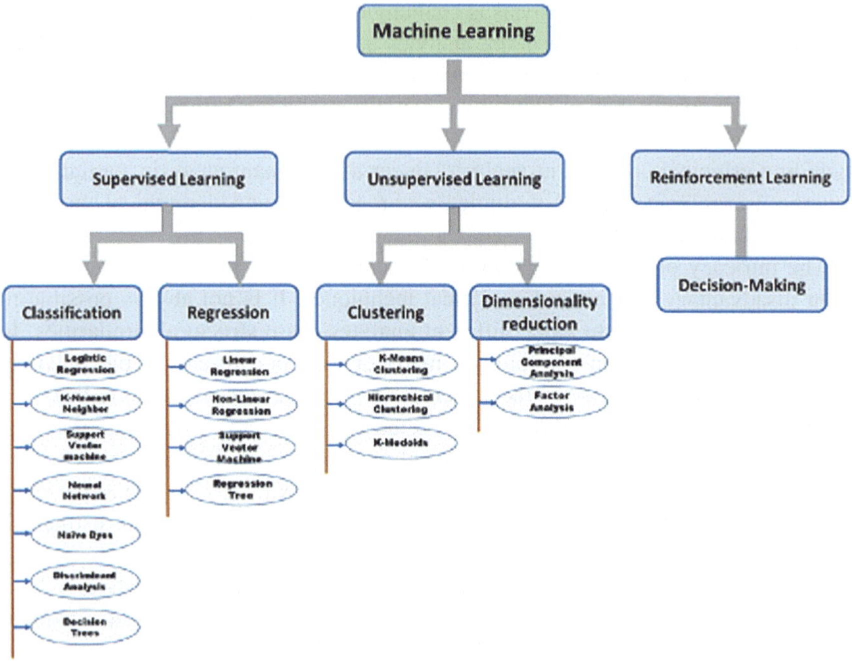

Fig. 15.2 Diagram showing several kinds of machine-learning methods. (Reprinted with permission (Ranbir et al. 2022))

dimension is reduced. Labeled data are used in supervised learning techniques like LDA, which predict outcomes based on input (Stewart et al. 2014).

Use of Machine Learning as a Tool in Food Forensics

In the last 30 years, research in optical chemical sensors has flourished, with many publications emphasizing the advantages of optical sensing compared to alternative transduction methods. The demand for efficient, rapid, and cost-effective identification of hazardous compounds in food is pressing. Array-based sensing methodologies provide unique response patterns, facilitating chemical detection through correlation with a designated library of responses. Array-based methodologies utilizing digital imaging techniques provide distinct interactions with analytes, producing unique fingerprint patterns for various analytes. The increasing prevalence of array-based sensing has resulted in the development of colorimetric and fluorescent sensor arrays for diverse analytes, enabling detection in both solution and vapor phases. It is also critical to design portable devices that can interact with numerous target analytes to generate distinct fingerprint patterns or that can be further examined using a variety of machine-learning techniques (Li et al. 2018).

For Meat and Protein-Rich Foods

In contemporary society, the foremost challenge is to provide nutritious food devoid of harmful residues, pesticides, allergies, and pathogen contamination. Microorganisms can infect food and influence its texture, flavor, aroma, and color. Sulfur compounds and biogenic amines are critical indicators for assessing meat freshness during storage. It was claimed that two fluorescent cucurbits probes were used to build a supramolecular assembly (Minami et al. 2012), while some researcher (Salinas et al. 2012) designed an optoelectronic nostrils using 16 pigments embedded in porous silica. Xiao-Weiet. al (Xiao-wei et al. 2016). fabricated a nanoporous color-responsive sensor consists of TiO2 film for the detection of trimethylamine gas. Microfluidic paper-based color-responsive sensor arrays have value due to their low cost, ease of use, and easy fabrication (Li et al. 2016). Researchers also found aldehydes as Warmed-Over Flavor (WOF) markers of cooked beef (Kim et al. 2016), and BP-ANN algorithm was used for better classified and prediction analysis (Li et al. 2014). Nguyen et al. (2019) created a cost-effective colorimetric sensor for the finding of gaseous ethanolamine, trimethylamine, and dimethylamine. Liu et al. (2020) formulated a color model for the developed probe and created a multicolor sensor array utilizing gold nanorods and nanospheres to differentiate and detect biogenic amines in meat and fish samples.

Regarding Drinks and Sweeteners

Sixty-two percent of sweeteners manufactured in the business are artificial sweeteners used in a vast range of products. Since 1977, the FDA has authorized non-nutritive sweeteners such as sucralose, aspartame, saccharin, and acesulfame-K. At biological pH, 14 different organic and artificial sweeteners were recognized using a color-responsive sensor array. By identifying various ethanol amounts and flavoring ingredients, the sensor was also able to distinguish between different commercial beers (Musto et al. 2009).

For Pesticides

Utilizing insecticides and organophosphates, farmers manage vermin, which can infiltrate the food via soil, air, or water. Several commercially available dyes, including 4-methylumbelliferyl phosphate, fluorescein, were utilized to identify and separate these pollutants. A fluorescent sensor array was developed by Liu and Bonizzoni (2014), which can identify glyphosate, methylphosphonate, and inorganic phosphates simultaneously. Another colorimetric sensor was developed using easily accessible, thiocholine, and H2O2 sensitive indicators for detection of organophosphates (Qian and Lin 2015). PCA and HCA, two unsupervised classification techniques, demonstrated that the suggested sensor array can identify and separate various organophosphates and carbamates in actual samples. Other researchers have developed color-responsive sensor arrays using citrate-capped gold nanoparticles, graphene quantum dots, and a laser-induced fluorescent device for identifying pesticide residues (Gao et al. 2017).

Concerning Dairy Products

Misuse of antibiotics in animal husbandry can lead to the accumulation of antibiotic residues in animal bodies and products derived from them, posing a risk to humans. While many scientists produced label-free ssDNA-AuNPs conjugates for AA detection and evaluation (Mungkarndee et al. 2016), Mungkardne et.al created a fluorescent array-based sensor for milk assessment (Yan et al. 2018). Zhao et al. developed a three-unit fluorescence array sensor to distinguish between powdered newborn formulae (Long et al. 2019), while Long et al. manufactured a quadruple fluorescence array-based sensor for tetracyclines (Zhao et al. 2021).

Instruments for Food Forensics that Are Inspired by Machine Learning

Gas or liquid chromatography and mass spectrometry are two popular laboratory techniques for assessing the chemical safety of food. These techniques are known to be the best quantitative research, but they also transporting the samples to a lab where they are examined by trained people and utilizing high valuable equipment. Handheld, mobile devices are becoming more and more popular as a means of providing fast, precise, and onsite food contamination screening. Using human understanding such as taste and smell, sensory analysis is an essential technique for identifying the type of food quality. Scientists have developed artificial senses to swiftly and accurately detect and quantify food contamination in place of human senses. Paper-based sensors, electronic tongues, electronic noses, and more are a few examples of them. However, with the abundance of such low-cost and portable sensors, other designs for the associated read-out devices need to be considered. For these sensing devices to be widely used, they must have properties like field mobility, affordability, and network connection. A handheld analyzer with a cartridge and a 20 element color-responsive sensor is used shown in Fig. 15.3.

An array-based instrument known as an electronic nose is used to rapidly distinguish between different types of pollutants based solely on their odor. Chemical sensor units in this device react with gaseous mixes to produce unique odor profiles called fingerprints. After comparing the obtained olfactory profiles to industry-standard odor profiles, they may be evaluated. To further confirm the results, machine learning-based methods were employed. First, air pollutants previously identified by the olfactory system were analyzed with an electronic nose. Since this equipment evaluates liquids and O2 gas combinations, its application is currently more prevalent (Sliwinska et al. 2014). In order to identify meals high in protein, Li et.al built a chemical sensor array that was combined with a handheld device to quickly assess and identify the freshness of meat products (Li and Suslick 2016). This portable device consists of a one-dimensional (1D) complementary metal oxide semiconductor (CMOS) camera to gather on-site color data and a non-board diaphragm pump, as seen in Fig. 15.3a. Figure 15.4 represents illustration demonstrating a color-responsive. At the PPB (parts per billion) level, this built sensor array exhibits a strong response to amines and sulfides (Fig. 15.3b) (Li and Suslick

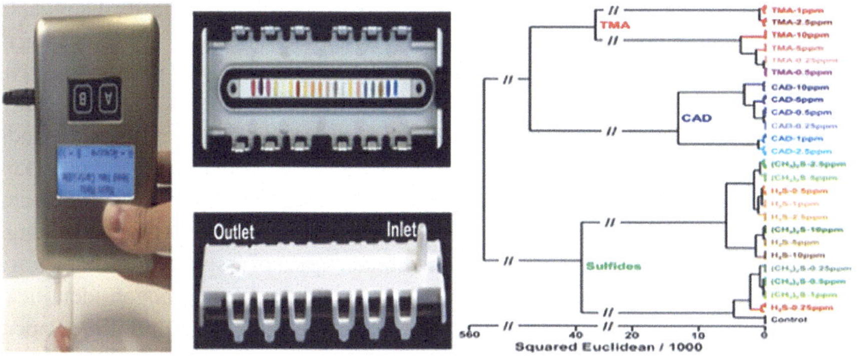

(A) (B)

Fig. 15.3 (**a**) Using a portable analyzer with a 20-element color-responsive sensor and a cartridge, sample beef. (**b**) Dendrogram showing the hydrogen ions released by rotting meat. (Reproduced with permission from the American Chemical Society (Li and Suslick 2016))

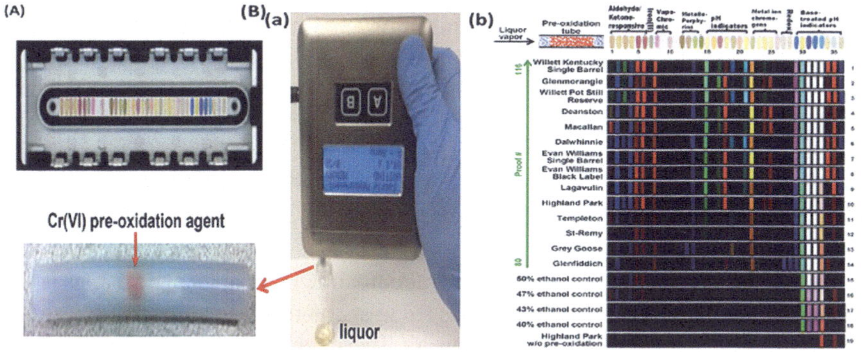

Fig. 15.4 (A) An illustration of a color-responsive array-based sensor that uses an oxidation tube agent and a handheld analyzer for sampling. Diagrammatically depicted in (B) (**a**), the preoxidation of liquid vapors is aimed toward target samples (**b**) in its colorimetric array. (Reproduced by permission of the American Chemical Society (Li and Suslick 2018))

2016). Additionally, they built a 36 element color-responsive array-based sensor for the quick and simple identification of alcoholic drinks, such as vodka, bourbon, scotch, and brandy (Li and Suslick 2018). This newly created sensor array is very sensitive to aldehydes, carboxylic acids, polyphenols, ethanol, and a variety of other important constituents of alcoholic drinks (Li and Suslick 2018). After being exposed to vapors for two minutes, each of the 14 alcoholic beverages in the control group exhibited distinct color differences that were visible even with the unaided eye (Fig. 15.4a, b). Only a tiny number of electronic noses have reached the market despite significant developments because of various recognition problems. Like, metal oxide-based electronic noses can need complex power sources in order to operate at heavy temperatures.

These moisture-sensitive metal oxide-based sensors are unable to distinguish between two chemically identical materials (Kiselev et al. 2018). Thus, commercializing electronic noses requires a system with a very strong array of cross-reactive sensors and power to reliably forecast fingerprint configuration. In this regard, researchers also developed a color-responsive barcode consisting of different kinds of porous nanocomposites implanted on cellulose acetate with dye-loaded nanoparticles. The created model can detect both intelligent fingerprints and fragrances simultaneously and is limited by a convolutional neural network (Deep Learning) (Guo et al. 2020). With an emphasis on the human sense of taste, liquid samples are analyzed utilizing electronic tongues. They are employed for pathogen identification in the medical sciences, food processing, and environmental monitoring. Paper-based analytical devices are being used more and more in food quality testing, evaluation of drug quality, clinical and health determination, and contaminant monitoring. A common method that makes use of color information included in digital photos is computerized picture-based colorimetric detection (Fuyal and Giri 2020; Sharma et al. 2018). Using a smartphone application, researchers have also developed an onsite method for identifying food deterioration. There are still attempts on to build portable and reasonably priced devices for the purpose of differentiating biogenic amines in food items, even in the face of rapidly accurating analytical techniques for biogenicamine detection in food products. Cloud computing and user-friendly designs have been included into commercial gadgets (Pokhrel et al. 2020).

The Significance of This Study

One of the most important things to think about while making meal decisions is the safety and freshness of the food. This has led to an interest in analytical applications for the development of portable food sensors. These sensors might be more advantageous than more sophisticated apparatus like gas chromatography-mass spectrometry (GC-MS) and high-performance liquid chromatography (HPLC). Many scientists have focused their research efforts on developing portable food forensics systems that are robust, user-friendly, process data quickly, and do not require specialized personnel to operate in light of the development of AI and machine-learning applications in this context. In light of this, we have discussed some recently developed portable gadgets that track the freshness and deterioration of food using machine-learning approaches. We believe it will be simpler to offer tailored solutions for certain problems involving onsite food forensics as advances in machine learning and electronics continue. Figure 15.5 displays ensuring the purity of food items is the aim of food authenticity (Saadat et al. 2022).

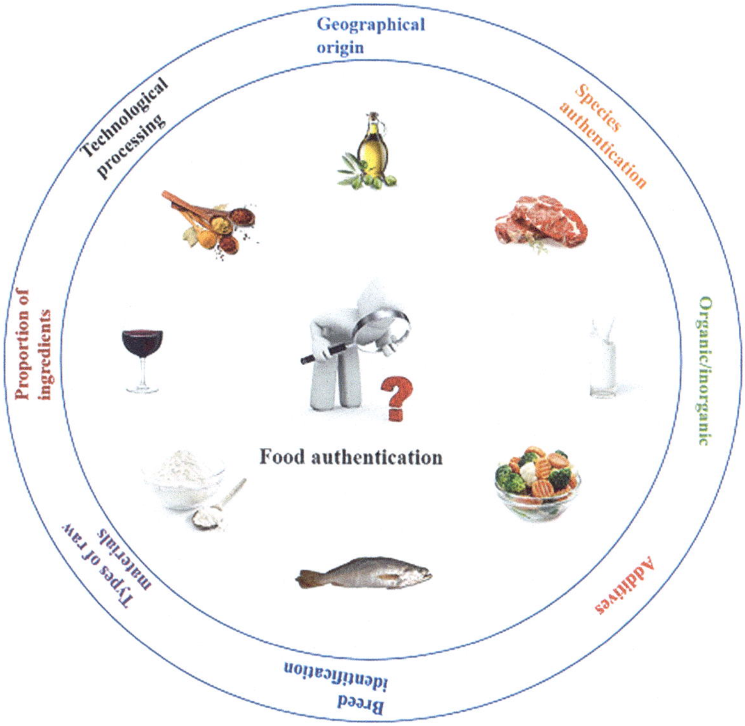

Fig. 15.5 Displayed ensuring the purity of food items is the aim of food authenticity (Saadat et al. 2022)

15.2 Techniques for Authentic Determination of Food Products

In the last several decades, there is been a rise in interest for precise and reliable analytical processes for food authenticity and traceability due to the high probability of food fraud driving excessive competition in the manufacturing of food items. This article discusses and applies several techniques, such as chromatographic, DNA-based, spectrometric, spectroscopic, lateral flow immunoassay techniques based on neutrons and protons, to forensic analysis of food products. These techniques are useful for food fraud investigations including a variety of legal concerns since they enable rapid food sample analysis and the identification of certain parameters. Thanks to these methodologies, food forensics can also find solutions to a number of important analytical issues regarding food traceability and authentication. Additionally, forensic study of colors in food, fish, plant-based meals, beverages, and viscera after death is studied (Saadat et al. 2022). The goal of food authenticity is to ensure that food items are displayed in a pure manner.

Figure 15.5 displays ensuring the purity of food items is the aim of food authenticity (Saadat et al. 2022). Food forensics refers to the likelihood of establishing a food product's legitimacy. Food forensics includes the potential application of forensic techniques to food item authenticity and traceability. Numerous areas, including biology, nanotechnology, geology, mathematics, and many more, are closely connected to this field (Bhatt et al. 2020).

Analytical Methods Used in Food Analysis for Forensics

The scientific field of food forensics is concerned with establishing the authenticity of food products, identifying adulteration, and addressing any additional legal issues. Food forensics examines the potential analytical techniques to ascertain the provenance and authenticity of food items. Food fraud is on the rise globally, which means that forensic food investigations require more accurate and reliable analytical techniques (Teletchea et al. 2005). Various analytical techniques, including chromatographic, spectroscopic, spectrometric, DNA-based, lateral flow immunoassay, and nuclear techniques based on neutrons and protons, are routinely used to evaluate for forensic investigations. How these tactics are applied is shown in Fig. 15.2. Numerous methodologies, such as recombinase polymerase amplification (RPA), real-time polymerase amplification (RT-PCR), are used. Food forensic analysis uses a variety of acronyms, including SERS (surface-enhanced Raman spectroscopy), GC (gas chromatography), MS (mass spectrometry), HPLC (high-performance liquid chromatography), TLC (thin-layer chromatography), ELISA (enzyme-linked immunosorbent assay), are used in food forensics.

Methods Based on DNA

DNA-based methods are very handful for food verification because of their high sensitivity, enhanced specificity, and resistance to deterioration. Figure 15.6 details the application of several analytical techniques to forensic food analysis. Amplification, species detection, and polymerase chain reaction (PCR) are the subsequent steps in DNA-based techniques. The foundation of PCR technologies used in food control to authenticate food is the tracing of unique sequences using DNA fingerprinting. Polymorphism is the term for the ability of PCR techniques to detect small differences in base-pair sequence. For food authentication, these techniques, which include single nucleotide polymorphisms (PCR-SNPs), restriction fragment length polymorphisms (PCR-RFLPs), randomly amplified polymorphic DNA (PCR-RAPD), amplified fragment length polymorphisms (AFLP) and randomly amplified polymorphic DNA (RAPD), are reviewed elsewhere (Abbas et al. 2018; Lo and Shaw 2018; Vucelić-Radović et al. 2019). For example, direct PCR has been used to identify the meat species food goods in Asian, and multi-plex real-time PCR has been utilized to analyze various species for the authenticity of processed food items (Silva and Hellberg 2021). Because direct PCR is sensitive and repeatable, it may be used in forensic investigations to detect a variety of meat species (Thanakiatkrai et al. 2019). Several analytical procedures are required for food authenticity, depending on the sample type, in order to generate adequate proof for use in legal proceedings. The capacity of the MEAT 5.0 kit to detect species of

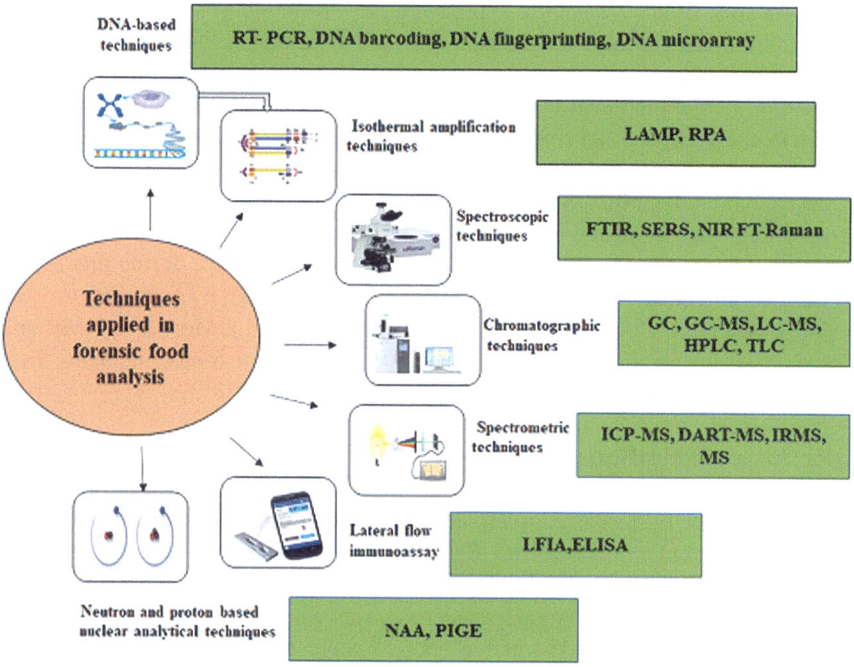

Fig. 15.6 Uses of various analytical methods in forensic food analysis (Saadat et al. 2022)

animals has been verified by Acutis et al. When tested on different types of milk and meat, the kit has shown excellent sensitivity and specificity. As a result, it is useful in food forensic investigations and for the detection of general contamination in food derived from animal sources (Beltramo et al. 2017).

Spectrometric Methods

Many studies examine the uses of different spectrometric methods for food authentication. These methods include isotope ratio mass spectrometry (IRMS), nuclear magnetic resonance (NMR), plasma optical emission spectrometry (ICP-OES), inductively coupled plasma mass spectrometry (ICP-MS), and others. Trace element identification and the chemical, biological, and geographical provenance of food ingredients can be accomplished with the aid of these analytical techniques. Particularly in the area of forensic food toxicology, these methods have grown in popularity for forensic food analysis (De Lima and Barbosa 2019). ICP-MS has been used by Pérez-Álvarez et al. to analyze the main, minor, and trace elements in wines (Pérez-Álvarez et al. 2019). ICP-MS has been utilized in a few additional investigations to identify harmful elements in plant-based diets (Cui et al. 2020). This method has low detection limits, sensitive, quick, and accurate. It can also detect many elements at once. This method's ability to provide information in multiplex samples without the need for standards is one of its major benefits, making it appropriate for forensic food analysis (Zhang et al. 2018). Regarding a number of legal concerns, IRMS is a helpful approach for determining the validity and attribution of food goods (Matos and Jackson 2019). For example, Park et al. employed this approach to analyze the isotope ratio of C, H, O, N, and S for the purpose of evaluating the geographical origin of onions (Park et al. 2019), while Peng et al. used it to analyze Brazilian coffees for their isotope ratio (Peng et al. 2019). Among the aforementioned techniques, MS has grown in popularity because of its exceptional analytical flexibility. Shiga toxins, which are protein toxins that can be found in food, and prions, which are infectious proteins, have recently been identified using mass spectrometry. Shiga and prions are rendered inactive during processing by this instrument, after which these proteins are broken down by a protease to yield peptides suitable for mass spectrometry analysis (Yuan n.d.). Moreover, NMR and UV-vis have been utilized to authenticate a wide range of foods, including fruit juices, soybeans, honey, edible oils, and spices (Schieber 2018).

Spectroscopic Methods

Non-destructive spectroscopic methods are increasingly being used as the analytical method of choice for forensic investigations due to their great sensitivity and ability to obtain a trace amount of material. Because of these techniques' speed, sensitivity, quantitative analysis, and non-destructive nature, analytical procedures can be preserved and reanalysis can be made easier in cases when analytical results are disputed. Consequently, non-destructive spectroscopic methods are preferred over destructive methods like GC-MS, such as Fourier transform infrared spectroscopy (FT-IR) and surface enhanced Raman spectroscopy (SERS) (Zhou et al. 2020). The use of NIR FT, Fourier Transform and visible micro-Raman spectroscopy, ultraviolet and visible spectroscopy, fluorescence spectroscopy, near-infrared (NIR) spectroscopy, and FT-IR for food analysis are just a few of the spectroscopic techniques that have been reviewed previously (He et al. 2019; Sun et al. 2020). There are two types of spectroscopic analysis: quantitative and qualitative. Determining the concentration of different ingredients in food products is the primary use of spectroscopy in food analysis. Forensic food analysis frequently uses FT-IR due to its excellent resolution and low cost. It can identify the authenticity and adulteration of a wide range of culinary items, including honey, beef, oil, spices, and herbs. Another method, called SERS, makes use of Raman scattering to determine the chemical makeup of food samples. It uses a laser light source to create Raman scattered light, which is detected using CCD camera (Ravindran et al. 2018; Zhang et al. 2021; Lu et al. 2019). SERS has been used to detect tropane alkaloids and titanium dioxide in food products, proving their presence at levels of 100 ppm or higher (Lim et al. 2018).

Chromatographic Methods

Chromatographic methods have several potential uses in food forensics. A number of studies (Kumar 2021; Stachniuk et al. 2021; Kalogiouri et al. 2020; Greer et al. 2021; Núñez and Lucci 2020; Pauk and Lemr 2018; Jansson et al. 2018) provide a thorough explanation of the applications of various techniques like liquid chromatography-mass spectrometry (LC-MS), with ultraviolet (LC-UV) for examining the authenticity of various food products, detecting adulteration, and identifying toxic materials from a forensic point of view. Gas chromatography (GC) is an analytical technique that may be very useful in food analysis, especially when working with volatile compounds. This allows flavor and aroma to be used as indicators of food quality. The enantiomeric ratio is necessary to determine the origins of fruit flavorings and essential oils. Furthermore, food quality and authenticity might be evaluated using GC enantioselective techniques, which correlate flavor and smell analysis with the meal's place of origin (Ribeiro et al. 2021). In order to increase analytical sensitivity, scientists have created a novel LC-MS approach for identifying aristolochic acids in dietary grains. There may be forensic investigations that benefit from this approach. Using nano-LC-HRMS, a multi-residue approach was developed for the quantitative identification of veterinary medicines in milk, eggs, and honey. Although it requires sample pretreatment, it could be handful in food forensic applications due to its lack of standard addition methods (Chan et al. 2019).

Other Methodologies

Two analytical methods for onsite target element identification in food analysis are enzyme-linked immunosorbent assay (ELISA) and lateral flow immunoassay (LFIA). Pregnancy test strips and other LFIAs are inexpensive, lightweight, and easy-to-use diagnostic instruments. Forensic food inspection can benefit from the quick, portable biomolecular detection offered by laser-induced breakdown spectroscopy (LIBS) and lanthanide-modified LFIA. Nevertheless, multiplexed LFIA increases efficiency and lowers costs while LFIA can only detect one target at a time. While fipronil in eggs and cucumbers has been detected using LFIA, ELISA has been developed to identify flunixin residues in milk. Neutron activation study has been used extensively in food analysis (NAA). Because food samples need to be analyzed with high precision and using the right approaches to determine their elemental compositions is therefore preferable, samples can be examined nondestructively by NATs to ascertain the elemental makeup in both qualitative and quantitative ways. When intrusive elements like manganese, salt, potassium, chlorine, and bromine are present, NAA has been used to determine both low and high iodine concentrations (Acharya and Pujari 2019).

Utilizing Forensics in Certain Food Categories

In spite of the difficulties in verifying the composition and origination of food products, this section addresses the use of forensic techniques for analysis, including seafood, drinks, and plant-based foods. It also addresses the use of postmortem viscera for forensic investigations. For hundreds of years, people have utilized plants as food and medicine. While they have many medicinal uses, some plants can be toxic if taken in excess or by someone who is allergic. These plants have the potential to help forensic scientists identify hazardous chemicals and uncover criminal activity. It is possible to obtain fingerprints from a variety of food surfaces, and the quality of the fingerprints changes with room temperature and refrigeration levels. In order to verify the authenticity of Jasmine rice, a genetic fingerprinting technique based on microsatellite markers may be used. DNA metabarcoding has been used for identifying processed food and contaminants. Using IC-MS analysis, the results of an enzyme inhibition assay that was developed to identify adulteration of edible herbal plants were validated. By using these techniques, food products' quality can be raised and possible pollutants can be found. Overall, the use of plants in forensic investigations can help identify criminal offenses and improve food safety (Li et al. 2020; Adhikari et al. 2021; Awuchi 2019; Ben-Shabat et al. 2020; Shedoeva et al. 2019).

Commonplace items such as shoes and clothing contain dyes that, if disposed of incorrectly, can poison the environment. By identifying dyes in complex matrices, appropriate detection techniques may help alleviate these issues. Currently, mass spectrometry and chromatography are used to identify colors in food. In forensic situations, textile fibers with dye found in carpets, furniture, and clothes can be an invaluable piece of evidence. Thin-layer chromatography and high-performance liquid chromatography are the two techniques used to identify artificial coloring in food products. To differentiate between animal species and human genomes, a unique five-color panel has been developed. But the overuse of permitted food coloring in Sri Lankan confections and drinks emphasizes the need for comprehensive laws, public awareness campaigns, and proper labeling to prevent adulteration with artificial food coloring (Terán et al. 2021; Schotman et al. 2017).

Maintaining equilibrium between food supply and demand is key to solving the global problem of food adulteration. Food adulterant detection methods are necessary for food analysis. Sanger sequencing of PCR is suitable for species detection of seafood, albeit it is not preferable for mixed species and needs to be canned. Human health may be impacted by aquatic organisms that develop toxicity or heavy metal poisoning. Fish identity is being determined more and more by molecular methods, and species identification can be accomplished quickly and reliably with DNA barcoding technology. Mitochondrial genomes have been used to assess the authenticity of frozen fillets from the northern Amazon coast, and forensic research based on DNA markers has proven to be successful in identifying species and spotting fillet fraud (Kappel et al. 2017).

Forensic investigations also heavily depend on the amount of ethanol found in alcoholic beverages, but traditional methods require sophisticated sample processing techniques. Directly analysis in real-time mass spectrometry offers a quick and accurate method with an accuracy of more than 98% (DART-MS). Meng et al. used GC-MS to determine the amount of gamma-hydroxybutyrate in beverages following the dispersion of organic solvents using ultrasonication. For sensitive GHB detection, dispersive liquid-liquid microextraction and GC-MS are the most efficient techniques. In real-world applications, Raman spectroscopy can be used to identify compounds in alcoholic beverages. Cyanide analysis in blood may now be done easily, quickly, and sensitively with flame ionization detection (GC-FID) and headspace GC (Ali and Edwards 2017; Meng et al. 2020; Sundermann et al. 2021).

Why It Is So Much Required?

Comprehensive information about food product qualities needs to be gathered from the supply chain in order for the food industry to deliver commodities that are trustworthy, secure, healthy, and sustainable. Customers also want to know if the food they eat is legitimate so they can make informed judgments. Data on the characteristics of food were previously gathered using specialist labs and cutting-edge forensic equipment. However, because of the development of small, inexpensive, and user-friendly forensic equipment, nanotechnology has revolutionized this process. By affixing these devices to food packaging or their smartphones, customers may promptly assess the meal's authenticity, composition, and other attributes. Poisons and artificial food coloring may also be detected by these devices. In order for the researchers to continue the process in the future time, the application of several analytical tactics for food product analysis in forensics investigations is underlined.

15.3 Detecting Protein Contaminants in Food by Mass Spectrometry, as Demonstrated by Prion Strains and Shiga Toxin Variants

A variety of tools are required by food forensicists to detect any potential food contamination. One such tool is mass spectrometry, which has a high degree of analytical versatility. Other applications of the multiple reaction monitoring technique include quantification and the identification of infectious proteins and protein toxins, such as Shiga toxins. The proteins are broken down to produce peptides suitable for multiple reaction monitoring-based analysis. Using this technique, the samples' prions and Shiga toxins are rendered inactive. Prion identification is possible because to differences in covalent modification and distinct physicochemical properties. Identifying peptides from the 5 binding B subunits that make up the toxin is the foundation of Shiga toxin analysis. 15 N-labeled standards are made using proteins that have been cloned (Silva 2018).

Multiple Reaction Monitoring (MRM) Method

It is possible to analyze peptides made from proteins efficiently using the MRM approach. Sector instruments are required for MRM due to their high cost and unsuitability for high-throughput sample processing. Since the MRM technique detects suitable peptides, a first qualitative proteomic analysis is necessary to identify peptides appropriate for MRM-based analysis. Preparing synthetic peptides and obtaining stable isotope internal standards are additional prerequisites (see infra). Nevertheless, MRM-based techniques clearly offer the advantages of greater accuracy and flexibility (Liebler and Zimmerman 2013). The MRM methodology provides more information than can be gained with other approaches and allows one to simultaneously detect numerous unique molecules (multiplex) and detect molecules directly. Molecular resonance microscopy (MRM) can be used to study proteins by cleaving them into shorter peptides. Trypsin or another protease is typically used for this kind of digestion (Chalker et al. 2009). In automated processes, disulfide linkages are reduced using a depleting agent such as dithiothreitol, and the corresponding carbamidomethyl group is obtained by reacting this reduced form of the compound with an alkylating agent (iodoacetamide). A set of (tryptic) peptides is the result of the reduced and alkylated protein being broken down by a protease such as trypsin. To examine certain proteins, additional reagents, alkylating agents, and reducing agents might be employed. To find peptides appropriate for MRM analysis, a qualitative analysis is conducted on the resultant set of peptides. The strongest signaling peptides are good candidates for MRM-based examination. The peptides' physicochemical characteristics are also taken into account. After a group of appropriate peptides has been found, each peptide's triple quadrupole instrument parameters can be adjusted. The analyte peptide with the strongest signal is identified by repeating this procedure for the remaining peptides. The peptides tailored for MRM analysis can easily have stable isotope-labeled standards produced for them using ordinary molecular biology procedures (Vit and Petrak 2017). In a suitable *E. coli* host, a intended gene is cloned into a suitable vector and overexpressed. The creation of internal standards and the practical manufacturing of proteins are made possible by this. When the *E. coli* host is cultured in a medium supplemented with a vital nutrient that is heavily enriched (99.7%) with 15 nitrogen, like 15 NH4Cl, all of the molecules that the *E. coli* synthesizes will also be heavily enriched with 15 nitrogen rather than the naturally occurring 14 natural nitrogen (Silva et al. 2014).

Prions

Transmission-spongiform encephalopathy (TSE) is a deadly neurological illness that is brought on by prions, which are abnormal proteins. Death occurs after a very brief time of symptomatic illness after a protracted asymptomatic incubation (Prusiner 1998). Disease is brought on by prions (PrP Sc), which induce the prion shape to be adopted by a normal cellular prion protein (Stahl et al. 1992; Silva et al. 2010). There is just a conformational difference between PrP C and PrP Sc, but their covalent structures are the same. PrP C is a monomeric protein that is highly conserved and has an unclear function (Schmitz et al. 2014a). A minor amount of β-sheet, unstructured areas, and α-helices make up its secondary structure, which is soluble in non-denaturing detergents (Schmitz et al. 2014b). Among the prion illnesses affecting humans are FFI (Llorens et al. 2016), GSS disease (Liberski 2012), Kuru (Liberski 2013), and Creutzfeldt-Jakob disease (CJD) (Manix et al. 2015). Cattle kept domestically contract BSE when they eat tainted feed (Harman and Silva 2009). vCJD, the human variant of BSE, is the only known zoonotic prion disease (Will et al. 1996). Transmissible mink encephalopathy (TME) is spread to farmed mink by contaminated feed, and sheep scrapie is the oldest known TSE (Detwiler and Baylis 2003; Marsh and Hadlow 1992). Only one prion disease,

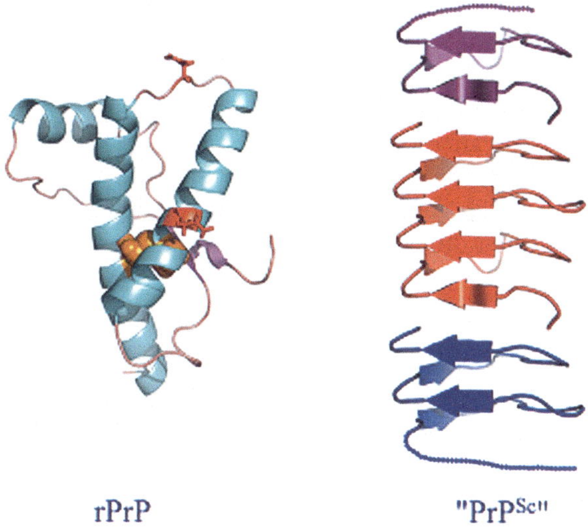

Fig. 15.7 The hamster rPrP NMR structures are illustrated. The four-rung solenoid, representing PrP Sc, is derived from a β-helical antifreeze protein (James et al. 1997; Leinala et al. 2002). PyMOL was used to build the two structures (1B10 and 1M8N) that were provided by the Protein Data Bank (PDB) (Berman et al. 2000). The color of α-helices in rPrP is blue, β-sheets are purple, the disulfide bond between the two cysteines is represented by orange spheres, and the location of the glycans in hamster PrP C is indicated by red sticks. The proposed "PrP Sc" multimer structure is based on the four-rung β-solenoid multimer model (Vázquez-Fernández et al. 2016)

chronic wasting disease (CWD), can naturally spread among untamed animals (Haley and Hoover 2015). Figure 15.7 represents β-solenoid multimer model is the foundation of the suggested "PrP Sc" multimer structure.

Detecting Prions Using Mass Spectrometry

Mass spectrometry has been used to study prions in great detail (Silva 2014). The study has been done on trypsin digests of PrP (Silva et al. 2017). The initial research revealed that PrPC and PrPSc had the same posttranslational alterations (Silva et al. 2010). The GPI anchor and the asparagine-bound glycan compositions were then found to differ similarly in PrPC and PrPSc using mass spectrometry. A good amount of β-sheet are thought to inhibit PK digestion (Silva et al. 2015a). Mass spectrometry has been extensively used to identify the remaining peptides after partial PK digestion, which has allowed for the determination of the boundaries of the β-sheet structure. It has been applied to ascertain the relative vulnerability of the two polymorphisms to the infrequent atypical scrapie in a heterozygous sheep or the naturally occurring scrapie. Pron diseases and prions can be identified by mass spectrometry, which eliminates the requirement for processing PK samples. The chemical environment of an amino acid within a protein can be significantly changed by the structural changes that take place when PrP C adopts the PrP Sc conformation (Silva et al. 2015a). Presumably, the PrP conformation affects the reaction that occurs between an amino acid and the same reagent. Theoretically, after PrP Sc has been denatured to PrP before analysis, a covalent difference originating from such a disparity in reactivity should remain. Using mass spectrometry, many prion strains have been found. Diverse prion strains often have distinct PK cleavage places. Mass spectrometry has been applied to pinpoint those spots and has also made it possible to differentiate scrapie strains in sheep. A further technique makes use of the ε-amino group of lysine's reactivity with the N-hydroxysuccinimide ester (Ac-NHS) of acetic acid. It is possible to identify different prion strains and detect prions using mass spectrometry (Silva 2014).

Shiga Toxins

In the nineteenth century, the bacteria Shigella dysenteriae was first identified as the source of the shigatoxin (Trofa et al. 1999). A great deal later, a toxin akin to this one was identified from *E. coli* cultures and named type 1 Shiga toxin (Trofa et al. 1999; Konowalchuk et al. 1978). Figure 15.8 represents two perspectives of Shiga toxin 2a's structure (4M1U). Researchers have identified unique but structurally and functionally comparable type 2 Shiga toxins from different Escherichia coli strains. At times, these toxins have also been noted to as verotoxins and Shiga-like toxins (Scheutz et al. 2012). Their amino acid sequence divides them into 2 groups of Shiga toxins. The two toxins are AB5 hexamers, where the A subunit has a catalytic domain that kills the eukaryotic host cell while the B subunits attach to the target eukaryotic cell (Yamasaki et al. 1991; Jacobson et al. 2014). Although both kinds of toxins have the potential to seriously harm people, second type of Shiga toxins are more frequently linked to the more severe signs and symptoms of Shiga poisoning. The main virulence factor linked to these foodborne infections is the Shiga toxins generated by STEC (Tarr et al. 2005). Hexameric (AB 5) proteins, which make up shiga toxins with a molecular weight of about 70 kDa. By attaching itself to the surface of eukaryotic cells and inducing endocytosis, the toxin known as Shiga

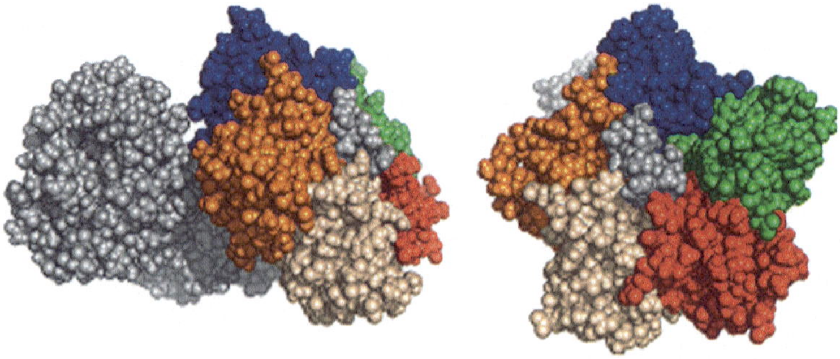

Fig. 15.8 Shiga toxin 2a's structure from two perspectives (4M1U). The atoms in the A subunit are colored black, but the five identical B subunits are colored red, green, blue, wheat, or orange. The structure is an image of the PyMOL-produced 4M1U structure found in the PDB (Berman et al. 2000; Jacobson et al. 2014)

toxin (STEC) destroys the cells. Finding the origins of these epidemics requires the application of food forensics. The Shigatoxin-producing lambdoid phage (stx phage), which regulates the actual manufacturing of Shiga toxins, infects an *E. coli* host. Antibiotics can trigger STEC phages to create Shiga toxins, which is why treating STEC-infected individuals with antibiotics is not advised (Huang et al. 1987; Newland et al. 1985; O'Brien et al. 1984; Freedman et al. 2016). This further renders PCR-based techniques for STEC detection questionable (see infra) (Koitabashi et al. 2006). The DNA required to create a single Shiga toxin is present in every single phage. An *E. coli* bacteria host, or one of the other bacteria, can be infected by many Stx phages (Casjens and Hendrix 2015). This suggests that multiple forms of Shiga toxin can be produced by a single host bacterial cell and that the amount of Shiga toxin produced by a specific STEC can vary greatly. Shiga toxin analysis is further complicated by the substances present in human serum. Some Shiga toxins, but not all of them, bind to certain proteins found in human serum and tissues. Human serum amyloid protein (HuSAP) in serum (Cox et al. 2014; Pepys et al. 1982). This protein is not bound by type 1 Shiga toxins, but it is by type 2 Shiga toxins (Bitzan et al. 1993; Caplioli 1994; Kimura et al. 2001). The use of animal bioassays, cell-based assays, and antibody-based assays to detect type 2 Shiga toxins is hindered by binding to HuSAP (Kimura et al. 2001, 2003; Armstrong et al. 2006).

Figure 15.9 represents PyMOL representation depicts the five B subunits of Shiga toxin type 1.

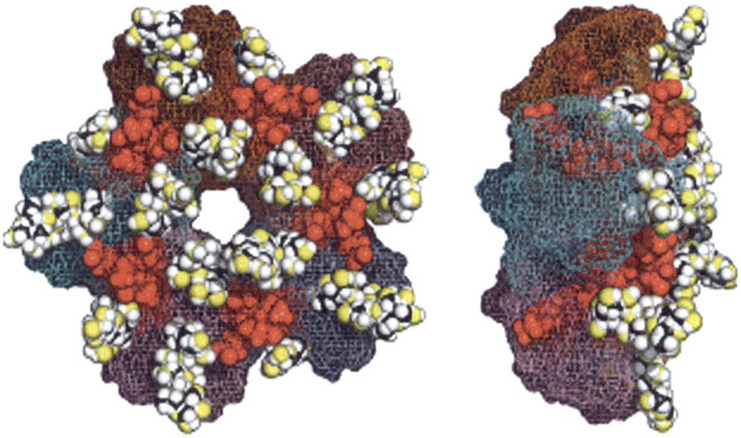

Fig. 15.9 The PyMOL representation of the five B subunits of Shiga toxin type 1 is a mesh of various colors, where the glycan component of the Gb3 gangliosides is represented by yellow, white, and black spheres, and the analyte peptides are represented by red spheres. The images are created from the 1BOS crystal structure in the RCSB PDB (Berman et al. 2000; Kitova et al. 2009) using PyMOL software

Identification of Shiga Toxins Using Mass Spectrometry

Shiga toxin analysis has been done extensively using mass spectrometry (Kitova et al. 2009; Alam et al. 2012; Fagerquist and Sultan 2010, 2011; Fagerquist and Zaragoza 2015; Kondo et al. 1997). It has been applied to research how the subunits interact (Kondo et al. 2003). Because the molecular weights of the intact proteins vary, it has also been used to identify and separate the different Shiga toxins (Alam et al. 2012; Fagerquist and Sultan 2010, 2011; Fagerquist and Zaragoza 2015; Kondo et al. 1997). Using the MRM-based mass spectrometry approach offers a more sensitive way to measure, identify, and separate the different Shiga toxins (Silva et al. 2015b). Since the B subunits of Shiga toxins are found in a 5:1 ratio to the A subunit and do not cause harm when isolated, they make a desirable objective for analytical techniques. The B subunits of recognized Shiga toxins were cloned, produced, and purified using their protein sequences. Qualitative MS method can detect various tryptic peptides suitable for MRM-based analysis, including YNDDDTFTVK (type 1 Shiga toxins), which has a low amount of detection limit (Silva et al. 2015b). This peptide has analogues in other Shiga toxins. To include these analyte peptides along with other peptides required for identifying and differentiating Shiga toxins, an artificial gene with a His tag was created. Analyte peptides are participated in binding B subunits to one another and to glycolipids on the target eukaryotic cell surface (Ling et al. 1998). If further new Shiga toxin variant discovered in the future is likely to have analogues or similar peptides, these areas are probably preserved. Differences in the phenotypic characteristics of the toxins are probably connected to variations in the B subunit region that contains analyte peptides (Fig. 15.9).Biosafety concerns also make the study of Shiga toxins more difficult. To treat shiga toxins, a BL2 laboratory is needed. In order to examine samples in mass spectrometry, they need to be aerosolized. Even though a very small amount of material is required for examination using mass spectrometry, toxin aerosolization still carries some risk. Additionally, the instrument itself must be kept in a BL2 laboratory, and instrument operators must handle toxins while assessing intact poisons. When examining intact poisons, mass spectroscopists may encounter many difficulties. It is interesting to note that trypsin only cleaves the furin site of Shiga toxins; the rest of the toxin is not broken down (MacLeod et al. 1991). Shiga toxins that were intact were subjected to the minimization and trypsin digestion processes required for an MRM study. Because domestic cattle lack endocytosis receptors, the mechanism of detecting Shiga toxins in them is still developing. This makes it possible for STEC to be amplified in cattle and for new serotypes to be produced when susceptible species or uninfected *E. coli* are infected by stx phages. At least 200 distinct serotypes of *E. coli* can be found in the STEC population in cattle, and even non-domestic *E. coli* serotypes have the potential to trigger large epidemics (Pruimboom-Brees et al. 2000).

Importance

Mass spectrometry can play a vital role for the identification and separation of known Shiga toxins. In fact, samples can be used with the MRM approach without needing to be fully purified, and it does not even require a whole colony. Instead, supernatants and complex matrices like human serum can be processed using it. More challenging than developing a novel MRM method is developing a novel antibody against a distinct Shiga toxin. Food forensic experts possess the ability to promptly detect and identify any novel subtypes of Shiga toxin, such as the recently discovered type 2 subtype. Using mass spectrometry, other foodborne protein toxins besides Shiga toxins have also been discovered. Toxins like those produced by Staphylococcus aureus enterotoxins, Bacillus anthracis, Botulinum toxin have all been identified using mass spectrometry-based techniques. A protease (usually trypsin) is required to decrease, alkylate, and degrade these extra toxins before analysis; thus, they might also be inactivated beforehand. By applying common molecular biology procedures, pertinent segments of the dangerous protein can be created in an internal standard gene that will contain the required internal standard peptides in a stable, non-toxicity protein. Similar to Shiga toxins, new variables of known toxins or freshly discovered new toxins can be swiftly and simply identified using the MRM approach. This provides forensic experts with an invaluable resource to assist them in solving current and future food contamination cases.

15.4 Detection of Hidden Toxins in Common Consumable Products

According to a recommendation from the Food Safety and Standards Authority of India, which also called FSSAI, simple chemical tests were used to detect common adulterants found in most common consumable items. Scientists looked into the following different food products: milk, sugar, asafoetida, jaggery, red chili powder, turmeric powder, ghee, honey, pulse, and ghee. Locally accessible goods that were easily acquired were shown to be most susceptible to adulteration (Manjunatha and Savia 2018).

Materials and Methods

(a) Glassware consists of droppers, test tube supports, glass rods, and test tubes.
(b) Chemicals: concentrated water, iodine, ethanol, concentrated H2SO4, ether, N-hexane, acetonitrile.
(c) Other: filter paper, red litmus paper, magnet, burner, and lactometer.

For each adulterant, the unidentified samples and the positive control samples were examined together. To select the unknown/test samples at random, locally made and branded products were used. Each of the ten unique products had ten or more variations, which were collected and put through the appropriate chemical test protocols (Xie and Wang 2013).

Experiment

Five milliliters of the sample were obtained and quantified for each test tube (Deconinck et al. 2015). The tabular column below Table 15.1 provides an illustration of the tests conducted for the chemical evaluation of several adulterants.

Outcomes of This Experiment

Two of the 10 different milk samples that they investigated turned out to contain detergent. Four of the eight tea leaf samples exhibited indications of synthetic

Table 15.1 Vanaspati is an Indian culinary product made by saponifying vegetable oils as opposed to animal fat. In Indian cuisine, it is typically eaten as a vegetarian dish (Manjunatha and Savia 2018)

SL	Food item	Adulterant	Test performed	Observation
1	Milk	Water	Lactometer(for reading value)	More than 26 (reading).
		Starch	Use of iodine solution to the sample,	A blue color formation.
		Vanaspati	HCl and sugar to the sample.	A red coloring.
		Formalin	Concentrated sulfuric acid introduced to the sample.	Violet/ blue ring formation at the intersection of the liquids.
2	Ghee	Coal tar dyes	1 gm material mixed with sulfuric acid and agitated.	Pink color with sulfuric acid and scarlet color with HCl.
		Vanaspati	After melting the sample, concentrated hydrochloric acid and sugar were mixing and shaking.	Crimson appears.
		Any starch	Iodine drops in the sample.	Blue color formation.
3	Sugar	Chalk powder	Sample dissolve in DW.	If any, sinks to the bottom.
		Urea	Dissolve in water.	The smell of ammonia.
4	Honey	Sugar solution	A sample of honey was used to soak and then burn cotton.	Either not burn at all or burn with a crackling sound.
		Invert sugar or jaggery	5 ml of ether and honey were added. The ether layer was decanted onto a petri dish after this was shook. After ether evaporation, a few drops of resorcinol were applied.	A cherry red appearance.

(continued)

Table 15.1 (continued)

SL	Food item	Adulterant	Test performed	Observation
5	Chili powder	Brick pieces	Sample blend in a glass of water.	Rubbing sediment with grit.
		Artificial color	Sample blend in a glass of water.	Water turns colored right away.
		Oil soluble coal tar dye 0.	2 ml ether mixed in sample. The ether layer was placed into a different tube and agitated with 2 ml of diluted HCl (1:1).	A pink appearance.
		Sudan III	After adding two milliliters of hexane, shaken. The clear solution was then poured into a different test tube after being given time to settle. 2 ml acetonitrile was added and agitated.	Red color appearance.
		Saw dust	On top of water, the sample was scattered.	If there is sawdust, it floats on the water.
6	Tea leaves	Artificial coal tar dye	A damp filter paper was used to disperse the tea leaves.	If the paper becomes discolored, it means artificial color has been added.
		Leather flakes	On top of a flaming paper ball were tea leaves.	+ ve means the scent of burnt leather.
7	Asafoetida	Soapstone	After shaking the sample in the water, it was let to settle.	If present, settle at the bottom
		Starch	The sample was blend with an iodine solution.	The presence of starch is indicated by the formation of a blue color.
		Foreign resin	Lights are used to ignite the sample.	When there is no residue remaining, the pure sample burns like camphor.
8	Turmeric powder	Metanil yellow	After collecting the sample, a few drops of HCl were placed in a test tube.	Produces a pink hue that fades with dilution but, if it remains, indicates the presence of metanil yellow.
		Chalk or yellow soapstone	Drops of HCl were added to a sample that was placed in a tube filled with water.	Should effervescence be noticed, means present.
9	Pulses	Lead chromate	Five grams of the pulse sample were obtained and mixed with five milliliters of water and a few drops of HCl.	The presence of lead chromate is indicated by a pink appearance.
10	Jaggery	Metanil yellow	After the sample was obtained, 10 drops of HCl and 3 milliliters of alcohol were put to a test tube.	The presence is indicated by a pink appearance.
		Washing soda	Drops of HCl and some water to the sample in a tube.	The presence of washing soda is indicated by effervescence.
		Chalk powder	In a test tube, the material was dissolved in water.	If there is chalk powder, it sinks to the tube's bottom.

coloring. Two of the 14 samples tested positive for Sudan III in the chili powder, and a minor number also included sawdust. Various ghee samples tested, one included vanaspati and two contained starch. One of the 10 samples of pulses that were tested included an unidentified hue. Four of the 12 samples that were examined included sugar and jaggery, suggesting that the honey had also been tampered with. There was dust and chalk powder in two of the sugar samples. Two samples contained washing soda, while five of the 10 jaggery samples that were examined contained chalk powder and stone. Five of the 10 asafoetida samples that were tested positively for the presence of dirt and other contaminants, while all other samples showed the presence of starch, which was probably caused by the addition of wheat flour during processing (Manjunatha and Savia 2018). Because locally available items are more frequently purchased than their more expensive counterparts, it is recommended to analyze that necessary measures be implemented to prevent adulteration in general. Overall, it was observed that domestically available items were more susceptible to adulteration.

15.5 Molecular Identification of Food and Forensics Also the Current State and Obstacles

Over the past few years, there has been a steady growth in the need for precise and trustworthy techniques of differentiating between different animal species. This is especially important in light of the recent food scares and the greater issue of biodiversity, which is mostly caused by the widespread illegal trade in threatened and endangered species. Promising answers can be found in the relatively young discipline of species molecular identification, which uses DNA amplification and analysis. The identification of animal species in practically any organic substrate is now possible thanks to a variety of technically sound processes, even though the extraction and analysis of DNA from processed items still poses significant challenges. But as this science enters a new phase, it has to rely increasingly on information from three core disciplines: systematics, molecular evolution, and paleogenetics (Teletchea et al. 2005).

Foodborne DNA and Forensic Samples

Many molecular or protein studies can be performed on fresh food products or unprocessed forensic evidence as mentioned in previous protein-based biochemical procedures employed immunological, chromatographic, or electrophoretic techniques (Civera 2003). Regretfully, the majority of consumables and forensic evidence go through processing that typically alters DNA. Still, this kind of DNA analysis had been used previously by several scientific disciplines: paleogenetics, which analyzes DNA from fossil bones human forensics, which analyzes DNA from crime scenes; non-invasive ecological studies, which analyzes DNA from animal feces or hairs found in the field; and, more recently, food authentication. The collective evidence from these domains has demonstrated that (i) DNA is more thermostable and resilient than proteins, even after modification; as a result, small DNA fragments can still be amplified by PCR (with sufficient information for identification); and (ii) since the DNA is present in almost all of an organism's cells, it might be possible to extract it from any substrate. Because of the degeneracy of the genetic code and the presence of several non-coding regions, phylogenetics and molecular evolution have also shown that DNA contains significantly more information. However, processed foods contain altered DNA, which has several distinct properties.

Substrates, DNA Purity, and Impurities

Brief Stretches of DNA

During the manufacturing process, food products may be subjected to various thermal treatments such as cooking, pasteurization, irradiation, high pressure, and drying. For example, food components are heated to up to thousand degree C for 1 hour while maintaining a pH of less than 4. As a result, very short DNA fragments, ideally between 100 and 200 base pairs, should be the basis of molecular detection techniques from severely degraded substrates (e.g., researchers were unable to amplify fragments longer than 200 bp from canned tunas (Quinteiro et al. 1998) or in processed animal meals (Frezza et al. 2003)).

Minimal Quantities of DNA

Apart from degradation, DNA is also present in trace amounts, which reduces the quantity of DNA fragments large enough for molecular analysis. PCR cycles must often be increased, sometimes up to 45 to 55 cycles, in order to extract sufficient copied DNA for additional investigation, as evidenced by paleogenetics and non-invasive molecular research. However, it is possible that minuscule amounts of DNA from members of minority groups and other elements were mistakenly or illegally inserted. Therefore, species detection algorithms need to be very particular in order to obtain correct results.

Contaminations

A foreign DNA molecule may occasionally be preferentially copied rather than the degraded one due to the sensitivity of PCR (Cooper and Wayne 1998). Thus, food and forensic materials must to be processed in a dedicated lab that is physically isolated from sections that handle amplified DNA and unmodified DNA from new samples prior to PCR. Every run should also include appropriate negative controls, like PCR blanks and extraction samples.

Inhibitors of PCR

Food and its component products, such as bacterial organisms, non-target DNA, organic chemicals, milk proteins, polysaccharides, glycogen, lipids, fulvic acids, collagen, iron, etc., can prevent DNA amplification (Wilson 1997). Rather than a lack of target DNA or inadequate quantity, the absence of amplification in food materials may be the result of PCR amplification inhibition (Hänni et al. 1995). Inhibitors to species detection in food items should be removed with DNA isolation techniques (Pääbo et al. 1988).

Artefacts from PCR and Chemical Alterations

Aspects of the environment that are probably the less known about DNA in forensic and culinary applications—such as moisture, acidity, and UV radiation—can chemically change DNA molecules, as paleogenetic research has shown. According to a review in (Poinar 2002), altered DNA molecules may exhibit breaks or fictitious mutations in this situation. Thus, especially if a food product has undergone significant processing, it is crucial to ensure that DNA degradation is absent before utilizing methods based on a limited number of variable sites. These DNA modifications can, in fact, lead to a misdetection of species since the acquired DNA sequence varies from the reference sequence. Food items have never been the subject of research on these chemical degradations, yet unknown differences were discovered in the DNA taken from tuna that had been canned by Ram et al. (Ram et al. 1996). These differences may have resulted from chemical changes made to the DNA molecule under study. Additionally, non-invasive and paleogenetic research has demonstrated that PCR amplification beginning with minute quantities of modified DNA may result in artifactual outcomes (such as allelic drop-out in microsatellite analysis (Goossens et al. 1998) or chimeric molecules created by PCR jumping (Pääbo et al. 1990)).

Combination among Individuals and Species

The DNA found in food products is a mixture of DNA from various origins, such as fungi, bacteria, plants, and animals. Furthermore, a food product could include a vast number of individuals from the identical species (typically from different lineages). Sadly, relatively few currently available methodologies allow for the simultaneous detection of a wide range of species present in food items. Undoubtedly, one of the most significant problems in the area remains unresolved: the simultaneous identification of many species. Cloning, sequencing, and DNA chips can all yield reliable and consistent results. DNA chips, which known to be a DNA macroarrays or DNA microarrays, allow for the study of complex mixtures of PCR results in addition to the simultaneous identification of vast number of species. These techniques are not widely applied for species identification in food and forensic materials, but they have previously been employed in many domains, including ecology (Prokhorova et al. 2010). It has been demonstrated that despite all of these technological limitations and challenges, several extraction procedures can be applied to extract and analyze DNA from a large variety of substrates with efficient size and information to assume for reliable species identification.

DNA Markers Use in Species Identification

DNA recovered from food items has unique characteristics that make detection more difficult. Because of its compact circular genome, many copies within a cell, maternal inheritance, haploid nature, and rapid evolution, mitochondrial DNA (mtDNA) is targeted for species recognition when compared to nuclear DNA. A DNA marker that is well studied for a broad variety of species and sufficiently variable between species is good for species identification, with variation across geographic distribution zones. The gene encoding cytochrome b is used in phylogenetic studies published in the past several years (Kocher 1989; Gongora et al. 2004; Nielsen et al. 2001). Markers with intra-specific variability, however, are helpful for breeds, regional origins, or individual assignments. Of the roughly 100 molecular identification experiments, cytochrome b was the target of around half (44%), followed by 16S, 12S, and D-loop.

Some Important Inputs

Despite these conceptual and technological challenges, the field of molecular species detection in food materials are predicted to expand quickly. In fact, a wide range of reliable methods exist now for recognizing species in almost any kind of substrate. Some of these techniques have even yielded noteworthy and fascinating outcomes, like in gourmet food (like species detection in caviar or foie gras, for example) (DeSalle and Birstein 1996; Rodriguez et al. 2003) and forensic samples derived from endangered species. Molecular identification has previously shown benefits in legal proceedings (Withler et al. 2004; Moore et al. 2003), and certain techniques are already in use in business (Pascal and Mahé 2001; Klein 2003). However, this biotechnology field will need to take into account the conceptual and technological understanding of the closest core domains in order to increase accuracy and dependability.

15.6 Current Developments in Mass Spectrometry and Capillary Electrophoresis Combined: From Component to Single-Cell Examination

This study includes the latest advancements in microfluidic technologies and capillary electrophoresis for MS detection. It highlights developments in the planning and application of three interfacing methods: ICP, ESI, and matrix-assisted desorption/ionization (MACP). Proteomics, forensics, pharmacology, food analysis, and single-cell analysis are a few examples of applications (Klepárník 2013).

Strong instruments in various disciplines, such as elemental studies, low molecular mass chemicals, big proteins, and superamolecular complexes, are electrophoresis in microfluidic devices hyphenated with MS detection (CE-MS). Extreme selectivity of MS detection is combined with the ease of use, rapid separation speed, and efficiency of electrophoresis (CE). With this combination, individual sample components in complex matrices can be identified in a comparatively short amount of time. The categories of CE-MS interfaces (Kaufmann 2012) include ESI, MALDI (Peš and Preisler 2010), and ICP; other, less frequently utilized methods include atmospheric pressure chemical ionization, photoionization (Axén et al. 2010), thermospray ionization, ESI (Barbula et al. 2011), and flowing atmospheric pressure afterglow ion source (Frazier et al. 1999). Electrophoretic separations in seconds are made possible by advancements in microfluidic devices. When combined, the CE and MS systems perform much better together, making it easier to select the ideal conditions for separation and yielding more precise results in complicated mixes. With established protocols for devices that are sold commercially, CE-MS is now a common technique (Kitagawa and Otsuka 2011; Yeo et al. 2011; Frazier et al. 2000; Huhn et al. 2010; Lindeberg 1996).

Instrumentation Development

A number of thorough evaluations of CE-MS interface have been published already. A summary of the most recent developments in the application of CE combined with high-resolution MS for the study of small compounds was provided by Klampfl and Buchberger (Heinemann and Zenobi 2011; Klampfl and Buchberger 2010). A thorough analysis of the soft ionization methods employed in CE-MS was provided (Kaufmann 2012). Along with describing different interfaces, they also discuss the ideas and layouts. Analogous evaluations emphasize the advancements in electrophoresis technique and instruments for microfluidic devices (Kitagawa and Otsuka 2011) and also their use in clinical analysis (Watanabe and Terabe 2000). The publication (Huhn et al. 2010) provides an understand of the relevance and use of capillary coatings in CE-MS applications. Figure 15.10 represents the random forest detection model's recognition accuracy for twelve diseases.

CE-ESI-MS

Electron transfer interaction (ESI) is the ionization method most frequently utilized for the CE-MS interface. Three essential elements are provided: a steady stream of spray or sheath liquids; an electrical connection to control the spray potential and CE output potential; and a suitable configuration to facilitate the rapid transfer of distinguish analytes to the spray. The three forms of interfaces between CE and ESI are unsheathed, sheath flow, and liquid-junction. Sheath flow interfaces are the most popular and most accessible; they combine analytes at the spray tip and apply voltage through a liquid sheath that travels through the coaxial tube and round about the capillary used for separation.

Figure 15.11 displays the linking diagram for CE, ESI, and MS. The stainless-steel sprayer is filled with the fused silica separation capillary. Sheathless interfaces are electrically coupled to a conductive membrane or an etched porous silica emitter that spans the separation and spray capillaries. A tiny gap between the separation and spray capillaries that are inundated with spray liquid and maintained at a spray

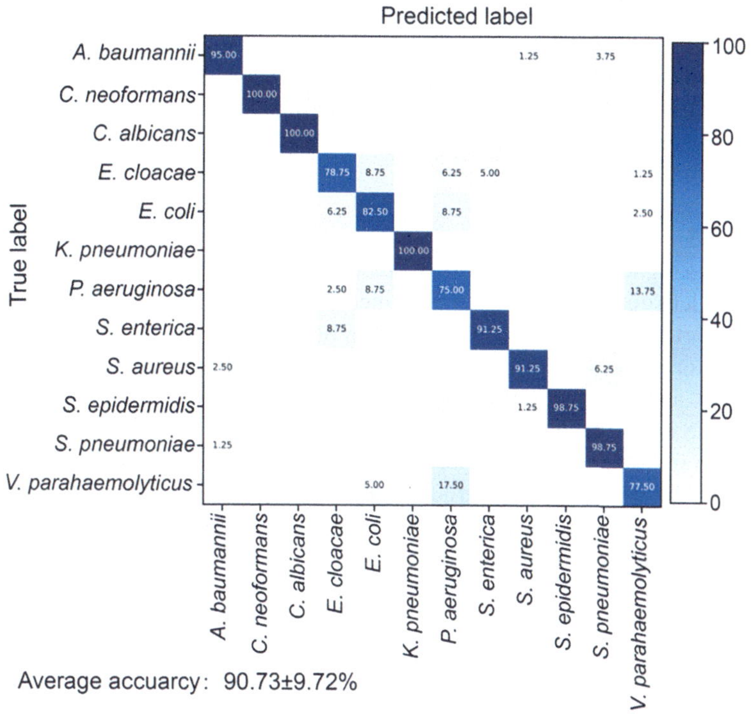

Fig. 15.10 The random forest detection model's recognition accuracy for 12 diseases. The percentage of accurate predictions for each disease is displayed in the confusion matrix. (Reproduced here with permission from (Lu et al. 2023))

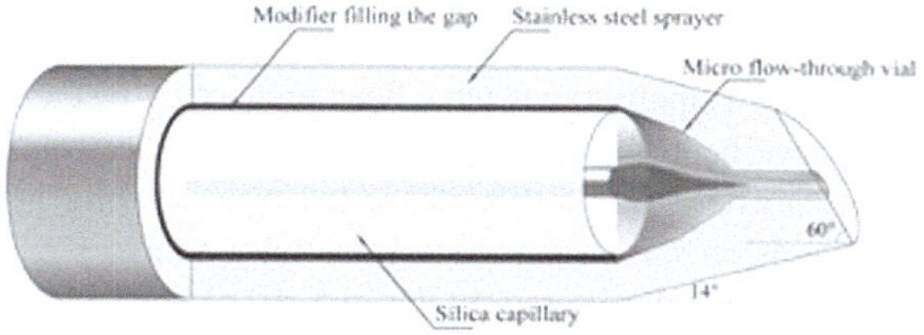

Fig. 15.11 Diagram illustrating the connections between CE, ESI, and MS. We install the fused silica separation capillary in the sprayer made of stainless steel. A tiny flow-through vial is formed by the space between the capillary's end and the inside surface of the stainless-steel tip. The modifier solution in the micro vial is mixed with the CE effluent. (Permission to adapt from (Zhong et al. 2011))

potential can result in an electric contact in liquid-junction interfaces. Analyte transport into the spray capillary is controlled by the electric field and spray liquid pressure, which lowers the dilution at liquid-junction interfaces. The CE-MS instrumentation's operating conditions have been developed and optimized with the help of theory (Al-Awadi et al. 2013; Zhong et al. 2011; Martín et al. 2012; Wojcik et al. 2010). Numerical models of mass transfer, hydrodynamic movement, and electric field strength can reveal a lot about how tiny devices operate. New instrumentations for the CE-ESI-MS interface have been studied experimentally. Current MS equipment can be readily integrated with lab-made or commercial research CE systems (Sorensen et al. 2007; Arceo 2017; Gorbatsova et al. 2012; Grundmann and Matysik 2011).

CE-ICP-MS

ICP-MS is a highly helpful detection method in chemical chromatography (CE) due to its excellent quantification potential, linearity, and low limit of detection (LOD). It works particularly well for elemental differentiation analysis and is particularly beneficial for metal-based drugs and their byproducts in body fluids (Pröfrock 2010; Meermann and Sperling 2012; Herrero et al. 2010). Recent developments in CE-ICP-MS apparatus include the development of a suction-free interface for electrophoresis in a microfluidic device and a demountable capillary microflow nebulizer for iodine speciation. Numerous analytical tasks have been conducted with it, including the speciation analysis of arsenic and selenium, rice samples, dogfish liver, and coal fly ash. A number of publications showed the potential of CE-ICP-MS for various applications (Cheng et al. 2012; Han et al. 2014; Wang et al. 2017; Willberger et al. 2019).

CE-MALDI-MS

Because of its distinct separation and detection processes, non-deterministic MS data gathering, and recurrent analysis of collected samples, MALDI detection is frequently carried out offline in microseparation methods. Applications in biology have made use of this strategy, including peptide and protein digest analysis and neuropeptidomic study. Additionally, utilizing CE-MALDI-MS systems, researchers have created novel capillary coatings for the investigation of peptides and protein digests. A trustworthy fraction collector is essential to the CEMALDI-MS offline combination. PTFE membranes have been created by researchers to collect proteins after they have been separated in denaturing CE electrolyte. It has been shown that sample can be continuously deposited on a moving MALDI sample plate interface. It has been shown that high-sensitivity analyses can benefit from field-enhanced sample injection. Protein LODs of 5–10 nM and neuropeptide LODs of 400 fmol have been attained by microfluidic sampling systems utilizing offline MALDI detection (Wang et al. 2010; Bachmann et al. 2011; Geiger et al. 2012; Said 2017; Chen et al. 2013; Zhong et al. 2012).

Analytical Applications

The CE-MS technology is a vital tool in systems, molecular, and cellular biology and has been extensively researched in proteomics and glycomics. Comprehensive knowledge of the posttranslational modifications of proteins, enzymes, antibodies, and biological systems is required. Three essential needs for proteomics analytical techniques are high throughput, sensitivity, and resolution. Both horizontal and vertical techniques include enzymatic cleavage of protein mixtures after solid proteins are sorted in liquid phase and fragmented using high-resolution mass spectrometers.

Unaltered Proteins

While distinguishing and describing whole biomolecules, like proteins, the combination of CE and MS is an effective analytical tool. The importance of intact protein analysis is highlighted by the huge potential of vertical proteomic investigations using modern high-resolution mass spectrometry equipment. In fact for large molecular mass proteins like antibodies, the thermo-LTQ Ultra ion-cyclotron MS may reach a resolution of 0.1 Da with its stated resolution of 10^6. The problems of using CE-MS to assess intact proteins for vertical proteomics were also examined (François et al. 2019). Various applications of CE-MS for the analysis of intact proteins, such as glycoproteins, biopharmaceuticals, protein-ligand complexes, biomarkers, and dietary proteins, are included in the review. The several electrospray-based ionization techniques shown are tailored for proteome studies. A high-resolution CE-TOF-MS approach was created for small protein changes, and an optimized CE-ESI-MS method was designed for biopharmaceutical analysis (Taichrib 2012; Haselberg et al. 2013; Behnken et al. 2014). Metallothioneins are low molecular mass proteins that are studied using a variety of techniques. For comprehensive characterization, a combination of CE-MS with ICP and ESI has been utilized (Ryvolova et al. 2012).

Peptides and Biomarkers

Two essential elements of the bottom-up approach are reliable MS identification of peptides and high-resolution separation. For examining biological markers, micro-column separation methods using mass spectrometry are effective analytical instruments (Li et al. 2012). CE-MS has been applied in proteins, peptides, and biomarker investigations, including tumor biomarker investigations and prostate cancer biomarker analysis (Desiderio et al. 2010; Gamagedara and Ma 2011). In clinical proteomics, it has difficulties in finding biomarker peptides. Certain studies have elucidated the potential of −1-Acid glycoprotein as a biomarker for bladder cancer. The potential of CE-MS for clinical research based on comprehensive isoform distribution of unmodified glycoproteins was demonstrated by its application to characterize more than 150 human isoforms (Mittermayr et al. 2013; Puerta et al. 2011; Ongay and Neusüβ 2010).

Single-Cell Analysis

Compact capillary or microfluidic devices offer an appropriate setting for the examination of the chemical composition of individual cells with volumes of many femtoliters, as has been shown throughout the course of SCAs almost 25-year history (Watanabe and Terabe 2000; Kortmann et al. 2011). Moreover, single-cell injection, component separation, cell lysis and MS or fluorescence detection can all be included in miniature devices. These devices may be used to identify individual cell components either offline using MALDI identification of fragment recovered after CE separation (Zhong et al. 2012; Amantonico et al. 2010) or online using interfaces with flow rates of several nL/min to link to MS detectors (Heinemann and Zenobi 2011; Amantonico et al. 2010). Research in biological and clinical analysis has tremendous potential thanks to the advancements in SCA technique and technology (Watanabe and Terabe 2000). The focus of the SCA research was on many aspects of physiology and biology. Understanding how the environment, different life situations, and the cell cycle affect many elements of the genome, such as the proteome, peptidome, metabolome, and so on, can be gained by studying individual cell organelles inside of them. Several writers (Heinemann and Zenobi 2011; Amantonico et al. 2010; Oikawa and Saito 2012; Liu et al. 2014) examined the evolution of diagnostic tools and procedure for single-cell metabolomics. Label-free techniques such as MS are necessary for a thorough study, even if fluorescent probes can be used for the focused analysis of certain components to detect them (Amantonico et al. 2010). Single-cell metabolite analysis is now possible thanks to advancements in ionization methods, which have also increased detection efficiency and sensitivity. This is especially helpful in comprehending how metabolites are affected by variations in epigenetics. Examining pharmacokinetics on a single-cell basis and a cell's impact on the fate of a medication or toxicant is made possible by microfluidics-based manipulation and mass spectrometry detection. A vital biological mechanism for cell-to-cell communication, single-cell exocytosis, has also been the subject of studies. These studies highlight the importance of cellular heterogeneity in the domains of neurology, developmental biology, and stem cell biology.

Food Analysis

Recent advancements in mass spectrometry (MS) and capillary electromigration methods have been beneficial for the food analysis fields and foodomics. These methods have opted to food analysis to check for impurities, alien substances, and dangerous components. The CEESI-MS combination was developed to detect melamine and related chemicals in milk and other food products. Because CE-ESI-MS is hyphenated, human milk oligosaccharides can be profiled in breast-milk and the feces of breastfed neonates. Almond skin extract, virgin olive oil, honey, seed oils, and soy drink all contain antioxidant phenolic compounds that can now be easily identified and separated. Within the expanding field of foodomics, complex peptide mixtures have also been analyzed using CE-MS (Kohler et al. 2011; Albrecht et al. 2011; Dominguez-Alvarez et al. 2011).

Forensic Applications

In biological specimens, CE-MS systems can detect, identify, and quantify the abused medications and their byproducts. These technologies offer many benefits for forensic applications. They have been used for determining therapeutic, abused medications and their metabolites, and for analyzing phosphorus consisting amino acid-type herbicides in beverages spiked with herbicides (Gottardo et al. 2012a, b; Kawai et al. 2011). Additionally, nonaqueous CE-TOF-MS has been used to determine 20 antidepressants in plasma samples, with a sensitivity 10–50 times higher than optical detectors (Sasajima et al. 2010).

Pharmacological Analyses

Recent advancements in chemical element separations (CE-MS) have proven beneficial for several applications such as pharmaceutical analyses, drug impurity testing, and the identification of medications in bioactive fluids. The enantioseparation of beta-blocker medications, metallodrug binding research, physicochemical evaluation of antibody-drug conjugates, and biopharmaceutical analysis utilizing noncovalently coated capillaries have all been accomplished with the aid of CE-MS. Additionally, it may be used to the study of complicated pharmaceutical structures, including drug-lysozyme conjugates for targeted cell delivery and selective drug separations employing nonaqueous base metal ions (BGEs) (Timerbaev et al. 2011; Haselberg et al. 2011a; Prokhorova et al. 2010; Sánchez-Hernández et al. 2010; Haselberg et al. 2011b; Anurukvorakun et al. 2010; Veizerová et al. 2012; Ganzera et al. 2010).

Future Development

The study of CE-MS instruments and methods has drawn more attention from a variety of applications due to the possible improvements in sensitivity and reliability. Advances in sample preparation methods, ionization strategies, and electrolyte systems have opened up new domains. Still, there are challenges: appropriate electrolytes are not available; also, an efficient ionization interface is needed for the transfer of analytes from separating columns into MS entry orifices. The creation of novel microfluidic devices is gaining more attention since disposable devices save time and money in environmental investigation, point-of-care diagnostics, forensic applications, food testing. Unexpected uses could arise from investigating new materials to make it easier to produce microfluidic devices with the right mechanical, electrical, and chemical properties. Continuous CE-MS analysis is expected to be frequently used in future cellular research.

15.7 Pathogen Identification and Antibiotic Susceptibility Assessment at the Single-Cell Level Using Raman Spectroscopy in Conjunction with Machine Learning

Clinical diagnostics face a technological challenge from single-cell resolution and fast, precise, label-free identification of pathogenic microorganisms and antibiotic resistance. The eradication of the time-consuming antibiotic susceptibility tests and the tedious pathogenic bacterial culture process will significantly improve early detection and the most effective use of antibiotics in clinics. Suitable for single-cell pathogen identification, Raman spectroscopy offers a label-free, culture-independent technique for acquiring pathogen molecular fingerprints. Through the use of machine learning and Raman spectroscopy, researchers have developed a rapid and accurate method for identifying harmful bacteria and detecting drug resistance at single-cell resolution. Their research indicates that the machine-learning method has an average detection accuracy of 90.73 ± 9.72% for 12 common pathogenic bacterial species. Using a machine-learning model, 99.92 ± 0.06% accuracy was achieved in differentiating between antibiotic-sensitive and resistant strains of Acinetobacter baumannii isolated from hospital patients. A greater nucleic acid/protein ratio was seen in sensitive bacteria, but amide II protein structures were more prevalent in antibiotic-resistant strains. Raman spectroscopy appears to be a viable technique for quickly identifying infections and determining their sensitivity to antibiotics, according to this study (Lu et al. 2023).

Materials and Methods

Strains of Yeast and Bacteria

Including two species of fungi (*Cryptococcus neoformans* and *Salmonella albicans*) and three gram-positive species (*Staphylococcus aureus*, *S. epidermidis*, and *Streptococcus pneumoniae*), this study used seven gram-negative species of bacteria (*A. baumannii*, *Enterobacter cloacae*, *E. coli*, *K. pneumonia*, *P. aeruginosa*, *Salmonella enterica*, and *Vibrio parahaemolyticus*) and three gram-positive species (*Staphylococcus aureus*, *S. epidermidis*, and *Streptococcus pneumoniae*). In the Cangzhou Central Hospital in Hebei Province, China, sputum from patients in the intensive care unit was found to have five clinical strains of *A. baumannii* (ST2 sequence type). Several previous reports (Woodford et al. 2006; Lu et al. 2023) state that PCR testing verified the presence of the carbapenem resistance gene oxa23 in all five clinical strains of *A. baumannii*.

Measurements of the Single-Cell Raman Spectra

Luria-Bertani medium and yeast extract peptone dextrose medium were used to cultivate both yeast and bacterial cells under various conditions, including varying culture times. After being harvested, the bacteria were suspended in a 0.85% NaCl solution. Two dozen suspended cells in μl were added to the sealed chamber in order to measure the Raman spectra. Raman spectroscopy (LTRS) was employed to measure the Raman spectra using laser tweezers, as was previously mentioned (Vieira et al. 2010). In order to calibrate the LTRS's Raman spectra, the 10 μm polystyrene spheres were set to 620.9 cm−1, 1001.4 cm−1, and 1602.3 cm⁄.1. The integration time of 60–90 s was selected. Following the collection of at least 300 single cells from various batches, one Raman spectra was produced for each species. The machine-learning algorithms were trained on these Raman spectra. To assess how accurately the machine-learning models classified each species, 80 individual cells were extracted from distinct culture batches that differed from the batches used to train the models. The Raman spectra of these single cells were used as testing data to evaluate the models.

Processing Raman Spectral Data and Building Machine-Learning Models

The spectra were pre-processed as follows: Min-max normalization is performed following Savitzky-Golay smoothing to eliminate noise and polynomial fitting to eliminate the fluorescence background. The spectral range chosen for the training and testing of the model was between 555 and 1815 cm−1. We built models for training Raman spectrum data based on a foundation of random forest, support vector machine, decision tree, bagging, and naive Bayes. The models' endurance was assessed by running a tenfold cross-validation on the training dataset. This study's five models were trained on an HP for around four hours. The effectiveness of each species' 80 single cells was evaluated using distinct spectra. The collection of Raman spectra (integration time: 60–90 s), species prediction (less than 30 s), and sample preparation are all part of the sample identification process (Lu et al. 2020).

The Model's Design for Detecting Antibiotic Susceptibility and Its Training Information

The VITEK 2 instrument was used in the investigation to ascertain the antibiotic susceptibilities of six A. baumannii strains. Imipenem, meropenem, ampicillin, cefoperazone, and cefepime were among the drugs that were tried. Antibiotic-free cultures of the reference strain and clinical isolates were obtained, and Raman spectra were recorded. Table 15.2 displays the various spectra from various A. baumannii strains that were used to create the RF antibiotic susceptibility identification model. To evaluate the prognosticative abilities of the detection models and the model, 1255 spectra were gathered for an independent testing dataset. The diagnostic accuracy was evaluated using the receiver operating characteristic curve, and the model that had the best prediction accuracy was chosen. The antimicrobial susceptibility test results were used to categorize the A. baumannii strains. Using several batches of Raman spectra, the model was trained on the four-antibiotic prediction task. For statistical analysis, PCA, building machine-learning models, and pre-processing Raman spectra, R ($\geq$3.6.1) was utilized as a programming language (Fig. 15.12).

Using Raman Spectra and Machine Learning, to Identify Pathogens

In the study, the identification accuracy of several machine-learning algorithms was examined, including decision trees (DT), random forests (RF), support vector machines (SVM), naïve Bayes (NB), and bagging. The models were trained using 3982 spectra from the Raman dataset. The discriminative power of these models was evaluated using tenfold cross-validation. The results showed that compared to DT, bagging, and NB, RF and SVM models outperformed them in precisely detecting pathogens at single-cell resolution. Additionally, RF, SVM obtained higher kappa values than the other remain models. The models were assessed using 960 spectra from 12 different diseases in an unbiased test dataset. Based on the data, the

Table 15.2 Raman band assignments in the difference spectra of Acinetobacter baumannii that are resistant to antibiotics and those that are susceptible to them

Strains	Wavenumber (cm−1)	Assignment/wavenumber (cm−1)	Reference
Resistant	860–918	Polysaccharides (868), tryptophan (880), protein (890), saccharide (891), glucose (913)	Hanlon et al. (2000); Krafft et al. (2005); Movasaghi et al. (2007); Kendall et al. (2011)
	1330–1367	CH2/CH3, twisting, collagen (1336), tryptophan (1359), glucose (1343), protein (1250–1350)	Krafft et al. (2005); Movasaghi et al. (2007); Cheng et al. (2005); Ramirez-Mora et al. (2019)
	1445	CH2/CH3 of carbonic acid, cholesterol (1440–1460), phospholipids (1445)	Maquelin et al. (2002); Töpfer et al. (2019); Park et al. (2021)
	1554	Amide II of protein	Faolain et al. (2005)
	1602	C–C, C–N of protein (1061)	Kuhar et al. (2021)
Sensitive	729	Adenine of DNA or RNA(728)	Neugebauer et al. (2010)
	783	Cytosine, uracil ring of DNA or RNA	Kendall et al. (2011); Maquelin et al. (2002)
	811	O–P–O of lipids	Notingher et al. (2003)
	1002	Phenylalanine of protein	Töpfer et al. (2019)
	1099	CC skeletal and C–O–C glycosidic link (1098)	Maquelin et al. (2002)
	1125	C–C skeletal of acyl backbone in lipid (1126)	Cheng et al. (2005)
	1281	Amide III of protein CH2 wagging vibrations from glycine backbone (1280)	Dukor (2001)
	1576	Guanine, adenine of DNA/RNA (ring stretching) (1575)	Maquelin et al. (2002)

average accuracy of RF was 90.73 ± 9.72%, which was marginally better than SVM. Gram-positive bacteria S. pneumoniae and S. epidermidis had a classification accuracy of 100%, whereas K. pneumoniae had an accuracy of 98.75% for identification. *E. coli* exhibited an accuracy of 82.5% for identification, compared to 75–83% for the other three gram-negative bacteria (Cohen 1960; García et al. 2009; Vieira et al. 2010).

Machine Learning-Based Identification of the Antibiotic-Resistant Strain of Acinetobacter Baumannii

Utilizing Raman spectroscopy to provide an antibiotic susceptibility test devoid of culturing is the primary objective of the study. Five drug-resistant A. baumannii strains were used as a proof-of-concept by the researchers, who got them from patients in the intensive care unit. Figure 15.13 shows A. baumannii single's Raman spectra, which show characteristics associated with antibiotic resistance. The bacteria are the cause of severe nosocomial infections, particularly in patients with weakened immune systems and those receiving treatment in an ICU. Antibiotic resistance

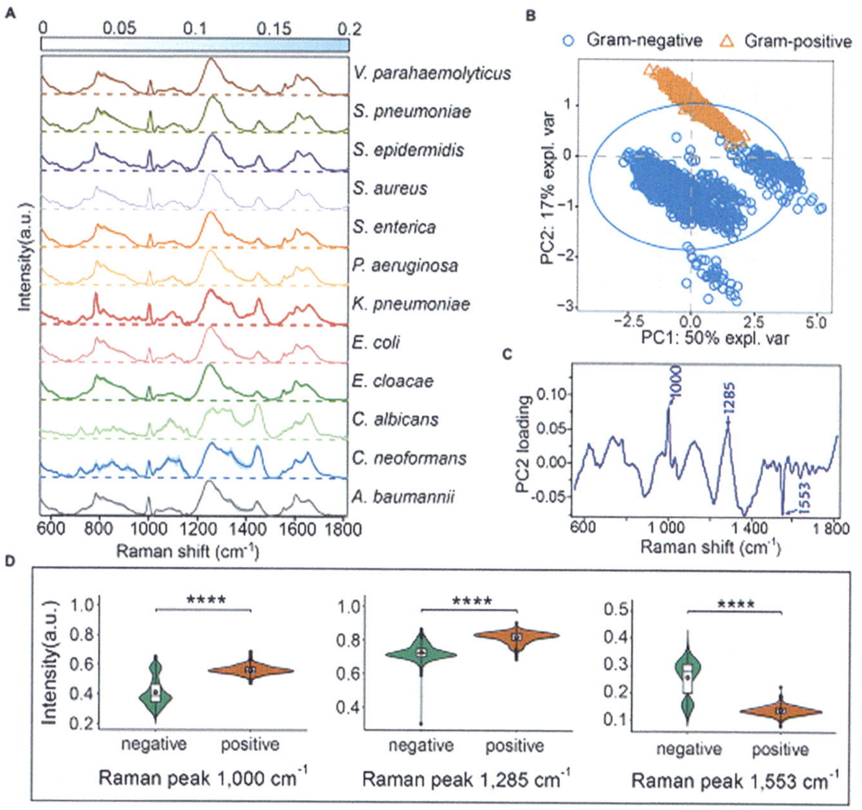

Fig. 15.12 Typical clinical diseases' mean Raman spectra are displayed. (**a**) The solid line displays the average Raman spectra for each species; the shadow shows the standard deviations. Heatmaps are used to display the standard deviations for each pathogen and the Raman band fluctuation. LTRS obtained the Raman spectra of over 300 different cells for each species. (**b**) The distinctions between gram-positive and negative microbes are shown by principal component analysis (PCA). The first two PCs explained 17% and 50% of the variances, respectively. For any strain, the confidence ellipse is 0.95. (**c**) PC2 loading throughout the PCA analysis process. Violin plots (**d**) display the Raman intensities of gram-positive and gram-negative bacteria at peaks of 1000, 1285, and 1553 cm−1. While embedded box plots show the median, first, and third quartiles, whiskers show the lowest and greatest values within 1.5 interquartile ranges from the first and third quartiles. The Raman intensity on average is shown by red dots. Two-sided t-tests were used to assess the statistical significance of both gram-positive and gram-negative microorganisms. P is not equal to 0.00001 (Lu et al. 2023). Figure 15.12 represents prevalent clinical infections' average Raman spectra

must be identified quickly and accurately in order to promptly deploy treatment strategies. The researchers acquired single-cell Raman spectra from the antibiotic-sensitive strain and five multiple drug-resistant bacteria without administering additional medications. Using the training dataset, the antibiotic susceptibility detection model was trained. The model demonstrated good specificity and sensitivity in distinguishing the antibiotic-resistant and susceptible strains of A. baumannii.

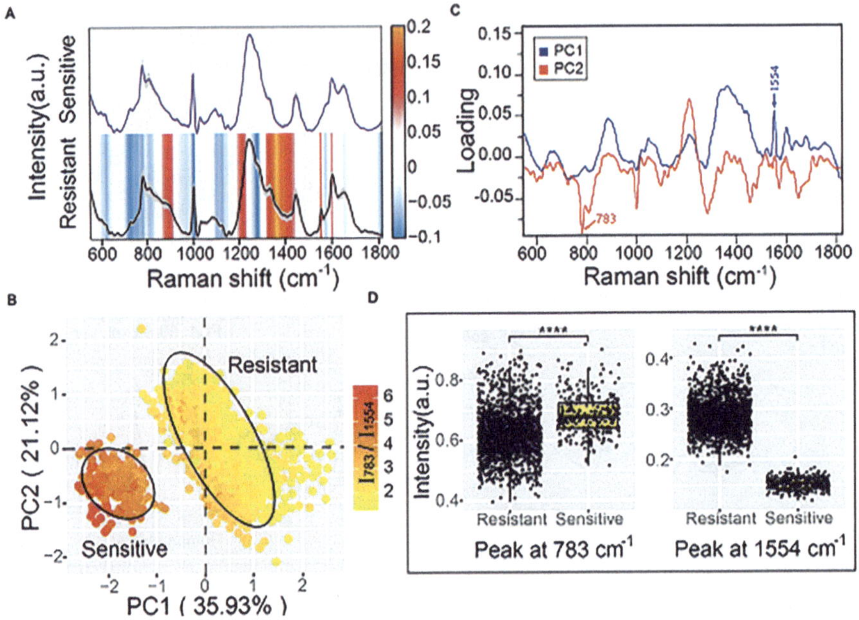

Fig. 15.13 A. baumannii single's Raman spectra illustrate the features of antibiotic resistance. (**a**) The Raman spectrum of A. baumannii, demonstrating both antibiotic resistance and susceptibility. A straight line represents the average Raman spectra, while a gray shadow represents the standard deviations. The spectrum differences between bacteria that are susceptible to and resistant to antibiotics are shown in the heatmaps. Principal component analysis (**b**) is used to compare drug-resistant and drug-sensitive isolates. 21.12% and 35.93, respectively, were the deviations that the first two PCs explained. Any strain has a confidence ellipse of 0.95. The Raman intensity ratio at 783 cm−1 and 1554 cm−1 (I783/I1554) is shown by the color gradient. The loading of PC1 and PC2 in PCA analysis (C). (D) The Raman intensity of bacteria that are resistant to and susceptible to antibiotics at the 783 cm−1 and 1554 cm−1 peaks. With the whiskers denoting these values, box plots show the first, third, and median quartiles as well as the lowest and greatest values within 1.5 interquartile ranges from the first and third quartiles. The statistical significance between microorganisms that were resistant to antibiotics and those that were susceptible to them was examined using two-sided t-tests. P is smaller than 0.00001. The Raman intensity of A. baumannii cell is indicated by each dot (Lu et al. 2023). The Raman spectra of A. baumannii single are shown in this figure, which shows characteristics associated with antibiotic resistance

Moreover, it achieved the highest possible detection accuracy of 95.66%. A mean accuracy of 99.92 ± 0.06% was estimated by the RF model. The study found that the application of Raman spectroscopy and machine learning can help estimate antibiotic susceptibility, accurately identify various bacterial species, and facilitate single-cell antibiotic selection. To determine the Raman spectrum distinction, the average spectrum of A. baumannii that is sensitive to antibiotics was subtracted from the average spectrum of A. baumannii that is resistant to antibiotics. Figure 15.4a and Table 15.1 illustrate the antibiotic-resistant bacteria's enhanced Raman bands at 860–918 cm−1 (polysaccharides and proteins), 1336–1367 cm−1 (carbohydrates and proteins), 1554 cm−1 (amide II of proteins), and 1602 cm−1 (C–C or C–N

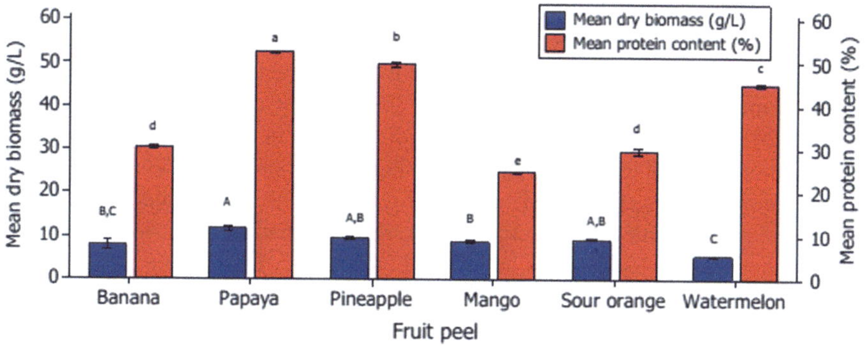

Fig. 15.14 Pineapple, watermelon, papaya, mango, sour orange, and banana peel effects on the dry biomass and protein composition of SCP obtained from a mixed palmyrah toddy culture. To compare means at $p < 0.05$, Tukey's multiple comparison test was employed. Significant variations in dry biomass and protein content are denoted by different capital and lowercase characters, respectively (Thiviya et al. 2022)

protein bonds). However, in the antibiotic-resistant bacteria, the Raman peaks at 729, 783, and 1576 cm−1 (DNA/RNA Raman bands) reduced. Moreover, phenylalanine at 1002 cm−1 and protein amide III at 1281 cm−1 in antibiotic-sensitive A. baumannii showed a little increase in Raman peaks (also seen in Fig. 15.4a and Table 15.1) (Talebi Bezmin Abadi et al. 2019; Vázquez-López et al. 2020; Butler et al. 2019).

Discussion

Antibiotic susceptibility testing and prompt pathogen identification are very much crucial for both effective clinical treatment and slowing the emergence of antimicrobial resistance (AMR). In this paper, researchers identified 12 common clinical infections at single-cell resolution using Raman spectroscopy and ML techniques. Additionally, the machine-learning model was shown to be capable of effectively identifying antibiotic-resistant A. baumannii cells using the Raman spectrum. Most significantly, Raman spectra and ML could assume bacterial medication resistance in the unlikely case that antibiotic therapy is not administered. They believe that developing a method for detecting drug-resistant viruses in practical environments may be essential. The use of an automated system and this methodology would facilitate the prompt and precise identification of all microbial cells present in clinical specimens, as well as the examination of their individual biochemical and metabolic characteristics. Furthermore, it would permit the investigation of pathogenic ecophysiology in natural settings. In this work, prediction models are constructed using 5 ML techniques: RF, SVM, DT, NB, and bagging. When comparing the accuracy of the various models, the SVM and RF models fared the better result. The RF antibiotic susceptibility identification model predicted the antibiotic susceptibility of A. baumannii with 99.92% accuracy, while the RF identification model had an average accuracy of 90.73% in identifying 12 prevalent clinical infections. Thus, for the purpose of developing the prediction model in this study, RF is the best ML technique. For approval on the diseases and generate category labels for those that earn the majority of votes, RF uses many independent decision trees (Biau and Scornet 2016; Shaikhina et al. 2019). According to Cui et al. (2020), this approach may be more reliable than the bagging and single DT approaches. Additionally, RF model training is faster than DT and SVM training which is advantageous when using vast size of data (Parmar et al. 2019). Bacterial pathogens have tremendous genetic flexibility, which results in antibiotic resistance. Developing ways to stop resistance from emerging and spreading requires a thorough understanding of the molecular mechanisms that lead to resistance. This work establishes the groundwork for using markers present in the Raman spectrum to decipher the mechanism of antibiotic resistance. The variations in the Raman spectrum were nearly consistent when comparing the antibiotic-sensitive A. baumannii strain to five antibiotic-resistant strains. The most prominent examples of this were found in the Raman peaks of proteins (1554 and 1602 cm−1), which exhibit markedly enhanced intensity. The higher protein Raman peaks could be related to the elevated oxacillinase production level in vivo, since all five strains carry the oxa23 gene, which codes for the enzyme. Additionally, the antibiotic-resistant A. baumannii may promote alterations to the bacterial cell membrane that would improve the development of biofilms due to the greater concentrations of phospholipids (1445 cm−1), polysaccharides, and proteins (860–918 cm−1 and 1330–1367 cm−1) (Park et al. 2021; García et al. 2009; Gieroba et al. 2020). Because Raman spectrum wavenumbers represent many different biomolecules, their complex and high-dimensional nature makes it challenging to correctly identify a particular biomolecule with a Raman spectrum wavenumber in order to infer the chemical mechanism of bacterial drug resistance. Further development of computational approaches is necessary to facilitate the mapping of Raman spectra to biomolecular profiles. By utilizing machine learning and merging massive amounts of data from transcriptomes, single-cell genomes, proteomes and metabolomes, the drug resistance mechanism of bacteria depending on the Raman spectra will be unveiled in the interim. The resistant strains of A. baumannii that make clinical anti-infective treatment difficult and seriously jeopardize public health are the subject of this investigation. Because the current technique, RF antibiotic susceptibility detection, can only distinguish between resistant and sensitive species, it is insufficient to identify specific antibiotic resistance. Subsequent research will employ deep learning techniques to discern resistance attributes. Although pathogen identification and antibiotic susceptibility assessment have not been accomplished in clinical settings using Raman spectroscopy, a standardized Raman spectral database of pathogenic microbes covering a greater variety of growth media and situations, resistant as well as susceptible strains, and antibiotic vulnerability profiles would be beneficial in bridging the knowledge gap between academic research and clinical practice. In order to analyze a single pathogen without the need for laborious laboratory analysis and time-consuming cultivation, Raman spectroscopy could be utilized in conjunction with technology like microscopy (Neugebauer et al. 2010). The clinical strains' Raman spectrum data would support field testing with a portable equipment. A single step, not harmful, unaffected by culture, free of labels, and quick method of detecting pathogenic microorganisms in patient samples in order to determine antibiotic susceptibility would be revolutionary. This would lead to better results for patients.

15.8 Single-Cell Protein Is Produced from Fruit Peel Wastes by Utilizing Palmyrah Toddy Yeast

Recently, there has been some interest in single-cell protein (SCP), which is produced from agricultural waste, as an inexpensive and different source of protein to address the expanding population's nutritional needs. Environmental problems are further decreased by the bioconversion of these wastes into SCP like value-added items. This study investigated whether peel wastes from locally available pineapple (Ananas comosus), watermelon (Citrullus lanatus), papaya (Carica papaya), sour orange (Citrus medica), banana (Musa acuminata), and mango (Mangifera indica) could be used to produce SCP. The system used for this was a liquid state fermentation system with a palmyrah (Borassus flabellifer) toddy that held a natural mixed yeast and bacteria culture. In an endeavor to enhance protein yield, this investigation also endeavored to determine the ideal substrate and process parameters for SCP synthesis. An analysis using physical and chemical methods was conducted on a specific set of fruit peels. After a 48-hour fermentation in triplicate in a shaking incubator at 100 rpm, the residues were collected by centrifugation at 1252 × g, oven-dried, and the dry weight was determined to ascertain the protein content. The sterilized peel extracts (10%, v/v) were then inoculated with 5 mL of palmyrah toddy. The biomass yield ranged from 5.3 to 11.7 g/L, based on the data. Peels from papayas produced the highest output, whereas peels from watermelon produced the least amount of biomass. With 52.4 percent protein, papaya peels were significantly higher than the protein contents of pineapple (49.7%), watermelon (45.2%), banana (30.4%), sour orange (29.5%), and mango (24.6%) peels. Papaya waste needs to ferment for 24 hours at 25 °C with a pH of 7 in order to get the best results. The dry weight and protein content of biomass are greatly decreased by nucleic acid reduction treatment. In summary, it can be said that the natural toddy blend culture of palmyrah is an excellent substrate for generating cell biomass high in protein from leftover papaya peels (Thiviya et al. 2022).

15.9 Remarks and Prospects for the Future

The waste from papaya, watermelon, pineapple, bananas, mangoes, and sour orange peels can be used as an alternative and productive carbon source to create SCP, according to the present research. Figure 15.14 shows how the protein content and dry biomass of SCP, which was produced from a mixed palmyrah toddy culture, were affected by pineapple, watermelon, papaya, mango, sour orange, and banana peel. Using a mixed culture of palmyrah toddy that was naturally formed through liquid fermentation, six peels were evaluated for their ability to produce SCP. Of these, papaya peel produced the highest yield of protein content. After optimization, there was a little increase in the papaya peel biomass protein content and a twofold increase in biomass. In perfect conditions, the SCP derived from the papaya peel medium had a greater protein content than the control medium. The nucleic acid reduction process has a major effect on the lowering of protein content in SCP and dry biomass. Furthermore, the amount of amino acids in SCP derived from papaya peel matches the recommendations for amino acids provided by the UNU, WHO, and FAO. Papaya peel extract may be the best choice for making SCP, based on the natural blend culture of palmyrah toddy culture in liquid fermentation, the study's findings indicate. The OFAT approach has several drawbacks because it is time-consuming and requires careful consideration of how each factor interacts with the formation of SCPs. Despite the fact that a high protein production was recorded under these enhanced circumstances, the use of statistical optimization techniques, such as response surface methodology (RSM), enhances the SCP manufacturing process by evaluating the correlations among inputs and responses. Since agricultural wastes usually have limited nitrogen content, increasing the amount of nitrogen from additional sources can improve the protein composition of SCP (Oshoma et al. 2019). By either effectively using and saccharifying the substrate or by getting rid of growth-inhibiting byproducts, co-cultures outperform monocultures in biomass productivity. Because co-culture eliminates the requirement for substrate treatment and shortens the fermentation duration, it may result in cost savings (Bajpai 2017). The hybrid cultivation of palmyrah toddy is therefore one possible source for the manufacture of SCP. The high content protein of the biomass harvested further demonstrates the commercial viability of the natural palmyrah toddy crop. Since no toxicological studies have been done in this field, more research is suggested.

References

Abbas O, Zadravec M, Baeten V, Mikuš T, Lešić T, Vulić A et al (2018) Analytical methods used for the authentication of food of animal origin. Food Chem 246:6–17

Acharya R, Pujari PK (2019) Potential of conventional and internal monostandard NAA and PGNAA and PIGE in forensic sciences: an overview. Forensic Chem 12:107–116

Adhikari B, Marasini BP, Rayamajhee B, Bhattarai BR, Lamichhane G, Khadayat K et al (2021) Potential roles of medicinal plants for the treatment of viral diseases focusing on COVID-19: A review. Phytother Res 35(3):1298–1312

Alam SI, Kumar B, Kamboj DV (2012) Multiplex detection of protein toxins using MALDI-TOF-TOF tandem mass spectrometry: application in unambiguous toxin detection from bioaerosol. Anal Chem 84(23):10500–10507

Al-Awadi SJ, Ghareeb AM, Salo WH (2013) Food dyes as an alternative tracking dye for DNA gel electrophoresis. Baghdad Sci J 10(4):1150–1156

Albrecht S, Schols HA, van Zoeren D, van Lingen RA, Jebbink LJG, van den Heuvel EG et al (2011) Oligosaccharides in feces of breast-and formula-fed babies. Carbohydr Res 346(14):2173–2181

Ali EM, Edwards HG (2017) The detection of flunitrazepam in beverages using portable Raman spectroscopy. Drug Test Anal 9(2):256–259

Amantonico A, Urban PL, Zenobi R (2010) Analytical techniques for single-cell metabolomics: state of the art and trends. Anal Bioanal Chem 398:2493–2504

Anurukvorakun O, Buchberger W, Himmelsbach M, Klampel CW, Suntornsuk L (2010) A sensitive non-aqueous capillary electrophoresis-mass spectrometric method for multiresidue analyses of β-agonists in pork. Biomed Chromatogr 24(6):588–599

Arceo J (2017) Metabolomic analysis of complex biological samples via capillary zone electrophoresis–electrospray ionization-mass spectrometry and MALDI imaging mass spectrometry (Doctoral dissertation, University of Notre Dame)

Armstrong GD, Mulvey GL, Marcato P, Griener TP, Kahan MC, Tennent GA et al (2006) Human serum amyloid P component protects against Escherichia coli O157: H7 Shiga toxin 2 in vivo: therapeutic implications for hemolytic-uremic syndrome. J Infect Dis 193(8):1120–1124

Awuchi CG (2019) Medicinal plants: the medical, food, and nutritional biochemistry and uses. Int J Adv Acad Res 5(11):220–241

Axén J, Malmström D, Axelsson BO, Petersson P, Sjöberg PJ (2010) Efforts to improve detection sensitivity for capillary electrophoresis coupled to atmospheric pressure photoionization mass spectrometry. Rapid Commun Mass Spectrom 24(9):1260–1264

Bachmann S, Bakry R, Huck CW, Polato F, Corradini D, Bonn GK (2011) Peptide mapping using capillary electrophoresis offline coupled to matrix-assisted laser desorption ionization time of flight mass spectrometry. Electrophoresis 32(20):2830–2839

Bajpai P (2017) Single cell protein production from lignocellulosic biomass. Springer, Singapore, pp 1–78

Barbula GK, Safi S, Chingin K, Perry RH, Zare RN (2011) Interfacing capillary-based separations to mass spectrometry using desorption electrospray ionization. Anal Chem 83(6):1955–1959

Behnken HN, Ruthenbeck A, Schulz JM, Meyer B (2014) Glycan analysis of Prostate Specific Antigen (PSA) directly from the intact glycoprotein by HR-ESI/TOF-MS. J Proteome Res 13(2):997–1001

Beltramo C, Riina MV, Colussi S, Campia V, Maniaci MG, Biolatti C et al (2017) Validation of a DNA biochip for species identification in food forensic science. Food Control 78:366–373

Ben-Shabat S, Yarmolinsky L, Porat D, Dahan A (2020) Antiviral effect of phytochemicals from medicinal plants: applications and drug delivery strategies. Drug Deliv Transl Res 10:354–367

Berman HM, Westbrook J, Feng Z, Gilliland G, Bhat TN, Weissig H et al (2000) The protein data bank. Nucleic Acids Res 28(1):235–242

Bhatt PV, Pandey G, Tharmavaram M, Rawtani D, Mustansar Hussain C (2020) Nanotechnology and taggant technology in forensic science. In: Technology in forensic science: Sampling, analysis, data and regulations, pp 279–301. https://doi.org/10.1002/9783527827688.ch14

Biau G, Scornet E (2016) A random forest guided tour. TEST 25:197–227

Bitzan M, Klemt M, Steffens R, Müller-Wiefel DE (1993) Differences in verotoxin neutralizing activity of therapeutic immunoglobulins and sera from healthy controls. Infection 21(3):140–145

Butler DA, Biagi M, Tan X, Qasmieh S, Bulman ZP, Wenzler E (2019) Multidrug resistant Acinetobacter baumannii: resistance by any other name would still be hard to treat. Curr Infect Dis Rep 21:1–17

Caplioli A (1994) Frequency and nature of verocytotoxin 2 (VT2) neutralizing activity (NA) in human and animal sera. In: Recent advances in verocytotoxin-producing Escherichia coli Infections. Elsevier Sciences, Amsterdam, pp 353–356

Casjens SR, Hendrix RW (2015) Bacteriophage lambda: early pioneer and still relevant. Virology 479:310–330

Chalker JM, Bernardes GJ, Lin YA, Davis BG (2009) Chemical modification of proteins at cysteine: opportunities in chemistry and biology. Chem Asian J 4(5):630–640

Chan CK, Pavlović NM, Chan W (2019) Development of a novel liquid chromatography-tandem mass spectrometric method for aristolochic acids detection: application in food and agricultural soil analyses. Food Chem 289:673–679

Chen HX, Busnel JM, Qiao L, Gasilova N, Zhang XX, Girault HH (2013) Compatible buffer for capillary electrophoresis and matrix-assisted laser desorption/ionization mass spectrometry. Anal Methods 5(16):4258–4262
Cheng WT, Liu MT, Liu HN, Lin SY (2005) Micro-Raman spectroscopy used to identify and grade human skin pilomatrixoma. Microsc Res Tech 68(2):75–79
Cheng H, Liu J, Yin X, Shen H, Xu Z (2012) Elimination of suction effect in interfacing microchip electrophoresis with inductively coupled plasma mass spectrometry using porous monolithic plugs. Analyst 137(13):3111–3118
Civera T (2003) Species identification and safety of fish products. Vet Res Commun 27:481
Cohen J (1960) A coefficient of agreement for nominal scales. Educ Psychol Meas 20(1):37–46
Cooper A, Wayne R (1998) New uses for old DNA. Curr Opin Biotechnol 9(1):49–53
Cox N, Pilling D, Gomer RH (2014) Serum amyloid P: a systemic regulator of the innate immune response. J Leukoc Biol 96(5):739–743
Cui H, Guo W, Jin L, Hu S (2020) Elemental screening of plant-based foods by slurry nebulization ICP-MS. J Anal At Spectrom 35(3):592–599
De Lima MD, Barbosa R (2019) Methods of authentication of food grown in organic and conventional systems using chemometrics and data mining algorithms: a review. Food Anal Methods 12:887–901
Deconinck E, Kamugisha A, Van Campenhout P, Courselle P, De Beer JO (2015) Development of a stationary phase optimised selectivity liquid chromatography based screening method for adulterations of food supplements for the treatment of pain. Talanta 138:240–246
DeSalle R, Birstein VJ (1996) PCR identification of black caviar. Nature 381:197
Desiderio C, Iavarone F, Rossetti DV, Messana I, Castagnola M (2010) Capillary electrophoresis-mass spectrometry for the analysis of amino acids. J Sep Sci 33(16):2385–2393
Detwiler LA, Baylis M (2003) The epidemiology of scrapie. Revue scientifique et technique (International Office of Epizootics) 22(1):121–143
Dominguez-Alvarez J, Rodríguez-Gonzalo E, Hernández-Méndez J, Carabias-Martinez R (2011) Programmed nebulizing gas pressure for efficient and stable capillary electrophoresis–mass spectrometry analysis of anionic compounds in positive separation mode. Anal Chem 83(7):2834–2839
Dukor RK (2001) Vibrational spectroscopy in the detection of cancer. In: Handbook of vibrational spectroscopy. Wiley, Hoboken
Fagerquist CK, Sultan O (2010) Top-down proteomic identification of furin-cleaved α-subunit of Shiga toxin 2 from Escherichia coli O157: H7 using MALDI-TOF-TOF-MS/MS. Biomed Res Int 2010(1):123460
Fagerquist CK, Sultan O (2011) Induction and identification of disulfide-intact and disulfide-reduced β-subunit of Shiga toxin 2 from Escherichia coli O157: H7 using MALDI-TOF-TOF-MS/MS and top-down proteomics. Analyst 136(8):1739–1746
Fagerquist CK, Zaragoza WJ (2015) Shiga toxin 2 subtypes of enterohemorrhagic E. coli O157: H-E32511 analyzed by RT-qPCR and top-down proteomics using MALDI-TOF-TOF-MS. J Am Soc Mass Spectrom 26(5):788–799
Faolain EO, Hunter MB, Byrne JM, Kelehan P, McNamara M, Byrne HJ, Lyng FM (2005) A study examining the effects of tissue processing on human tissue sections using vibrational spectroscopy. Vib Spectrosc 38(1–2):121–127
Frades I, Matthiesen R (2010) Overview on techniques in cluster analysis. In: Bioinformatics methods in clinical research. Humana Press, pp 81–107
François YN, Marie AL, Ruel C, Gahoual R, Tran NT, Taverna M (2019) Capillary electrophoresis coupled to mass spectrometry as a powerful tool to investigate heterogeneities, conformers, and oligomers of intact proteins. In: Advances in chromatography. CRC Press, pp 137–179
Frazier RA, Ames JM, Nursten HE (1999) The development and application of capillary electrophoresis methods for food analysis. Electrophoresis 20(15–16):3156–3180
Frazier RA, Nursten HE, Ames JM (2000) Capillary electrophoresis for food analysis: method development. Royal Society of Chemistry

Freedman SB, Xie J, Neufeld MS, Hamilton WL, Hartling L, Tarr PI et al (2016) Shiga toxin–producing Escherichia coli infection, antibiotics, and risk of developing hemolytic uremic syndrome: a meta-analysis. Clin Infect Dis 62(10):1251–1258

Frezza D, Favaro M, Vaccari G, Von-Holst C, Giambra V, Anklam E et al (2003) A competitive polymerase chain reaction–based approach for the identification and semiquantification of mitochondrial DNA in differently heat-treated bovine meat and bone meal. J Food Prot 66(1):103–109

Fuyal M, Giri B (2020) A combined system of paper device and portable spectrometer for the detection of pesticide residues. Food Anal Methods 13(7):1492–1502

Gamagedara S, Ma Y (2011) Biomarker analysis for prostate cancer diagnosis using LC–MS and CE–MS. Bioanalysis 3(18):2129–2142

Ganzera M, Krüger A, Wink M (2010) Determination of quinolizidine alkaloids in different Lupinus species by NACE using UV and MS detection. J Pharm Biomed Anal 53(5):1231–1235

Gao L, Ju L, Cui H (2017) Chemiluminescent and fluorescent dual-signal graphene quantum dots and their application in pesticide sensing arrays. J Mater Chem C 5(31):7753–7758

García S, Fernández A, Luengo J, Herrera F (2009) A study of statistical techniques and performance measures for genetics-based machine learning: accuracy and interpretability. Soft Comput 13:959–977

Geiger M, Hogerton AL, Bowser MT (2012) Capillary electrophoresis. Anal Chem 84(2):577–596

Gieroba B, Krysa M, Wojtowicz K, Wiater A, Pleszczyńska M, Tomczyk M, Sroka-Bartnicka A (2020) The FT-IR and Raman spectroscopies as tools for biofilm characterization created by cariogenic streptococci. Int J Mol Sci 21(11):3811

Gongora J, Fleming P, Spencer PB, Mason R, Garkavenko O, Meyer JN et al (2004) Phylogenetic relationships of Australian and New Zealand feral pigs assessed by mitochondrial control region sequence and nuclear GPIP genotype. Mol Phylogenet Evol 33(2):339–348

Goossens B, Waits LP, Taberlet P (1998) Plucked hair samples as a source of DNA: reliability of dinucleotide microsatellite genotyping. Mol Ecol 7(9):1237–1241

Gorbatsova J, Borissova M, Kaljurand M (2012) Electrowetting-on-dielectric actuation of droplets with capillary electrophoretic zones for off-line mass spectrometric analysis. J Chromatogr A 1234:9–15

Gottardo R, Fanigliulo A, Sorio D, Liotta E, Bortolotti F, Tagliaro F (2012a) Monitoring compliance to therapy during addiction treatments by means of hair analysis for drugs and drug metabolites using capillary zone electrophoresis coupled to time-of-flight mass spectrometry. Forensic Sci Int 216(1–3):101–107

Gottardo R, Mikšík I, Aturki Z, Sorio D, Seri C, Fanali S, Tagliaro F (2012b) Analysis of drugs of forensic interest with capillary zone electrophoresis/time-of-flight mass spectrometry based on the use of non-volatile buffers. Electrophoresis 33(4):599–606

Greer B, Chevallier O, Quinn B, Botana LM, Elliott CT (2021) Redefining dilute and shoot: the evolution of the technique and its application in the analysis of foods and biological matrices by liquid chromatography mass spectrometry. TrAC Trends Anal Chem 141:116284

Grundmann M, Matysik FM (2011) Fast capillary electrophoresis–time-of-flight mass spectrometry using capillaries with inner diameters ranging from 75 to 5 μm. Anal Bioanal Chem 400:269–278

Guo L, Wang T, Wu Z, Wang J, Wang M, Cui Z et al (2020) Portable food-freshness prediction platform based on colorimetric barcode combinatorics and deep convolutional neural networks. Adv Mater 32(45):2004805

Haley NJ, Hoover EA (2015) Chronic wasting disease of cervids: current knowledge and future perspectives. Annu Rev Anim Biosci 3(1):305–325

Han C, Sun J, Cheng H, Liu J, Xu Z (2014) Speciation analysis of urine iodine by ion-pair reversed-phase liquid chromatography and inductively coupled plasma mass spectrometry. Anal Methods 6(14):5369–5375

Hanlon EB, Manoharan R, Koo T, Shafer KE, Motz JT, Fitzmaurice M et al (2000) Prospects for in vivo Raman spectroscopy. Phys Med Biol 45(2):R1

Hänni C, Brousseau T, Laudet V, Stehelin D (1995) Isopropanol precipitation removes PCR inhibitors from ancient bone extracts. Nucleic Acids Res 23(5):881
Harman JL, Silva CJ (2009) Bovine spongiform encephalopathy. J Am Vet Med Assoc 234(1):59–72
Haselberg R, Brinks V, Hawe A, De Jong GJ, Somsen GW (2011a) Capillary electrophoresis-mass spectrometry using noncovalently coated capillaries for the analysis of biopharmaceuticals. Anal Bioanal Chem 400:295–303
Haselberg R, Harmsen S, Dolman MEM, De Jong GJ, Kok RJ, Somsen GW (2011b) Characterization of drug-lysozyme conjugates by sheathless capillary electrophoresis–time-of-flight mass spectrometry. Anal Chim Acta 698(1–2):77–83
Haselberg R, de Jong GJ, Somsen GW (2013) Low-flow sheathless capillary electrophoresis–mass spectrometry for sensitive glycoform profiling of intact pharmaceutical proteins. Anal Chem 85(4):2289–2296
He H, Sun DW, Pu H, Chen L, Lin L (2019) Applications of Raman spectroscopic techniques for quality and safety evaluation of milk: a review of recent developments. Crit Rev Food Sci Nutr 59(5):770–793
Heinemann M, Zenobi R (2011) Single cell metabolomics. Curr Opin Biotechnol 22(1):26–31
Herrero M, García-Cañas V, Simo C, Cifuentes A (2010) Recent advances in the application of capillary electromigration methods for food analysis and Foodomics. Electrophoresis 31(1):205–228
Huang A, Friesen J, Brunton JL (1987) Characterization of a bacteriophage that carries the genes for production of Shiga-like toxin 1 in Escherichia coli. J Bacteriol 169(9):4308–4312
Huhn C, Ramautar R, Wuhrer M, Somsen GW (2010) Relevance and use of capillary coatings in capillary electrophoresis–mass spectrometry. Anal Bioanal Chem 396:297–314
Jacobson JM, Yin J, Kitov PI, Mulvey G, Griener TP, James MN et al (2014) The crystal structure of Shiga toxin type 2 with bound disaccharide guides the design of a heterobifunctional toxin inhibitor. J Biol Chem 289(2):885–894
James TL, Liu H, Ulyanov NB, Farr-Jones S, Zhang H, Donne DG et al (1997) Solution structure of a 142-residue recombinant prion protein corresponding to the infectious fragment of the scrapie isoform. Proc Natl Acad Sci 94(19):10086–10091
Jansson D, Lindström SW, Norlin R, Hok S, Valdez CA, Williams AM et al (2018) Part 2: forensic attribution profiling of Russian VX in food using liquid chromatography-mass spectrometry. Talanta 186:597–606
Kalogiouri NP, Aalizadeh R, Dasenaki ME, Thomaidis NS (2020) Application of high resolution mass spectrometric methods coupled with chemometric techniques in olive oil authenticity studies-a review. Anal Chim Acta 1134:150–173
Kappel K, Haase I, Käppel C, Sotelo CG, Schröder U (2017) Species identification in mixed tuna samples with next-generation sequencing targeting two short cytochrome b gene fragments. Food Chem 234:212–219
Kaufmann A (2012) The current role of high-resolution mass spectrometry in food analysis. Anal Bioanal Chem 403(5):1233–1249
Kawai M, Iwamuro Y, Iio-Ishimaru R, Chinaka S, Takayama N, Hayakawa K (2011) Analysis of phosphorus-containing amino acid-type herbicides by sheathless capillary electrophoresis/electrospray ionization–mass spectrometry using a high sensitivity porous sprayer. Anal Sci 27(8):857–857
Keith JA, Vassilev-Galindo V, Cheng B, Chmiela S, Gastegger M, Muller KR, Tkatchenko A (2021) Combining machine learning and computational chemistry for predictive insights into chemical systems. Chem Rev 121(16):9816–9872
Kendall C, Hutchings J, Barr H, Shepherd N, Stone N (2011) Exploiting the diagnostic potential of biomolecular fingerprinting with vibrational spectroscopy. Faraday Discuss 149(1):279–290
Kim SY, Li J, Lim NR, Kang BS, Park HJ (2016) Prediction of warmed-over flavour development in cooked chicken by colorimetric sensor array. Food Chem 211:440–447
Kimura T, Tani S, Matsumoto YI, Takeda T (2001) Serum amyloid P component is the Shiga toxin 2-neutralizing factor in human blood. J Biol Chem 276(45):41576–41579

Kimura T, Tani S, Motoki M, Matsumoto YI (2003) Role of Shiga toxin 2 (Stx2)-binding protein, human serum amyloid P component (HuSAP), in Shiga toxin-producing Escherichia coli infections: assumption from in vitro and in vivo study using HuSAP and anti-Stx2 humanized monoclonal antibody TMA-15. Biochem Biophys Res Commun 305(4):1057–1060

Kiselev I, Sysoev V, Kaikov I, Koronczi I, Adil Akai Tegin R, Smanalieva J et al (2018) On the temporal stability of analyte recognition with an e-nose based on a metal oxide sensor array in practical applications. Sensors 18(2):550

Kitagawa F, Otsuka K (2011) Recent progress in microchip electrophoresis–mass spectrometry. J Pharm Biomed Anal 55(4):668–678

Kitova EN, Mulvey GL, Dingle T, Sinelnikov I, Wee S, Griener TP et al (2009) Assembly and stability of the Shiga toxins investigated by electrospray ionization mass spectrometry. Biochemistry 48(23):5365–5374

Klampfl CW, Buchberger W (2010) Recent advances in the use of capillary electrophoresis coupled to high-resolution mass spectrometry for the analysis of small molecules. Curr Anal Chem 6(2):118–125

Klein U (2003) Experiments, models, paper tools: cultures of organic chemistry in the nineteenth century. Stanford University Press

Klepárník K (2013) Recent advances in the combination of capillary electrophoresis with mass spectrometry: from element to single-cell analysis. Electrophoresis 34(1):70–85

Kocher TD (1989) Dynamics of mitochondrial DNA evolution in animals: amplifications and sequencing with conserved primers. Proc Natl Acad Sci USA 86:189–191

Kohler I, Cognard E, Marchi I, Ortelli D, Edder P, Veuthey JL et al (2011) Single-run separation of closely related cationic and anionic compounds by CE-ESI-MS: application to the simultaneous analysis of melamine and its analogs in milk. Chimia 65(6):389–389

Koitabashi T, Vuddhakul V, Radu S, Morigaki T, Asai N, Nakaguchi Y, Nishibuchi M (2006) Genetic characterization of Escherichia coli O157: H7/–strains carrying the stx2 gene but not producing Shiga toxin 2. Microbiol Immunol 50(2):135–148

Kondo F, Kobayashi S, Matsumoto M, Yamada S, Saito M, Suzuki Y et al (1997) Analysis of Vero toxins 1 and 2 by high-performance liquid chromatography/electrospray ionization mass spectrometry. J Mass Spectrom 32(10):1140–1142

Kondo F, Saito H, Hayashi R, Onda H, Kobayashi S, Matsumoto M et al (2003) Identification of Shiga toxins in Shiga toxin-producing Escherichia coli using immunoprecipitation and high-performance liquid chromatography-electrospray ionization mass spectrometry. Analyst 128(11):1360–1364

Konowalchuk J, Dickie N, Stavric S, Speirs J (1978) Properties of an Escherichia coli cytotoxin. Infect Immun 20(2):575–577

Kortmann H, Blank LM, Schmid A (2011) Single cell analytics: an overview. In: High resolution microbial single cell analytics. Springer, pp 99–122

Krafft C, Neudert L, Simat T, Salzer R (2005) Near infrared Raman spectra of human brain lipids. Spectrochim Acta A Mol Biomol Spectrosc 61(7):1529–1535

Kuhar N, Sil S, Umapathy S (2021) Potential of Raman spectroscopic techniques to study proteins. Spectrochim Acta A Mol Biomol Spectrosc 258:119712

Kumar S (2021) Recent applications of hyphenated liquid chromatography techniques in food forensics. Eurasian J Anal Chem 16(2):10

Leinala EK, Davies PL, Doucet D, Tyshenko MG, Walker VK, Jia Z (2002) A β-helical antifreeze protein isoform with increased activity: structural and functional insights. J Biol Chem 277(36):33349–33352

Li Z, Suslick KS (2016) Portable optoelectronic nose for monitoring meat freshness. Acs Sensors 1(11):1330–1335

Li Z, Suslick KS (2018) A hand-held optoelectronic nose for the identification of liquors. ACS sensors 3(1):121–127

Li Y, Champion MM, Sun L, Champion PAD, Wojcik R, Dovichi NJ (2012) Capillary zone electrophoresis-electrospray ionization-tandem mass spectrometry as an alternative proteomics

platform to ultraperformance liquid chromatography-electrospray ionization-tandem mass spectrometry for samples of intermediate complexity. Anal Chem 84(3):1617–1622
Li H, Chen Q, Zhao J, Ouyang Q (2014) Non-destructive evaluation of pork freshness using a portable electronic nose (E-nose) based on a colorimetric sensor array. Anal Methods 6(16):6271–6277
Li Z, Li H, LaGasse MK, Suslick KS (2016) Rapid quantification of trimethylamine. Anal Chem 88(11):5615–5620
Li Z, Askim JR, Suslick KS (2018) The optoelectronic nose: colorimetric and fluorometric sensor arrays. Chem Rev 119(1):231–292
Li Y, Kong D, Fu Y, Sussman MR, Wu H (2020) The effect of developmental and environmental factors on secondary metabolites in medicinal plants. Plant Physiol Biochem 148:80–89
Liberski PP (2012) Gerstmann-Sträussler-Scheinker disease. Neurodegener Dis 724:128–137
Liberski PP (2013) Kuru: a journey back in time from Papua New Guinea to the Neanderthals' extinction. Pathogens 2(3):472–505
Liebler DC, Zimmerman LJ (2013) Targeted quantitation of proteins by mass spectrometry. Biochemistry 52(22):3797–3806
Lim JH, Bae D, Fong A (2018) Titanium dioxide in food products: quantitative analysis using ICP-MS and Raman spectroscopy. J Agric Food Chem 66(51):13533–13540
Lindeberg J (1996) Capillary electrophoresis in food analysis. Food Chem 55(1):73–94
Ling H, Boodhoo A, Hazes B, Cummings MD, Armstrong GD, Brunton JL, Read RJ (1998) Structure of the Shiga-like toxin I B-pentamer complexed with an analogue of its receptor Gb3. Biochemistry 37(7):1777–1788
Liu Y, Bonizzoni M (2014) A supramolecular sensing array for qualitative and quantitative analysis of organophosphates in water. J Am Chem Soc 136(40):14223–14229
Liu JX, Aerts JT, Rubakhin SS, Zhang XX, Sweedler JV (2014) Analysis of endogenous nucleotides by single cell capillary electrophoresis-mass spectrometry. Analyst 139(22):5835–5842
Liu T, Yang L, Feng W, Liu K, Ran Q, Wang W et al (2020) Dual-mode photonic sensor array for detecting and discriminating hydrazine and aliphatic amines. ACS Appl Mater Interfaces 12(9):11084–11093
Llorens F, Thüne K, Schmitz M, Ansoleaga B, Frau-Mendez MA, Cramm M et al (2016) Identification of new molecular alterations in fatal familial insomnia. Hum Mol Genet 25(12):2417–2436
Lo YT, Shaw PC (2018) DNA-based techniques for authentication of processed food and food supplements. Food Chem 240:767–774
Long D, Peng J, Peng H, Xian H, Li S, Wang X et al (2019) A quadruple-channel fluorescent sensor array based on label-free carbon dots for sensitive detection of tetracyclines. Analyst 144(10):3307–3313
Lu J, Cai Z, Zou Y, Wu D, Wang A, Chang J et al (2019) Silver nanoparticle-based surface-enhanced Raman spectroscopy for the rapid and selective detection of trace tropane alkaloids in food. ACS Appl Nano Mater 2(10):6592–6601
Lu W, Chen X, Wang L, Li H, Fu YV (2020) Combination of an artificial intelligence approach and laser tweezers Raman spectroscopy for microbial identification. Anal Chem 92(9):6288–6296
Lu W, Li H, Qiu H, Wang L, Feng J, Fu YV (2023) Identification of pathogens and detection of antibiotic susceptibility at single-cell resolution by Raman spectroscopy combined with machine learning. Front Microbiol 13:1076965
MacLeod DL, Gyles CL, Valdivieso-Garcia A, Clarke RC (1991) Physicochemical and biological properties of purified Escherichia coli Shiga-like toxin II variant. Infect Immun 59(4):1300–1306
Manix M, Kalakoti P, Henry M, Thakur J, Menger R, Guthikonda B, Nanda A (2015) Creutzfeldt-Jakob disease: updated diagnostic criteria, treatment algorithm, and the utility of brain biopsy. Neurosurg Focus 39(5):E2
Manjunatha G, Savia FG (2018) Food forensics: detection of hidden toxins present in common household consumables. J Forensic Sci Crim Inv. ISSN, 2476-1311

Maquelin K, Kirschner C, Choo-Smith LP, van den Braak N, Endtz HP, Naumann D, Puppels GJ (2002) Identification of medically relevant microorganisms by vibrational spectroscopy. J Microbiol Methods 51(3):255–271
Marsh RF, Hadlow WJ (1992) Transmissible mink encephalopathy. Revue scientifique et technique (International Office of Epizootics) 11(2):539–550
Martín A, Vilela D, Escarpa A (2012) Food analysis on microchip electrophoresis: an updated review. Electrophoresis 33(15):2212–2227
Matos MP, Jackson GP (2019) Isotope ratio mass spectrometry in forensic science applications. Forensic Chem 13:100154
Meermann B, Sperling M (2012) Hyphenated techniques as tools for speciation analysis of metal-based pharmaceuticals: developments and applications. Anal Bioanal Chem 403:1501–1522
Meng L, Chen S, Zhu B, Zhang J, Mei Y, Cao J, Zheng K (2020) Application of dispersive liquid-liquid microextraction and GC–MS/MS for the determination of GHB in beverages and hair. J Chromatogr B 1144:122058
Minami T, Esipenko NA, Zhang B, Kozelkova ME, Isaacs L, Nishiyabu R et al (2012) Supramolecular sensor for cancer-associated nitrosamines. J Am Chem Soc 134(49):20021–20024
Mittermayr S, Bones J, Guttman A (2013) Unraveling the glyco-puzzle: glycan structure identification by capillary electrophoresis. Anal Chem 85(9):4228–4238
Moore MK, Bemiss JA, Rice SM, Quattro JM, Woodley CM (2003) Use of restriction fragment length polymorphisms to identify sea turtle eggs and cooked meats to species. Conserv Genet 4:95–103
Movasaghi Z, Rehman S, Rehman IU (2007) Raman spectroscopy of biological tissues. Appl Spectrosc Rev 42(5):493–541
Mungkarndee R, Techakriengkrai I, Tumcharern G, Sukwattanasinitt M (2016) Fluorescence sensor array for identification of commercial milk samples according to their thermal treatments. Food Chem 197:198–204
Musto CJ, Lim SH, Suslick KS (2009) Colorimetric detection and identification of natural and artificial sweeteners. Anal Chem 81(15):6526–6533
Neugebauer U, Clement JH, Bocklitz T, Krafft C, Popp J (2010) Identification and differentiation of single cells from peripheral blood by Raman spectroscopic imaging. J Biophotonics 3(8–9):579–587
Newland JW, Strockbine NA, Miller SF, O'Brien AD, Holmes RK (1985) Cloning of Shiga-like toxin structural genes from a toxin converting phage of Escherichia coli. Science 230(4722):179–181
Nguyen TTT, Huy BT, Lee YI (2019) Disposable colorimetric paper-based probe for the detection of amine-containing gases in aquatic sediments. ACS omega 4(7):12665–12670
Nielsen EE, Hansen MM, Schmidt C, Meldrup D, Grønkjaer P (2001) Population of origin of Atlantic cod. Nature 413(6853):272–272
Notingher I, Verrier S, Haque S, Polak JM, Hench LL (2003) Spectroscopic study of human lung epithelial cells (A549) in culture: living cells versus dead cells. Biopolymers 72(4):230–240
Núñez O, Lucci P (2020) Application of liquid chromatography in food analysis. Food Secur 9(9):1277
O'Brien AD, Newland JW, Miller SF, Holmes RK, Smith HW, Formal SB (1984) Shiga-like toxin-converting phages from Escherichia coli strains that cause hemorrhagic colitis or infantile diarrhea. Science 226(4675):694–696
Oikawa A, Saito K (2012) Metabolite analyses of single cells. Plant J 70(1):30–38
Ongay S, Neusüβ C (2010) Isoform differentiation of intact AGP from human serum by capillary electrophoresis–mass spectrometry. Anal Bioanal Chem 398:845–855
Oshoma CE, Eguakun-Owie SO, Obuekwe IS (2019) Utilization of banana peel as a substrate for single cell protein and Amylase production by Aspergillus Niger. African Sci 18(3):143–150
Pääbo S, Gifford JA, Wilson AC (1988) Mitochondrial DNA sequences from a 7000-year old brain. Nucleic Acids Res 16(20):9775–9787

Pääbo S, Irwin DM, Wilson AC (1990) DNA damage promotes jumping between templates during enzymatic amplification. J Biol Chem 265(8):4718–4721

Park JH, Choi SH, Bong YS (2019) Geographical origin authentication of onions using stable isotope ratio and compositions of C, H, O, N, and S. Food Control 101:121–125

Park J, Kim M, Shin B, Kang M, Yang J, Lee TK, Park W (2021) A novel decoy strategy for polymyxin resistance in Acinetobacter baumannii. elife 10:e66988

Parmar A, Katariya R, Patel V (2019) A review on random forest: an ensemble classifier. In: International conference on intelligent data communication technologies and internet of things (ICICI) 2018. Springer International Publishing, pp 758–763

Pascal, G., & Mahé, S. (2001). Identity, traceability, acceptability and substantial equivalence of food Cell Mol Biol (Noisy-le-Grand, France), 47(8), 1329–1342

Pauk V, Lemr K (2018) Forensic applications of supercritical fluid chromatography–mass spectrometry. J Chromatogr B 1086:184–196

Peng CY, Zhang YL, Song W, Cai HM, Wang Y, Granato D (2019) Characterization of Brazilian coffee based on isotope ratio mass spectrometry (δ13C, δ18O, δ2H, and δ15N) and supervised chemometrics. Food Chem 297:124963

Pepys MB, Baltz ML, De Beer FC, Dyck RF, Holford S, Breathnach SM et al (1982) Biology of serum amyloid P component. Ann N Y Acad Sci 389(1):286–298

Pérez-Álvarez EP, García R, Barrulas P, Dias C, Cabrita MJ, Garde-Cerdán T (2019) Classification of wines according to several factors by ICP-MS multi-element analysis. Food Chem 270:273–280

Peš O, Preisler J (2010) Off-line coupling of microcolumn separations to desorption mass spectrometry. J Chromatogr A 1217(25):3966–3977

Poinar HN (2002) The genetic secrets some fossils hold. Acc Chem Res 35(8):676–684

Pokhrel P, Jha S, Giri B (2020) Selection of appropriate protein assay method for a paper microfluidics platform. Pract Laborat Med 21:e00166

Pröfrock D (2010) Progress and possible applications of miniaturised separation techniques and elemental mass spectrometry for quantitative, heteroatom-tagged proteomics. Anal Bioanal Chem 398:2383–2401

Prokhorova AF, Shapovalova EN, Shpigun OA (2010) Chiral analysis of pharmaceuticals by capillary electrophoresis using antibiotics as chiral selectors. J Pharm Biomed Anal 53(5):1170–1179

Pruimboom-Brees IM, Morgan TW, Ackermann MR, Nystrom ED, Samuel JE, Cornick NA, Moon HW (2000) Cattle lack vascular receptors for Escherichia coli O157: H7 Shiga toxins. Proc Natl Acad Sci 97(19):10325–10329

Prusiner SB (1998) Prions. Proc Natl Acad Sci 95(23):13363–13383

Puerta A, Díez-Masa JC, Martín-Álvarez PJ, Martín-Ventura JL, Barbas C, Tuñón J et al (2011) Study of the capillary electrophoresis profile of intact α-1-acid glycoprotein isoforms as a biomarker of atherothrombosis. Analyst 136(4):816–822

Qian S, Lin H (2015) Colorimetric sensor array for detection and identification of organophosphorus and carbamate pesticides. Anal Chem 87(10):5395–5400

Quinteiro J, Sotelo CG, Rehbein H, Pryde SE, Medina I, Pérez-Martín RI et al (1998) Use of mtDNA direct polymerase chain reaction (PCR) sequencing and PCR− restriction fragment length polymorphism methodologies in species identification of canned tuna. J Agric Food Chem 46(4):1662–1669

Ram JL, Ram ML, Baidoun FF (1996) Authentication of canned tuna and bonito by sequence and restriction site analysis of polymerase chain reaction products of mitochondrial DNA. J Agric Food Chem 44(8):2460–2467

Ramirez-Mora T, Dávila-Pérez C, Torres-Méndez F, Valle-Bourrouet G (2019) Raman spectroscopic characterization of endodontic biofilm matrices. J Spectrosc 2019(1):1307397

Ranbir, Kumar M, Singh G, Singh J, Kaur N, Singh N (2022) Machine learning-based analytical systems: food forensics. ACS Omega 7(51):47518–47535

Ravindran A, Nesamani FP, Nirmal D (2018) A study on the use of spectroscopic techniques to identify food adulteration. In: 2018 International Conference on Circuits and Systems in Digital Enterprise Technology (ICCSDET). IEEE, pp 1–6

Ribeiro C, Gonçalves R, Tiritan ME (2021) Separation of enantiomers using gas chromatography: application in forensic toxicology, food and environmental analysis. Crit Rev Anal Chem 51(8):787–811

Rodriguez MA, Garcia T, González I, Asensio L, Mayoral B, López-Calleja I et al (2003) Identification of goose, mule duck, chicken, Turkey, and swine in foie gras by species-specific polymerase chain reaction. J Agric Food Chem 51(6):1524–1529

Ryvolova M, Adam V, Kizek R (2012) Analysis of metallothionein by capillary electrophoresis. J Chromatogr A 1226:31–42

Saadat S, Pandya H, Dey A, Rawtani D (2022) Food forensics: techniques for authenticity determination of food products. Forensic Sci Int 333:111243

Said N (2017) Characterization of therapeutic proteins by capillary electrophoresis (CE) coupled to mass spectrometry (MS) (Doctoral dissertation, Université de Strasbourg)

Salinas Y, Ros-Lis JV, Vivancos JL, Martinez-Manez R, Marcos MD, Aucejo S, Lorente I (2012) Monitoring of chicken meat freshness by means of a colorimetric sensor array. Analyst 137(16):3635–3643

Sánchez-Hernández L, García-Ruiz C, Luisa Marina M, Luis Crego A (2010) Recent approaches for enhancing sensitivity in enantioseparations by CE. Electrophoresis 31(1):28–43

Sasajima Y, Lim LW, Takeuchi T, Suenami K, Sato K, Takekoshi Y (2010) Simultaneous determination of antidepressants by non-aqueous capillary electrophoresis-time of flight mass spectrometry. J Chromatogr A 1217(48):7598–7604

Scheutz F, Teel LD, Beutin L, Piérard D, Buvens G, Karch H et al (2012) Multicenter evaluation of a sequence-based protocol for subtyping Shiga toxins and standardizing Stx nomenclature. J Clin Microbiol 50(9):2951–2963

Schieber A (2018) Introduction to food authentication. In: Modern techniques for food authentication. Academic Press, pp 1–21

Schmitz M, Greis C, Ottis P, Silva CJ, Schulz-Schaeffer WJ, Wrede A et al (2014a) Loss of prion protein leads to age-dependent behavioral abnormalities and changes in cytoskeletal protein expression. Mol Neurobiol 50:923–936

Schmitz M, Zafar S, Silva CJ, Zerr I (2014b) Behavioral abnormalities in prion protein knockout mice and the potential relevance of PrPC for the cytoskeleton. Prion 8(6):381–386

Schotman TG, Xu X, Rodewijk N, van der Weerd J (2017) Application of dye analysis in forensic fibre and textile examination: case examples. Forensic Sci Int 278:338–350

Shaikhina T, Lowe D, Daga S, Briggs D, Higgins R, Khovanova N (2019) Decision tree and random forest models for outcome prediction in antibody incompatible kidney transplantation. Biomed Signal Process Contr 52:456–462

Sharma N, Barstis T, Giri B (2018) Advances in paper-analytical methods for pharmaceutical analysis. Eur J Pharm Sci 111:46–56

Shedoeva A, Leavesley D, Upton Z, Fan C (2019) Wound healing and the use of medicinal plants. Evid Based Complement Alternat Med 2019(1):2684108

Silva CJ (2014) Applying the tools of chemistry (mass spectrometry and covalent modification by small molecule reagents) to the detection of prions and the study of their structure. Prion 8(1):42–50

Silva CJ (2018) Food forensics: using mass spectrometry to detect foodborne protein contaminants, as exemplified by Shiga toxin variants and prion strains. J Agric Food Chem 66(32):8435–8450

Silva AJ, Hellberg RS (2021) DNA-based techniques for seafood species authentication. In: Advances in food and nutrition research, vol 95. Academic Press, pp 207–255

Silva CJ, Onisko BC, Dynin I, Erickson ML, Vensel WH, Requena JR et al (2010) Assessing the role of oxidized methionine at position 213 in the formation of prions in hamsters. Biochemistry 49(9):1854–1861

Silva CJ, Erickson-Beltran ML, Skinner CB, Dynin I, Hui C, Patfield SA et al (2014) Safe and effective means of detecting and quantitating Shiga-like toxins in attomole amounts. Anal Chem 86(10):4698–4706

Silva CJ, Vázquez-Fernández E, Onisko B, Requena JR (2015a) Proteinase K and the structure of PrPSc: the good, the bad and the ugly. Virus Res 207:120–126

Silva CJ, Erickson-Beltran ML, Skinner CB, Patfield SA, He X (2015b) Mass spectrometry-based method of detecting and distinguishing type 1 and type 2 Shiga-like toxins in human serum. Toxins 7(12):5236–5253

Silva CJ, Erickson-Beltran ML, Hui C, Badiola JJ, Nicholson EM, Requena JR, Bolea R (2017) Quantitating PrP polymorphisms present in prions from heterozygous scrapie-infected sheep. Anal Chem 89(1):854–861

Sliwinska M, Wisniewska P, Dymerski T, Namiesnik J, Wardencki W (2014) Food analysis using artificial senses. J Agric Food Chem 62(7):1423–1448

Sorensen H, Sorensen S, Bjergegaard C, Michaelsen S (2007) Chromatography and capillary electrophoresis in food analysis. Royal Society of Chemistry

Stachniuk A, Sumara A, Montowska M, Fornal E (2021) Liquid chromatography–mass spectrometry bottom-up proteomic methods in animal species analysis of processed meat for food authentication and the detection of adulterations. Mass Spectrom Rev 40(1):3–30

Stahl N, Baldwin MA, Teplow D, Hood L, Beavis R, Chait B et al (1992) Cataloguing post-translational modifications of the scrapie prion protein by mass spectrometry. Ellis Horwood Limited, Chichester

Stewart S, Ivy MA, Anslyn EV (2014) The use of principal component analysis and discriminant analysis in differential sensing routines. Chem Soc Rev 43(1):70–84

Su Z, Li T, Wu D, Wu Y, Li G (2022) Recent progress on single-molecule detection technologies for food safety. J Agric Food Chem 70(2):458–469

Sun T, Wang X, Cong P, Xu J, Xue C (2020) Mass spectrometry-based lipidomics in food science and nutritional health: a comprehensive review. Compr Rev Food Sci Food Saf 19(5):2530–2558

Sundermann TR, Schmitt G, Bartel M (2021) Modified congener analysis: quantification of cyanide in whole blood, other body fluids, and diverse beverages. Drug Test Anal 13(2):412–418

Taichrib A (2012) Analysis of peptides and proteins by capillary electrophoresis-mass spectrometry: method optimization for the improved characterization of peptides and intact proteins in the pharmaceutical area (Doctoral dissertation, Jena, Friedrich-Schiller-Universität Jena, Diss., 2012)

Talebi Bezmin Abadi A, Rizvanov AA, Haertlé T, Blatt NL (2019) World Health Organization report: current crisis of antibiotic resistance. BioNanoScience 9(4):778–788

Tarr PI, Gordon CA, Chandler WL (2005) Shiga-toxin-producing Escherichia coli and haemolytic uraemic syndrome. Lancet 365(9464):1073–1086

Teletchea F, Maudet C, Hänni C (2005) Food and forensic molecular identification: update and challenges. Trends Biotechnol 23(7):359–366

Terán JE, Millbern Z, Shao D, Sui X, Liu Y, Demmler M, Vinueza NR (2021) Characterization of synthetic dyes for environmental and forensic assessments: a chromatography and mass spectrometry approach. J Sep Sci 44(1):387–402

Thanakiatkrai P, Dechnakarin J, Ngasaman R, Kitpipit T (2019) Direct pentaplex PCR assay: an adjunct panel for meat species identification in Asian food products. Food Chem 271:767–772

Thiviya P, Gamage A, Kapilan R, Merah O, Madhujith T (2022) Production of single-cell protein from fruit peel wastes using palmyrah toddy yeast. Fermentation 8(8):355

Timerbaev AR, Pawlak K, Gabbiani C, Messori L (2011) Recent progress in the application of analytical techniques to anticancer metallodrug proteomics. TrAC Trends Anal Chem 30(7):1120–1138

Töpfer N, Müller MM, Dahms M, Ramoji A, Popp J, Slevogt H, Neugebauer U (2019) Raman spectroscopy reveals LPS-induced changes of biomolecular composition in monocytic THP-1 cells in a label-free manner. Integr Biol 11(3):87–98

Trofa AF, Ueno-Olsen H, Oiwa R, Yoshikawa M (1999) Dr. Kiyoshi Shiga: discoverer of the dysentery bacillus. Clin Infect Dis 29(5):1303–1306

Vázquez-Fernández E, Vos MR, Afanasyev P, Cebey L, Sevillano AM, Vidal E et al (2016) The structural architecture of an infectious mammalian prion using electron cryomicroscopy. PLoS Pathog 12(9):e1005835

Vázquez-López R, Solano-Gálvez SG, Juárez Vignon-Whaley JJ, Abello Vaamonde JA, Padró Alonzo LA, Rivera Reséndiz A et al (2020) Acinetobacter baumannii resistance: a real challenge for clinicians. Antibiotics 9(4):205
Veizerová L, Piešťanský J, Maráková K, Galba J, Rauová D, Dokupilová S et al (2012) Comparison of column-coupled electrophoresis with liquid chromatography methods in food analysis of quinine. Eur Pharma J 59(1):67–80
Vieira SM, Kaymak U, Sousa JM (2010) Cohen's kappa coefficient as a performance measure for feature selection. In: International conference on fuzzy systems. IEEE, pp 1–8
Vit O, Petrak J (2017) Integral membrane proteins in proteomics. How to break open the black box? J Proteome 153:8–20
Vucelić-Radović B, Lazić D, Nikšić M (2019) Application of molecular methods and Raman microscopy. In: Spectroscopy in agricultural sciences and food technology. Ubiquity Press, p 219
Wang J, Zhang Y, Xiang F, Zhang Z, Li L (2010) Combining capillary electrophoresis matrix-assisted laser desorption/ionization mass spectrometry and stable isotopic labeling techniques for comparative crustacean peptidomics. J Chromatogr A 1217(26):4463–4470
Wang H, Liu X, Nan K, Chen B, He M, Hu B (2017) Sample pre-treatment techniques for use with ICP-MS hyphenated techniques for elemental speciation in biological samples. J Anal At Spectrom 32(1):58–77
Watanabe T, Terabe S (2000) Analysis of natural food pigments by capillary electrophoresis. J Chromatograp A 880(1–2):311–322. [257] Kleparnik, K., Otevrel, M., Electrophoresis 2010,31,879–885
Will RG, Ironside JW, Zeidler M, Estibeiro K, Cousens SN, Smith PG et al (1996) A new variant of Creutzfeldt-Jakob disease in the UK. Lancet 347(9006):921–925
Willberger C, Leichtfuß D, Amayri S, Reich T (2019) Determination of the stability constants of the acetate complexes of the actinides Am (III), Th (IV), Np (V), and U (VI) using capillary electrophoresis-inductively coupled plasma mass spectrometry. Inorg Chem 58(8):4851–4858
Wilson IG (1997) Inhibition and facilitation of nucleic acid amplification. Appl Environ Microbiol 63(10):3741–3751
Withler RE, Candy JR, Beacham TD, Miller KM (2004) Forensic DNA analysis of Pacific salmonid samples for species and stock identification. In: Genetics of subpolar fish and invertebrates. Springer, pp 275–285
Wojcik R, Dada OO, Sadilek M, Dovichi NJ (2010) Simplified capillary electrophoresis nanospray sheath-flow interface for high efficiency and sensitive peptide analysis. Rapid Commun Mass Spectrom 24(17):2554–2560
Woodford N, Ellington MJ, Coelho JM, Turton JF, Ward ME, Brown S et al (2006) Multiplex PCR for genes encoding prevalent OXA carbapenemases in Acinetobacter spp. Int J Antimicrob Agents 27(4):351–353
Xiao-wei H, Zhi-Hua L, Xiao-bo Z, Ji-yong S, Han-ping M, Jie-wen Z et al (2016) Detection of meat-borne trimethylamine based on nanoporous colorimetric sensor arrays. Food Chem 197:930–936
Xie J, Wang DQ (2013) Textual research on adulteration of Chinese materia medica in ancient China. Zhonghuayishi za zhi (Beijing, China: 1980) 43(5):262–265
Yamasaki S, Furutani M, Ito K, Igarashi K, Nishibuchi M, Takeda Y (1991) Importance of arginine at position 170 of the A subunit of Vero toxin 1 produced by enterohemorrhagic Escherichia coli for toxin activity. Microb Pathog 11(1):1–9
Yan S, Lai X, Du G, Xiang Y (2018) Identification of aminoglycoside antibiotics in milk matrix with a colorimetric sensor array and pattern recognition methods. Anal Chim Acta 1034:153–160
Yeo LY, Chang HC, Chan PP, Friend JR (2011) Microfluidic devices for bioapplications. Small (Weinheim an der Bergstrasse, Germany) 7(1):12–48
Yuan TZ. Food forensics: can non-targeted mass spectrometry be a simple and robust solution for detecting food adulteration?

Zhang N, Shen K, Yang X, Li Z, Zhou T, Zhang Y et al (2018) Simultaneous determination of arsenic, cadmium and lead in plant foods by ICP-MS combined with automated focused infrared ashing and cold trap. Food Chem 264:462–470

Zhang W, Ma J, Sun DW (2021) Raman spectroscopic techniques for detecting structure and quality of frozen foods: principles and applications. Crit Rev Food Sci Nutr 61(16):2623–2639

Zhao XY, Du N, Huang Y, Shen Y, Tan Y, Tan C (2021) Discrimination of powdered infant formula according to species, country of origin, and brand using a fluorescent sensor array. ACS Food Sci Technol 1(8):1392–1398

Zhong X, Maxwell EJ, Chen DD (2011) Mass transport in a micro flow-through vial of a junction-at-the-tip capillary electrophoresis-mass spectrometry interface. Anal Chem 83(12):4916–4923

Zhong M, Lee CY, Croushore CA, Sweedler JV (2012) Label-free quantitation of peptide release from neurons in a microfluidic device with mass spectrometry imaging. Lab Chip 12(11):2037–2045

Zhou L, Zhang C, Qiu Z, He Y (2020) Information fusion of emerging non-destructive analytical techniques for food quality authentication: a survey. TrAC Trends Anal Chem 127:115901

Industry Voices on Biofilm Innovations

16

Sarika S. Suryawanshi, Pranali P. Patil, Vipul M. Patil, Rajnikant B. Ghotane, Pramod B. Patil, Poournima S. Sankpal, and Shrirang V. Kharmate

Abstract

Biofilms are widely recognized as a contributing factor in significant problems currently facing human health and industry. The following paper summarizes a round table forum held at the 2021 International Biodeterioration and Biodegradation Symposium which discussed the potential role of standards in biofilm research and industry innovation. Standards and other forms of best practice guidance are reviewed in an academic research context as well as in relation to industry impacts and product development.

The understanding of fundamental aspects of biofilms is rapidly evolving, driven in part by new analytical methods. However, the complex and multidisciplinary nature of biofilm-associated problems and typically limited training available for industry personnel tackling the associated issues often reduces the ability to provide best practice solutions. As such it is argued that more effort needs to be made by both academia and industry experts to provide consensus and associated documentation on standard test methods or guidance documents related to studying and combating biofilms.

Keywords

Food authentication · Traceability · Fraud detection · Analytical techniques · Safety compliance

S. S. Suryawanshi (✉) · P. P. Patil · V. M. Patil · R. B. Ghotane · P. B. Patil · S. V. Kharmate
AshokraoMnae College of Pharmacy, Peth Vadgaon, India
e-mail: sarikasuryawanshi@amcoph.org

P. S. Sankpal
Bharati Vidyapeeth College of Pharmacy, Palus, Palus, India

T. Sarkar et al. (eds.), *Biofilm Applications to Revolutionize Food Technology*,
https://doi.org/10.1007/978-3-031-85205-3_16

16.1 Introduction

Biofilms are self-organized, single- or multispecies communities of microorganisms that are typically attached to a surface. Biofilm research has a long history of publication and both nationally and internationally, there are now a large number of research institutes and collaborative networks devoted to this field (Fields et al. 2020; Flemming et al. 2000). A wide range of industries, including medicine, oil and gas production and storage, marine vessels and constructions, water supply and treatment facilities, and oil and gas storage, depend heavily on biofilms and their effects on health and the economy (Cámara et al. 2022). This makes the employment of biocides, antibiotics, antimicrobials, and clean-in-place procedures that focus on biofilm control necessary.

Antonie van Leeuwenhoek, a dry-goods merchant, first noticed "animalcules" swarming over live and dead things in the seventeenth century. Leeuwenhoek's resourcefulness and curiosity were astounding; even after carefully cleaning his teeth, the remaining opaque deposits between them were "as thick as if it were batter." He found these "animalcules" in the tartar on his own teeth. These deposits included a mat of several kinds of "animalcules," which we now understand to be dental plaque bacteria. It is plausible to propose that the earliest known proof of microbial biofilms was this early tooth plaque investigation. Currently, microbial communities that are attached to a substrate and covered in an extracellular polymeric substance (EPS) made by the microbial cells themselves. After van Leuwenhoek's early work, it was not until 1940 that the so-called bottle effect in marine microorganisms was first observed (Heukelekian and Heller 1940). This demonstrated that when bacteria were adhered to a surface, their growth was greatly accelerated. When Zobell discovered in 1943 that the quantity of bacteria on surfaces was higher than that of the surrounding saltwater, it further advanced our understanding of biofilms. Based on his research, Zobell also proposed that bacterial adhesion was a two-step process that started with reversible adhesion and ended with irreversible adherence. Although the aforementioned studies are the earliest ones that have been documented on biofilms, it was not until the late 1960s and early 1970s that bacterial biofilms were extensively analyzed physically and chemically, after a few researchers realized how common they were (Characklis 1973; Costerton) In 1969, Jones et al. employed scanning and transmission electron microscopy to examine biofilms on trickling filters in a wastewater treatment plant. This research demonstrated that the matrix material, or EPS, was mainly made up of polysaccharides and that biofilms were made up of a wide range of distinct microorganisms. The use of electron microscopy, which revealed details on both the presence and structure of EPS in biofilms, was crucial in aiding the exploration of biofilms at this time. Biofilms were shown to be persistent and extremely resistant to the antimicrobial effects of chlorine, according to Characklis's 1973 investigation into microbial slimes in industrial water. Studies revealing that many bacteria lived most of their lives in surface-attached, sessile communities led to the first thorough investigation of biofilms as a whole, which was recognized in 1978.

A biofilm is made up of live, reproducing microorganisms that live in a colony, or community, such as bacteria. Put another way, biofilms are living things with a complicated social structure that continues to elude biologists and engineers but serves to both shield and promote their growth.

When specific microorganisms (such some types of bacteria) stick to an object's surface in a damp environment and start to multiply, a biofilm is created. The microbes attach themselves to the object's surface by secreting a glue-like, slimy material. Biofilms can develop on nearly any surface that can be imagined, including metals, polymers, natural materials (like rocks), medical implants, kitchen countertops, contact lenses, hot tub or pool walls (have you ever noticed that the sides of hot tubs and swimming pools seem a little slimy?), human and animal tissue, and a host of other surfaces. In fact, biofilms are likely to be present anywhere there is a surface, moisture, and nutrients.

One type of microbe can form a biofilm community, but in the natural world, biofilms are nearly invariably made up of a variety of bacterial species mixed with fungus, algae, yeasts, protozoa, and other microorganisms, as well as non-living detritus and corrosion products. For instance, the biofilms of dental plaque typically contain approximately 500 different kinds of bacteria! Only a few cell layers thick, biofilms can be so thin as to be invisible to the unaided eye. For example, most biofilms on your kitchen counter are invisible to the naked eye (unless, like some college students, you don't wash your counters very regularly). They have the potential to grow to be many inches thick; however, we doubt they would do so on a countertop.

16.2 Prevalence and Importance of Biofilms in Animals and Humans

Biofilms may form on a wide variety of surfaces, including natural aquatic systems living tissues, indwelling medical devices, and industrial/potable water system piping. The vast majority of microbes grow as biofilms in aqueous environments. Introduction to Biofilms 43, these biofilms can be benign or pathogenic, releasing harmful products and toxins, which become encased within the biofilm matrix. Biofilm formation is a phenomenon that occurs in both natural and manmade environments under diverse conditions, occurring on most moist surfaces, plant roots, and nearly every living animal. Biofilms may exist as beneficial epilithic communities in rivers and streams, wastewater treatment plant trickling beds or in the alimentary canal of mammals. Biofilms are not, however, confined to solid/liquid interfaces and can also be found at solid/air or liquid/liquid interfaces (e.g., airborne pathogens and deteriogens have been shown to be important factors in the biodeterioration of surface coatings; biofilms at liquid/liquid interfaces have been implicated in hydrocarbon degradation, including fuels, oils, and industrial coolants). In humans, an estimated 65% of all hospital infections are of biofilm origin (Costerton et al. 1999; Donlan 2001; Donlan and Costerton 2002; Douglas 2003; Ramage et al. 2009). Once established, biofilm infections are very difficult to eradicate due to their resilience to removal by host defense mechanisms and antimicrobials. Table 1 outlines several examples of clinically significant biofilm infections. Of particular note are biofilm infections in patients with medical devices, Pseudomonas aeruginosa infection in cystic fibrosis patients, tooth decay and periodontal disease, and chronic wound infections. As well as increasing patient morbidity and mortality, there is also an economic burden to biofilm infections with Table 1 Comparison of biofilm and planktonic lifestyles Characteristic Lifestyle Physiological Free swimming, often in an aqueous environment Planktonic Metabolic products continuously removed from the milieu, stresses of physical, chemical and biological nature Planktonic Slow growth Biofilm Found ubiquitously Biofilm Generally higher tolerance to antibiotics, hydrogen peroxide, phagocytosis Biofilm Stress conditions: lower metabolic activity (e.g., decreased nutrient accessibility, limitation of oxygen, increased osmotic pressure, pH variation, problems with accumulation of waste metabolites) Biofilm Bacteria develop stress responses Biofilm Source of infections Planktonic and biofilm Display unique gene expression patterns Biofilm Attachment can influence metabolic activities Biofilm Express virulence factors Biofilm and planktonic Structural Attached to a solid substrate Biofilm Many cells are dormant, likely to be smaller, not actively engaged in cell division Biofilm Flagella function in transport and initiation of cell-to-surface interactions and detachment Biofilm 44 S.L. Percival et al. those involving medical devices estimated to cost $20 billion in the USA alone. Furthermore, biofilms in air-conditioning and other water retention systems can cause human infection if ingested or inhaled, and examples include those biofilms caused by *Legionella* species, which persist in water storage tanks despite chlorination (Percival and Walker 1999) (Fig. 16.1).

1. Free-floating, or *planktonic*, bacteria encounter a submerged surface and within minutes can become attached. They begin to produce slimy extracellular polymeric substances (EPS) and to colonize the surface.

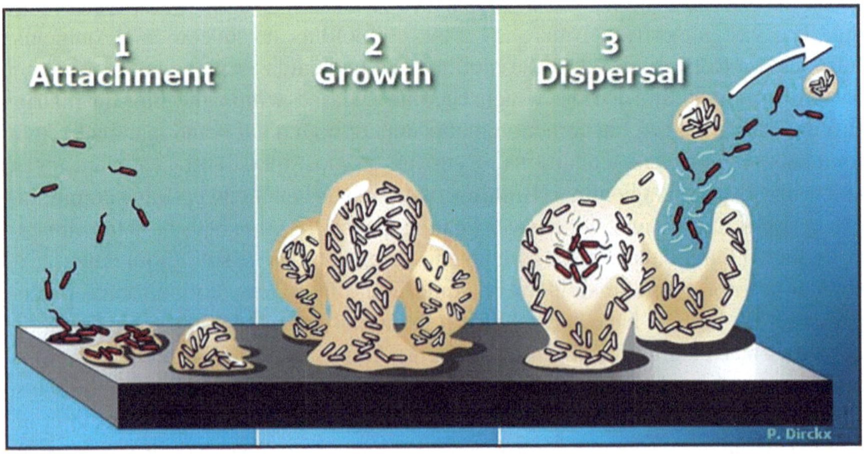

Fig. 16.1 The biofilm life cycle

2. EPS production allows the emerging biofilm community to develop a complex, three-dimensional structure that is influenced by a variety of environmental factors. Biofilm communities can develop within hours.
3. Biofilms can propagate through detachment of small or large clumps of cells, or by a type of "seeding dispersal" that releases individual cells. Either type of detachment allows bacteria to attach to a surface or to a biofilm downstream of the original community.

The truth is that most treatment formulas available today, including mouthwashes, antibiotics, disinfectants, and household cleaners, were tested on harmful microorganisms in a planktonic state—a condition in which the microorganisms float in a solution as individuals rather than as a part of a biofilm colony attached to a surface. Therefore, the effectiveness of these treatments is significantly limited when applied to real-world situations where the microbes to be killed are actually members of a biofilm, even though they may certainly kill millions of bacteria on contact.

16.3 Stages in the Formation of Biofilms

The process of biofilm formation is complex, but generally recognized as consisting of five stages (Palmer and White 1997): (1) Development of a surface conditioning film (2) Movement of microorganisms into close proximity with the surface (3) Adhesion (reversible and irreversible adhesion of microbes to the conditioned surface) (4) Growth and division of the organisms with the colonization of the surface, microcolony formation and biofilm formation; phenotype and genotype changes (5) Biofilm cell detachment/dispersal Each of these processes will be considered in turn. Introduction to Biofilms 47 6.1 Development of Conditioning Films and Substratum Effects Within the natural environment, microorganisms do not adhere directly to a substratum per se, but actually adhere to a conditioning film, which is known to form on most substrata. The composition of the conditioning film is complex and results in chemical modification of the original surface, thus influencing the rate and extent of microbial adhesion (Mittelman 2019). The first documented evidence that a "conditioning" film existed was by Loeb and Neihof (1975), although whether a conditioning film is a prerequisite for bacterial attachment is still debatable with evidence of this still elusive in the scientific literature. In aquatic or terrestrial environments, the conditioning layer has been shown to consist of complex polysaccharides, glycoproteins, and humic compounds (Chamberlain 1992; Marshall and Blainey 1991; Baier 1980; Rittle et al. 1990). In comparison, human host conditioning films may also be complex and determined by the site being conditioned. An area of "biofilmology" that has been studied extensively is the role of dental plaque in oral disease. In this regard, the enamel of teeth is conditioned by a proteinaceous "pellicle" composed of albumin, glycoproteins, lipids, lysozyme, phosphoproteins, and other components of saliva and gingival crevicular fluids (Marsh 1995). Other types of conditioning films have also been reported, particularly on biomaterials used for human use. In these situations, the components of blood, tears, urine, saliva, intravascular fluid, and respiratory secretions may all contribute to the conditioning film. The role of the conditioning film in biofilm development is in its ability to modify the physico-chemical properties of the substratum, as well as providing a concentrated nutrient source and important trace elements. It is important to note that conditioning films may actually inhibit rather than promote the adhesion of certain bacteria. The topography of the surface to which a microbial cell attaches is also fundamental to biofilm formation. Generally, as the roughness of a surface increases, bacterial adhesion will also increase (Characklis et al. 1990a). There are several possible reasons for this, including most specifically the provision of shelter for the microbes from the effects of shear forces. Apart from increasing the available interfacial area, a rough surface enhances mass transfer coefficients and allows cells to "anchor" to micro-irregularities, where they are better protected from possible desorption (Characklis et al. 1990b). Regardless of surface roughness, the attachment of living particles is energetically favorable if the change in the free energy during the process is negative. Despite metallic surfaces being energetically favorable to the attachment of the pioneer colonizers, the chemical composition of surfaces may interfere with adhesion, cellular metabolism, and production of exopolymers (Beech and Gaylarde 1989). Vieira et al. (1993) reported that the surface properties of certain metals influence bacterial adhesion. Specifically, they observed that Pseudomonas fluorescens preferentially adhered to aluminum surfaces within a few hours, followed by copper and brass. This observation, along with other related findings, is also discussed by S.L. Percival et al. Beyond surface roughness, the physicochemical properties of a surface significantly affect bacterial adhesion. Specifically, microorganisms exhibit a higher rate of attachment to hydrophobic, non-polar surfaces, such as Teflon and plastics, than to hydrophilic surfaces like glass or metals (Pringle and Fletcher 1986; Bendinger et al. 1993; Percival and Thomas 2009). However, such results need to be addressed with care, as many studies have proved contradictory, primarily due to non-availability of standardized methods for surface hydrophobicity measurements.

This new understanding of how harmful microorganisms actually present in our environment—as biofilms—means that the entire study of treating them needs to be reexamined. This is the conclusion that biofilm researchers around the world are currently pursuing with great vigor. It is necessary to comprehend the composition, development, defensive mechanisms, reproduction, and every other aspect of biofilms in order to create new therapies and assess their efficacy.

These are the major obstacles that people who work in the biofilm industry must overcome. Even though there has been a lot of progress, there is still more to learn and accomplish in terms of creating therapies that work and educating the public about biofilms and their effects. For this reason, we recommend this profession as a viable career option for students pursuing degrees in biology, microbiology, environmental engineering and science, biochemistry, and other related fields.

Every year, biofilms actually cost the US billions of dollars in energy losses, equipment damage, contaminated products, and medical diseases. However, biofilms also present a great deal of promise for the bioremediation of hazardous waste sites, the biofiltration of wastewater and municipal and industrial water, and the creation of biobarriers to prevent contamination of soil and groundwater. The intricacy of biofilm activity and behavior necessitates research contributions from numerous fields, including microbiology, biochemistry, engineering, and mathematics. Numerous science and engineering journals are continually publishing new findings that shed light on the secrets of biofilm.

Since van Leeuwenhoek's initial findings in the late seventeenth century, research, testing, and understanding of biofilms have significantly advanced. Modern research is being done in this topic, which spans several scientific and engineering specialties, to address issues related to biofilms. Research and monitoring on biofilms can be very specialized, and testing frequently needs to be done in settings with inadequate infrastructure. A small number of cross-disciplinary recommendations that offer guidance for decision-making by the corrosion specialist, field engineer, and plant operator, for example, exacerbate this problem. Therefore, it is evident that standards for methodology and reporting, as well as guidelines for developing new ones, are needed in both the research and industry. For all research domains, the question of reporting and test method standards is crucial. The issue of a third party not being able to replicate a study result is one that is frequently discussed in regard to reproducibility in scientific investigations (Baker 2016; Ioannidis 2014; National Academies of Sciences, Engineering, and Medicine 2019). The publication of Minimum Information Guidelines, which outline the details an author must provide for a research experiment to be repeatable in a different laboratory, is one way the biofilm research community is attempting to address this "crisis" in reproducibility (Allkja et al. 2020; Lourenço et al. 2015). This essay will examine related initiatives to create policies and procedures in the energy industry; yet, more work in this area is desperately needed. On the other hand, the industry's creation and application of private testing techniques that are considered "trade secrets" and may provide them a competitive edge, may restrict the exchange of crucial information.

This paper presents the talks from a round table panel session at the 18th International Biodeterioration and Biodegradation Symposium, held in July 2021, with the theme "Closing the gap: the role of regulatory standards in biofilm research and industry innovation." Ten presenters made up the panel (see Table 1), which featured talks and presentations on the value of conventional biofilm test procedures from both academic and industrial viewpoints in the fields of corrosion, oil and gas, and medicine. Important audience comments were made during the conversations, and when applicable, these pertinent issues have been integrated into the text.

It is obvious that the viewpoints offered by industrial and academic laboratories would not always coincide. Actually, minimal information criteria for publications may be more beneficial to academia than industry, which would benefit more from standard test procedures or best practices. Remarkably, though, by the end of the panel discussion, it was evident that industry and academia both require unanimity.

The paper has been broken up into the main topics of discussion included in the panel session, with content provided by presenters and additional comments as appropriate. At this stage, the authors would like to acknowledge the many pioneers of biofilm research for their hard work starting from scratch, which included developing many of the test methods that are routinely used today but were completely novel at the time. One of the aims of this paper was to help build on their efforts, rather than reinventing the wheel or repeating past mistakes. As the quote often associated with Isaac Newton states "*If I have seen further it is by standing on the shoulders of Giants.*" Finally, to paraphrase a senior biofilm researcher from a talk discussion during the IBBS18 Symposium "*If I was starting in this field from scratch, what would I like to know so that I didn't make silly mistakes and focused instead on research that was useful.*"

16.4 Section Snippets

Are Academic Research and Standardization Clashing Paradigms?

The rush to publish a major scientific discovery in a high-impact publication often overshadows the benefits of utilizing a conventional test technique in fundamental academic research. An academic may be persuaded to add standard methods to their list of preferred options when designing experiments if it is made clear what constitutes a standard test method, how they are used in industry and regulatory decision-making, and how a research laboratory may benefit as a result.

The Role of Standards in Driving Innovation

One essential component of the infrastructure needed to enable industrial expansion can be appropriate regulations and standards. As an important illustration, it is impossible to make product claims for cutting-edge biofilm technology in the absence of the necessary standards. Standards also assist businesses in de-risking research and development by highlighting the advantages of innovation and minimizing the potential for misuse.

The co-authors recently conducted an international road-mapping exercise with academia and industry.

Standards Related to Microbiologically Influenced Corrosion

A number of pertinent historical standards are now available regarding biofilms and MIC, and numerous new or updated standards are in the process of being produced. As previously mentioned, the ancient standards were not very detailed and frequently relied on test procedures based on culture to identify microorganisms. In the past few decades, test procedures and related technologies have advanced quickly, and costs have decreased.

Current Biofilm Standard Development

As was previously mentioned, MIC and biofouling have an effect on assets in a variety of sectors. The most effective biocides must be carefully chosen before being used in order to manage and control MIC and biofouling. Nevertheless, there are not any industry guidelines in place yet that outline how to choose biocides for these applications by assessing how they affect the prevention and reduction of biofilms. The NACE/AMPP standard TM0194 concentrates on the use of culture-based testing for biocide selection in.

IBSTG and Biofilm Center Perspectives and Outlook

There are chances to learn from best practices and guarantee coordination of efforts because the needs and issues surrounding the establishment of biofilm standards are global in nature. In light of this, the Center for Biofilm Engineering, the National Biofilms Innovation Center, the Singapore National Biofilms Center, and an EU Cooperation in Science and Technology (COST) activity have formed the International Biofilm Standards Tasks Group (IBSTG).

16.5 Discussion

The panel concluded by saying that, in order to inform business and academia about biofilms, related biologically induced degradation, and methods for investigating these subjects in the lab and field, there is a need for guidance documentation (regardless of whether it takes the form of standards, best practices, guidelines, etc.) and for discussions about these documents (e.g., workshops, conferences, and collaborations). These publications provide a set of checks and balances that the commercial or not-for-profit sectors can use to provide ethical guidelines.

References

Allkja J, Bjarnsholt T, Coenye T, Cos P, Fallarero A, Harrison JJ, Azevedo NF (2020) Minimum information guideline for spectrophotometric and fluorometric methods to assess biofilm formation in microplates. Biofilm 2:100010

Baier RE (1980) Substrate influence on adhesion of microorganisms and their resultant new surface properties. In: Bitton G, Marshall KC (eds) Adsorption of microorganisms to surfaces. Wiley, New York, pp 59–104

Baker BJ, Von Salm JL, Witowski CG, Fleeman RM, McClintock JB, Amsler CD, Shaw LN (2016) Darwinolide, a new diterpene scaffold that inhibits methicillin-resistant Staphylococcus aureus biofilm from the Antarctic sponge Dendrilla membranosa. Organic Letters 18(11):2596–2599

Beech IB, Gaylarde CC (1989) Adhesion of Desulfovibrio desulfuricans and Pseudomonas fluorescens to mild steel surfaces. J Appl Bacteriol 67:2017

Bendinger B, Rijnaarts HHM, Altendorf K, Zehnder AJB (1993) Physicochemical cell surface and adhesive properties of coryneform bacteria related to the presence and chain length of mycolic acids. Appl Environ Microbiol 59:3973–3977

Cámara M, Green W, MacPhee CE, Rakowska PD, Raval R, Richardson MC, Webb JS (2022) Economic significance of biofilms: A multidisciplinary and cross-sectoral challenge. npj Biofilms and Microbiomes 8(1):42

Chamberlain AHL (1992) The role of adsorbed layers in bacterial adhesion. In: Melo LF, Bott TR, Fletcher M, Capdeville B (eds) Biofilms-science and technology. Kluwer Academic, Dordrecht, pp 59–67

Characklis WG (1973) Attached microbial growths-II. Frictional resistance due to microbial slimes. Water Res 7:1249–1258

Characklis WG, McFeters GA, Marshall KC (1990a) Physiological ecology of biofilm systems. In: Characklis WG, Marshall KC (eds) Biofilms. Wiley, New York, pp 341–393

Characklis WG, Turakhia MH, Zelver N (1990b) Transfer and interfacial transport phenomena. In: Characklis WG, Marshall KC (eds) Biofilms. Wiley, New York, pp 265–340

Donlan R (2001) Biofilms and device-associated infections. Emerg Infect Dis 7:277–281

Donlan R, Costerton JW (2002) Biofilms: survival mechanisms of clinically relevant microorganisms. Clin Microbiol Rev 15:167–193

Douglas LJ (2003) Candida biofilms and their role in infection. Trends Microbiol 11:30–36

Flemming H-C, Wingender J, Griegbe T, Mayer C (2000) Physico-chemical properties of biofilms. In: Evans LV (ed) Biofilms: recent advances in their study and control. Harwood Academic, Amsterdam, pp 19–34

Fields B, Riquelme, SA, Liimatta, K, Lung TWF, Ahn D, Chen D, Prince A (2020) Pseudomonas aeruginosa utilizes host-derived itaconate to redirect its metabolism to promote biofilm formation. Cell Metabolism 31(6):1091–1106

Habash M, Reid G (1999) Microbial biofilms: their development and significance for medical device-related infections. J Clin Pharmacol 39:887–898

Hamilton WA (1987) Biofilm: microbial interaction and metabolic activities. In: Fletcher M, Gray TRG, Jones JG (eds) Ecology of microbial communities. Society for general microbiology symposium 41. Cambridge University Press, Cambridge, pp 361–387

Heukelekian H, Heller A (1940) Relation between food concentration and surface for bacterial growth. J Bacteriol 40:547–558

Khardori N, Yassien M (1995) Biofilms in device-related infections. J Ind Microbiol Biotechnol 15:141–147

Mittelman MW (2019) Bacterial biofilms in pharmaceutical water systems. In Filtration and Purification in the Biopharmaceutical Industry, Third Edition (pp. 587-607). CRC Press

Marshall KC, Blainey BL (1991) Role of bacterial adhesion in biofilm formation and biocorrosion. In Biofouling and Biocorrosion in Industrial Water Systems: Proceedings of the International Workshop on Industrial Biofouling and Biocorrosion, Stuttgart, September 13–14, 1990 (pp. 29–46). Berlin, Heidelberg: Springer Berlin Heidelberg

Marsh PD, Bradshaw DJ (1995) Dental plaque as a biofilm. Journal of Industrial Microbiology and Biotechnology 15(3):169–175

Ioannidis A, Kyratsa A, Ioannidou V, Bersimis S, Chatzipanagiotou S (2014) Detection of biofilm production of Yersinia enterocolitica strains isolated from infected children and comparative antimicrobial susceptibility of biofilm versus planktonic forms. Molecular Diagnosis and Therapy 18:309–314

Loeb GI, Neihof RA (1975) Marine conditioning films. Adv Chem Ser 145:319–335

Lourenço TGB, Heller D, Souto RMD, Silva-Senem MXE, Varela VM, Torres MCB, Colombo, APV (2015) Long-term evaluation of the antimicrobial susceptibilityand microbial profile of subgingival biofilms in individuals with aggressive periodontitis. Brazilian Journal of Microbiology 46(2):493–500

Palmer RJ, White DC, Kirkegaard RD, Flemming CA, Chen G, Leung KT, Arrage AA (1997) The biofilm ecology of microbial biofouling, biocide resistance andcorrosion (No. CONF-9704166-1). Oak Ridge National Lab.(ORNL), Oak Ridge, TN (United States)

Percival SL, Thomas JG (2009) Transmission of helicobacter pylori and the role of water and biofilms. Journal of Water and Health 7(3):469–477
Percival SL, Walker JT (1999) Potable water and biofilms: a review of the public health implications. Biofouling 14(2):99–115
Pringle JH, Fletcher M (1986) Adsorption of bacterial surface polymers to attachment substrata. Microbiology 132(3):743–749
Ramage G, Mowat E, Jones B, Williams C, Lopez-Ribot J (2009) Our current understanding of fungal biofilms. Critical Reviews in Microbiology 35(4):340–355
Rittle KH, Helmstetter CE, Meyer AE, Baier RE (1990) Escherichia coli retention on solid surfaces as functions of substratum surface energy and cell growth phase. Biofouling 2(2):121–130
Vieira MJ, Melo LF, Pinheiro MM (1993) Biofilm formation: Hydrodynamic effects on internal diffusion and structure. Biofouling 7(1):67–80

Navigating the Future: Biofilm Trajectories

17

Sonali Nikam, Pranali Pangam, Swapnali Patil, and Poournima Sankpal

Abstract

The study of biofilms, intricate communities of microorganisms thriving in a matrix of extracellular polymeric substances, has captivated scientific interest across diverse disciplines. Biofilms, ubiquitous in natural and engineered environments, exhibit remarkable resilience and adaptive capabilities. It is essential to comprehend the dynamic nature of biofilms in order to predict how they will develop in response to shifting environmental conditions, advances in technology, and new difficulties. Biofilm-associated infections present major medical concerns that motivate study, developing tailored medication delivery systems, antimicrobial coatings, and innovative therapeutic approaches. A balance between preventing antibiotic resistance and guaranteeing patient safety and effectiveness is required for the future of biofilms in healthcare. Biofilms present chances for biocatalysis, biofouling reduction, and bioremediation in industrial environments. But in order to maximize their potential and reduce the risks, significant consideration must be given to how they affect industrial processes, product quality, and environmental sustainability. Additionally, biofilm paths interact with cutting-edge technologies, including artificial

S. Nikam (✉)
Department of Pharmaceutical Chemistry, Appasaheb Birnale College of Pharmacy, Sangli, Tal-Miraj, Maharashtra, India

P. Pangam
Department of Pharmaceutical Quality Assurance, Vijayrao Naik college of pharmacy, Shirval, Tal-Kankavli, Maharashtra, India

S. Patil
Department of Pharmaceutical Quality Assurance, Vijayrao Naik college of pharmacy, Shirval, Tal-Kankavli, Maharashtra, India

P. Sankpal
Department of Pharmaceutical Chemistry, Bharati Vidyapeeth College of Pharmacy, Palus, Maharashtra, India

T. Sarkar et al. (eds.), *Biofilm Applications to Revolutionize Food Technology*,
https://doi.org/10.1007/978-3-031-85205-3_17

intelligence, nanotechnology, and synthetic biology. Using the synergies between biofilms and these state-of-the-art technologies could lead to creative solutions in environmental management, biomanufacturing, and healthcare. Future developments must take ethical issues and societal ramifications into account as biofilms continue to influence fields of industry, human health, and the environment. In order to navigate the intricate dynamics of biofilm use and guarantee fair outcomes for people and communities, responsible innovation, ethical governance, and stakeholder involvement are crucial. By embracing the challenges and opportunities presented by biofilms, we can chart a course toward sustainable and resilient bio-based solutions that benefit society and the planet.

Keywords

Biofouling mitigation · Future trajectories · Biofilms · Synthetic biology · Ethical governance

17.1 Introduction

Biofilms, which are structured communities of microorganisms encased in a self-produced matrix of extracellular polymeric substances (EPS), have been extensively studied due to their widespread presence across both natural and artificial environments (Flemming and Wingender 2010). These complex microbial communities thrive in diverse habitats, ranging from natural ecosystems such as rivers and soils to artificial settings including medical devices, industrial pipelines, and food processing environments (Arruda et al. 2022). In the context of food technology, biofilms have historically been viewed as a formidable challenge (Socransky and Haffajee 2002). Their ability to adhere to surfaces and persist despite rigorous cleaning and disinfection protocols presents significant risks to food safety (Ezzatpanah et al. 2022). Biofilms can grow on a variety of surfaces, including rubber, plastic, and stainless steel, thanks to the moist, nutrient-rich conditions found in food processing plants (Chmielewski and Frank 2003). Biofilms are especially troublesome in the food business because of their adherence and tenacity (Mazaheri et al. 2021).

Biofilms present a variety of difficulties for food technology (Galié et al. 2018). Their resistance to sanitization and cleaning is a major issue (Bland et al. 2022). The biofilms are protected from disinfectants and mechanical cleaning procedures by the EPS matrix that surrounds them (Shineh et al. 2023). Because of this resistance, it is more difficult to maintain hygienic standards in food processing plants, where even minute biofilm leftovers can pose serious contamination concerns (Abebe 2020). Furthermore, harmful microbes like *Salmonella* species, *Escherichia coli*, and *Listeria monocytogenes* can be found in biofilms (El-Liethy et al. 2020). Food safety is constantly at risk from these bacteria because, once established in a biofilm, they become more resistant to antimicrobial treatments and can linger in the environment (Ghosh et al. 2021). The risk of cross-contamination, in which germs from one batch of food products can spread to another and perhaps cause major outbreaks of foodborne illness, is also increased when biofilms are present on processing equipment. This perseverance has serious economic ramifications for the food sector in addition to posing a threat to human health. Despite these obstacles, new developments in microbiology and biotechnology are starting to change the way that biofilms are thought of—from merely harmful to possibly helpful. Scientists are looking for novel ways to use the special qualities of biofilms for purposes that could boost food production, increase food safety, and support more environmentally friendly methods.

As biocontrol agents, for example, biofilms made up of advantageous bacteria can be used to outcompete and stop the growth of dangerous pathogens. By providing a natural substitute for chemical preservatives, this strategy—known as competitive exclusion—may increase food goods' shelf lives while lowering their dependency on artificial chemicals (Barberis et al. 2018). Furthermore, biofilm-based biosensors are being developed by utilizing the metabolic processes of bacteria that create biofilms. Compared to conventional testing methods, these sensors offer a more quick and more sensitive way to ensure food safety since they can identify toxins or track environmental conditions in real time (Bashir et al. 2022).

17.2 The Dual Nature of Biofilms in Food Technology

Because of their intricate structure, biofilms provide bacteria with a greater chance of surviving in harsh environments. The EPS matrix, which is made up of proteins, lipids, polysaccharides, and extracellular DNA, acts as a barrier to protect the biofilm from external stressors such as immune reactions and antimicrobial chemicals (Singh et al. 2021).

Biofilms as a Challenge in Food Technology

Because of their propensity to remain on surfaces, host pathogens, and withstand antimicrobial treatments, biofilms pose a number of serious issues in the field of food technology. These elements make it more difficult to keep food processing areas hygienic and free from infection (Bryan 1988). Here is a thorough analysis of these difficulties (Table 17.1).

The Potential of Biofilms in Food Technology

Biofilms present a number of encouraging prospects for advancement in food technology, nevertheless their difficulties. Biofilms' intrinsic qualities, such as their capacity to create ordered communities and their resilience in a range of environments, can be used to their advantage (Li et al. 2022) (Table 17.2).

Table 17.1 Challenges of biofilms in food technology

Challenge	Details	Impact on food technology
Persistence on surfaces	Biofilms can adhere to various surfaces, including stainless steel, plastic, and rubber, leading to persistent contamination	Continuous risk of contamination, difficulty in maintaining hygiene, and potential for cross-contamination
Pathogen reservoirs	Pathogenic bacteria such as *Listeria monocytogenes*, *Salmonella* spp., and *E. coli* can thrive within biofilms	Increased risk of foodborne illness outbreaks, and challenges in eliminating pathogens from processing environments
Resistance to antimicrobials	Biofilms exhibit higher resistance to antimicrobial treatments due to the protective EPS matrix and altered microbial physiology	Complicates food safety management, necessitates more effective and innovative disinfection strategies

Table 17.2 Opportunities of biofilms in food technology

Opportunity	Details	Impact on food technology
Biocontrol agents	Beneficial biofilms can outcompete pathogens and produce antimicrobial compounds, acting as natural preservatives	Reduces reliance on chemical preservatives, enhances food safety, and supports cleaner food processing practices
Biofilm-based fermentation	Biofilm reactors improve efficiency and consistency, and reduce contamination risks in fermentation processes	Increases product yields, ensures consistent quality, and supports more sustainable food production methods
Biosensors	Biofilm-based biosensors enable real-time monitoring of food safety and quality parameters	Provides rapid and sensitive detection of contaminants, improves quality control, and enhances food safety

17.3 Emerging Applications of Biofilms in Food Technology

Communities of bacteria that stick to surfaces, known as biofilms, have fascinating potential for revolutionizing food technology. With encouraging advancements in biocontrol methods, biosensor technology, and environmentally friendly food packaging, biofilms have a wide range of potential uses (Coughlan et al. 2016).

Biofilm-Based Biocontrol

In food technology, biofilms—structured populations of microbes affixed to surfaces—offer creative ways to manage infections and spoilage organisms. These advantageous biofilms, which are composed of nonpathogenic bacteria, have become useful instruments for improving food sustainability and safety via two main processes: the generation of antimicrobial chemicals and competitive exclusion (Asma et al. 2022).

The deliberate deployment of advantageous biofilms to stop the growth of dangerous germs is known as competitive exclusion. Beneficial biofilms form a protective layer that fills available niches and resources when they are applied to surfaces or surroundings used in food processing. This occupation successfully prevents harmful or spoiling microbes from adhering and growing (Yuan et al. 2020). Beneficial biofilms contribute to a cleaner, safer food processing environment by outcompeting these dangerous organisms for nutrients and space. This strategy lessens the need for chemical disinfectants and antibiotics, which are being examined more closely due to their effects on the environment and their role in the development of antibiotic resistance. In addition to encouraging a more environmentally responsible and sustainable approach to food safety, competitive exclusion supports more general objectives of lowering the use of chemicals in food production (Jia et al. 2023).

A key component of biofilm-based biocontrol is the synthesis of antibacterial chemicals. The ability of some biofilms to produce bioactive compounds with antibacterial qualities increases their efficiency in reducing infections and spoiling organisms. Bacteriocins, which are proteins made by microorganisms that selectively target and prevent the growth of other bacteria, are among these substances. These proteins are useful instruments for enhancing food safety since they work well against a range of infections and spoiling organisms (Ibrahim et al. 2021). Furthermore, by reducing the pH of their surroundings, organic acids like lactic acid, which are produced by biofilms, are important in preventing the proliferation of microorganisms. These organic acids produce an acidic environment that inhibits the growth of undesirable microbes. Moreover, biofilms have the ability to produce lysozymes and other enzymes that degrade bacterial cell walls, enhancing their antibacterial efficacy. By controlling microbial populations, these enzymatic activities improve the food safety and preservation potential of biofilm-based systems. By utilizing these naturally occurring antimicrobial compounds, biofilms provide a safe and sustainable substitute for artificial preservatives, extending food goods' shelf lives and enhancing their safety. Direct application of these antimicrobial agents to food items or processing surfaces is possible. When utilized as natural preservatives, they provide a good substitute for artificial ones, extending food safety and shelf life (Sharif et al. 2017). The food industry can improve food preservation and quality in a way that is both environmentally friendly and health-conscious by utilizing the inherent antibacterial qualities of materials formed from biofilms.

Biofilm-Based Biosensors

One of the most recent developments in the field of food safety and quality monitoring is the use of biofilm-based biosensors. These cutting-edge sensors provide notable advancements over conventional techniques by using the metabolic activity of biofilm-forming bacteria to identify pollutants, poisons, and other important quality factors in real time (Jalilian 2020).

One of the main uses of biofilm-based biosensors is pathogen detection. These sensors can be designed to precisely detect harmful microbes in food products and processing settings, including Salmonella or E. coli (Adley 2014). These biosensors can offer a quick, accurate, and affordable substitute for traditional microbiological testing techniques by harnessing the metabolic activities of biofilm-forming microbes. In contrast to biofilm-based biosensors, which provide faster detection and response capabilities, traditional approaches frequently entail laborious culture processes and intricate laboratory procedures (Saccomano et al. 2021). This quick turnaround lowers the risk of foodborne diseases and improves overall food safety by enabling prompt interventions in the event of contamination (Kovac et al. 2017).

Biofilm-based biosensors are very useful for environmental monitoring in food processing plants in addition to pathogen detection. Numerous environmental factors, including pH, temperature, and nutrition levels, can be measured by these sensors. In order to maintain ideal processing conditions and guarantee the safety and quality of food products, real-time monitoring of these variables is essential. For example, keeping the pH and temperature at the right conditions can stop the growth of dangerous microbes and biofilms that could contaminate food or endanger human health (Rawat 2015). Biofilm-based biosensors aid in adjusting processing parameters, increasing production efficiency, and halting the formation of harmful biofilms by continuously delivering data on environmental conditions.

Biofilm-Derived Materials for Food Packaging

Materials made from biofilms are starting to appear as environmentally friendly substitutes for traditional food packaging. Microorganisms manufacture these compounds, which have special qualities that improve food products' safety and shelf life.

- *Edible Films*: Edible films made from biofilms can be used to wrap food items, providing a barrier of protection that prolongs shelf life and is completely biodegradable. By inhibiting bacteria growth and spoiling, antimicrobial chemicals can be added to these films to further improve food safety (Sung et al. 2013).
- *Active Packaging*: Active packaging technologies that interact with the food product to prevent microbial development and spoiling might incorporate biofilm-based materials. Packaging materials, for example, can be made to release antimicrobial chemicals in reaction to temperature or moisture changes in the environment. This flexible strategy guarantees that the food will stay safe and fresh for the duration of its shelf life (Nayak and Waterson 2019).

17.4 Biofilms in Fermentation and Food Production

One of the earliest and most common uses of biofilms in food production is fermentation. When different food products ferment, biofilms are essential for developing their flavor, texture, and nutritional qualities (Gong et al. 2021).

Traditional Fermentation Systems

Biofilms spontaneously develop within the fermenting substrate or on the surfaces of fermenting containers in conventional fermentation procedures. These biofilms aid in the formation of fermented foods' desired qualities (Gong et al. 2021).

Biofilms affect a number of sensory and nutritional aspects of fermented foods, which is why they are important to their growth. Development of flavor and scent is significantly influenced by the microbial communities in biofilms. These microbes create a wide range of metabolites, such as alcohols, esters, and organic acids, which are essential in determining the flavor and fragrance characteristics of fermented foods (Zhang et al. 2023). These substances add to the nuanced sensory characteristics that distinguish various fermented food varieties. Additionally, biofilms produce extracellular polysaccharides that affect the textural characteristics of fermented foods. By altering the finished product's viscosity and mouthfeel, these polysaccharides can improve its texture and overall sensory appeal (Yang et al. 2020). Furthermore, by generating vitamins, bioactive peptides, and other advantageous substances, biofilm-associated microbes improve nutrition. These improvements raise the nutritional profile of fermented foods, increasing their nutritional value in addition to their flavor and texture appeal (Shiferaw Terefe and Augustin 2020).

Biofilm-Based Fermentation Reactors

Biofilm-based fermentation reactors, which provide a number of noteworthy benefits over conventional fermentation systems, are the result of recent developments in bioprocess engineering (Gunes 2021). Compared to traditional planktonic fermentation systems, biofilm-based reactors can attain larger cell densities and more metabolic activity, which leads to improved fermentation efficiency. This improvement makes the process more economical and efficient by increasing product yields and reducing fermentation periods. Furthermore, biofilm-based reactors can lower the danger of contamination (Gunes 2021). By acting as a barrier, the biofilm matrix improves the fermentation process's consistency and safety by reducing the possibility of contamination by unwanted microbes. The stability and dependability of the fermentation environment are guaranteed by this protective feature. Furthermore, by lowering the need for chemical additions like pH regulators and preservatives, biofilm-based fermentation systems promote sustainability (Mahdinia and Demirci 2020). In line with the larger objectives of sustainable food production, this reduction results in the creation of more natural and eco-friendly food products. All of these developments show how biofilm-based fermentation reactors improve fermentation processes' safety and efficiency while encouraging more environmentally friendly behaviors (Elsayad et al. 2024).

17.5 Future Directions and Technological Advancements

With a number of possible developments in the works, the future of biofilm applications in food technology is bright. This section examines a few of the major fields in which biofilm research is anticipated to have a major impact in the years to come (Cámara et al. 2022).

Synthetic Biology and Biofilm Engineering

Exciting opportunities exist for the engineering of biofilms with specialized characteristics for certain food technology applications thanks to synthetic biology.

Targeted Antibacterial Activity: A dynamic and adaptable approach to food safety is possible by engineering biofilms to produce particular antibacterial chemicals in response to the presence of pathogens (Chitlapilly Dass and Wang 2022).

- Enhanced Nutrient Synthesis: Engineered biofilms can be utilized to increase the nutritional profile of food products by creating high levels of vitamins, critical amino acids, or bioactive substances (Ghosh et al. 2022).
- Bioremediation: To make food production safer and cleaner, biofilms can be engineered to break down or detoxify dangerous materials found in food processing settings, such as mycotoxins or pesticide residues (Awuchi et al. 2021).

Integration with Smart Technologies

Food safety and quality monitoring could be completely transformed by combining biofilm-based technologies with smart sensors and data analytics platforms. Recent developments in biofilm engineering have produced a number of creative uses that improve the quality and safety of food. One important advancement is targeted antimicrobial activity, which allows biofilms to be modified to release particular antibiotic substances in reaction to infections (Koo et al. 2017). This strategy offers a flexible and adaptable way to manage food safety, improving the general safety of food products and enabling exact control over microbial contamination. Another advantage of tailored biofilms is improved nutrient production. High concentrations of vital nutrients, such vitamins and amino acids, or bioactive substances that enhance the nutritional profile of food products can be produced by these biofilms. This feature promotes more nutrient-dense meals in addition to increasing the food's health benefits. Additionally, biofilms can be used for bioremediation. They can be designed to break down or purify toxic materials found in food processing settings, like pesticide residues or mycotoxins (Gavahian et al. 2020). Through the efficient removal of pollutants that may jeopardize food safety, this application helps to produce food in a safer and cleaner manner. When taken as a whole, these developments show how engineered biofilms can be used to solve particular issues with food safety and nutrition, opening the door to more efficient and environmentally friendly food production methods (Neethirajan et al. 2018).

17.6 Regulatory and Ethical Considerations

It is imperative to address the ethical and regulatory issues that arise with the increasing integration of biofilm-based technologies into food production and processing (Mishra et al. 2021).

Regulatory Frameworks

Instead of focusing on the possible advantages of biofilms, current food safety rules are largely intended to address the problems they pose. Regulatory frameworks will need to be modified as biofilm applications in food technology develop further in order to guarantee the technologies' safe and efficient use (Highmore et al. 2022).

- Safety Assessment: To analyze the possible dangers connected to the use of modified biofilms in food production, thorough safety assessments will be necessary. This involves determining if compounds formed from biofilms have the potential to add new toxins or allergies to the food supply (Elfaky et al. 2024).
- Labeling and Transparency: To educate customers about the nature and advantages of biofilm-based goods and packaging, it will be crucial to have clear labeling and transparency as these materials become more widely used (Mishra et al. 2021).

Ethical Considerations

A number of ethical issues are also brought up by the use of biofilms in food technology, especially when it comes to genetic engineering and synthetic biology.

- Consumer Acceptance: How the general public views biofilm-based technologies will have a big impact on how widely they are used. Gaining the trust of consumers will need open and honest communication about the advantages and security of these technologies (Lam et al. 2020).
- Environmental Impact: Although biofilm-based technologies may lessen the environmental impact of food production, it is crucial to take into account any

potential unforeseen repercussions, such as how engineered biofilms may affect natural ecosystems (Maurya et al. 2023).

17.7 Conclusion

Food technology's investigation of biofilm trajectories indicates a promising and innovative topic. The dual nature of biofilms—both a difficulty and an opportunity—will continue to influence food technology as research advances. By harnessing the beneficial properties of biofilms, the food industry can develop novel solutions that enhance food safety, improve production efficiency, and contribute to sustainability.

The future of biofilm applications in food technology will be driven by advances in synthetic biology, bioprocess engineering, and smart technologies, all of which promise to unlock new possibilities for the food industry. However, realizing this potential will require careful consideration of the regulatory and ethical implications, as well as a commitment to transparency and consumer engagement.

The integration of biofilms into food technology represents a significant shift in how we approach food production and safety. As we navigate this future, continued research and innovation will be essential in ensuring that biofilms become a force for positive change in the global food system.

References

Abebe GM (2020) The role of bacterial biofilm in antibiotic resistance and food contamination. Int J Microbiol 2020(1):1705814

Adley CC (2014) Past, present and future of sensors in food production. Food Secur 3(3):491–510

Arruda V, Simões M, Gomes IB (2022) Synthetic musk fragrances in water systems and their impact on microbial communities. Water 14(5):692

Asma ST, Imre K, Morar A, Herman V, Acaroz U, Mukhtar H, Arslan-Acaroz D, Shah SRA, Gerlach R (2022) An overview of biofilm formation–combating strategies and mechanisms of action of antibiofilm agents. Life 12(8):1110

Awuchi CG, Ondari EN, Ogbonna CU, Upadhyay AK, Baran K, Okpala COR, Korzeniowska M, Guiné RP (2021) Mycotoxins affecting animals, foods, humans, and plants: types, occurrence, toxicities, action mechanisms, prevention, and detoxification strategies—A revisit. Food Secur 10(6):1279

Barberis S, Quiroga HG, Barcia C, Talia JM, Debattista N (2018) Natural food preservatives against microorganisms. In: Food safety and preservation. Academic Press, pp 621–658

Bashir O, Bhat SA, Basharat A, Qamar M, Qamar SA, Bilal M, Iqbal HM (2022) Nano-engineered materials for sensing food pollutants: technological advancements and safety issues. Chemosphere 292:133320

Bland R, Brown SR, Waite-Cusic J, Kovacevic J (2022) Probing antimicrobial resistance and sanitizer tolerance themes and their implications for the food industry through the Listeria monocytogenes lens. Compr Rev Food Sci Food Saf 21(2):1777–1802

Bryan FL (1988) Risks of practices, procedures and processes that lead to outbreaks of foodborne diseases. J Food Prot 51(8):663–673

Cámara M, Green W, MacPhee CE, Rakowska PD, Raval R, Richardson MC, Slater-Jefferies J, Steventon K, Webb JS (2022) Economic significance of biofilms: a multidisciplinary and cross-sectoral challenge. NPJ Biofilms Microbiomes 8(1):42

Chitlapilly Dass S, Wang R (2022) Biofilm through the looking glass: a microbial food safety perspective. Pathogens 11(3):346

Chmielewski R, Frank J (2003) Biofilm formation and control in food processing facilities. Compr Rev Food Sci Food Saf 2(1):22–32

Coughlan LM, Cotter PD, Hill C, Alvarez-Ordóñez A (2016) New weapons to fight old enemies: novel strategies for the (bio) control of bacterial biofilms in the food industry. Front Microbiol 7:1641

Elfaky MA, Ibrahim SR, Obaid NA, Wasfi R (2024) Biofilms of pathogenic bacteria. In: Perspectives of quorum quenching in new drug development. CRC Press, pp 1–33

El-Liethy MA, Hemdan BA, El-Taweel GE (2020) Prevalence of E. coli, Salmonella, and Listeria spp. as potential pathogens: a comparative study for biofilm of sink drain environment. J Food Saf 40(4):e12816

Elsayad RM, Sharshir SW, Khalil A, Basha AM (2024) Recent advancements in wastewater treatment via anaerobic fermentation process: a systematic review. J Environ Manag 366:121724

Ezzatpanah H, Gómez-López VM, Koutchma T, Lavafpour F, Moerman F, Mohammadi M, Raheem D (2022) Risks and new challenges in the food chain: viral contamination and decontamination from a global perspective, guidelines, and cleaning. Compr Rev Food Sci Food Saf 21(2):868–903

Flemming H-C, Wingender J (2010) The biofilm matrix. Nat Rev Microbiol 8(9):623–633

Galié S, García-Gutiérrez C, Miguélez EM, Villar CJ, Lombó F (2018) Biofilms in the food industry: health aspects and control methods. Front Microbiol 9:898

Gavahian M, Pallares N, Al Khawli F, Ferrer E, Barba FJ (2020) Recent advances in the application of innovative food processing technologies for mycotoxins and pesticide reduction in foods. Trends Food Sci Technol 106:209–218

Ghosh S, Sarkar T, Chakraborty R (2021) Formation and development of biofilm-an alarming concern in food safety perspectives. Biocatal Agric Biotechnol 38:102210

Ghosh S, Nag M, Lahiri D, Sarkar T, Pati S, Kari ZA, Nirmal NP, Edinur HA, Ray RR (2022) Engineered biofilm: innovative nextgen strategy for quality enhancement of fermented foods. Front Nutr 9:808630

Gong C, He Y, Tang Y, Hu R, Lv Y, Zhang Q, Tardy BL, Richardson JJ, He Q, Guo J (2021) Biofilms in plant-based fermented foods: formation mechanisms, benefits and drawbacks on quality and safety, and functionalization strategies. Trends Food Sci Technol 116:940–953

Gunes B (2021) A critical review on biofilm-based reactor systems for enhanced syngas fermentation processes. Renew Sust Energ Rev 143:110950

Highmore C, Melaugh G, Morris R, Parker J, Direito S, Romero M, Soukarieh F, Robertson S, Bamford N (2022) Translational challenges and opportunities in biofilm science: a BRIEF for the future. NPJ Biofilms Microbiomes 8(1):68

Ibrahim SA, Ayivi RD, Zimmerman T, Siddiqui SA, Altemimi AB, Fidan H, Esatbeyoglu T, Bakhshayesh RV (2021) Lactic acid bacteria as antimicrobial agents: food safety and microbial food spoilage prevention. Food Secur 10(12):3131

Jalilian F (2020) Biofilm sensing tool development for environmental monitoring

Jia F, Shahzadi G, Bourlakis M, John A (2023) Promoting resilient and sustainable food systems: a systematic literature review on short food supply chains. J Clean Prod 435:140364

Koo H, Allan RN, Howlin RP, Stoodley P, Hall-Stoodley L (2017) Targeting microbial biofilms: current and prospective therapeutic strategies. Nat Rev Microbiol 15(12):740–755

Kovac J, den Bakker H, Carroll LM, Wiedmann M (2017) Precision food safety: a systems approach to food safety facilitated by genomics tools. TrAC Trends Anal Chem 96:52–61

Lam TK, Heales J, Hartley N, Hodkinson C (2020) Consumer trust in food safety requires information transparency. Australas J Inf Syst 24. https://doi.org/10.3127/ajis.v24i0.2219

Li Z, Wang X, Wang J, Yuan X, Jiang X, Wang Y, Zhong C, Xu D, Gu T, Wang F (2022) Bacterial biofilms as platforms engineered for diverse applications. Biotechnol Adv 57:107932

Mahdinia E, Demirci A (2020) Biofilms in fermentation for the production of value-added products. In: Microbial biofilms. CRC Press, pp 73–108

Maurya A, Kumar R, Raj A (2023) Biofilm-based technology for industrial wastewater treatment: current technology, applications and future perspectives. World J Microbiol Biotechnol 39(5):112

Mazaheri T, Cervantes-Huamán BR, Bermúdez-Capdevila M, Ripolles-Avila C, Rodríguez-Jerez JJ (2021) Listeria monocytogenes biofilms in the food industry: is the current hygiene program sufficient to combat the persistence of the pathogen? Microorganisms 9(1):181

Mishra B, Varjani S, Parida M, Iragavarapu GP, Awasthi MK, Awasthi SK, Zhang Z (2021) Film based packaging for food safety and preservation: issues and perspectives. In: Environmental microbiology and biotechnology: Volume 2: Bioenergy and environmental health. Springer, Singapore, pp 309–336

Nayak R, Waterson P (2019) Global food safety as a complex adaptive system: key concepts and future prospects. Trends Food Sci Technol 91:409–425

Neethirajan S, Ragavan V, Weng X, Chand R (2018) Biosensors for sustainable food engineering: challenges and perspectives. Biosensors 8(1):23

Rawat S (2015) Food spoilage: microorganisms and their prevention. Asian J Plant Sci 5(4):47–56

Saccomano SC, Jewell MP, Cash KJ (2021) A review of chemosensors and biosensors for monitoring biofilm dynamics. Sens Actuators Rep 3:100043

Sharif ZM, Mustapha F, Jai J, Zaki N (2017) Review on methods for preservation and natural preservatives for extending the food longevity. Chem Eng Res Bull 19:145

Shiferaw Terefe N, Augustin MA (2020) Fermentation for tailoring the technological and health related functionality of food products. Crit Rev Food Sci Nutr 60(17):2887–2913

Shineh G, Mobaraki M, Perves Bappy MJ, Mills DK (2023) Biofilm formation, and related impacts on healthcare, food processing and packaging, industrial manufacturing, marine industries, and sanitation–a review. Appl Microbiol 3(3):629–665

Singh S, Datta S, Narayanan KB, Rajnish KN (2021) Bacterial exo-polysaccharides in biofilms: role in antimicrobial resistance and treatments. J Genet Eng Biotechnol 19:1–19

Socransky SS, Haffajee AD (2002) Dental biofilms: difficult therapeutic targets. Periodontol 28(1):12–55

Sung S-Y, Sin LT, Tee T-T, Bee S-T, Rahmat A, Rahman W, Tan A-C, Vikhraman M (2013) Antimicrobial agents for food packaging applications. Trends Food Sci Technol 33(2):110–123

Yang X, Li A, Li X, Sun L, Guo Y (2020) An overview of classifications, properties of food polysaccharides and their links to applications in improving food textures. Trends Food Sci Technol 102:1–15

Yuan L, Hansen MF, Røder HL, Wang N, Burmølle M, He G (2020) Mixed-species biofilms in the food industry: current knowledge and novel control strategies. Crit Rev Food Sci Nutr 60(13):2277–2293

Zhang K, Zhang T-T, Guo R-R, Ye Q, Zhao H-L, Huang X-H (2023) The regulation of key flavor of traditional fermented food by microbial metabolism: a review. Food Chem X 19:100871

Biofilm Revisited: Enhancing Food Safety Using Phytonanotechnology

18

Anusuya Debnath and Hemanta Koley

Abstract

Food spoilage caused by bacterial biofilm leads to substantial economic losses for the food industry worldwide with probable losses amounting to billions of dollars annually. Biofilm acts as a protective shield that enables these organisms to survive the action of chemical disinfectants, sanitizers, and antimicrobials. Most foodborne pathogens, such as *Escherichia coli*, *Vibrio* spp., *Clostridium* spp., *Staphylococcus aureus*, *Pseudomonas* spp., *Listeria monocytogenes*, and *Bacillus cereus*, are capable of producing antimicrobial resistant biofilm. The inadequacies of chemical methods for biofilm removal raised the need for more robust and effective technology to handle the menace created by biofilm-producing pathogens to ensure food safety in the industry. As an alternative, bioactive compounds from plants having inhibitory action on biofilm formation are used as preservatives in the food industry. However, their usage is restricted due to poor availability, stability during food processing and storage, and lack of standardized protocol. These limitations can be overcome by the intervention of nanotechnology. There are a number of benefits in nanotechnology-enhanced interventions that target achieving optimal use of phytochemical constituents among which include enhanced penetration through biofilm matrix, sustained release, site-specific delivery, improved bioavailability, efficacy as well as stability. The surface-functionalized or encapsulated phyto-fabricated nanoparticles represent an attractive strategy for tackling highly resilient microbial biofilms. This chapter provides a comprehensive detail of recent developments in the field

A. Debnath (✉)
Department of Biotechnology, Brainware University, Barasat, India
e-mail: asd.bt@brainwareuniversity.ac.in

H. Koley
National Institute for Research in Bacterial Infections (NIRBI), Kolkata, India

T. Sarkar et al. (eds.), *Biofilm Applications to Revolutionize Food Technology*,
https://doi.org/10.1007/978-3-031-85205-3_18

of phytonanotechnology used for the prevention and elimination of biofilms formed by foodborne pathogens.

Keywords

Biofilm evolution · Technological advancements · Industrial applications · Sustainability · Regulatory challenges

18.1 Introduction to Foodborne Pathogen Biofilms and Their Impact on Food Industry and Human Health

The food industry needs to adapt new methods of cleaning and disinfection procedures due to the substantial risk of food cross-contamination posed by the existence of microbial biofilms. The biofilm serves as an organized habitat for one or more species of microorganisms, firmly anchored to either biotic or abiotic surfaces, enveloped by a shielding matrix comprising extracellular polysaccharides, nucleic acids, and proteins (Fig. 18.1). Biofilms consist of intricately arranged water channels facilitating the passage of both large and small molecules (Flemming and Wingender 2010; Lewandowski et al. 1994). Biofilms having multiple bacterial species show significant ecological benefits by increasing the adherence to the solid substratum. These multispecies biofilms demonstrate increased resilience against disinfectants like quaternary ammonium compounds and other antimicrobials (Giaouris et al. 2015). In general, this sessile form is an excellent survival strategy under unfavorable conditions that confers many advantages to the microbial cells in food industry as well, such as physical resistance against desiccation, mechanical resistance against liquid streams in pipelines and chemical protection against

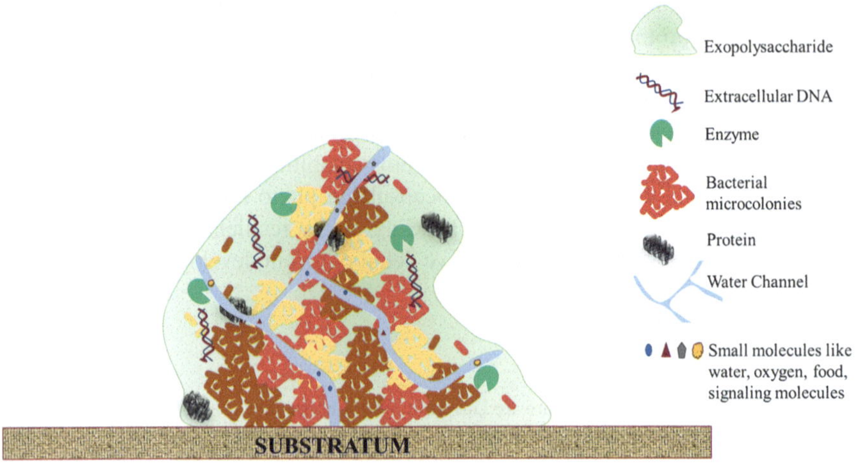

Fig. 18.1 The structure of mature biofilm (Flemming and Wingender 2010; Lewandowski et al. 1994)

chemicals, antimicrobials, and disinfectants used in the industry (Flemming et al. 2016). Biofilms have demonstrated heightened resistance to disinfection measures across various sectors of the food industry, encompassing seafood processing, brewing, dairy processing, and meat and poultry processing (Srey et al. 2013).

Biofilms can quickly grow in food industry settings due to favorable conditions. The process consists of several sequential steps (Fig. 18.2): the first step involves conditioning the surface to which the bacterial community binds reversibly. After that cells irreversibly attach to the surface which leads to the formation of microcolonies. Eventually, an intricate three-dimensional elaborate ecosystem is established ready for dispersal to spread further (Nikolaev and Plakunov 2007; Srey et al. 2013). The biofilm-forming ability of certain human pathogens like *Bacillus cereus*, *Listeria monocytogenes*, Staphylococcus spp., Salmonella spp., *Escherichia coli* O157:H7, *Campylobacter jejuni*, and *Yersinia enterocolitica* puts food industry under a continuous threat of becoming a source of an epidemic (Anand et al. 2014). These pathogens demonstrate the ability to create biofilm formations on various synthetic substrates commonly found in food processing, such as stainless steel, polyethylene, wood, glass, polypropylene, rubber, and others (Abdallah et al. 2014; Colagiorgi et al. 2017).

Bacillus cereus is a gram-positive bacterium capable of forming spores and surviving a wide range of temperatures (4 °C–50 °C). *B. cereus* is resistant to several

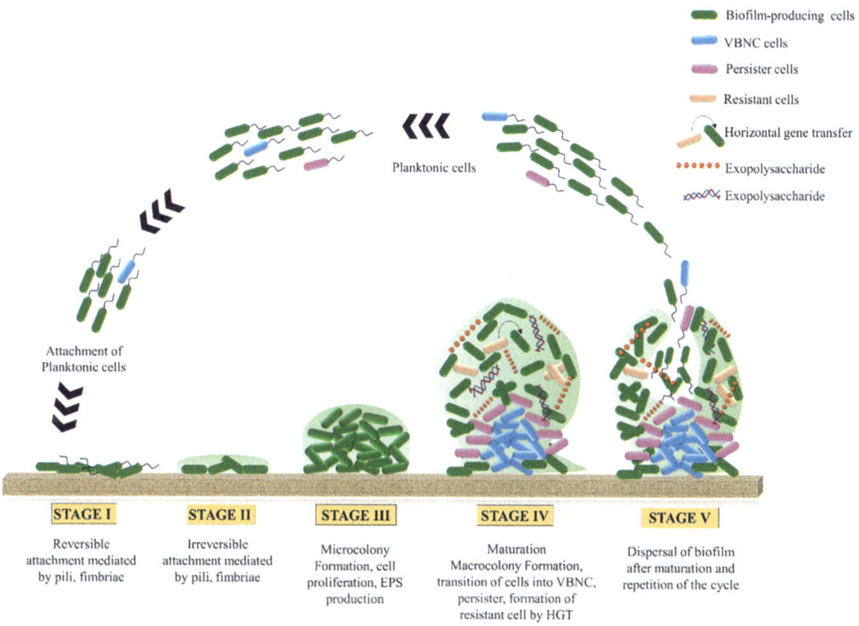

Fig. 18.2 Illustration showing the step-wise development of biofilm (1) reversible attachment, (2) irreversible attachment, (3) microcolonies formation, (4) biofilm maturation, (5) dispersal (Kirmusaoglu 2016)

stress-inducing factors such as chemicals, heat, and radiation (Bottone 2010). It is commonly found in soil and food products like rice, dairy, vegetables, and meat. The toxins secreted by B. cereus are responsible for manifesting diarrhea in humans. It can form biofilm on food contact surfaces such as stainless steel pipes and conveyor belts and affects the food quality by releasing bacteriocins, metabolites, surfactants, and enzymes like proteases and lipases (Grigore-Gurgu et al. 2019). The bacterial motility facilitated by flagella plays a crucial role in biofilm formation by allowing access to the surface and rapid spread but does not participate directly in surface attachment (Houry et al. 2010). *B. cereus* caused numerous foodborne illnesses in the EU between 2007 and 2014, leading to 413 outbreaks with 6657 cases reported. These outbreaks were associated with a variety of food types, such as buffet meals, cereal products, red meat products, poultry meat products, and vegetables or juices (EFSA 2016). Another gram-positive bacterium, *Listeria monocytogenes* poses a widespread and serious threat as a foodborne pathogen. Various food products have been identified as potential sources of L. monocytogenes transmission, including seafood, dairy products, meat, ready-to-eat products, fruits, unpasteurized milk, frozen vegetables, and poultry (CDC 2017; Rothrock Jr et al. 2017). *L. monocytogenes* forms biofilms primarily composed of teichoic acids on surfaces like polypropylene, steel, rubber, or glass within food industry facilities, allowing it to contaminate food batches and proliferate even under refrigeration temperatures (Silva et al. 2008). Particularly vulnerable populations such as pregnant women, infants, the elderly, and immunocompromised individuals are at risk of developing listeriosis—a severe illness characterized by septicemia and meningitis (Silk et al. 2012).

Poultry meat serves as a common reservoir for *Salmonella enterica* serovar Enteritidis, a gram-negative bacterium in processed food. Its significance as a foodborne pathogen was established as early as 1966 due to its biofilm-producing capability (Duguid et al. 1966). *S. enterica* easily attaches to meat and other food substrates, facilitating cross-contamination between food batches in processing plants or supermarkets (Wang et al. 2013a, b). Biofilm formation on infrastructure used in precooked food manufacturing is the primary source of contamination, leading to widespread outbreaks, some of which have been lethal, affecting thousands of individuals, as observed in Spain in 2005 (Lenglet et al. 2005; Wang et al. 2013a, b). Another gram-negative organism known for causing outbreaks of bloody diarrhea and hemolytic uremic syndrome (HUS) worldwide is EHEC serotype O157:H7, a predominant strain responsible for infections in humans (Gould et al. 2013). This pathogen can be transmitted through various sources including raw milk, drinking water, fresh meat, and a range of fruits and vegetables such as melons, tomatoes, parsley, coriander, spinach, and lettuce (Galié et al. 2018). *E. coli* adheres to the surfaces via pili, flagella, and membrane proteins and subsequently forms biofilms that enhance its resistance toward disinfectants (Lim et al. 2019). Effective strategies for preventing or treating EHEC infections are still lacking despite attempts, since using antibiotics can make problems like HUS and renal failure severe (Lee et al. 2007). As the infectious dose for *E. coli* O157:H7 is as low as 50 ingested colony-forming units (CFUs), this further emphasizes the necessity for stringent

Table 18.1 List of biofilm-forming pathogens in the food industry

Pathogen	Characteristics	Contaminated food	Negative impact	References
Bacillus cereus	Gram-positive, spore-forming, anaerobic, facultative anaerobic	Dairy products, rice, vegetables, meat	Diarrhea and vomiting symptoms	Bottone (2010); Grigore-Gurgu et al. (2019)
Listeria monocytogenes	Gram-positive, rod-shaped, facultative anaerobic	Dairy products, meat, ready-to-eat products, fruit, unpasteurized milk, candied apples, frozen vegetables, poultry	Listeriosis in the elderly, pregnant women, and immune-compromised patients	CDC (2017); Rothrock Jr et al. (2017)
Staphylococcus aureus	Gram-positive, nonspore-forming, nonmotile, facultative anaerobic	Meat products, poultry, egg products, dairy products, salads, bakery products, especially cream-filled pastries and cakes, and sandwich fillings	Methicillin resistance can cause vomiting and diarrhea	Giaouris et al. (2015); Kadariya et al. (2014)
Salmonella Enterica	Gram-negative, rod-shaped, flagellate, facultative aerobic	Poultry meat, bovine, ovine, porcine, fish	Can cause gastroenteritis or septicemia	Nguyen et al. (2014); Duguid et al. (1966)
Escherichia coli	Gram-negative, rod-shaped	Raw milk, fresh meat, fruits, and vegetables	Diarrhea outbreaks and hemolytic uremic syndrome	Galié et al. (2018)
Campylobacter jejuni	Gram-negative, aerobic, and anaerobic	Animals, poultry, unpasteurized milk	Bloody diarrhea, fever, stomach cramp, nausea, and vomiting	Klančnik et al. (2021); Chlebicz and Śliżewska (2018)

control measures in food production facilities (Marouani-Gadri et al. 2010) (Table 18.1).

18.2 Conventional Antimicrobial Methods and Their Limitations

In the food industry, maintenance of proper sanitation protocols is of utmost importance to ensure food safety. The process of sanitation involves two main steps, i.e., cleaning and sanitizing. Cleaning is the process of applying detergents to surfaces to remove residues and debris. Sanitizing, on the other hand, employs chemical agents to decrease the microbial load on surfaces to levels that are safe for food production (Srey et al. 2013). Presently, a diverse array of chemical disinfectants finds application in the food industry. These disinfectants are broadly classified into three main groups based on their modes of action: (i) oxidizing agents, such as chlorine-based compounds (including hypochlorite), hydrogen peroxide, ozone, and peracetic acid, which hinder protein synthesis and enzymes involved in carbohydrate metabolism and glucose oxidation, causing irreversible inactivation; (ii) surface-active compounds, including quaternary ammonium compounds and alcohols, which disrupt cell membrane bonds, leading to membrane rupture; and (iii) iodophors, which introduce iodine into cellular components, resulting in cell death (Fraise et al. 2013). The effectiveness and affordability of hypochlorites make it the most preferred and frequently used sanitizer. However, there are several limitations to using hypochlorites such as equipment and surface corrosion, potential human toxicity, and environmental contamination. As a less-potent alternative solution, quaternary ammonium-based sanitizers are a good option due to their properties like biodegradability, less harmful for humans and as well as equipment (Wang et al. 2017). Even though effective, few of these chemicals may inadvertently produce by-products like trihalomethanes and haloacetic acids, which may pose potential carcinogenic risks, environmental hazard or have the potential to foster development of antibiotic-resistance mechanism in microorganism (Wang et al. 2017; Coroneo et al. 2017; Davidson and Harrison 2002; Doyle 2006; Halden 2014; Marques et al. 2007; Xue et al. 2017; Falcó et al. 2019).

The molecular mechanism behind antibiotic resistance is still not clear because of complex multifactorial regulation, particularly structural barriers within the biofilm and genetic adaptations. Research outcome of several groups led to three probable explanations through different hypotheses as potential reasons to elaborate this mechanism. According to the first hypothesis, antibiotics diffuse either slowly or partially into the inner core of the biofilm (Marić and Vraneš 2007). The second hypothesis advocates for the changes in the microfilm microenvironment, where a certain population of bacterial cells enters into a slow-growth phase either due to a lack of nutrients or accumulation of toxic metabolites of physiological processes. This slow growth rate increases their chances of survival under harsh conditions (Beveridge et al. 1997). The third hypothesis suggests that a subpopulation of the cells within the biofilm gave rise to spore formation. These cells exhibit a distinct and highly resistant phenotype, shielding them from antibiotic effects, and represent a biologically programmed response to the sessile lifestyle of bacteria (Flemming 1993). Hence, various substitutes for conventional disinfectants like plant-based antimicrobials, development of surface material composition aimed at deterring bacterial adhesion, and bacteriophage-targeted elimination of biofilm-producing microorganisms.

18.3 Nanotechnology: An Emerging Intervention in Food Safety Through Nano-enhanced Solutions

Natural antimicrobials are hydrophobic and/or have distinct odors, thus rendering them difficult to employ in the food industry. They are also typically sensitive to harsh environmental conditions such as high temperature or pressure (Cacciatore et al. 2021). Nanotechnology is suggested as a solution to these problems in order to make use of natural ingredients in food. Nanoparticulate systems offer protection to sensitive compounds, preventing premature inactivation, and enabling controlled release of antimicrobials, thus prolonging their effectiveness (Lopes and Brandelli 2018; Niaz et al. 2019; Brandelli et al. 2017; Lee et al. 2015). The precise antibacterial process of nanoparticles is not fully understood yet. However, the potential mechanism might start with nanoparticle absorption into the cell membrane and its disintegration. After this, the nanoparticles enter the cell and disrupt bacterial growth by causing leakage of intracellular contents, producing reactive oxygen species, disrupting the electron transport system, deactivating efflux pumps, and importantly, interfering with the cell's enzymatic and metabolic functions (Slavin et al. 2017). In terms of food safety, nanotechnology offers two ways to control harmful and spoiling microorganisms: (i) encapsulation of an antimicrobial agent in a nanocarrier-based system, or (ii) transformation of an antimicrobial agent into a particle of nanoscale (Srividya et al. 2017).

Nanoencapsulation of Natural Compounds with Antibiofilm Activity

Encapsulation is the methodology that can be used to protect the antimicrobial against adverse conditions. This technique enables the use of a low dosage of the active compound under controlled release that helps maintain the antimicrobial activity for a longer period of time. Extensive research has been carried out on nanocarriers used in food systems including nanoemulsions, nanoliposomes, polymer nanoparticles, and nanofibers (Srividya et al. 2017; Blanco-Padilla et al. 2014; Lopes et al. 2017). Figure 18.3 illustrates certain structural characteristics of these nanocarriers. Antimicrobial compounds enclosed in nontoxic matrices have been employed as nanocarrier systems in food, a concept that could also apply to crafting nanostructured disinfectants.

Lipid Nanoparticles: Nanoliposomes

Liposomes are smart nanocarriers that can extend the duration of their presence in the blood circulation without triggering immune clearance or showing any toxicity. Thus, can be utilized for targeting bacterial pathogens (Ahmed et al. 2001; Wang et al. 2020). Nanoliposomes are composed of an aqueous nucleus and hydrophobic phospholipid bilayer that can encapsulate water-soluble (localized in the nucleus), lipo-soluble (phospholipid bilayer), or amphipathic compounds (can be present both in the nucleus as well as in the bilayer) (Lopes and Brandelli 2018). Nanoliposomes are one of the most studied lipid-based nanoformulations, due to their ease of production, biocompatibility, low toxicity, enhanced stability, and ability to entrap both lipophilic and hydrophilic bioactive compounds (Filipczak et al. 2020).

Liposomes preferentially adsorb onto biofilm surfaces and penetrate the exopolysaccharide. This leads to the inhibition of bacterial growth and its ability to

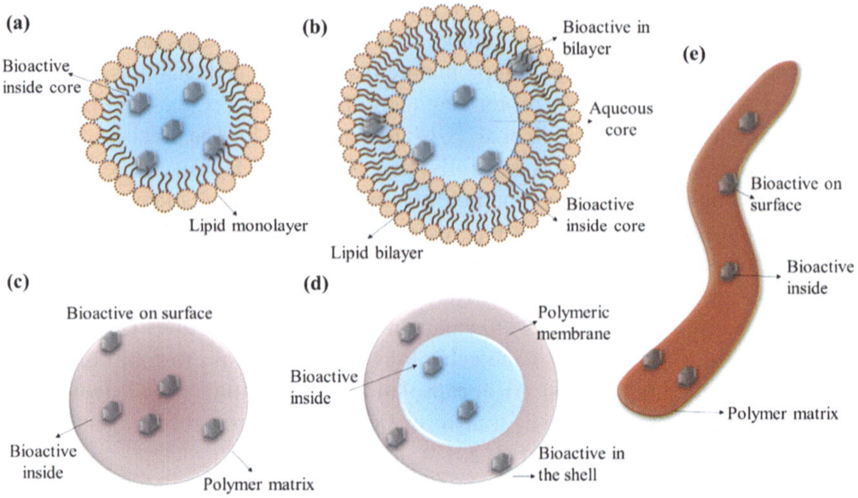

Fig. 18.3 Schematic representation of the structural characteristics of more commonly used nanostructures in food systems. (**a**) Nanoemulsions, (**b**) nanoliposomes, (**c**) polymeric nanoparticles—nanospheres, (**d**) polymeric nanoparticles—nanocapsules, (**e**) nanofibers

recognize the drug and thus, hinders the development of drug resistance through horizontal gene transfer (Wang 2021; Dos Santos Ramos et al. 2018; Rukavina and Vanić 2016). Some of the examples where nanoliposome-based targeting of biofilm-producing bacterial pathogens was used are mentioned in this chapter. Liposomes derived from a combination of soy lecithin and cholesterol (at a ratio of 5:1) prolonged the stability of cinnamon oil for up to 96 hours and enhanced its antibiofilm effectiveness against MRSA by approximately tenfold (Cui et al. 2016a, b). An effective nanoliposome against *Staphylococcus aureus* was prepared by encapsulating a mixture of cinnamon oil and colistin that improves the chemical stability of the essential oil (Ellboudy et al. 2023). In another study, preestablished biofilms of *L. monocytogenes* and *S. enteritidis* were successfully targeted by cinnamaldehyde (CIN)-loaded nanoliposomes (CIN-NLs) (Hasanzadeh Baboli et al. 2023). The biofilm of *Staphylococcus aureus* on containers is the primary cause of microbial contamination in milk. The nanoliposomes encapsulating salvia oil were used to trigger an antibacterial effect on *S. aureus* biofilm in a milk container (Cui et al. 2016a, b). There were numerous other examples to establish the significance of nanoliposomes in food safety against biofilm-forming microorganisms.

Lipid Nanoparticles: Nanoemulsion

Nanoemulsions can be defined as nano-sized droplets of stable dispersions, comprised of surfactant-stabilized aqueous and oily phases. These are extensively used for the encapsulation of hydrophobic compounds like essential (Blanco-Padilla et al. 2014). Nanoemulsions are able to penetrate porous matrices and interact with exopolysaccharide (EPS), enabling the targeted administration of antibacterial drugs at extremely high doses. Their proficient penetration into porous matrices and close interaction with the biofilm surface render nanoemulsions highly effective for disrupting biofilms (Rossi et al. 2022; Pircalabioru and Chifiriuc 2020). Now, let us look at some of the studies reported by several groups working on nanoemulsion-based control of biofilm-producing pathogens.

A cumin oil-based nanoemulsion was able to inhibit biofilm formation in *E. coli* and *S. enterica* by reducing the production of EPS and quorum sensing (QS)-related traits (Amrutha et al. 2017). Similarly, another study examined the effectiveness of nanoemulsions containing methyl salicylate (0.0025%) extracted from *Gaultheria fragrantissima* essential oil and eugenol (0.0005%). This nanoemulsion demonstrated a reduction in the surface adhesion and expression of the biofilm-related genes in *E. coli* O15:H7, ultimately inhibiting biofilm formation (Prateeksha et al. 2019). Another study used a citral nanoemulsion on the pineapple surface as a treatment against *S. enterica* Typhimurium that reduced 2.57 log CFU/g of bacterial load (Prakash et al. 2020). A recently published report showed the significance of garlic essential oil nanoemulsion in inhibiting planktonic growth and biofilm formation of *L. monocytogenes* (Liu et al. 2024).

Polymeric Nanoparticles

Polymeric nanoparticles are commonly categorized as nanospheres and nanocapsules. In the nanosphere, the active compound is either adsorbed on the polymer surface or encapsulated, without including oils in its composition. However, in nanocapsules, the active compound is dissolved within an oily core contained within the polymer particle (Schaffazick et al. 2003). The use of polymeric nanoparticles is recommended when the active compound needs to be released slowly (Blanco-Padilla et al. 2014). These polymeric materials are either natural polysaccharides such as chitosan, cellulose, and dextran and polypeptides such as albumin, gelatin, and legumin, or synthetic polymers like polyglycolide, polylactides (PLA), and copolymers including polylactide co-glycolide (PLGA), and others (Michalak et al. 2016). These nanoparticles function by engaging in electrostatic interactions with the negatively charged extracellular polymeric substances (EPS) present in the outer layer of the biofilms (Da Costa et al. 2021).

Numerous studies have highlighted the promise of biopolymeric nanosystems in combating the growth and formation of biofilms by foodborne pathogens. Chitosan is a naturally derived amino polysaccharide polymer known for its inherent antibacterial and antibiofilm properties and thus an attractive choice in the packaging industry. A recently published work claimed that chitosan nanoparticles loaded with usnic acid are highly efficient in removing the persister cells of foodborne pathogens like *S. aureus*, *E. coli*, and *L. monocytogenes* residing within the biofilm (Khan et al. 2020; Pompilio et al. 2016). Synthetic polymers like PLA and PLGA are frequently employed as nanocarriers across various applications. They efficiently transport antimicrobials deep into the inaccessible core of biofilms, eliminating preformed biofilms and impeding the growth and formation of foodborne bacteria (Gaspar et al. 2018). It has been reported that xylitol typically struggles to penetrate bacterial biofilms and is vulnerable to degradation by bacterial beta-lactamase enzymes. However, incorporating xylitol into PLGA nanoparticles facilitated its penetration into the EPS which leads to the disruption of biofilms in *S. aureus* (Anjum et al. 2019). Nowadays, research has been conducted to explore the potential of encapsulated natural antimicrobials as sanitizing agents for direct application on food or food-contact surfaces. Chitosan-tripolyphosphate nanoparticles infused with carvacrol were able to keep the levels of bacterial load 2 to 3 logs lower than carrots disinfected with hypochlorite (Martínez-Hernández et al. 2017). The polymeric Eudragit® nanocapsules loaded with carvacrol were tested against *S. aureus, L. monocytogenes, E. coli*, and *S. enterica* adhered to stainless steel. This nanocapsule was able to reduce the viable counts for *E. coli* and *S. enterica* below the detection limit within 1-minute of contact time, while approximately 10^4 CFU/cm^2 reduction was observed for *S. aureus* and *L. monocytogenes* (Cacciatore et al. 2020).

Nanofibers

Nanofibers are ultra-thin fibers of diameter below 100 nm, composed of carbohydrates, proteins, or lipids. Due to its large surface area, it is a promising material for active food packaging and edible films (Blanco-Padilla et al. 2014). In addition, it can be used for delivering soluble or insoluble bioactive compounds to control foodborne pathogens (Rezaei et al. 2019). It has been demonstrated that poly (ethylene oxide) nanofibers containing cinnamon essential oil showed antibacterial activity against *Bacillus cereus* in beef (Lin et al. 2017). Nanofibers composed of polycaprolactone/gelatin nanofibers containing Lawsone (2-hydroxy-1,4-naphthoquinone), the active ingredient of *Lawsonia inermis* L. (commonly known as Henna), showed antibacterial and antibiofilm activity against Methicillin-resistant *S. aureus* (MRSA), *S. aureus, P. aeruginosa* (Adeli-Sardou et al. 2018). Two prominent challenges in the food industry are extending product shelf life and preventing microbial contamination. These issues can be addressed by introducing active food packaging made of nanofibrous material infused with antimicrobial agents. Such few examples include a recently developed polyamide-based nanofiber functionalized with natamycin/rosemary extract/green tea extract that showed antibiofilm properties against *Escherichia coli, Listeria monocytogenes, Salmonella enterica*, and *Staphylococcus aureus*. These nanofibers also increased chicken breast's shelf life by reducing the pathogens' total viable count (Lencova et al. 2022). Another example where polylactic acid (PLA) film coated with gelatin-chitosan nanofibers (GC-NF) incorporated with tannic acid and chitin oligosaccharides showed high antibacterial and antioxidant activity. Thus, could be a potential active packaging alternative for food applications (Gulzar et al. 2022).

Nanoparticles with Antibiofilm Activity Using Surface Functionalization with Phytochemicals

The fusion of phytochemicals in metal nanoparticles amplifies the bioactivity of functionalized plant bioactives. This surface functionalization onto the nanoparticle can enhance the solubility of the phytochemicals and penetrance power, thus augmenting their antibiofilm capabilities (Barros et al. 2020). Plant extracts containing bioactive phytochemicals can be used as a capping agent for the stable synthesis of metallic nanoparticles such as gold, silver, copper, and zinc. So, both metallic nanoparticles and phytochemicals complement each other's stability, solubility, and activity (Table 18.2). The biofabrication of metallic nanoparticles has garnered significant attention due to its straightforward single-step process that avoids the generation of toxic chemicals and gave rise to a new field of phytonanotechnology. Plant biomolecules, such as organic acids, proteins, vitamins, amino acids, and secondary metabolites such as flavones, ketones, alkaloids, terpenoids, phenolics, saponins, aldehydes, tannins, and polysaccharides, assist in the reduction of metal ions into nanoparticles. According to numerous studies, these metabolites derived from plants can prevent the agglomeration and aggregation of metallic nanoparticles by reducing and stabilizing the reaction in a nontoxic manner (Das et al. 2021).

There are several mechanisms reported for nanoparticles-mediated antimicrobial activity against biofilm-forming pathogens that can be mainly categorized into three types as illustrated in Fig. 18.4 (Shkodenko et al. 2020).

- Generation of ROS (Reactive oxygen species) induced oxidative stress by the synthesis of hydrogen peroxide (H_2O_2), superoxide (O_2^-), and hydroxyl radical (.OH),
- The release of metal ions causes alterations in enzyme activity, disruption of cell structures, and protein denaturation.
- The mechanical destruction of the cell wall occurs due to electrostatic interactions.

18.4 Conclusion and Future Perspectives

The use of naturally derived compounds with inhibitory effects on the growth of foodborne bacterial pathogens and their ability to remove biofilms has been acknowledged, but there are certain drawbacks. These include low levels of availability, activity, solubility, and stability of the recognized biological active molecules and possible alterations of their functions in microbial biofilms owing to changes in microenvironments. To most free drugs, it is recommended that these natural bioactives be encapsulated or combined with nanoparticles as this generally improves their effectiveness. This chapter elucidates the recent developments in exploiting nanotechnology for potentiating natural products against foodborne bacterial pathogen biofilms which is still an emerging field but of increasing importance. However we need to address some issues like the interaction of nanoparticles with food components, toxicity assessment, application-focused research to overcome the limitations to ensure consumer safety. Additional studies are required in order to fully evaluate these aspects and increase the safe and efficient application of nanoparticles in food systems probably together with the development of enhanced active packaging technologies.

Table 18.2 Recent studies on the bioactivity of phyto-fabricated nanoparticles against foodborne pathogen biofilms

Type of nanoformulation	Plant source and bioactive agents	Size of nanoparticles	Biological activity against foodborne pathogens	Effective concentration	References
Silver nanoparticles (GA-AgNPs)	*Gmelina arborea* leaf extract Flavonoids such as luteolin, kaempferol, quercetin, isoquercetin, rutin, and astragalin	34–40 nm	*B. cereus*, *E. coli*, *P. aeruginosa*, and *S. aureus*	20–1000 μg/ml	Chandrasekharan et al. (2022)
Silver nanoparticles (J-AgNPs)	*Jacaranda mimosifolia* flower aqueous extract	33.02 nm	*P. aeruginosa* and *S. aureus*	5.3 μg/ml	Verduzco-Chavira et al. (2023)
Gold nanoparticles (Au NPs)	*Tinospora cordifolia* stem extract	16.1 nm	*Pseudomonas aeruginosa* PAO1	150 μg/ml	Ali et al. (2020)
Gold nanoparticles (Au NPs)	*Trachyspermum ammi* seed extract	16.63 nm	*S. marcescens and L. monocytogenes*	16 and 32 μg/ml	Perveen et al. (2021)
Zinc oxide nanoparticle (ZnONPs)	*Veronica multifida* leaf extract	10–100 nm	*P. aeruginosa and S. aureus*	10–100 μg/ml	Doğan and Kocabaş (2020)
Zinc oxide nanoparticle (ZnONPs)	*Nigella sativa* seed extract	24 nm	*L. monocytogenes, E. coli, C. violaceum, and P. aeruginosa PAO1*	128–512 μg/ml	Al-Shabib et al. (2016)
Copper oxide nanoparticles (CuONPs)	*Eucalyptus globulus* leaf extract	16.78–22.5 nm	*E. coli-336, P. aeruginosa-621 and MRSA-1*	125–2000 μg/ml	Ali et al. (2019)
CuNPs	*Cardiospermum halicacabum* leaf extract	30–40 nm	*P. aeruginosa, S. aureus, and E. coli*	100 μg/ml	Punniyakotti et al. (2020)
TiO_2	*Ocimum sanctum* leaf extract	90–100 nm	*P. aeruginosa, E. coli, S. aureus and C. albicans*	250–650 μg/ml	Dan and Khan (2019)
TiO_2	*Rosa davurica* leaf extract	146 nm	*S. aureus, B. cereus, S. enterica, and E. coli*	15.62 μg/ml	Jin et al. (2022)
Calcium oxide nanoparticles (FC-CaONPs)	*Ficus carica* (an edible fruit) aqueous extract	84.87 nm	*E. coli, P. aeruginosa, S. aureus, K. pneumoniae*, and *P. vulgaris*	62.5 μg/ml	Khan et al. (2023)
Paladium nanoparticles (PdNPs)	*Orthosiphon stamineus* leaf extract	110 nm	Methicillin-sensitive *Staphylococcus aureus* (MSSA) strain	15.25 μg/ml	Prakashkumar et al. (2021)

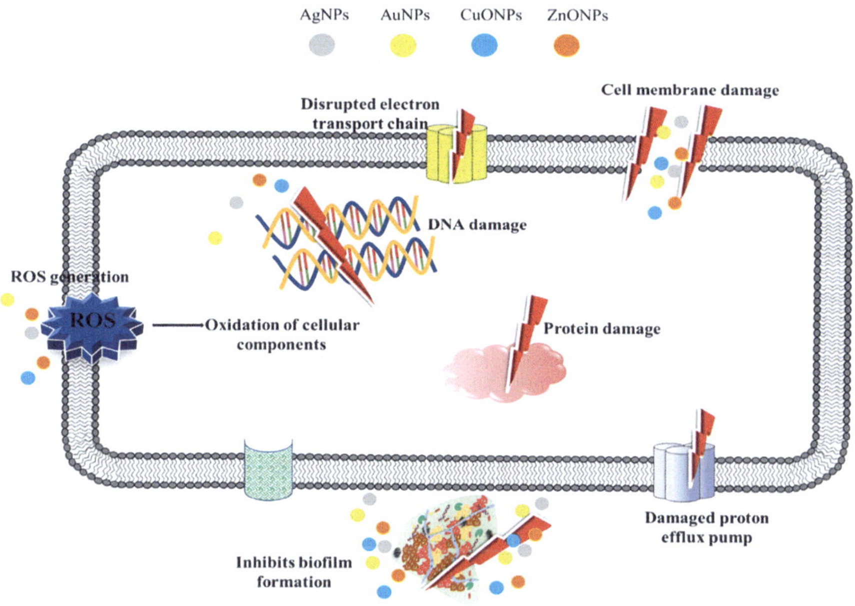

Fig. 18.4 Diagrammatic illustration to represent the mode of action of various metal/metal oxide nanoparticles on a bacterial cell (Das et al. 2021)

References

Abdallah M, Benoliel C, Drider D, Dhulster P, Chihib NE (2014) Biofilm formation and persistence on abiotic surfaces in the context of food and medical environments. Arch Microbiol 196(7):453–472

Adeli-Sardou M, Torkzadeh-Mahani M, Yaghoobi MM, Dodel M (2018) Antibacterial and anti-biofilm investigation of electrospun PCL/gelatin/Lawsone nano fiber scaffolds against biofilm-producing bacteria. Biom J 4(1):46–57

Ahmed K, Muiruri PW, Jones GH, Scott MJ, Jones MN (2001) The effect of grafted poly (ethylene glycol) on the electrophoretic properties of phospholipid liposomes and their adsorption to bacterial biofilms. Colloids Surf A Physicochem Eng Asp 194(1–3):287–296

Ali K, Ahmed B, Ansari SM, Saquib Q, Al-Khedhairy AA, Dwivedi S et al (2019) Comparative in situ ROS mediated killing of bacteria with bulk analogue, Eucalyptus leaf extract (ELE)-capped and bare surface copper oxide nanoparticles. Mater Sci Eng C 100:747–758

Ali SG, Ansari MA, Alzohairy MA, Alomary MN, AlYahya S, Jalal M et al (2020) Biogenic gold nanoparticles as potent antibacterial and antibiofilm nano-antibiotics against *Pseudomonas aeruginosa*. Antibiotics 9(3):100

Al-Shabib NA, Husain FM, Ahmed F, Khan RA, Ahmad I, Alsharaeh E et al (2016) Biogenic synthesis of zinc oxide nanostructures from Nigella sativa seed: prospective role as food packaging material inhibiting broad-spectrum quorum sensing and biofilm. Sci Rep 6(1):36761

Amrutha B, Sundar K, Shetty PH (2017) Spice oil nanoemulsions: potential natural inhibitors against pathogenic *E. coli* and *Salmonella* spp. from fresh fruits and vegetables. LWT-Food Sci Technol 79:152–159

Anand S, Singh D, Avadhanula M, Marka S (2014) Development and control of bacterial biofilms on dairy processing membranes. Compr Rev Food Sci Food Saf 13(1):18–33

Anjum A, Chung PY, Ng SF (2019) PLGA/xylitol nanoparticles enhance antibiofilm activity via penetration into biofilm extracellular polymeric substances. RSC Adv 9(25):14198–14208
Barros CH, Devlin H, Hiebner DW, Vitale S, Quinn L, Casey E (2020) Enhancing curcumin's solubility and antibiofilm activity via silica surface modification. Nanoscale Adv 2(4):1694–1708
Beveridge TJ, Makin SA, Kadurugamuwa JL, Li Z (1997) Interactions between biofilms and the environment. FEMS Microbiol Rev 20(3–4):291–303
Blanco-Padilla A, Soto KM, Hernández Iturriaga M, Mendoza S (2014) Food antimicrobials nanocarriers. Sci World J 2014(1):837215
Bottone EJ (2010) Bacillus cereus, a volatile human pathogen. Clin Microbiol Rev 23(2):382–398
Brandelli A, Lopes NA, Boelter JF (2017) Food applications of nanostructured antimicrobials. In: Food preservation. Academic Press, pp 35–74
Cacciatore FA, Dalmás M, Maders C, Isaía HA, Brandelli A, da Silva Malheiros P (2020) Carvacrol encapsulation into nanostructures: characterization and antimicrobial activity against foodborne pathogens adhered to stainless steel. Food Res Int 133:109143
Cacciatore FA, Brandelli A, Malheiros PDS (2021) Combining natural antimicrobials and nanotechnology for disinfecting food surfaces and control microbial biofilm formation. Crit Rev Food Sci Nutr 61(22):3771–3782
CDC (2017) Listeria (Listeriosis). Available at: https://www.cdc.gov/listeria/outbreaks/index.html
Chandrasekharan S, Chinnasamy G, Bhatnagar S (2022) Sustainable phyto-fabrication of silver nanoparticles using *Gmelina arborea* exhibit antimicrobial and biofilm inhibition activity. Sci Rep 12(1):156
Chlebicz A, Śliżewska K (2018) Campylobacteriosis, salmonellosis, yersiniosis, and listeriosis as zoonotic foodborne diseases: a review. Int J Environ Res Public Health 15(5):863
Colagiorgi A, Bruini I, Di Ciccio PA, Zanardi E, Ghidini S, Ianieri A (2017) *Listeria monocytogenes* biofilms in the wonderland of food industry. Pathogens 6(3):41
Coroneo V, Carraro V, Marras B, Marrucci A, Succa S, Meloni B et al (2017) Presence of Trihalomethanes in ready-to-eat vegetables disinfected with chlorine. Food Addit Contami Part A 34(12):2111–2117
Cui H, Zhou H, Lin L (2016a) The specific antibacterial effect of the Salvia oil nanoliposomes against *Staphylococcus aureus* biofilms on milk container. Food Control 61:92–98
Cui H, Li W, Li C, Vittayapadung S, Lin L (2016b) Liposome containing cinnamon oil with antibacterial activity against methicillin-resistant *Staphylococcus aureus* biofilm. Biofouling 32(2):215–225
Da Costa D, Exbrayat-Héritier C, Rambaud B, Megy S, Terreux R, Verrier B, Primard C (2021) Surface charge modulation of rifampicin-loaded PLA nanoparticles to improve antibiotic delivery in *Staphylococcus aureus* biofilms. J Nanobiotechnol 19:1–17
Dan SO, Khan SH (2019) Green coalescence and characterization of Tio2 nanoparticles and evaluation of its Antibiofilm activity. Rasāyan J Chem 12(4):2252–2259
Das P, Ghosh S, Nayak B (2021) Phyto-fabricated nanoparticles and their anti-biofilm activity: progress and current status. Front Nanotechnol 3:739286
Davidson PM, Harrison MA (2002) Resistance and adaptation to food antimicrobials, sanitizers, and other process controls. Food Technol 56(11):69–78
Doğan SŞ, Kocabaş A (2020) Green synthesis of ZnO nanoparticles with Veronica multifida and their antibiofilm activity. Hum Exp Toxicol 39(3):319–327
Dos Santos Ramos MA, Da Silva PB, Spósito L, De Toledo LG, Bonifacio BV, Rodero CF et al (2018) Nanotechnology-based drug delivery systems for control of microbial biofilms: a review. Int J Nanomedicine 13:1179–1213
Doyle MP (2006) Dealing with antimicrobial resistance. Food Technol 60(8):22–29
Duguid JP, Anderson ES, Campbell I (1966) Fimbriae and adhesive properties in Salmonellae. J Pathol Bacteriol 92(1):107–137
EFSA Panel on Biological Hazards [BIOHAZ] (2016) Risks for public health related to the presence of *Bacillus cereus* and other *Bacillus* spp. including *Bacillus thuringiensis* in foodstuffs. EFSA J 14:e04524

Ellboudy NM, Elwakil BH, Shaaban MM, Olama ZA (2023) Cinnamon oil-loaded nanoliposomes with potent antibacterial and antibiofilm activities. Molecules 28(11):4492

Falcó I, Verdeguer M, Aznar R, Sánchez G, Randazzo W (2019) Sanitizing food contact surfaces by the use of essential oils. Innovative Food Sci Emerg Technol 51:220–228

Filipczak N, Pan J, Yalamarty SSK, Torchilin VP (2020) Recent advancements in liposome technology. Adv Drug Deliv Rev 156:4–22

Flemming HC (1993) Biofilms and environmental protection. Water Sci Technol 27(7–8):1–10

Flemming H, Wingender J (2010) The biofilm matrix. Nat Publish Group 8:623–633

Flemming HC, Wingender J, Szewzyk U, Steinberg P, Rice SA, Kjelleberg S (2016) Biofilms: an emergent form of bacterial life. Nat Rev Microbiol 14(9):563–575

Fraise AP, Maillard JY, Sattar S (2013) Russel, Hugo & Ayliffe's principles and practice of disinfection, preservation and sterilization, 5th edn. Wiley-Blackwell

Galié S, García-Gutiérrez C, Miguélez EM, Villar CJ, Lombó F (2018) Biofilms in the food industry: health aspects and control methods. Front Microbiol 9:315815

Gaspar LMDAC, Dórea ACS, Droppa-Almeida D, de Mélo Silva IS, Montoro FE, Alves LL et al (2018) Development and characterization of PLGA nanoparticles containing antibiotics. J Nanopart Res 20(11):289

Giaouris E, Heir E, Desvaux M, Hébraud M, Møretrø T, Langsrud S et al (2015) Intra-and inter-species interactions within biofilms of important foodborne bacterial pathogens. Front Microbiol 6:144151

Gould LH, Mody RK, Ong KL, Clogher P, Cronquist AB, Garman KN et al (2013) Increased recognition of non-O157 Shiga toxin–producing *Escherichia coli* infections in the United States during 2000–2010: epidemiologic features and comparison with E. coli O157 infections. Foodborne Pathog Dis 10(5):453–460

Grigore-Gurgu L, Bucur FI, Borda D, Alexa EA, Neagu C, Nicolau AI (2019) Biofilms formed by pathogens in food and food processing environments. In: Bacterial biofilms. IntechOpen, pp 1–32

Gulzar S, Tagrida M, Nilsuwan K, Prodpran T, Benjakul S (2022) Electrospinning of gelatin/chitosan nanofibers incorporated with tannic acid and chitooligosaccharides on polylactic acid film: characteristics and bioactivities. Food Hydrocoll 133:107916

Halden RU (2014) On the need and speed of regulating Triclosan and Triclocarban in the United States. Environ Sci Technol 48(7):3603–3611

Hasanzadeh Baboli N, Hosseini SF, Gharsallaoui A (2023) Antibacterial and anti-biofilm properties of cinnamaldehyde-loaded nanoliposomes against *Listeria monocytogenes* and *Salmonella enteritidis* adhered to stainless steel. Int J Food Sci Technol 58(10):5275–5282

Houry A, Briandet R, Aymerich S, Gohar M (2010) Involvement of motility and flagella in *Bacillus cereus* biofilm formation. Microbiology 156(4):1009–1018

Jin Y, Li B, Saravanakumar K, Hu X, Wang MH (2022) Phytogenic titanium dioxide (TiO2) nanoparticles derived from *Rosa davurica* with anti-bacterial and anti-biofilm activities. J Clust Sci 33(4):1435–1443

Kadariya J, Smith TC, Thapaliya D (2014) Staphylococcus aureus and staphylococcal food-borne disease: an ongoing challenge in public health. Biomed Res Int 2014(1):827965

Khan F, Yu H, Kim YM (2020) Bactericidal activity of usnic acid-chitosan nanoparticles against persister cells of biofilm-forming pathogenic bacteria. Mar Drugs 18(5):270

Khan AU, Hussain T, Abdullah, Khan MA, Almostafa MM, Younis NS, Yahya G (2023) Antibacterial and Antibiofilm activity of Ficus carica-mediated calcium oxide (CaONPs) Phyto-nanoparticles. Molecules 28(14):5553

Kırmusaoğlu S (2016) Staphylococcal biofilms: Pathogenicity, mechanism and regulation of biofilm formation by quorum sensing system and antibiotic resistance mechanisms of biofilm embedded microorganisms. Microbial biofilms: importance and applications. IntechOpen 189–209

Klančnik A, Šimunović K, Sterniša M, Ramić D, Smole Možina S, Bucar F (2021) Anti-adhesion activity of phytochemicals to prevent *Campylobacter jejuni* biofilm formation on abiotic surfaces. Phytochem Rev 20:55–84

Lee J, Bansal T, Jayaraman A, Bentley WE, Wood TK (2007) Enterohemorrhagic *Escherichia coli* biofilms are inhibited by 7-hydroxyindole and stimulated by isatin. Appl Environ Microbiol 73(13):4100–4109

Lee BK, Yun YH, Park K (2015) Smart nanoparticles for drug delivery: boundaries and opportunities. Chem Eng Sci 125:158–164

Lencova S, Stiborova H, Munzarova M, Demnerova K, Zdenkova K (2022) Potential of polyamide nanofibers with natamycin, rosemary extract, and green tea extract in active food packaging development: interactions with food pathogens and assessment of microbial risks elimination. Front Microbiol 13:857423

Lenglet A, National Epidemiological Surveillance Network of Spain (2005) E-alert 9 August: over 2000 cases so far in Salmonella Hadar outbreak in Spain associated with consumption of precooked chicken, July-August, 2005. Weekly Releases (1997–2007) *10*(32):2770

Lewandowski Z, Stoodley P, Altobelli S, Fukushima E (1994) Hydrodynamics and kinetics in biofilm systems-recent advances and new problems. Water Sci Technol 29(10–11):223

Lim ES, Koo OK, Kim MJ, Kim JS (2019) Bio-enzymes for inhibition and elimination of *Escherichia coli* O157: H7 biofilm and their synergistic effect with sodium hypochlorite. Sci Rep 9(1):9920

Lin L, Dai Y, Cui H (2017) Antibacterial poly (ethylene oxide) electrospun nanofibers containing cinnamon essential oil/beta-cyclodextrin proteoliposomes. Carbohydr Polym 178:131–140

Liu M, Guo W, Feng M, Bai Y, Huang J, Cao Y (2024) Antibacterial, anti-biofilm activity and underlying mechanism of garlic essential oil in water nanoemulsion against *Listeria monocytogenes*. LWT 196:115847

Lopes NA, Brandelli A (2018) Nanostructures for delivery of natural antimicrobials in food. Crit Rev Food Sci Nutr 58(13):2202–2212

Lopes NA, Pinilla CMB, Brandelli A (2017) Pectin and polygalacturonic acid-coated liposomes as novel delivery system for nisin: preparation, characterization and release behavior. Food Hydrocoll 70:1–7

Marić S, Vraneš J (2007) Characteristics and significance of microbial biofilm formation. Period Biol 109:115–121

Marouani-Gadri N, Firmesse O, Chassaing D, Sandris-Nielsen D, Arneborg N, Carpentier B (2010) Potential of Escherichia coli O157: H7 to persist and form viable but non-culturable cells on a food-contact surface subjected to cycles of soiling and chemical treatment. Int J Food Microbiol 144(1):96–103

Marques SC, Rezende JDGOS, Alves LADF, Silva BC, Alves E, Abreu LRD, Piccoli RH (2007) Formation of biofilms by Staphylococcus aureus on stainless steel and glass surfaces and its resistance to some selected chemical sanitizers. Braz J Microbiol 38:538–543

Martínez-Hernández GB, Amodio ML, Colelli G (2017) Carvacrol-loaded chitosan nanoparticles maintain quality of fresh-cut carrots. Innovative Food Sci Emerg Technol 41:56–63

Michalak G, Głuszek K, Piktel E, Deptuła P, Puszkarz I, Niemirowicz K, Bucki R (2016) Polymeric nanoparticles–a novel solution for delivery of antimicrobial agents. Medical Studies/Studia Medyczne 32(1):56–62

Nguyen HDN, Yang YS, Yuk HG (2014) Biofilm formation of *Salmonella Typhimurium* on stainless steel and acrylic surfaces as affected by temperature and pH level. LWT-Food Sci Technol 55(1):383–388

Niaz T, Shabbir S, Noor T, Imran M (2019) Antimicrobial and antibiofilm potential of bacteriocin loaded nano-vesicles functionalized with rhamnolipids against foodborne pathogens. LWT 116:108583

Nikolaev YA, Plakunov VK (2007) Biofilm- "City of microbes" or an analogue of multicellular organisms? Microbiology 76:125–138

Perveen K, Husain FM, Qais FA, Khan A, Razak S, Afsar T et al (2021) Microwave-assisted rapid green synthesis of gold nanoparticles using seed extract of Trachyspermum ammi: ROS mediated biofilm inhibition and anticancer activity. Biomol Ther 11(2):197

Pircalabioru GG, Chifiriuc MC (2020) Nanoparticulate drug-delivery systems for fighting microbial biofilms: from bench to bedside. Future Microbiol 15(8):679–698

Pompilio A, Riviello A, Crocetta V, Di Giuseppe F, Pomponio S, Sulpizio M et al (2016) Evaluation of antibacterial and antibiofilm mechanisms by usnic acid against methicillin-resistant Staphylococcus aureus. Future Microbiol 11(10):1315–1338

Prakash A, Baskaran R, Nithyanand P, Vadivel V (2020) Effect of nanoemulsification on the antibacterial and anti-biofilm activities of selected spice essential oils and their major constituents against Salmonella enterica Typhimurium. J Clust Sci 31:1123–1135

Prakashkumar N, Vignesh M, Brindhadevi K, Phuong NT, Pugazhendhi A, Suganthy N (2021) Enhanced antimicrobial, antibiofilm and anticancer activities of biocompatible neem gum coated palladium nanoparticles. Prog Org Coat 151:106098

Prateeksha, Barik SK, Singh BN (2019) Nanoemulsion-loaded hydrogel coatings for inhibition of bacterial virulence and biofilm formation on solid surfaces. Sci Rep 9(1):6520

Punniyakotti P, Panneerselvam P, Perumal D, Aruliah R, Angaiah S (2020) Anti-bacterial and anti-biofilm properties of green synthesized copper nanoparticles from Cardiospermum halicacabum leaf extract. Bioprocess Biosyst Eng 43:1649–1657

Rezaei A, Fathi M, Jafari SM (2019) Nanoencapsulation of hydrophobic and low-soluble food bioactive compounds within different nanocarriers. Food Hydrocoll 88:146–162

Rossi C, Chaves-López C, Serio A, Casaccia M, Maggio F, Paparella A (2022) Effectiveness and mechanisms of essential oils for biofilm control on food-contact surfaces: an updated review. Crit Rev Food Sci Nutr 62(8):2172–2191

Rothrock MJ Jr, Davis ML, Locatelli A, Bodie A, McIntosh TG, Donaldson JR, Ricke SC (2017) Listeria occurrence in poultry flocks: detection and potential implications. Front Vet Sci 4:125

Rukavina Z, Vanić Ž (2016) Current trends in development of liposomes for targeting bacterial biofilms. Pharmaceutics 8(2):18

Schaffazick SR, Guterres SS, Freitas LDL, Pohlmann AR (2003) Physicochemical characterization and stability of the polymeric nanoparticle systems for drug administration. Química Nova 26:726–737

Shkodenko L, Kassirov I, Koshel E (2020) Metal oxide nanoparticles against bacterial biofilms: perspectives and limitations. Microorganisms 8(10):1545

Silk BJ, Date KA, Jackson KA, Pouillot R, Holt KG, Graves LM et al (2012) Invasive listeriosis in the Foodborne Diseases Active Surveillance Network (FoodNet), 2004–2009: further targeted prevention needed for higher-risk groups. Clin Infect Dis 54(suppl_5):S396–S404

Silva S, Teixeira P, Oliveira R, Azeredo J (2008) Adhesion to and viability of *Listeria monocytogenes* on food contact surfaces. J Food Prot 71(7):1379–1385

Slavin YN, Asnis J, Hńfeli UO, Bach H (2017) Metal nanoparticles: understanding the mechanisms behind antibacterial activity. J Nanobiotechnol 15:1–20

Srey S, Jahid IK, Ha SD (2013) Biofilm formation in food industries: a food safety concern. Food Control 31(2):572–585

Srividya N, Ghoora MD, Padmanabh PR (2017) Antimicrobial nanotechnology: research implications and prospects in food safety. In: Food preservation. Academic Press, pp 125–165

Verduzco-Chavira K, Vallejo-Cardona AA, González-Garibay AS, Torres-González OR, Sánchez-Hernández IM, Flores-Fernández JM, Padilla-Camberos E (2023) Antibacterial and Antibiofilm activity of chemically and biologically synthesized silver nanoparticles. Antibiotics 12(7):1084

Wang Y (2021) Liposome as a delivery system for the treatment of biofilm-mediated infections. J Appl Microbiol 131(6):2626–2639

Wang H, Ding S, Dong Y, Ye K, Xu X, Zhou G (2013a) Biofilm formation of Salmonella serotypes in simulated meat processing environments and its relationship to cell characteristics. J Food Prot 76(10):1784–1789

Wang H, Ding S, Wang G, Xu X, Zhou G (2013b) In situ characterization and analysis of Salmonella biofilm formation under meat processing environments using a combined microscopic and spectroscopic approach. Int J Food Microbiol 167(3):293–302

Wang H, Tay M, Palmer J, Flint S (2017) Biofilm formation of *Yersinia enterocolitica* and its persistence following treatment with different sanitation agents. Food Control 73:433–437

Wang DY, Van der Mei HC, Ren Y, Busscher HJ, Shi L (2020) Lipid-based antimicrobial delivery-systems for the treatment of bacterial infections. Front Chem 7:872

Xue R, Shi H, Ma Y, Yang J, Hua B, Inniss EC et al (2017) Evaluation of thirteen haloacetic acids and ten trihalomethanes formation by peracetic acid and chlorine drinking water disinfection. Chemosphere 189:349–356

Index

T. Sarkar et al. (eds.), *Biofilm Applications to Revolutionize Food Technology*, https://doi.org/10.1007/978-3-031-85205-3

The manufacturer's authorised representative in the EU is Springer Nature Customer Service Centre GmbH, Europaplatz 3, 69115 Heidelberg, Germany. If you have any concerns regarding our products, please contact ProductSafety@springernature.com

Printed and bound by CPI Group (UK) Ltd, Croydon, CR0 4YY
30/09/2026
02225568-0015